Breeding for Disease Resistance
in Farm Animals

Breeding for Disease Resistance in Farm Animals

Edited by

J.B. Owen *and* R.F.E. Axford

School of Agricultural and Forest Sciences,
University of Wales, Bangor,
Gwynedd LL57 2UW, UK

C·A·B INTERNATIONAL

C·A·B International   Tel: Wallingford (0491) 32111
Wallingford   Telex: 847964 (COMAGG G)
Oxon OX10 8DE   Telecom Gold/Dialcom: 84: CAU001
UK   Fax: (0491) 33508

British Library Cataloguing in Publication Data
Breeding for disease resistance in farm animals.
  1. Livestock. Breeding
  I. Owen, John B. (John Bryn) *1931–* II. Axford, R.F.E.
  636.0821

ISBN 0-85198-710-9 ✓

Typeset by Leaper & Gard Limited
Printed and bound in the UK by Redwood Press Ltd., Melksham

Contents

Preface

This book is dedicated to the memory of our friend and colleague, the late Dr Ian Herbert, who was involved from an early stage in the genesis and development of the idea for the book and the Symposium upon which it is based. Despite his serious illness which Ian fought from Christmas 1987 to his death in June 1989, he played an active and enthusiastic part in the organization.

The Symposium, which was held at Bangor on 13 and 14 September 1990, brought together for the first time the views of a diverse group of scientists, working on all aspects of genetic resistance to disease, on the application of current knowledge to the problems of present day animal improvement. The topics discussed included the problems of controlling gastrointestinal worms, tsetse-fly transmitted trypanosomes, lameness in sheep and cattle, mastitis, scrapie and BSE. The striking implications of major genes for animal health and welfare were considered in relation to examples like double muscling in cattle and inherited resistance to diarrhoea in young pigs. Contributors ranged from those working in the fundamental areas of molecular biology to the veterinarians, geneticists and agriculturalists working with stockmen at the practical end of animal care.

This book which contains the reviews presented at the Symposium highlights a looming problem of world-wide dimensions of which many in the industry are subliminally aware but which has not been sufficiently appreciated by all concerned. The inventiveness of the pharmaceutical industry has lulled commercial producers and pedigree breeders to a possibly false hope that the challenges addressed in this book can be indefinitely postponed.

The book not only chronicles problems but also indicates promising solutions, even though there is realism about the practical and monetary implications involved.

We believe the book should be read by students of animal breeding whether in veterinary, agricultural or applied biology schools of Universities and Higher Education Institutes. It also provides an up-to-date review of the many aspects involved for a wide range of researchers, advisers and breeders whose activities impinge in some way on the general theme of the book. We even hope that the forceful message emerging throughout the book may help more positive action in breeding for resistance in the future.

As Editors we were delighted with the response from the contributors and we are satisfied that our original aims have been well realized. We are much indebted to all who contributed and particularly to Dr Hussein Omed who ably handled the business end of the organizing group.

The editors acknowledge with thanks, the encouragement and financial assistance received from the main sponsor, Holstein Friesian Society of Great Britain and Ireland: and other sponsors, National Sheep Association, Royal Welsh Agricultural Society, Centre for Arid Zone Studies, Bangor, Farmers Union of Wales, Welsh Black Cattle Society, Royal Agricultural Society of England, Gwynedd County Council, Welsh Tourist Board, Lleyn Sheep Society, Welsh Quality Lamb and United Meat Packers towards the expenses of the Symposium.

J.B. Owen and R.F.E. Axford
Bangor
October 1990

Section 1
Principles of Breeding for Disease Resistance

Current systems of disease prevention and control put little emphasis on the innate animal mechanisms evolved to maintain a healthy status. Moreover, they tend to favour diversification of the pathogens to more resistant forms creating serious disease problems. In this section the animal's genetic mechanisms for combatting disease are discussed, particularly the role that the important complex of genes – the major histocompatibility complex (MHC) – can play in some cases. It is evident that no genes have been discovered that confer universal resistance to disease; indeed animals resistant to some pathogens can be more susceptible to even closely related organisms, as exemplified by the response of chickens to infection by different species of *Eimera*. However, it is shown that the immune system is regulated by sets of genes which not only control innate immunity but can alter the specificity and quality of acquired immunity.

Many of the principles of the genetic basis of disease that are applicable to farm animals are derived from model systems based on laboratory and other animals. A particularly important example is described of the use of mice with respect to parasite resistance mechanisms.

Chapter 1

Strategies for Disease Control

R.F.E. Axford and J.B. Owen

*School of Agricultural and Forest Sciences, University
of Wales, Bangor, Gwynedd LL57 2UW, UK*

Summary

The spread of diseases among domestic animals can be limited by eradication of infected stock, isolation and quarantine of susceptible groups or individuals, the application of principles of hygiene or vaccination. Many diseases, however, cannot be practically controlled by these means and increasing use is made of drugs to eliminate or destroy pathogens or their vectors. The regular use of therapeutic agents minimizes the selective advantage of natural resistance in the animals, while at the same time pathogens are encouraged to diversify into forms that are resistant to many drugs. The scale of antibiotic and pesticide use to control many common diseases could have serious effects on the environment of the developed world, and is prohibitively expensive for third world countries.

Urgent consideration is needed to develop a balanced strategy in which drug use is minimized where possible by raising resistant stock. The identification of disease-resistant animals promises major benefits in a number of specific areas such as trypanotolerance in African livestock breeds, mastitis of cattle and colibacillosis in pigs. This gives the promise of major benefits on limited and fruitful fronts once the high initial costs of research and development are met. In this process the new tools of molecular biology, particularly the exploitation of genes of the major histocompatibility complex (MHC), are relevant.

The complete eradication of a few specific pathogens of major importance may be economically justified. However, the eradication of one member of a pathogen family may simply lead to its substitution by another. In a balanced strategy of disease control a more realistic view of the optimum intensity of animal production could lessen the dependence on artificial aids.

Determinants of disease

The determinants of disease have often been represented as the triad shown in Fig. 1.1 in which animal and pathogen coexist in an environment which may favour either of the protagonists. The infectivity and virulence of the pathogen is opposed by the susceptibility and infectiousness of the animal, and the incidence of health or disease depends on the balance of these factors. For some major diseases, such as foot and mouth disease, the environmental component is not obvious, while for others of multifactorial origin such as scours and mastitis the nature of the environment has a profound effect (Thrusfield, 1986). Control measures against diseases may be directed towards any component of the triad or a combination of components.

Current preventive measures

Preventive methods have been reviewed many times. Figure 1.2 (Davies, 1985) illustrates the attention given to environmental and therapeutic control methods. The possible increase of resistance obtainable by the use of genetically resistant stock is rarely discussed. For example, in a useful series of reviews recently published on mastitis (Booth, 1988; Francis, 1989; Grindal, 1988) the culling of chronically infected cattle is mentioned, but not the selection for genetic improvement of resistance to mastitis.

Eradication is perhaps the ideal solution to the problem of disease. This can be attained only for a few diseases with favourable characteristics, particularly those found only in domestic species and capable of being readily and promptly diagnosed. Regional eradication and quarantine schemes have produced areas of the world which are usually free from foot and mouth disease, apart from occasional costly incursions, capitalizing on the short stability of the foot and mouth virus outside its host, and the ready detection of incursions (Brown, 1986). Some diseases like swine fever can be eradicated under an umbrella of vaccination, but this is only possible because immunity is solid enough to preclude the development of carriers and subclinical cases which could harbour and excrete the pathogen. Most pathogens, however, are not susceptible to complete eradication and their disease manifestations have to be controlled on a herd and individual animal basis with the recognition that these are under continuous challenge.

The chief defences employed against these pathogens are environmental measures, vaccination, and the use of pharmaceuticals.

The first two of these defences are generally strictly beneficial in their effects, although it is possible that the regular and routine use of vaccines protects genotypes that are slow in developing natural resistance from the

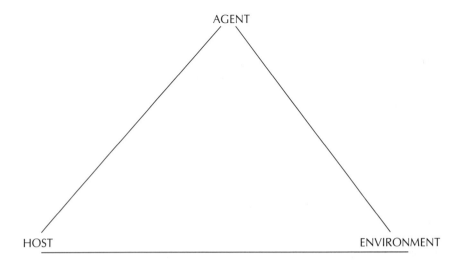

Fig. 1.1. The determinants of disease (Thrusfield, 1986).

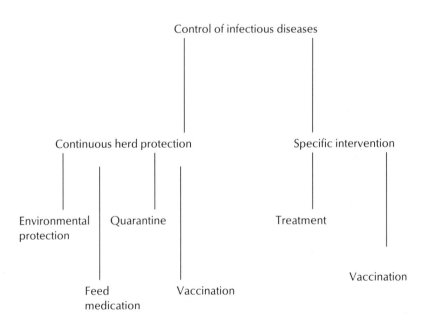

Fig. 1.2. Options for controlling disease in farm animals (Davies, 1985).

process of natural selection. In contrast the use of pharmaceuticals, which is currently the dominant method of preventing and treating disease, carries several major disadvantages.

The first is cost, as an initial dependence on drug use usually leads to the need for its continuous application as in the control of mastitis in cattle (Dodd, 1987). This may be prohibitively expensive for poorer parts of the world (Vogel and Stephens, 1989).

The second is environmental damage caused by drug residues. These may only become apparent after considerable use. The adverse effects of ivermectin residues on invertebrate fauna in animal dung has been described (Wall and Strong, 1987).

Third, drug use, especially if it is continued over a long period encourages the development of drug-resistant strains among pathogenic organisms (Falkow, 1975) to the detriment of the treated animal. Once developed, these organisms might invade other livestock (Jones, 1987) or humans, raising problems when they in turn are treated with antibiotics (Falkow, 1975) or pesticides (Waller, 1985). This is perhaps the greatest drawback to the use of drugs to maintain farm animal health.

Fourth, several of the major diseases causing economic loss, e.g. many so-called production diseases like mastitis and lameness, have proved recalcitrant to drug therapy.

Finally, a pharmaceutical umbrella reduces the selective advantage of natural disease resistance so much so that in some cases, breeds are being developed that are increasingly susceptible to infection. Mastitis is a good example of this as selection for milk yield probably results in a higher incidence of the condition (Emanuelson, 1988) and this will lead to ever greater drug dependence.

Natural resistance

World populations have naturally developed a broad-based resistance to the disease organisms endemic in their environment. In most cases potentially debilitating or pathogenic organisms are not completely resisted but a symbiotic genetic compromise is set up allowing both protagonists to survive in a delicate equilibrium. Changes in the environment, including invasion by foreign pathogenic organisms can lead to a breakdown in the balance, leading to epidemics. The threatened species, if not exterminated, reach a new genetic equilibrium. In some situations, e.g. myxomatosis in rabbits, it may take a period of cyclic ebb and flow before a more stable pattern is reached, possibly involving genetic changes (Fenner and Ratcliffe, 1965); the pathogen decreases in virulence and the host increases in resistance or tolerance.

With domestication the environment has been radically altered, and further changes to the equilibrium have occurred through intensification of farm animal enterprises. Man's reaction could be one or two strategies, either to do nothing, be patient and wait for the natural process of equilibrium to occur as in the wild, or to intervene to protect his livestock possessions. With the advent of immunization and powerful synthetic drugs, the latter choice was inevitable and in Western developed countries intensive drug therapy is employed, especially in the pedigree breeding herds and flocks that are the main genetic source for those species.

The practical exploitation of natural disease resistance has only recently been recognized as possible, and its application is on a very limited scale.

Future strategies

The choice for the future is stark and difficult. It lies between three options:

1. Continuation of a policy of maximum intervention in the hope that new drugs and vaccines can keep pace with new pathogen genotypes even though natural animal resistance would continue to weaken progressively.
2. A deliberate policy of non-intervention with the use of drugs and vaccines only *in extremis*, the policy advocated by the 'organic' farmer. This would maximize the development of specific and general natural resistance and tolerance mechanisms. Embarking on such a policy could be hazardous where drugs and vaccines have been intensively used for decades.
3. A cautious balanced policy of reduced intervention. This would be coupled with the development and introduction of breeds in which increased resistance to important endemic disorders could be demonstrated. A deliberate policy of lowering the intensity of animal production systems would be a useful adjunct so that stocking densities could be lowered in conjunction with multiple land use systems.

The first of these options, that of maximum intervention may not be sustainable much longer even in the rich developed countries, for reasons discussed earlier. This would first become apparent for those herds and flocks at the top of the breed society hierarchies. Their products may eventually be refused by ordinary commercial farmers who cannot cope with the cost of keeping such animals healthy for a reasonable length of productive life. Such animals will certainly be quite inappropriate for other parts of the world and exports will decrease.

The second policy of non-intervention is an option not to be lightly discarded. It is promoted by the organic lobbies in developed countries (Soil Association, 1989), and is especially relevant for underdeveloped countries where animal production is still dominated by native, adapted stock. In such situations a very careful policy decision should be made before any foreign genotypes from contrasting environments are introduced. It is too often assumed that native species of farm livestock are inferior to highly developed breeds because of their apparent low level of productivity. It may well be possible to exploit genetic variance for production traits within such stock without losing the desired disease resistance, and avoid introducing exotic breeds even in cautious cross-breeding ventures. There are many examples of native adapted stock that may be capable of more efficient production in the purebred form than any crosses with exotic European or North American breeds. It may be more prudent to introduce for product testing, stock from similar neighbouring environments for any such cross-breeding adventures.

The third option, that of actively selecting and breeding animals adapted to specific pathogens and diseases endemic in their environment and thus reducing the need for drugs to control these would seem the most strongly indicated for general use. For this to bear economic fruit and yield environmental benefits as well, large amounts of highly speculative research and a long time period for its establishment would be required. It has been estimated (Eriksson, 1991) that at least 5.5 years will be needed to evaluate the sons of a bull-sire for mastitis. Production of resistant animals is likely to be expensive, and more of a high technology procedure than a 'green' one. Support for research and development of this order can only come from governmental and international agencies. The individual breeder will be tempted to use the safer, short-term pharmaceutical solution to disease control unless persuaded or directed otherwise, particularly as disease-resistant animals are generally not the highest producers.

Naturally resistant animals are already proving their worth in areas where trypanosome infection is a problem (Trail *et al.*, 1991). There are possibilities of emulating this success and reducing the incidence of scours, lameness, mastitis and internal parasites by similar means. An important objective of this book is to increase awareness at all levels in the possibilities of exploiting natural and induced immunities by reviewing what is already being achieved and considering fruitful areas for further investigation.

References

Booth, J.M. (1988) Update on mastitis: control measures in England and Wales. How have they influenced incidence and aetiology? *British Veterinary Journal* 144, 316–22.

Brown, F. (1986) Review lecture. Foot and mouth disease – one of the remaining great plagues. *Proceedings of the Royal Society London B* 229, 216–26.

Davies, G. (1985) Art, science and mathematics: new approaches to animal health problems in agriculture. *Veterinary Record* 117, 263–7.

Dodd, F.H. (1987) The role of therapy in mastitis control. *Proceedings of the International Mastitis Symposium*, Canada 1987, 161–75.

Emanuelson, U., Danell, B. and Philipsson, J. (1988) General parameters for clinical mastitis, somatic cell counts and milk production estimated by multiple trait restricted maximum likelihood. *Journal of Dairy Science* 71, 467–76.

Eriksson, J.A. (1991) Mastitis in Cattle. In: J.B. Owen and R.F.E. Axford (eds), *Breeding for Disease Resistance in Farm Animals.* CAB International, Wallingford, pp. 394–411.

Falkow, S. (1975) In: *Infectious Multiple Drug Resistance.* Pion Ltd., Chapter 4.

Fenner, F. and Ratcliffe, F.N. (1965) *Myxomatosis.* Cambridge University Press, Cambridge.

Francis, P.G. (1989) Update on mastitis: therapy. *British Veterinary Journal* 145, 302–11.

Grindal, R.J. (1988) Update on mastitis: the role of the milking machine in mastitis. *British Veterinary Journal* 145, 524.

Jones, T.O. (1987) Intramammary antibiotic preparations and cephalosporin resistance in *Salmonella typhimurium* 204C. *Veterinary Record* 120, 399.

Soil Association (1989) *Standards for Organic Agriculture.*

Thrusfield, M. (1986) *Veterinary Epidemiology.* Butterworths, London.

Trail, J.C.M., d'Ieteren, G.D.M. and Murray, M. (1991) Practical aspects of developing genetic resistance to Trypanosomiasis. In: J.B. Owen and R.F.E. Axford (eds), *Breeding for Disease Resistance in Farm Animals.* CAB International, Wallingford, pp. 224–34.

Vogel, R.J. and Stephens, B. (1989) Availability of pharmaceuticals in sub-Saharan Africa. Roles of the public, private and church mission sectors. *Social Science and Medicine* 29, 479–86.

Wall, R. and Strong, M. (1987) Effects of ivermectin on fauna in dung. *Australian Veterinary Journal* 64, 9.

Waller, P.J. (1985) Resistance to antihelminthics and the implications for animal production. In: N. Anderson and P.J. Waller (eds), *Resistance in Nematodes to Anthelminthic Drugs.* CSIRO Australia.

Chapter 2

Genetic Basis of Disease Resistance in Chickens

N. Bumstead, B.M. Millard, P. Barrow and J.K.A. Cook
The Institute for Animal Health, Houghton Laboratory,
Houghton, Huntingdon, Cambridgeshire PE17 2DA, UK

Summary

Variation in resistance to disease is a widespread phenomenon common in all species, and the genetic mechanisms which contribute to resistance are correspondingly varied. Roughly genetic resistance can be divided into three categories: first, genes involved specifically in host resistance to disease, such as the class I and class II genes of the major histocompatibility complex. A second category of genes are those possessing structural or metabolic functions which incidentally vary in ways which affect disease resistance, for example the genes coding for retroviral receptors. Third, genes derived from pathogens themselves may confer resistance, either as a consequence of natural processes, as in blocking of the E subgroup avian leukosis receptor by endogenous viral envelope genes, or following artificial introduction of genes, as in the equivalent blocking of A subgroup virus in transgenic animals.

It is clear that no genes conferring universal resistance to disease exist, since animals resistant to some pathogens are more susceptible to others, a situation particularly evident in the response to infection by different species of *Eimeria* in chickens. It seems likely that in some cases polymorphism in resistance genes is selectively maintained, notably among major histocompatibility complex haplotypes. However, in most cases there are insufficient data to say whether observed polymorphisms are stable or transient as rather little work has been done to assess gene frequencies in commercial (i.e. non-laboratory) environments. Fortunately, present developments in techniques for identifying and isolating genes should lead to a dramatic improvement in our understanding of this area as it becomes possible to discern the molecular mode of action of the resistance genes and to detect the genes efficiently on a commercial scale.

Introduction

Genetic resistance to disease is a widespread phenomenon common in all species, and it is possible to detect differences in susceptibility to most diseases. However, the mechanisms by which different pathogens reproduce and propagate themselves are very diverse and it would be surprising if the genes affecting disease resistance were not equally diverse. At present a number of factors limit our understanding of disease resistance: even in model species such as mice only some of the genes involved in the immune response have been isolated, and the structure or normal function of even fewer non-immune genes has been identified. Since the frequency of these genes within commercial flocks is generally unknown, and since studies of the epidemiology of infections are generally lacking it is difficult to determine the selective pressures which bear on resistance genes.

As an approach to these problems in chickens we have compared the resistance of a number of inbred lines to a range of diseases, in order to look for common resistance elements and to identify defined resistant and susceptible lines. These have then been used to investigate differences in pathology and to provide crosses in which to test candidate resistance genes. Although this approach has potential dangers, principally in that inbred lines may be fixed for recessive deleterious genes which would not normally be expressed in outbred populations, its potential as an initial step to gene identification has been amply demonstrated in mice.

Resistance to coccidiosis

Seven species of *Eimeria* can cause coccidiosis in chickens: *Eimeria maxima, E. tenella, E. praecox, E. necatrix, E. brunetti, E. acervulina* and *E. mitis.* Ingestion of oocysts leads to intracellular infection of the gut wall, in which the parasites undergo several cycles of asexual replication before entering a sexual phase which is followed by the release of oocysts into the lumen of the gut; these are excreted with the faeces and become infective for other chickens. Infection appears to be self-limiting, if ingestion of further oocysts is prevented, and in most species a strong protective immune response is developed which prevents the establishment of subsequent reinfections. Relatively few birds are initially infected, by low numbers of oocysts. Multiplication and excretion of parasites from these birds causes a rapid increase in oocyst numbers in the litter which results in the remainder of the flock becoming heavily infected by increasingly large

numbers of parasites, until eventually the infection ceases when all birds
have been infected and are immune.

Infection causes reduced growth rate and reduced feed efficiency and
may cause mortality at high dose levels, particularly in young chicks. The
consequences of infection are dependent on the dose and are also affected
by age of the chicken. The biology and pathogenesis of coccidial infections
have been extensively reviewed (Long, 1982).

Variation in resistance to coccidial infection has been reviewed (Jeffers,
1978; Wakelin and Rose, 1990) but most of this work has involved diverse
and often transient chicken stocks, and used a variety of measures of infection including mortality, lesion scores and weight gain.

More systematic experiments have been carried out by Bumstead and
Millard (1987), Clare *et al.* (1985) and Lillehoj (1988). In particular,
oocyst production is preferred as a measure of resistance, since this
determines the degree of contamination of the environment and hence the
consequences of infection.

The initial experiments consisted of inoculating eight widely available
inbred lines of chickens with a standard dose of sporulated oocysts at three
weeks of age. The numbers of oocysts present in the faeces were counted
on a daily basis throughout the subsequent infection. The results showed
that there were considerable differences in susceptibility between the inbred
lines; for all seven species there were highly significant differences in
numbers of oocysts produced, with differences between the most and least
susceptible lines being at least twofold (Table 2.1). Although some lines
produced oocysts earlier or for longer than other lines, these differences
were relatively small and in general the time course of infections was similar

Table 2.1.   Resistance to *Eimeria* spp. in various inbred lines of chickens.

Line	Eimeria acervulina	Eimeria praecox	Eimeria necatrix	Eimeria maxima	Eimeria mitis	Eimeria brunetti	Eimeria tenella
BRL	72.7[a]	64.4	7.13	18.3	45.5	24.2	18.3
WL	99.0	36.1	3.45	15.6	16.6	13.6	29.4
15I	112.0	77.3	8.89	29.3	53.7	36.3	8.9
7	97.3	77.0	2.95	18.2	45.3	43.5	7.1
6	109.0	78.9	3.76	24.6	47.1	34.7	13.9
C	45.8	35.4	3.18	13.9	37.4	36.2	26.9
N	71.1	41.9	3.66	21.9	45.2	13.8	9.6
O	45.2	70.6	4.67	19.2	43.3	17.4	11.0

[a]Millions of oocysts produced/bird resulting from the inoculation of 100 oocysts
at 3 weeks of age.

Values are means for groups of nine birds.

in different lines. There was no indication of a major difference such as that found in mice where asexual cycling of infection by *E. vermiformis* appears to be controlled by the immune response and the duration of oocyst production differs greatly between different strains of mouse (Rose *et al.*, 1984).

Comparison of the relative resistance of the different lines to the parasite species should indicate whether common resistance mechanisms operate with respect to the different species. Although the limited number of lines used restricts the sensitivity of such comparisons, in a number of cases there were significant correlations between the resistance of lines to the different species of *Eimeria* (Table 2.2). Even more interestingly, it is possible to relate these correlations so as to group species with relatively high or low common resistance. This seems to indicate a commonality of resistance between six of the species, with resistance to *E. maxima*, *E. mitis* and *E. praecox* being particularly highly correlated. However, all comparisons between resistance to *E. tenella* and the other six species produced negative correlations. This presumably indicates that genes which make the birds more resistant to, for example, *E. maxima* actually make the birds more susceptible to *E. tenella*.

Lillehoj *et al.* (1989) have reported that resistance to *E. tenella* is due to both MHC-associated and non-MHC-associated genes. We have attempted to assess the role of MHC genes in resistance to other species of *Eimeria*. In experiments in which F2 progeny of a cross between line N and line 15I were inoculated with *E. maxima* a consistent association of resistance to this species with the B^{21} haplotype relative to the B^{15} haplotype was found. The MHC-associated difference was approximately half the difference seen in the oocyst output of the parental lines. Interestingly this remained the case for differing strains of *E. maxima*, even where these were antigenically dissimilar and not fully cross-protective.

In similar experiments with *E. praecox* we have also found evidence of

Table 2.2. Rank correlation of oocyst output across lines.

	E. acervulina	*E. praecox*	*E. necatrix*	*E. maxima*	*E. mitis*	*E. brunetti*
E. praecox	0.57[a]					
E. necatrix	0.24	0.43				
E. maxima	0.45	0.76	0.71			
E. mitis	0.60	0.83	0.60	0.79		
E. brunetti	0.29	0.48	−0.10	0.10	0.55	
E. tenella	−0.17	−0.62	−0.10	−0.55	−0.57	−0.55

[a]All correlations have 6 d.f.

MHC association, in this case accounting for rather less of the difference in susceptibility between these two lines. For *E. brunetti*, on the other hand, resistance appears to be wholly unassociated with the MHC.

Since these results indicated a significant role of MHC genes, and hence presumably of immune response mechanisms, we have carried out further experiments in which the immune response was ablated by treatment with β-methasone before and during challenge. This drug has a broad activity but its effect on immune response in the challenged birds was manifest by their lowered response to human red blood cells and to mitogens. However, the results, both for *E. maxima* and *E. mitis*, showed that the difference in susceptibility was only marginally affected by this treatment. Interestingly, the drug did have a very clear effect in postponing the excretion of the oocysts, perhaps due to an effect of the drug on the growth of enterocytes and hence on the rate at which the parasites within the enterocytes are carried up the villi before release. There was no evidence of a continuation of excretion in the treated birds such as might be expected if the reproduction of the parasite was terminated by immunity. This would seem to confirm the fixed nature of the parasitic life cycle in these species.

Resistance to eimerian infection therefore seems to be due to both MHC and non-MHC genes: the relative magnitude of the contributions of these genes differing for the different species. Resistance is not impaired by treatment with β-methasone, which would seem to exclude mechanisms involving participation of T and B cells and perhaps other aspects of the immune response. However, although there may be common factors in the resistance to six of the seven species of *Eimeria*, resistance to these species implies susceptibility to *E. tenella*.

Resistance to salmonellosis

Oral ingestion of *Salmonella* by chickens leads to colonization of the gut but also to systemic infections of the host following penetration of the gut wall by the bacteria. Although the presence of the bacteria in the gut may have little effect on the bird, systemic infections can cause morbidity and mortality, depending on the age of the bird and the pathogenicity of the invading serotype. In mice a high level of resistance to *Salmonella typhimurium* infection (which also confers resistance to other facultative intracellular bacteria) is determined primarily by the *Ity* gene (Lissner *et al.*, 1983; Swanson and O'Brien, 1983). In order to see whether similar resistance can be detected in chickens we have compared responses to oral or intramuscular infection in inbred lines of chickens, generally using mortality as an indicator of susceptibility.

In the initial experiments chicks were inoculated orally on the day of hatch with a standard dose of *S. typhimurium*; this revealed large differences in mortality in the different inbred lines (Bumstead and Barrow, 1988). These differences were also present when birds were challenged intramuscularly, and hence were not due to aspects of penetration of the gut wall. Indeed when birds were challenged intravenously similar differences in resistance were observed. Under these circumstances virtually all bacteria are rapidly trapped by the phagocytic cells of the spleen and liver, and studies of pathogenesis in these birds showed that bacteria continued to be localized within these organs until numbers of bacteria either declined in resistant birds or caused total disruption of these organs and release of bacteria into other parts of the body, leading rapidly to death. This suggests that resistance is a function of the phagocytic cells of the spleen and liver (and possibly also those present in other tissues). However, challenge of cultured adherent spleen cells *in vitro* did not show differences in the ability of cells from resistant and susceptible birds to kill bacteria although there was some indication that resistant cells may survive ingestion of the bacteria better. It was also evident that resistant lines of chickens gave a greater yield of phagocytic cells per spleen, though whether this contributes significantly to the differences in resistance of the different lines is unclear.

Challenge of birds with *Salmonella* of other serotypes showed considerable consistency of resistance across serotypes. Challenged with *S. pullorum*, *S. gallinarum* or *S. enteriditis* lines 6_1, WL and N were resistant to all four serotypes, 15I, 7_2 and C being relatively susceptible (Table 2.3).

Crosses and backcrosses between pairs of resistant and susceptible lines indicated that this resistance is fully dominant, not sex-linked or associated with the MHC and may be due to a single gene. Within our experiments there was little evidence of genetic differences in time course of mortality

Table 2.3. LD$_{50}$ values following i.m. inoculation of flocks with differing serotypes of *Salmonella*.

Line	*S. typhimurium*	*S. gallinarum*	*S. pullorum*	*S. enteriditis*
6_1	4.53 ± 0.20[a]	> 5 —	6.23 ± 0.25	5.82 ± 0.15
WL	5.0 ± 0.43	4.54 ± 0.76	> 6 —	3.51 ± 0.28
N	3.0 ± 0.24	> 5 —	5.29 ± 0.34	3.23 ± 0.28
7_2	1.7 ± 0.36	1.74 ± 0.49	4.50 ± 0.38	4.22 ± 0.24
15I	< 1.0 —	< 2 —	5.93 ± 0.37	<2
C	0.84 ± 0.23	0.36 ± 1.1	4.80 ± 0.28	1.76 ± 0.22
φ				2.42 ± 0.25
Br1				3.11 ± 0.27

[a]LD$_{50}$ ± standard error (log$_{10}$ bacteria).

but this would not preclude a secondary role of MHC genes in speeding elimination of infection, such as that described in mice by Nauciel *et al.* (1990).

Resistance to infectious bronchitis

Infectious bronchitis is a highly contagious respiratory disease of chickens caused by a coronavirus, infectious bronchitis virus (IBV). The disease causes a respiratory infection in young birds and in older birds results in reduced egg production and poor egg quality. Although the primary infection takes place via the trachea, viral infection of visceral organs also takes place. The consequences of infection are exacerbated by secondary infection by other microorganisms present in the trachea at the time of infection and bacterial infection of internal organs is common (Cunningham, 1970; Smith *et al.*, 1985).

In order to mimic the natural infection we have used the protocol developed by Smith *et al.* (1985) in which birds are challenged intranasally with a mixed infection of virulent strains of IBV together with pathogenic strains of *Escherichia coli*. Birds were challenged at two weeks of age and resistance assessed on the basis of mortality, although the extent of bacterial infection was also assessed both in birds that died and those that survived to the end of the experiment. Birds of different lines showed very considerable differences in mortality, ranging from 3% in the BrL line to 87% in line 7_2 (Table 2.4).

Table 2.4. Resistance of inbred lines to infectious bronchitis.

Line	Mortality: mixed infection (%)	Mortality: virus alone (%)	Mortality: *E. coli* alone LD_{50}	s.e.
LS	16.7[a]	0.0[a]	5.76	0.13
RIR	55.2	9.9	6.21	0.15
BrL	3.3	3.3	5.67	0.14
N	40.0	0.0	5.96	0.16
WL	43.3	3.3	5.41	0.13
C	10.7	0.0	5.62	0.14
15I	80.0	46.2	5.31	0.13
6_1	60.0	9.9	6.37	0.13
7_2	86.7	82.5	6.51	0.20

[a]30 birds challenged per line.
Source: Bumstead *et al.* (1989).

When birds of these lines were challenged intranasally with the bacterial component of the inoculum alone no mortality was observed. Intravenous inoculation of the bacteria did cause mortality but only at very high dose levels and although there were statistically significant differences in LD_{50} dose levels between lines, the pattern of mortality across lines did not correlate with that seen for the mixed or viral infections. Hence differences in susceptibility to the bacteria are probably not responsible for the differences in resistance to the mixed infection (Table 2.4).

Crosses between resistant and susceptible lines were fully resistant and there was no evidence of maternal inheritance of resistance or of sex linkage.

The intermediate levels of mortality associated with some lines (for example line WL) may be due to a uniformly intermediate phenotype within these birds, but may also be a consequence of these lines segregating for resistant and susceptible genotypes, since although these are relatively inbred lines of chickens they were not inbred to the levels attained in mice.

Segregating F_2 and backcross populations showed no evidence of association of resistance with MHC haplotype, and levels of mortality were formally consistent with a single autosomal resistance gene. However, this is unlikely to be the case, since the pathology of infection differed in different resistant lines and between different susceptible lines. Particularly striking was the fact that line 7_2 showed very high mortality when challenged by virus alone; while a variety of microorganisms would have been present in the trachea there was also little evidence of visceral bacterial invasion in this line. In line 7_2 it appears, therefore, that the majority of death may be a direct consequence of viral infection.

In other lines omitting the bacterial component of the challenge reduced mortality, and in birds challenged with the mixed infection visceral bacterial lesions were present in many birds. In particular it was striking that despite very low mortality most birds of the BrL line showed evidence of visceral bacterial infection, whereas birds of other less resistant lines did not.

Otsuki *et al.* (1990) have shown that resistant birds of line C do undergo infection both in terms of clinical signs and of viral infection of internal organs. In comparisons between line C and line 15I both lines appeared equally susceptible to initial infection; however, the duration of viral infection was shorter in the resistant line (9 days) than in the susceptible line (18 days or longer). When tissues of the two lines were challenged *in vitro* no differences in infectivity or ability of the virus to replicate were observed.

Since resistant birds do show signs of infection, and virus can be recovered from visceral organs of these birds it seems unlikely that resistance is a consequence of the presence or absence of a cellular receptor as in the case of the murine coronavirus, mouse hepatitis virus (Boyle *et al.*, 1987). The difference in duration, rather than onset of infection suggests that differences in immune response are responsible for resistance. However Cook *et al.* (1990) have shown that systemic levels of IBV-specific antibody do not differ between lines C and 15I during infection, and as noted above it seems that genes of the MHC are also not involved.

Resistance to other diseases

In addition to the resistance to coccidiosis, salmonellosis and infectious bronchitis described above genetic resistance to avian leukosis and to Marek's disease have long been extensively documented. In the former case resistance is principally due to the absence of cellular receptors for the virus. Five subgroups of the virus can be rigidly defined by their host range, which reflects the presence of differing cellular receptors specific for each subgroup (reviewed in Weiss *et al.*, 1984). At present none of these receptors has been isolated biochemically despite considerable efforts (Moldow *et al.*, 1979), and it is not known whether resistance is due to the total absence of the receptor molecules or an expression of polymorphism. A putative receptor gene for murine leukaemia virus has now been cloned (Albritton *et al.*, 1989), and this may open up new possibilities of identifying the equivalent avian receptor genes.

In addition to resistance at the level of the receptor genes both MHC and non-MHC genes are involved in control of tumour development in those birds in which the site of integration of the retrovirus causes transformation (Bacon *et al.*, 1980).

Resistance to Marek's disease is principally a reflection of MHC haplotype, first described by Hansen *et al.* (1967), although other genes may also have an effect (Gilmour *et al.*, 1976). The mechanism of resistance has received much study (reviewed in Calnek, 1985), and appears to act by limiting the development of tumours in infected birds, rather than by preventing initial infection or viraemia. More virulent strains of Marek's disease virus have been identified recently, but these appear to be generally more pathogenic and birds with, for example, the B^{21} haplotype remain relatively more resistant.

Lamont *et al.* (1987) have recently described variation in resistance to *Pasteurella multocida*. In a cross between lines they found differences in survival following intramuscular inoculation. Comparing segregating MHC haplotypes they found small but statistically significant differences in survival between MHC haplotypes.

Conclusions

The results discussed above indicate the very limited state of our knowledge of the genes involved in disease resistance and the mechanisms by which they operate. However, it is possible to draw some limited conclusions from a comparison of the resistance of the different lines to the various pathogens (Table 2.5). It is clear that no line is consistently resistant to all of even the limited set of diseases for which we have information, strongly suggesting the absence of universal resistance genes. This is particularly well shown by the inverse relationship between resistance to *E. tenella* and resistance to the other coccidial species.

Line 15I is generally highly susceptible; this may well be a consequence of its derivation, since the line was selected for susceptibility to both lymphoid leukosis and Marek's disease. This process may have simultaneously selected for many susceptibility genes and, by the process of selection and inbreeding, fixed a number of what are, in effect, sublethal genes for immune functions.

The genetic resistance described above ranges from an apparent lack of involvement of MHC genes in the case of salmonellosis through a mixture of both MHC and non-MHC genes involved in resistance to the coccidial species to the highly MHC-associated resistance to Marek's disease. The labour of comparing the very numerous MHC haplotypes in a wide range of genetic backgrounds makes it impracticable to arrive at a precise estimate of the relative contributions of MHC and non-MHC components, though it seems likely that some element of MHC involvement will be present in the response to any disease due to its central role in the immune system.

Functionally, the chicken MHC resembles those of mammals in determining presentation of foreign antigens by class I and class II molecules (Dietert *et al.*, 1990) and also codes for at least some components of the complement system (Lee and Nordskog, 1982). The molecular organization of the chicken MHC does differ in some respects from those of humans or mice, principally by being much more compact and by the apparent absence of class IIA genes (Guillemot *et al.*, 1989). It is not known whether chicken equivalents of other genes such as those coding for the tumour necrosis factors are also located in or near the complex. Unfortunately the lack of readily occurring recombinational events within the chicken MHC (presumably a reflection of its small chromosomal size) makes it difficult to ascribe the disease resistance properties of the complex as a whole to individual genes within the MHC. Possibly the progress being made in defining the MHC at the molecular level may allow recombinants to be detected or alternatively allow identification of gene products and hence function.

With the exception of the immunoglobulin loci little is presently known of the location of other genes of the immune system or of the non-immune genes involved in disease resistance in chickens. Increasing numbers of genes of the immune response have been identified in mice and it may in

Table 2.5. Summary of known resistance properties of the inbred lines.

	Eimeria acervulina	Eimeria praecox	Eimeria necatrix	Eimeria maxima	Eimeria mitis	Eimeria brunetti	Eimeria tenella	Salmonellosis	E. coli	MD	IBV	Avian leukosis A	B	C	D	E
BrL	M	S	S	M	S	M	M	R	M	S	R	S	S	S	S	M
WL	S	R	R	R	R	R	S	R	R	S	M					R
151	S	S	S	S	S	S	R	S	R	S	S	S	S	S	S	R
7	S	S	R	M	S	S	R	M	S	R	S	R	R	S	R	R
6	S	S	R	S	S	S	R	R	S	R	M	S	S	S	S	R
C	R	R	R	R	S	S	S	S	M	R	R	R	R	S	S	R
N	M	R	R	M	R	R	R	R	M	R	M	S	S	S	S	
O	R	S	M	M	S	R	R	S		R		S	S	S	S	R

S: Relatively susceptible.
M: Intermediate.
R: Relatively resistant.

time be possible to identify the corresponding avian genes on the basis of their DNA homology to the murine genes. The recent isolation of a retroviral receptor in mice (Albritton *et al.*, 1989) also offers some indication that progress can be made using novel techniques. However, the *Ity* gene has not yet been identified at the molecular level despite much work, and this highlights the difficulty of identifying individual resistance genes at the molecular level.

Perhaps the most promising prospect for progress both in chickens and in the other domestic species lies in the development of linkage maps based on molecular markers. This should allow a generalized approach to the identification of resistance genes and by using linked markers to assess their individual properties and characteristics.

References

Albritton, L.M., Tseng, L., Scadden, D. and Cunningham, J.M. (1989) A putative ecotropic retrovirus receptor gene encodes a multiple membrane-spanning protein and confers susceptibility to virus infection. *Cell* 57, 659–66.

Bacon, L.D., Witter, R.L., Crittenden, L.B., Fadly, A. and Motta, J. (1980) β-haplotype influence on Marek's disease, Rous sarcoma, and lymphoid leukosis virus-induced tumours in chickens. *Poultry Science* 60, 1132–9.

Boyle, J.F., Weissmiller, D.G. and Holmes, K.V. (1987) Genetic resistance to mouse hepatitis virus correlates with absence of virus-binding activity on target tissues. *Journal of Virology* 61, 185–9.

Bumstead, N. and Barrow, P.A. (1988) Genetics of resistance to *Salmonella typhimurium* in newly hatched chicks. *British Poultry Science* 29, 521–30.

Bumstead, N., Huggins, M.B. and Cook, J.K.A. (1989) Genetic differences in susceptibility to a mixture of avian infectious bronchitis virus and *Escherichia coli. British Poultry Science* 30, 39–48.

Bumstead, N. and Millard, B.M. (1987) Genetics of resistance to coccidiosis: response of inbred chicken lines to infection by *Eimeria tenella* and *Eimeria maxima. British Poultry Science* 28, 705–16.

Calnek, B.W. (1985) Genetic resistance. In: Payne, L.N. (ed.), *Marek's Disease.* Martinus Nijhoff, Boston.

Clare, R.A., Strout, R.G., Taylor, R.L. Jr., Collins, W.M. and Briles, W.E. (1985) Major histocompatibility (B) complex effects on acquired immunity to cecal coccidiosis. *Immunogenetics* 22, 593–9.

Cook, J.K.A., Otsuki, K., Huggins, M.B. and Bumstead, N. (1990) Investigations into the resistance of chicken lines to infection with infectious bronchitis virus. *Advances in Experimental Medicine and Biology* 276 (in press).

Cunningham, C.H. (1970) Avian infectious bronchitis. *Advances in Veterinary Science* 14, 105–48.

Dietert, R.R., Taylor, R.L. Jr and Dietert, M.F. (1990) The chicken major histocompatibility complex: structure and impact on immune function, disease resistance and productivity. *Monographs in Animal Immunology* 1, 7–26.

Gilmour, D.G., Brand, A., Conelly, N. and Stone, H.A. (1976) Bu-1 and Thy-1, two loci determining surface antigens of B lymphocytes or T lymphocytes in the chicken. *Immunogenetics* 3, 549–63.

Guillemot, F., Kaufman, J.F., Skjoet, K. and Auffray, C. (1989) The major histo-compatibility complex in the chicken. *Trends in Genetics* 5, 300–4.

Hansen, H.P., van Zandt, N. and Law, G.R.J. (1967) Differences in susceptibility to Marek's disease in chickens carrying two different B locus blood group alleles. *Poultry Science* 46, 1268.

Jeffers, T.K. (1978) Genetics of coccidia and the host response. In: Long, P.L., Boorman, K.N. and Freeman, B.M. (eds), *Avian Coccidiosis.* British Poultry Science Limited, Edinburgh, pp. 51–125.

Lamont, S.J., Bolin, C. and Cheville, N. (1987) Genetic resistance to fowl cholera is linked to the major histocompatibility complex. *Immunogenetics* 25, 284–9.

Lee, W.H. and Nordskog, A.W. (1982) Complement activity and the MHC in the chicken. *Poultry Science* 61, 1500.

Lillehoj, H.S. (1988) Influence of inoculation dose, inoculation schedule, chicken age, and host genetics on disease susceptibility and development of resistance to *Eimeria tenella* infection. *Avian Diseases* 32, 437–44.

Lillehoj, H.S., Ruff, M.D., Bacon, L.D., Lamont, S.J. and Jeffers, T.K. (1989) Genetic control of immunity to *Eimeria tenella.* Interaction of MHC genes and non-MHC linked genes influences levels of disease susceptibility in chickens. *Veterinary Immunology and Immunopathology* 20, 135–48.

Lissner, C.R., Swanson, R.N. and O'Brien, A.D. (1983) Genetic control of the innate resistance of mice to *Salmonella typhimurium*: expression of the *Ity* gene in peritoneal and splenic macrophages isolated *in vitro. Journal of Immunology* 131, 3006–13.

Long, P.L. (ed.) (1982) *The Biology of the Coccidia*, University Park Press, Baltimore.

Moldow, C.F., Reynolds, F.H.Jr, Lake, J., Lundberg, K. and Stephenson, J.R. (1979) Avian sarcoma virus envelope glycoprotein (gp85) specifically binds chick embryo fibroblasts. *Virology* 97, 448–53.

Nauciel, C., Ronco, E. and Pla, M. (1990) Influence of different regions of the H-2 complex on the rate of clearance of *Salmonella typhimurium. Infection and Immunity* 58, 573–4.

Otsuki, K., Huggins, M.B. and Cook, J.K.A. (1990) Comparison of the suscept-ibility to avian infectious bronchitis virus infection of two inbred lines of white leghorns. *Avian Pathology* 19, 467–75.

Rose, M.E., Owen, D.G. and Hesketh, P. (1984) Susceptibility to coccidiosis: effect of strain of mouse on reproduction of *Eimeria vermiformis. Parasitology* 88, 45–54.

Smith, H.W., Cook, J.K.A. and Parsell, Z.E. (1985) The experimental infection of chickens with mixtures of infectious bronchitis virus and *Eschericia coli. Journal of General Virology* 66, 777–86.

Swanson, R.N. and O'Brien, A.D. (1983) Genetic control of the innate resistance of mice to *Salmonella typhimurium*: *Ity* gene is expressed *in vivo* by 24 hours after infection. *Journal of Immunology* 131, 3014–20.

Wakelin, D. and Rose, M.E. (1990) Immunity to coccidiosis. In: P.L. Long (ed.),

Coccidiosis of Man and Domestic Animals. CRC Press, Boca Raton, Florida, pp. 281–306.

Weiss, R., Teich, N., Varmus, H. and Coffin, J. (eds) (1984) *RNA Tumour Viruses,* 2nd edn. Cold Spring Harbour Laboratory.

Chapter 3

Genetic Improvement of the Immune System: Possibilities for Animals

M.J. Doenhoff[1] and A.J.S. Davies[2]

[1]School of Biological Sciences, University of Wales, Bangor, Gwynedd, LL57 2UW, UK and [2]The Institute of Cancer Research: Royal Cancer Hospital, The Haddow Laboratories, 15 Cotswold Road, Belmont, Sutton, Surrey SM2 5NG, UK

Summary

A genetic basis to the differences that exist between individuals of a species with respect to their capacity to resist infection has been demonstrated in many instances. However, discussion of the evidence is complicated by the widely different connotations of the words 'resistance' and 'susceptibility' as used in the literature, and also because resistance/susceptibility profiles are often attributable to the effects of more than one gene. The effects of one gene may be augmented or cancelled out by those of another.

Three sets of genes which govern the response of vertebrate hosts to infection can be distinguished; namely, those which control 'innate immunity', those which determine the specificity of acquired immune responses, and those which affect the quality of acquired immunity. The genetically controlled effects of innate and acquired immunity have at times been difficult to distinguish between.

The *Bcg/Lsh/Ity* gene, which prevents establishment of intracellular pathogens in mice, is one of the most extensively studied models of genetic control of infection. The gene product is involved in macrophage physiology, and it thus, not surprisingly, also seems to modify the quality of acquired immune responses. The extent to which this gene provides selective advantage in terms of enhanced survival of the murine host has still to be defined, and its homologue has not yet been found in other host species.

The specificity of an acquired immune response is determined by the histocompatibility–antigen complex, the T cell receptor and the immunoglobulin molecule. As a result of a multiplicity of genetic elements, and the processes of their rearrangement during T and B lymphocyte maturation, the potential for diversity of the antigen combining sites of both the T cell receptor and immunoglobulin molecules is considerably larger than the number of lymphocytes in the body. There would thus seem to be little scope for enhancing the specificity of acquired immune responses.

The amount and isotype of antibody produced during an immune response, the type and quantity of cytokine production, and the longevity of immunological memory are some of the factors affecting the 'quality' of immune responsiveness and progress is being made in identifying the relevant genes. Unfortunately, selection of a particular genetic trait (for example, high antibody production) does not necessarily lead to uniformly high levels of resistance to different infections.

Consideration of some of these paradoxes, together with the fact that infecting organisms are seldom rapidly and completely eliminated from the body, would seem to conflict with commonly held notions of the hegemonical role of the acquired immune response in controlling infectious diseases.

Introduction

Evidence from many sources indicates that individuals of vertebrate species vary in their capacity to resist, control and/or reject infections, and that these differences are under genetic control; trypanotolerance (Murray *et al.*, 1982) and differential resistance of sheep breeds to infection with *Haemonchus contortus* (Altaif and Dargie, 1978a,b; Gray, 1987) are two examples of importance in agriculture. It is, on the other hand, also clear that overall the complex and dynamic interactions between the host's immune response and its pathogens are controlled by many genes. Wakelin (1989), for example, has listed 16 distinct parameters of the immune response to infection with *Trichinella spiralis* that are known to vary between strains of mice.

Unfortunately, the relationships between the genes controlling these differences and those that govern the specificity or quality of acquired immune responses have not often been established. Thus, strains selected for particular responses, either in respect of resistance to one pathogen species (Brindley and Dobson, 1983; Windon and Dineen, 1984), or for a more general enhanced immune response potential (e.g. mouse lines reared for high antibody production (Biozzi *et al.*, 1968) are not necessarily resistant against all pathogens.

A problem of terminology

Our intention is to review the various facets of resistance to infection that are mediated by innate and acquired mechanisms of immunity, indicating for each what is known about the genetic determination of the response.

Unfortunately different connotations of the words 'resistance' and 'susceptibility' may give rise to confusion: the discrepancy between usage of these words in experimental and clinical settings is particularly wide. Thus, experimental host phenotypes defined as resistant range from those which are completely impervious to entry, let alone growth of pathogens, to those which allow pathogen survival and replication, but in which the infection is eventually controlled. In the instance of African trypanosome infections of mice, host strains which merely survive an infection longer than others are considered 'resistant', even though all infected animals eventually die. It is also a common observation that a more virulent strain of a pathogen will overcome the defences of hosts that have been typed resistant to isolates of lesser virulence.

In the public health context, Benenson (1985) considers man generally 'susceptible' to most of the communicable diseases with which he is afflicted, and little account is taken of differences in disease patterns that may exist between individuals or subpopulations consequent upon infection. There is often no relationship between the pathology associated with infection and the persistence, by replication or otherwise, of large numbers of the pathogen: the disease manifestations of the chronic phases of Chagas' disease (American trypanosomiasis) and schistosomiasis exemplify this.

Resistance and susceptible epithets can thus vary in their meaning, and are often dependent on the particular host–pathogen combination under study.

Control of infections by three different sets of genes

Genes which affect the host response to a pathogen may be distinguished according to whether they: (a) control the susceptibility or resistance of an animal to acquisition of an infection, i.e. are concerned with what has traditionally been called 'innate immunity'; (b) govern the specificity of acquired 'adaptive' immune responses: the known products of these genes are histocompatibility antigens, the antigen receptor on T cells, and immunoglobulins; and (c) affect the 'quality' of specific immune responses, including, for example, the isotype, rate of production, quantity, and affinity of antibody, and the activities of numerous cellular and humoral effector mechanisms which are used by the host in the expression of a state of acquired immunity.

Distinction between these various genes does not preclude the possibility that the overall interaction of a particular host organism with a potential pathogen is the outcome of expression of the products of all these sets of genes. The potential that might exist for genetic improvement of immunological resistance to disease is enormous, and it can only be exemplified here in outline. The recent explosive increase in our knowledge of the genetic basis of the specificity of acquired immune responses is also only touched upon.

Innate immunity

Factors traditionally listed as contributing to innate resistance include impenetrable barriers such as the skin, mucus and acidic secretions, ciliary activity in the respiratory tract, lysozyme in sweat and tears, and expropriation of habitable niches by normally harmless bacteria in the intestine. (In his immunology lectures W.J. Herbert is said to have included cows' tails in this category, for the defence they provide against disease-carrying insects: see Kennedy, 1990.) Innate resistance can thus be conferred by genes which have no obvious effect on acquired immunity: sickle trait and thalassaemia haemoglobinopathies which protect against malaria are an example (Allison, 1954; Miller and Carter, 1976; Weatherall *et al.*, 1988). In most instances, however, little is known about the degree of specificity of innate mechanisms in terms of the range of pathogens against which they are protective, nor about the controlling genes and their products.

The evolutionary development of innate resistance predates that of the adaptive immune response, but as specific immunity evolved its effectors have come to combine forces with innate mechanisms. The complement system, and phagocytic cells such as neutrophils and macrophages, are considered part of the innate system, but their activities are essential to the expression of certain types of acquired immune responses.

Difficulties in distinguishing between innate and acquired resistance

Schistosome infections of strain 129 mice

A criterion sought by many research programmes on immunity to infection, and particularly those in quest of a non-living vaccine, is a measurable adaptive immune response that correlates with good control of the infection in question, but studies on strain 129 mice infected with schistosomes illustrate the need for great rigour in interpretation of empirical evidence of resistance conferred by an acquired immune response. About 50% of 129/J infected mice were found to self-cure *Schistosoma japonicum* infections, and the cured mice had circulating antibody to a 26 kD polypeptide, subsequently shown by gene cloning and DNA sequencing to be glutathione *S*-transferase (GST) (Mitchell, 1989). Little anti-GST antibody was found in non-curing 129 strain mice or in other non-self-curing mouse strains such as BALB/c.

Subsequent work has attributed the differential handling of schistosome infections by 129 strain mice to within-strain variation in hepatic portal vasculature, with larvae in resistant mice having a greater chance of being shunted to the lungs rather than of becoming trapped in the liver where normal maturation could commence (Coulson and Wilson, 1989; Elsaghier *et al.*, 1989). Thus, it appears that instead of anti-GST antibody being an effector of anti-schistosome resistance in self-curing animals, it is produced as a consequence of the persistence and/or death of the parasite in an ectopic site, which in turn is due to an anatomical difference in the host. Earlier work showing that resistance to reinfection in *S. mansoni*-infected mice was due more to egg granuloma-induced portacaval shunting of hepatic portal blood than to a specific immune response (Harrison *et al.*, 1982) also drew attention to the fact that specific immunity was not the only means by which hosts acquired resistance.

Trypanosome virulence

The susceptibility/resistance response phenotype of different strains of mice to infection with *Trypanosoma brucei* also seemed at first to be related to production of antibody specific for the parasite variable surface glyco-protein antigens (VSG); thus, relative virulence (determined in terms of longevity of infected mice) of different clones appeared to be linked to VSG expression by the parasite (Inverso and Mansfield, 1983), and in resistant mice there was a temporal coincidence between clearance of blood parasites with the VSG of the infecting clone and the production of IgM antibody with specificity for that VSG (Levine and Mansfield, 1984).

Recent work on subclones of parasite selected for different degrees of virulence is indicative of a more complicated interplay between host and parasite genes, since the mouse genes controlling antibody-mediated clear-ance of parasitaemia (which were dominant) segregated independently of (recessive) genes controlling resistance to the parasite's lethal effects (De Gee *et al.*, 1988). Parasite-specific antibody responses were therefore not directly associated with host resistance against infection lethality.

The Bcg/Lsh/Ity gene

Probably the best known example of one gene controlling susceptibility to infection is the *Bcg/Lsh/Ity* gene that appears to regulate a property of macrophages in mice. Our present understanding about this gene (reviewed in Blackwell, 1989a) derives from work which showed that inbred mouse strains differed in their susceptibility to *Salmonella typhimurium* (Plant and Glynn, 1976), *Leishmania donovani* (Bradley, 1974) and infections of Bacille Calmette Guerin or BCG (Gros *et al.*, 1981), all of which are intracellular parasites of macrophages. In the spleens and livers of inoculated mice of resistant genotype, replication of the infecting organisms was slow or non-existent in the period following infection. After testing numerous inbred mouse strains and their hybrids the different research groups agreed that there is complete concordance between *Ity, Lsh* and *Bcg* with regard to resistance/susceptibility typing of mice, and resistance against infection by these different intracellular pathogens is most likely to be controlled by a single gene. The gene product has not been identified, but there is speculation that it is either a cytokine/interleukin or cyto-skeletal protein which maintains macrophages in a heightened activation state (Blackwell *et al.*, 1989; Buschman *et al.*, 1989).

Specificity of action of the Bcg gene

BCG-resistant mouse strains are also insusceptible to infection by other mycobacteria species, including *Mycobacterium lepraemurium, M. intra-cellulare* and *M. smegmatis* (Schurr *et al.*, 1990). Relative specificity of the action of the *Bcg/Lsh/Ity* gene is indicated, however, by the observation that mice with Lsh^r phenotype were no more protected against either cutaneous or viscerotropic infections of *Leishmania major* than Lsh^s mice (Blackwell, 1989b; Mock *et al.*, 1985). Innate resistance and susceptibility to *L. major*, and possibly also to *L. mexicana*, appears to be under the control of a distinct gene, *Scl* (Blackwell, 1989b).

The *Lsh/Ity/Bcg* gene has been located on the proximal end of mouse chromosome 1, and a linkage group of five other loci that have been shown to be congenic to *Lsh/Ity/Bcg* in the mouse are conserved in the same order on the distal region of human chromosome 2q (Schurr *et al.*, 1990). It is not known whether the putative human homologue of the *Bcg* gene influences the outcome of attempted vaccination of humans with BCG; if so, therein might lie an explanation for the great variation in efficacy of human vaccination programmes in different parts of the world (Fine, 1989). That the *Bcg* gene controls susceptibility to the consequences of infection by virulent mycobacteria, is doubtful. Pioneering experiments by Lurie *et al.* (1952) did indeed demonstrate that resistance to *Myco-bacterium tuberculosis* could be selected for in rabbits, and mouse strains have also been shown to differ in their capacity to control this pathogen (Gray *et al.*, 1960), but a clear relationship between the *Bcg* gene and resistance to *M. tuberculosis* has not been unequivocally demonstrated; rather, the opposite, since (a) in experiments in which resistance to infec-tion with BCG has been demonstrated the bacteria were introduced intra-venously and not via the respiratory tract, though the latter is the more usual route of entry of pathogenic tubercle bacilli; (b) the gene had no influence when BCG substrains of high virulence were used (Orme *et al.*, 1985); and (c) a lack of concordance between resistance to BCG and resistance to *M. tuberculosis* was found in the Biozzi strain of mice (Gheorghiu *et al.*, 1985).

Co-ordination between innate and acquired resistance

A single dominant autosomal gene gives C57BL strain mice innate protection against doses of ectromelia virus that are between 10^2 and 10^7 times greater than those that will kill other mouse strains (Schell, 1960), and the gene product appears to impose an early barrier to viral penetration and its spread via lymphatics or blood (O'Neill and Blanden, 1983). It was considered that the delay which ensued after virus inoculation of innately resistant mice allowed time for development of the specific cell-mediated cytotoxic immune response which controlled the pathogenicity of the infection.

If an ectromelia infection was imposed on innately resistant congenic B10 mouse strains by inoculation with a more virulent isolate of virus, the subsequent pattern of infection was governed by genes linked to the H-2 histocompatibility locus (O'Neill *et al.*, 1983) which, as will be seen, have crucial regulatory effects on acquired immune responses. As with ectromelia infections, the H-2 complex affects whether late phase infections of *L. donovani* in Bcg^s strains of mice are self-cured or not (Blackwell, 1989b).

Acquired immunity

Antigens which come into the intact mammalian body are exposed to a variety of degradative processes (antigen processing), from which emerge a welter of epitopic determinants that become associated with histocompatibility antigens for presentation to reactive lymphocytes. Helper T cells have a pivotal role in acquired immunity: they can react to antigen via a receptor of genetically predetermined structure, provided the antigen is presented to them in a complex with histocompatibility antigens on the surface of the antigen presenting cells. Helper T cells also react with other cells, including other T cell subpopulations, and with B cells, the latter leading to synthesis of antibody by the B cell, but all such interactions can occur only between cells of the same histocompatibility type (MHC restriction).

Specificity of the immune response

Specificity for the inducing antigen is one of the hallmarks of acquired immunity, and this specificity is governed by the interaction of the products of three gene complexes which code, respectively, for histocompatibility antigens, T cell receptors and B cell immunoglobulin receptors.

The major histocompatibility complex (MHC)

Structure and function   Histocompatibility antigens occur in two classes: class I antigens, the classic transplantation antigens, are found on most cells in the body, and are composed of a membrane-bound polypeptide chain of approximately 43 kD invariably associated with a molecule of β_2-micro-globulin of 12 kD. Distribution of class II histocompatibility antigens is more restricted, particularly to cells of the reticuloendothelial system, and they each consist of a heterodimer of a 33 kD alpha chain and a 28 kD beta chain.

The antigen presenting function of histocompatibility antigens was recently put into clearer perspective by crystallographic analysis of a purified human class I antigen (Bjorkman *et al.*, 1987). A groove formed by the two terminal domains of the 43 kD polypeptide is assumed to accommodate foreign peptides derived from 'processed' antigen molecules, and which can elicit responses from T cells. Modelling indicates a similar peptide-binding groove structure is formed by the terminal domains of each of the two polypeptides constituting class II heterodimers (Brown *et al.*, 1988).

Convention has it that class I and class II histocompatibility molecules receive antigenic epitopes via two different processing pathways, and their antigen presentation activities are respectively directed to two different classes of T cells – killers and helpers. The division makes some biological sense, since the activities of cytotoxic $CD8^+$ T cells can thereby be directed toward any class I-bearing cells in the body that may be carrying endogenously processed antigen derived from intracellular parasites (e.g., viruses). $CD4^+$ helper T cells on the other hand react with epitopes borne by class II positive cells such as macrophages and B cells which themselves require signals from T cells to become activated or to produce antibody.

Histocompatibility antigens can present a wide variety of epitopes to T cells because multiple genetic loci exist for both class I and class II antigens, both alleles are expressed in heterozygotes, and the genes are highly polymorphic. Gaps in the antigen presenting repertoire of the MHC were detected when synthetic peptides were used for testing immune responsiveness (Benacerraf, 1981). Recent studies indicate that although throughout a species individuals vary with respect to the particular T cell epitopes in a protein antigen which their MHC haplotypes can present, the class II molecules of any individual will be able to interact with and present at least one epitope in most antigens to T cells (Roy *et al.*, 1989).

Histocompatibility antigens and disease   During the course of an infection numerous different antigens will be processed and presented by class II-bearing cells for recognition by helper T cells. Studies in mice infected with *Ascaris suum* (Kennedy *et al.*, 1990), *Schistosoma mansoni*

(Kee *et al.*, 1986) and *Trichuris muris* (Else and Wakelin, 1989) have indeed shown that there is a MHC-restricted failure to produce antibody against some, as yet uncharacterized, antigens of each of these parasites, but there is so far little evidence that a particular gap in the antibody repertoire affects antiparasite resistance.

Conversely, there have been numerous demonstrations of MHC-linked effects on the pathogenesis of infectious and various idiopathic diseases, but in many cases the underlying mechanisms have not been elucidated. In some instances there is a negative or positive correlation between particular MHC haplotypes and the incidence of disease; an early example was the linking of susceptibility of mice to lymphocytic choriomeningitis virus infections to H-2 type (Oldstone *et al.*, 1973), and another clear example is the relative resistance of chickens with the B21 haplotype to Marek's disease virus (Pazderka *et al.*, 1975). Similarly, bovine lymphocyte antigen (BoLA) w12 confers enhanced susceptibility to lymphocytosis caused by bovine leukaemia virus (Lewin and Bernoco, 1986), and in humans associations occur between particular HLA haplotypes and leprosy and other infections (Pollack and Rich, 1985; Van Eden *et al.*, 1985). Since most of these diseases have an immunopathological basis the 'resistant' phenotype is a reflection of the capacity of the host to react in a way which limits the pathological consequences of infection.

Mouse strains that are congenic for MHC antigens, with a constant background of all other gene products, are now providing the most sensitive system for analysis of the relationships between inheritance of MHC haplotype and patterns of infectious disease, and there is an extensive literature on the influence of H-2 genes on bacterial, protozoan and helminth infections (Wakelin and Blackwell, 1989) as well as on viral infections (Maudsley *et al.*, 1989). Taking *Leishmania donovani* infections again as an example, on a B10 genetic background they rapidly resolve in H-2r,s,b mice (cure phenotype), while in H-2d,q,f (non-cure) mice parasite densities in spleen and liver are maintained at high levels for long periods (Blackwell *et al.*, 1980). This difference seems attributable to differential expression of the heterodimeric class II molecules coded by the IA and IE subregions of the H-2 complex, H-2d mice expressing both IA and IE gene products, whereas H-2b mice express only IE (Blackwell and Roberts, 1987).

Differential susceptibility of mice to helminth infections of *Trichinella spiralis* and *Nematospiroides dubius* has also been attributed to class II antigen expression, with relative susceptibility correlating with expression of both I-A and I-E, and resistance correlating with expression of I-A only (Wakelin, 1989; Wassom *et al.*, 1987). Thus, for both a protozoan and a helminth infection antigen presentation by I-E exacerbates disease by a process in which the activation of 'suppressor' cells has traditionally been implicated.

The control exerted by the MHC gene products may be modulated or even overridden by the effects of other background genes, or by parameters of the infection. For example, mouse strains that are identical in MHC genotype may demonstrate very different phenotypes with respect to patterns of rejection of *T. spiralis* infections, due to genetic control of the inflammatory reactions mediating expulsion of the parasite from the intestine (Wakelin, 1985); and cure and non-cure phenotypes of mice infected with *L. donovani* vary according to the size of initial inoculum, an effect which has been attributed to alteration in the balance of disease-controlling and disease-enhancing T cell subpopulations (Blackwell, 1989b).

T cell receptor genes

T cells recognize antigen via a disulphide linked heterodimeric receptor consisting of an alpha and beta peptide chain, each of approximately 40 kD. The heterodimer is found on the cell surface invariably associated with four to five other peptides forming the CD3 complex. Two subpopulations of T lymphocytes, helper cells and cytotoxic cells, are identified by antigenically distinct CD4 and CD8 accessory molecules. These determine that the reactivity of the receptors on the helper and cytotoxic T cell subpopulations is directed, respectively, towards class II and class I histocompatibility antigens. Another population of T cells with gamma/delta heterodimer receptors, and no CD4 or CD8 molecules, is present in skin epidermis and intestinal epithelium, and in low concentration in the circulation.

The mechanisms by which T lymphocyte populations are educated to be unresponsive to self-antigens, and to recognize foreign antigens in the restricted context of self-MHC antigens, are currently matters of intense speculation (Schwartz, 1989), and a model has been proposed which relates the structure of the T cell receptor to its function of recognition of antigen in the context of MHC (Davis and Bjorkman, 1988).

By a process of gene rearrangement that occurs during the maturation of T cells in the thymus up to four different elements are incorporated in the gene coding the primary structure of T cell receptor peptides. In addition to the element for the carboxyterminal constant domain, receptor peptide genes contain variable (V), joining (J), and in the case of beta and delta chains, diversity (D) regions, which code for the variable aminoterminal, antigen-recognition portion of receptor peptide. Random addition of nucleotides at V-J, V-D, and D-J junctions during rearrangement (N-region addition) creates further structural diversity.

The rearrangement process for receptor gene elements allows a mixture of V-region and J-region gene elements to combine randomly such that in the mouse for example, 100 alpha-chain V regions and 50 J regions can form up to 5,000 different alpha-chain genes. Similarly about 500 different beta-chain genes could be made up from 20 beta V genes, 2 D regions and 12 beta J genes. It is considered unlikely that all the different gene elements have equal likelihood of being used, but theoretically random combination of alpha and beta chains can thus give rise to 2.5×10^6 alpha–beta heterodimers, and when all mechanisms for creating diversity in T cell receptor structure are taken into account it is calculated that up to 10^{15} different alpha–beta and 10^{18} gamma–delta T cell receptors are possible (Davis and Bjorkman, 1988).

The potential for diversity of T cell receptors is thus considerably larger than the number of T cells present in an organism, and it is perhaps not surprising that gaps in the T cell receptor repertoire have not so far been held to account for a failure in immune responsiveness. Indeed, SJL and related strains of mice, and NZW mice, have a deletion in the beta-chain gene region (Behlke *et al.*, 1986; Noonan *et al.*, 1986), but do not have any recorded increased propensity to infection, though (NZW $\times$ NZB) F_1 mice do have an increased incidence of auto-immunity (Kotzin and Palmer, 1987).

A connection may yet be found between usage of particular T cell receptor genes or gene combinations and immune responses to some infections, particularly as in autoimmune encephalomyelitis that has been induced experimentally in mice and rats by injection of myelin basic protein or its peptides, disease seems to be due to reactivity of T cells which use a limited range of V alpha and V beta gene combinations (Heber-Katz, 1990). In multiple sclerosis patients T cells found in the brain similarly have limited heterogeneity of T receptor V-alpha chains (Oksenberg *et al.*, 1990). It should also be noted that the capacity of certain exotoxins of streptococci, staphylococci and mycoplasmas to induce mitosis in large numbers of T cells may depend on expression of a particular TCR gene repertoire (Fleischer, 1989).

Immunoglobulins, antibodies and B cells

The genes coding for immunoglobulin molecules are generated by gene segment rearrangement processes similar to those for T cell receptor genes (Tonegawa, 1983), with immunoglobulin heavy chain diversity resulting from recombination of V, D, and J gene segments, and N region additions, and light chain diversity from recombination of V and J regions (Max, 1989). Theoretically there is here again potential for up to 10^{11} different immunoglobulin variable domain sequences for the total antibody repertoire (Davis and Bjorkman, 1988), and somatic mutation is a further mechanism, not so far found in T cell receptor genes, which gives additional antibody sequence diversity that contributes to increasing affinity of antibody produced during the course of immune responses (French *et al.*,

1989). Also during an immune response Ig variable domain genes can recombine successively with different constant region genes coding for the various immunoglobulin isotypes (class or isotype switching), a process that is modulated by lymphokines produced by helper T cells (Mosmann and Coffman, 1989). Rearrangement of both T cell receptor and immuno-globulin genes may be controlled by recombinase enzymes with similar properties (Schatz *et al.*, 1989), but the genes which control affinity maturation and class switching of antibodies have not yet been identified.

Factors controlling effectuation of the immune response

Immune-deficiency syndromes

Mice with severe combined immunodeficiency (SCID mice; Bosma *et al.*, 1983), and which lack both T and B cell function, have a defect in the recombinase enzyme that is responsible for T cell receptor and immuno-globulin gene rearrangement (Okazaki *et al.*, 1988). Horses (McGuire *et al.*, 1975) and humans (Hitzig, 1968) are other species in which SCID is found, with severe consequences in terms of susceptibility to overwhelming infection. Progress has been made towards identification of the recom-binase gene (Schatz *et al.*, 1989).

'Nude' (nu/nu) mice congenitally fail to develop a thymus (Pantelouris, 1968), and because of the T-dependence of nearly all immune responses these and analogous animals with induced T cell deficiencies have been extremely useful for investigating the interface between infectious agents and acquired immunity. Nude mice fail to generate cell-mediated immune responses, and specific antibodies, mainly of the IgM isotype, are induced only by polysaccharides and other antigens with multiple repetitive epitopes, the so-called T-independent antigens.

As might be expected from the evolutionary advantages assumed to be conferred by the immune response, the course of many infections is exacer-bated in nu/nu, neonatally thymectomized and otherwise T cell deprived mice, with consequent higher infection loads and/or failure to recover. In some diseases, however, in which pathology is immunologically mediated, an absence of immune response potential results in better prognosis (see p. 39). Also, nude mice are more resistant than their heterozygous litter-mates to the early stages of some intracellular microbial infections (*Listeria monocytogenes*, Nickoll and Bonventre, 1977; *Brucella abortus*, Cheers and Waller, 1975; *Salmonella typhimurium*, Fauve and Hevin, 1974). This is suspected to be a result of macrophages in nude mice being in an heightened activation state, and thus effecting the same phenotype as the *Bcg/Lsh/Ity* gene.

T cell subpopulations

It is not yet clear whether gamma–delta T cells have an immunological role that is totally distinct from that of alpha–beta T cells, but their localization in epidermis and intestinal epithelium indicates specialization, and they have been shown to be selectively activated by mycobacterial antigens (Janis *et al.*, 1989). Genes controlling variations in the ratios of, respectively, gamma–delta : alpha–beta T cells, and CD4$^+$ (helper) : CD8$^+$ (cytotoxic) alpha/beta T cells, have not been identified so far.

Two populations of helper T cells have been distinguished by *in vitro* lymphokine secretion profiles of cloned cell lines (Mosmann and Coffman, 1989), and the outcome of some infectious diseases may depend on which of these subpopulations is activated. Thus, phenotypic resistance to cutaneous leishmaniasis caused by *L. major* in mice is attributed to TH1 cells, which are more effective inducers of delayed-type hypersensitivity (DTH) reactions. TH2 cells are considered mediators of the susceptibility phenotype, their lymphokine secretion pattern being appropriate for cooperative interaction with β cells (Scott, 1989). The proposed distinction may not be absolute, since some clones, ostensibly of TH1 cells, can effect DTH reactions and also exacerbate the growth of *L. major* lesions (Muller *et al.*, 1989). Nevertheless further experimentation along these lines may point the way to selection of animals with T cell subpopulation profiles which are appropriately defined to give increased protection against infection.

Quantity and quality of the antibody response

Mutations which affect antibody production in mice have been useful in characterizing immune effector mechanisms which regulate infections. Many are X-linked and affect B lymphocyte function: the *Xid* immunodeficiency of CBA/N strain mice results in low concentrations of serum IgM and IgG$_3$, but not of other isotypes (Perlmutter *et al.*, 1979). The *Xid* gene can counter the *Ity*^{r} phenotype in so far as CBA/N mice are susceptible to, and die from, a late phase of *S. typhimurium* replication as a result of impaired production of antibody against bacterial polysaccharide (O'Brien *et al.*, 1979; Wicker and Scher, 1986). (A third gene, *Lps*, also controls mouse *S. typhimurium* infections, and in this case increased susceptibility correlates with non-responsiveness to the biological effects of bacterial endotoxin (O'Brien *et al.*, 1980).)

Two selective breeding programmes have resulted in mouse lines with, respectively, high and low antibody responsiveness (Biozzi *et al.*, 1968), and high and low antibody affinity (Steward *et al.*, 1979). The Biozzi lines in particular have been examined for their infection handling capacity, and the high responder mice have generally, but not always, proved to exert better immunological control over protozoan and helminth infections than low responders (Biozzi *et al.*, 1985).

High responder mice were more resistant to primary *Trichinella spiralis* infections than low responder mice, but were only partially protected against reinfection, whereas low responders were completely immune (Perrudet-Badoux *et al.*, 1978). Against *Nematospiroides dubius* both lines responded equivalently to primary infections, with high responders in this instance showing more resistance to reinfections (Jenkins and Carrington, 1981).

There was no difference in the resistance of the two Biozzi lines to the lethal effects of a strain of *Plasmodium yoelii* (Heath *et al.*, 1989), but early rises in antibody titres seemed significant in controlling infections as diverse as *Trypansoma cruzi* (Kierszenbaum and Howard, 1976) and *Taenia* spp., (Joysey, 1986; Mitchell, 1982). Rate of antibody synthesis as a determinant factor in the outcome of infection with *T. taeniaeformis* was also indicated by earlier work comparing different mouse strains (Mitchell *et al.*, 1980).

The low responder mice are resistant to BCG (Gheorghiu *et al.*, 1985) and to *S. typhimurium* (Plant and Glynn, 1982), traits that are related to macrophage physiology, and the defect in antibody production may be accounted for by the high activation state of macrophages in these animals causing antigens to be broken down so rapidly and completely that the antigen presentation process leading to antibody production is impaired (Biozzi *et al.*, 1984).

It is not yet known whether the genetic difference between mouse lines responding with high and low affinity antibody is connected with genes controlling somatic mutation of immunoglobulin V domains, but CD8$^+$ T cells appear to have an important regulatory role in the process (Holland and Steward, 1989). High affinity mice have a lower propensity to develop chronic antibody–antigen immune complex disease than low affinity responders (Steward, 1979), but are more susceptible to autoimmunity in experimentally induced allergic encephalomyelitis (Devey *et al.*, 1990). High affinity mice reduced their *P. yoelii* parasitaemias somewhat earlier than low affinity mice, but peak parasitaemias were similar (Heath *et al.*, 1989).

Accessory immune effector mechanisms

Complement The alternate pathway of complement activation is an innate resistance mechanism that assists control of microbial infections. Its potent toxic activities have also been harnessed as an effector mechanism in acquired immunity through the capacity of some immunoglobulin isotypes to activate the classical complement pathway after complexing with antigen. Complement activation and inflammation are often associated phenomena.

Individuals with deficiencies of complement components are generally more susceptible to infection, particularly those of bacterial origin (Goldstein and Marder, 1983; Lachman, 1984). The relationship is not always as expected, however, and the complement system sometimes appears to facilitate infection. Thus, mice with C5 deficiency were found less susceptible to primary infections of *Schistosoma mansoni* (Ruppel *et al.*, 1982), and in the case of *Leishmania major*, complement receptor CR1, the ligand for C3b, is exploited by the parasite as a means of gaining entry into macrophages (Da Silva *et al.*, 1989).

One of the two major loci which governs resistance/susceptibility during the very early phases of *Listeria monocytogenes* infections in mice (Cheers and McKenzie, 1978; Skamene *et al.*, 1979), now designated *Lsr-1*, co-maps with the *Hc* locus which specifies the C5 component of the complement cascade. Traits for C5-deficiency and susceptibility to *Listeria* infection are concordant in most strains of mice, the exceptions being due to the effects of a second major locus governing *Listeria* susceptibility (Kongshavn, 1986). The defect in *Lsr-1* susceptible mice as regards the bacterial infection is due to reduced macrophage inflammatory responsiveness (Gervais *et al.*, 1984); a correlation between *Listeria* resistance and inflammatory responses mediated by C5a has been demonstrated (Stevenson *et al.*, 1985).

Effector cells Macrophages, eosinophils, mast cells, neutrophils and platelets can all act as effectors in acquired immune responses. They may react relatively non-specifically, as in the case of mycobacterium-activated macrophages being cytotoxic also for unrelated bacteria and tumour cells, or be targetted more specifically by antibody in antibody-dependent cell-mediated cytotoxicity (ADCC) reactions.

Studies on P strain mice illustrate the potential importance of macrophages in controlling infections, since these animals were poorly resistant to reinfection with *Schistosoma mansoni* (James *et al.*, 1983) and failed to inhibit growth of *L. major* (Fortier *et al.*, 1984). The defect has been attributed to a genetically controlled impaired production of, and macrophage response to activation by, interferon-gamma (James *et al.*, 1984).

Eosinophils have been particularly implicated as effector cells in the control of helminth infections, and a correlation was found between eosinophilia and resistance to *Trichostrongylus colubriformis* in guinea pigs (Handlinger and Rothwell, 1981). Levels of eosinophilia induced by infections are under genetic control (Vadas, 1982; Wakelin and Donachie, 1983), and differences in this response are primarily reflections of precursor cell numbers rather than of T cell derived lymphokine production (Lammas *et al.*, 1989). Nevertheless lymphokines are as important for eosinophil as for macrophage physiology, as shown by complete blockage of eosinophilia *in vivo* with an anti-IL-5 monoclonal antibody (Coffman *et al.*, 1989). Similarly, control of the mast cell response phenotype, which regulates infestation of mice by the tick *Haemaphysalis longicornis* (Matsuda *et al.*, 1987), is expressed at the level of both precursor cell number and mast cell proliferation rate (Reed *et al.*, 1988).

Cytokines Understanding of how immune responses are controlled by cytokines is increasing rapidly, and the severity of immunopathological disease is probably largely a reflection of a spectrum of cytokine production and action. Experimental pathology can already be dramatically alleviated by eliminating individual cytokines with specific antibodies (e.g., with antitumour necrosis factor/cachectin antibodies; Grau *et al.*, 1987; Piguet *et al.*, 1987), and in due course genetic regulation of cytokines could be a realistic means of controlling disease. The consequences of actions which ameliorate one disease may of course lead to exacerbation of others (Kindler *et al.*, 1989).

Also relevant is the likelihood that genes which affect resistance/ susceptibility phenotypes will code for cytokines. Thus, the gene locus *Ts-2* which affects the time of expulsion of *Trichinella spiralis* in mice appears to map to a cluster encoding both tumour necrosis factor-alpha and lymphotoxin (Grencis and Pinkerton, 1990).

Inflammation and immunopathology Few, if any immune responses are effected without some concomitant inflammation, and the inflammatory response may serve in a host-protective capacity. Thus, expulsion of *T. spiralis* adults from the small intestine of mice is believed to be due to acute inflammation mediated by CD4[+] T cells (Reidlinger *et al.*, 1986), and in strains which expel adults earlier there is a tendency for reduced female worm fecundity and lower densities of muscle larvae (Wakelin, 1989).

Immunopathology can on the other hand sometimes be almost wholly responsible for infection morbidity. This is the case in some virus infections where disease is ameliorated in the absence of T cells (Cole *et al.*, 1971; Ytterberg *et al.*, 1988). In schistosomiasis the parasite egg-induced granulomas are immune hypersensitivity reactions that, when sufficiently numerous to inhibit blood circulation, cause many of the disease symptoms (Warren, 1968); T cell deprived mice infected with *Schistosoma bovis* accordingly survived longer than immunologically intact controls (Agnew *et al.*, 1989). Ablation of pathology in *S. mansoni*-infected mice on the other hand curtailed host survival due to onset of an acute parasite egg-induced hepatotoxicity reaction (Doenhoff *et al.*, 1986).

Potential exists for selecting against disease-induced pathology, as indicated by variation in pathogenesis of *S. mansoni* infections in inbred mouse strains (Dean *et al.*, 1981), and a linkage between HLA haplotype and T cell responsiveness of *S. japonicum* egg antigens (Hirayama *et al.*, 1987).

Immunological memory, vaccination and chemotherapy

An important distinguishing feature of the acquired immune response, and a parameter that could be of consequence in respect of intercurrent infections, is its anamnestic quality. The means by which immunological memory is retained is still under debate (Beverley, 1990), and there is as yet no information about its genetic control.

Vaccines and drugs are front line measures used to protect against many infectious diseases. Dineen and colleagues clearly demonstrated a genetic basis to variation in the response of sheep to vaccination with irradiated larvae against *Trichostrongylus colubriformis* (Windon and Dineen, 1980), and went on to show an apparent association between high frequency of an ovine lymphocyte histocompatibility antigen and good response to vaccination (Outteridge *et al.*, 1985). In mice a genetic locus, *Rsm-1*, which controls protective immunity induced by vaccination with irradiation-attenuated *S. mansoni* larvae, was identified through the use of P strain animals that have a defective response (Correa-Oliveira *et al.*, 1986). The same defective gene product may be responsible for poor leishmanicidal and tumouricidal responses (Fortier *et al.*, 1984). That cytokine deficiency in some instances may be responsible for poor responses to vaccination is indicated by the capacity of interferon-gamma to act as an adjuvant in vaccination of both Biozzi low antibody responder mice and low affinity antibody mice against *Plasmodium yoelii* infection (Heath *et al.*, 1989).

There is now good evidence to indicate that the outcome of chemotherapy of different diseases is immune-dependent: this has been shown for schistosomiasis (Doenhoff, 1989), malaria (Lwin *et al.*, 1987) and malignancy (Carter *et al.*, 1973). There may also be an immunological basis to the potential that exists for selecting cattle that respond better to chemotherapy of trypanosomiasis (Murray *et al.*, 1982) since treatment of *T. brucei-*infected mice with either alpha-difluoromethylornithine (Bitonti *et al.*, 1986; De Gee *et al.*, 1983) or melarsoprol (Frommel, 1988) is antibody-dependent.

The role of the immune response

As this chapter indicates, efforts to improve the immune system of animals are unlikely to be immediately rewarded by a generalized enhancement of disease resistance. To begin with, the genetic repertoire coding for the specificity of acquired immune responses has a potential that is greater than the cellular capacity actually available in the body to express the full range of T cell receptor or immunoglobulin variants. Opportunities for genetic improvement of immune responses seem to lie more in the domain of traits affecting quantity and quality, rather than specificity of immune responses.

Unfortunately heritable intraspecific variation in resistant/susceptibility phenotypes is rarely under the exclusive control of a single gene, and as yet relatively little is known about the nature and the mode of action of the gene products regulating resistance/susceptibility traits, or of their inter-action with each other and with factors such as age and nutrition. Animal lines selected for a particular immune response may not be more resistant to infection, nor indeed show improved overall 'fitness', and by the same token, selection for 'resistance' to a particular infection may not coincide with 'resilience' (Albers and Gray, 1986).

An improvement in our understanding of the role of acquired immunity at the interface between hosts and potential pathogens is perhaps the most important prerequisite in the continuing effort to limit the damage caused by infectious diseases.

The immune response as an eliminator of pathogens

There is a widespread belief (adhered to perhaps more especially by 'pure' immunologists than by immunoparasitologists) that an hegemonic immune response can dispose of infections rapidly and completely, but this is hardly tenable. Benensen (1985) and Mimms (1987) list numerous infections that persist or exist in the carrier state in man, and the overdispersed nature of helminth infections, whereby the majority of infected people carry long-term, low intensity and asymptomatic infections (Anderson and May, 1985) is a situation that deserves comparison with the latency of persistent virus infections (Oldstone, 1989).

Conventional reasoning fails also to explain satisfactorily how overt immunological repression can substantially reduce the intensity of many infections, yet in the end there is a failure to clear completely the offending organism from the body. To take one experimental example, *Trypanosoma musculi* infection of mice appeared to be an instance where generation of an effective T-dependent response culminated in sterile immunity (Viens and Targett, 1971): upon closer examination small numbers of parasites were in fact found surviving in capillary loops in the kidneys (Viens *et al.*, 1972).

Much has been said about strategies available to pathogens to enable them to survive the immune response (reviewed in Zschiesche, 1987), but many of the hypotheses inherent in such work, initiated from an intellectual stance of immune responses that are being subverted, have lacked sub-stantiation in the form of either identification of the responsible genes and gene products, or the promise of better measures for disease control.

The high frequency of the carrier state, and the commonness of persistent low levels of infection in the face of demonstrably protective immune responses leads inexorably to the idea that the adaptive immune response is but a modulator of the interface between pathogens and their hosts (Davies, 1985, 1986). Though no immunity often means death as a consequence of infection, good immunity, it may be said, allows life to continue with infection.

Do host–parasite relationships tend to become symbiotic?

A common suspicion about host–parasite relationships is that the 'successful' ones inevitably evolve toward a state of relative parasite harmlessness and eventual symbiosis, a situation which the contrasting pathogenicities of African trypanosome infections of indigenous and imported ruminant host species are considered to exemplify.

Epidemiological principles and modelling studies indicate rather that the pathways taken in coevolution of host and pathogen are not predictable, and depend to a great extent on the interplay between virulence and transmissibility of the parasite (Anderson and May, 1982). Introduction of the myxomatosis virus from South America into Australia and England in the early 1950s provided an excellent opportunity to study a developing relationship. Although the virus did undergo some early attenuation of virulence, with time there was a marked increase in the resistance of rabbits to the lethality of the infection, and after 30 years a balance appeared to have been struck at an intermediate to high level of virus virulence (Fenner, 1983). In Britain an increase of resistance in the rabbit population may latterly have induced a compensatory further relative increase in virus virulence (Ross and Sanders, 1987), an observation which suggests that if used as a strategy in isolation (independently of other measures of good animal husbandry) selection for increased resistance to infection may not meet with great success.

The necessity of maintaining an intermediate level of virulence by the pathogen is understandable: chances of successful transmission would be diminished by either too high virulence which leads to very early death of the host, or too low virulence with consequent low virus titres and/or very quick recovery from infection (Anderson and May, 1982).

The characteristics of African trypanosome infections are also indicative of the existence of a 'delicate balance' between the mechanisms responsible for, and the order of, VSG gene switching in the parasite, and the kinetics of the immune response, such that there are neither too many parasites, which would kill the host, nor too few, resulting in decreased opportunity for ingestion and transmission by the vector (Barry, 1989). Prehn and Prehn (1989) have similarly argued that adaptation of a level of immunogenicity which maximized growth accounted for the survival of some tumours.

The immune response as another niche for exploitation

That pathogens can exploit even the immune response as a means of potentiating their transmission (Damian, 1987) is illustrated by observations on the immune-dependence of schistosome egg excretion (Doenhoff *et al.*, 1986), and the enhancement of cutaneous leishmania lesions by specific T cell lines (Titus *et al.*, 1984). Consistent with the latter observation is the radical demonstration that *L. tropica* growth can be suppressed by rendering mice immunologically tolerant to parasite antigens (Muller *et al.*, 1989), and induction of tolerance might thus prove as effective as conventional vaccination in limiting the growth and transmission of pathogens.

Stable host–parasite relationships as instances of co-operation

The foregoing begs the question whether acquired immunity is a device deliberately enabling hosts to retain infections over long periods: if so, host–parasite relationships would be more appropriately studied in the context of co-operative interactions (Axelrod, 1984; Dawkins, 1989, Chapter 12).

There is evidence that genes conferring susceptibility to infection indirectly confer enhanced fitness on the host (e.g., a positive association between liability to *N. dubius* and progeny litter size; Brindley and Dobson, 1981), but at its extreme, this paradigm shift would require the immune response to be investigated as a device which accommodates a parasite long-term for the benefits which it directly confers on the host (Desowitz, 1979, Chapters 9 and 10). Firm evidence for such a postulate is sparse, not the least reason for this being, we feel, a failure to look for it.

In conclusion, note has to be taken of the enormous economic potential offered by transgenic animals (Westphal, 1989). After the not insignificant hurdle of identification of useful genes has been overcome, methods for their transfection, in conjunction with techniques of embryo implantation, will rapidly yield progeny of appropriate disease resistance status.

Bibliography and further reading

Blackwell, J. (organizer) (1989) 27th Forum in Immunology: The macrophage resistance gene *Lsh/Ity/Bcg*. *Research in Immunology* 140, 767–828.

Briles, D.E. (ed.) (1986) Genetic control of the susceptibility to bacterial infection. *Contemporary Topics in Microbiology and Immunology*, Volume 124, Springer-Verlag, Berlin.

Kaufmann, S.H.E. (ed.) (1990) T-cell paradigms in parasitic and bacterial infections. *Contemporary Topics in Microbiology and Immunology*, Volume 155, Springer-Verlag, Berlin.

Morrison, W.I. (ed.) (1986) *The Ruminant Immune System in Health and Disease,*

Cambridge University Press, Cambridge.

Paul, W.E. (ed.) (1989) *Fundamental Immunology*, 2nd edn. Raven Press, New York.

Skamene, E. (ed.) (1985) Genetic control of host resistance to infection. *Progress in Leukocyte Biology*, Volume 3, Alan R. Liss, New York.

Wakelin, D.M. and Blackwell, J.M. (eds) (1989) *Genetics of Resistance to Bacterial and Parasitic Infection*. Taylor and Francis, London.

Wakelin, D. and Blackwell, J.M. (1990) Genetic variations in immunity to parasite infection. In: Warren, K.S. and Agabian, N. (eds), *Immunology and Molecular Biology of Parasitic Infections*. Blackwell Scientific Publications, Oxford (in press).

References

Agnew, A.M., Murare, H.M., Lucas, S.B. and Doenhoff, M.J. (1989) *Schistosoma bovis* as an immunological analogue of *S. haematobium*. *Parasite Immunology* 11, 329–40.

Albers, G.A.A. and Gray, G.D. (1986) Breeding for worm resistance: a perspective. In: Howell, M.J. (ed.) *Parasitology – Quo vadit*. Proceedings VI International Congress of Parasitology. Australian Academy of Sciences, Canberra, pp. 559–66.

Allison, A.C. (1954) Protection afforded by sickle cell trait against subtertian malarial infection. *British Medical Journal* 1, 290–4.

Altaif, K.I. and Dargie, J.D. (1978a) Genetic resistance to helminths: the influence of breed and haemoglobin type on the response of sheep to primary infections with *Haemonchus contortus*. *Parasitology* 77, 161–75.

Altaif, K.I. and Dargie, J.D. (1978b) Genetic resistance to helminths: the influence of breed and haemoglobin type on the response of sheep to reinfection with *Haemonchus contortus*. *Parasitology* 77, 177–87.

Anderson, R.M. and May, R.M. (1982) Coevolution of hosts and parasites. *Parasitology* 85, 411–26.

Anderson, R.M. and May, R.M. (1985) Helminth infections of humans: mathematical models, population dynamics and control. *Advances in Parasitology* 85, 1–101.

Axelrod, R. (1984) *The Evolution of Cooperation*. Basic Books, New York.

Barry, J.D. (1989) African trypanosomes: an elusive target. In: McAdam, K.P.W.J. (ed.) *New Strategies in Parasitology*. Churchill Livingstone, Edinburgh, pp. 101–16.

Behlke, M.A., Chou, A.S., Huppi, K. and Loh, D.Y. (1986) Murine T-cell receptor mutants with deletions of beta-chain variable region genes. *Proceedings of the National Academy of Sciences*, USA, 83, 767–71.

Benacerraf, B. (1981) Role of MHC gene products in immune regulation. *Science* 212, 1229–38.

Benenson, A.S. (ed.) (1985) *Control of Communicable Diseases in Man*. 14th edn. American Public Health Association.

Beverley, P.C. (1990) Is T-cell memory maintained by cross-reactive stimulation. *Immunology Today* 11, 203–5.

Biozzi, G., Mouton, D., Siqueira, M. and Stiffel, C. (1985) Effect of genetic modification of immune responsiveness on anti-infection and anti-tumour resistance. *Progress in Leukocyte Biology* 3, 3–18.

Biozzi, G., Mouton, D., Stiffel, C. and Bouthillier, Y. (1984) Macrophage in regulation of immunoresponsiveness. *Advances in Immunology* 36, 189–234.

Biozzi, G., Stiffel, C., Mouton, D., Bouthillier, Y. and Decreusefond, C. (1968) Selection artificielle pour la production d'anticorps chez la souris. *Annales d'Institute Pasteur* 115, 965–7.

Bitonti, A.J., McCann, P.P. and Sjoerdsma, A. (1986) Necessity of antibody response in the treatment of African trypanosomiasis with alpha-difluoromethylornithine. *Biochemical Pharmacology* 35, 331–4.

Bjorkman, P.J., Saper, M.A., Samraoui, B., Bennett, W.S., Strominger, J.L. and Wiley, D.C. (1987) Structure of the human class I histocompatibility antigen, HLA-A2. *Nature* 329, 506–12.

Blackwell, J. (organizer) (1989a) 27th Forum in Immunology: The macrophage resistance gene *Lsh/Ity/Bcg*. *Research in Immunology* 140, 767–828.

Blackwell, J.M. (1989b) Protozoan infections. In: Wakelin, D.M. and Blackwell, J.M. (eds), *Genetics of Resistance to Bacterial and Parasitic Infection*. Taylor and Francis, London, pp. 103–51.

Blackwell, J., Freeman, J. and Bradley, D. (1980) Influence of H-2 complex on acquired resistance to *Leishmania donovani* infection in mice. *Nature* 283, 72–4.

Blackwell, J.M., Roach, T.I.A., Kiderlen, A. and Kaye, P.M. (1989) Role of *LSH* in regulating macrophage priming/activation. *Research in Immunology* 140, 798–805.

Blackwell, J.M. and Roberts, M.B. (1987) Immunomodulation of murine visceral leishmaniasis by administration of monoclonal anti-I-A vs anti-I-E antibodies. *European Journal of Immunology* 17, 1669–72.

Bosma, G.C., Custer, R.P. and Bosma, M.J. (1983) A severe combined immunodeficiency mutation in the mouse. *Nature* 301, 527–30.

Bradley, D.J. (1974) Genetic control of natural resistance to *Leishmania donovani*. *Nature* 250, 353–4.

Brindley, P.J. and Dobson, C. (1981) Genetic control of liability to infection with *Nematospiroides dubius* in mice: selection of refractory and liable populations of mice. *Parasitology* 83, 51–65.

Brindley, P.J. and Dobson, C. (1983) Host specificity in mice selected for innate immunity to *Nematospiroides dubius*: infection with *Nippostrongylus brasiliensis*, *Mesocestoides corti* and *Salmonella typhimurium*. *Zeitschrift fur Parasitenkunde* 69, 797–805.

Brown, J.H., Jardetzky, T., Saper, M.A., Samraoui, B., Bjorkman, P.J. and Wiley, D.C. (1988) A hypothetical model of the foreign antigen binding site of class II histocompatibility molecules. *Nature* 332, 845–50.

Buschman, E., Taniyama, T., Nakamura, R. and Skamene, E. (1989) Functional expression of the *Bcg* gene in macrophages. *Research in Immunology* 140, 793–7.

Carter, R.L., Connors, T.A., Weston, B.J. and Davies, A.J.S. (1973) Treatment of a mouse lymphoma by L-asparaginase: success depends on the host's immune response. *International Journal of Cancer* 11, 345–57.

Cheers, C. and MacKenzie, I.F.C. (1978) Resistance and susceptibility of mice to bacterial infections: genetics of listeroses. *Infection and Immunity* 19, 755–63.

Cheers, C. and Waller, R. (1975) Activated macrophages in congenitally athymic "nude" mice. *Journal of Immunology* 115, 844–7.

Coffman, R.L., Seymour, B.W.P., Hudak, S., Jackson, J. and Rennick, D. (1989) Antibody to interleukin-5 inhibits helminth-induced eosinophilia in mice. *Science* 245, 308–10.

Cole, G.A., Gilden, D.H., Monjan, A.A. and Nathanson, N. (1971) Lymphocytic choriomeningitis virus: pathogenesis of acute central nervous system disease. *Federation Proceedings* 30, 1831–41.

Correa-Oliveira, R., James, S., McCall, D. and Sher, A. (1986) Identification of a genetic locus, *Rsm-1*, controlling protective immunity against *Schistosoma mansoni. Journal of Immunology* 137, 2014–19.

Coulson, P.S. and Wilson, R.A. (1989) Portal shunting and resistance to *Schistosoma mansoni* in 129 strain mice. *Parasitology* 99, 383–9.

Damian, R.T. (1987) The exploitation of host immune responses by parasites. *Journal of Parasitology* 73, 3–13.

Da Silva, R.P., Hall, B.F., Joiner, K.A. and Sacks, D.L. (1989) CR1, the C3B receptor, mediates binding of infective *Leishmania major* metacyclic promastigotes to human macrophages. *Journal of Immunology* 143, 617–22.

Davies, A.J.S. (1985) The possible biological significance of the immune response. In: Y.A.Ouchimnikov (ed.) *Proceedings of the 16th FEBS Congress.* Part C, pp. 33–41.

Davies, A.J.S. (1986) Evolution and functions of the immune system. In: Morrison, W.I. (ed.) *The Ruminant Immune System in Health and Disease*, Cambridge University Press, Cambridge, pp. 1–31.

Davis, M.M. and Bjorkman, P.J. (1988) T-cell antigen receptor genes and T-cell recognition. *Nature* 334, 395–402.

Dawkins, R. (1989) *The Selfish Gene.* Oxford University Press, Oxford.

Dean, D.A., Bukowski, M.A. and Cheever, A.W. (1981) Relationship between acquired resistance, portal hypertension and lung granulomas in ten strains of mice infected with *Schistosoma mansoni. American Journal of Tropical Medicine and Hygiene* 30, 806–14.

De Gee, A.L.W., Levine, R.F. and Mansfield, J.M. (1988) Genetics of resistance to the African trypanosomes. VI. Heredity of resistance and variable surface glycoprotein-specific immune responses. *Journal of Immunology* 140, 283–8.

De Gee, A.L.W., McCann, P.P. and Mansfield, J.M. (1983) Role of antibody in the elimination of trypanosomes after DL-alpha-difluoromethylornithine chemotherapy. *Journal of Parasitology* 69, 818–22.

Desowitz, R.S. (1979) *New Guinea Tapeworms and Jewish Grandmothers: Tales of Parasites and People.* W.W. Norton, New York.

Devey, M.E., Major, P.J., Bleasdale-Barr, K.M., Holland, G.P., Dal Canto, M.C. and Paterson, P.Y. (1990) Experimental allergic encephalomyelitis (EAE) in mice selectively bred to produce high affinity (HA) or low affinity (LA) antibody responses. *Immunology* 69, 519–24.

Doenhoff, M.J. (1989) The immune-dependence of chemotherapy in experimental schistosomiasis. *Memorias do Instituto Oswaldo Cruz* 84 (Supp. 1), 31–7.

Doenhoff, M.J., Hassounah, O., Murare, H.M. and Lucas, S. (1986) The schistosome egg granuloma: immunopathology in the cause of host protection or parasite survival? *Transactions of the Royal Society of Tropical Medicine and Hygiene* 80, 503–14.

Doenhoff, M., Musallam, R., Bain, J. and McGregor, A. (1979) *Schistosoma mansoni* infections in T-cell deprived mice, and the ameliorating effect of administering homologous chronic infection serum. I. Pathogenesis. *American Journal of Tropical Medicine and Hygiene* 28, 260–73.

Elsaghier, A.A.F., Knopf, P.M., Mitchell, G.F. and McLaren, D.J. (1989) *Schistosoma mansoni*: evidence that 'non-permissiveness' in 129/Ola mice involves worm relocation and attrition in the lungs. *Parasitology* 99, 365–75.

Else, K. and Wakelin, D. (1989) Genetic variation in the humoral immune responses of mice to the nematode *Trichuris muris*. *Parasite Immunology* 11, 77–90.

Fauve, R.M. and Hevin, B. (1974) Resistance paradoxale des souris thymoprives a l'infection par *Listeria monocytogenes* et *Salmonella typhimurium* et action immunostimulante d'un bacterien phospholipidique (EBP). *Comptes Rendus Academie Scientifique Paris* 279, 1603–5.

Fenner, F. (1983) Biological control, as exemplified by smallpox eradication and myxomatosis. *Proceedings of the Royal Society, Series B*, 218, 259–85.

Fine, P.E.M. (1989) The BCG story: lessons from the past and implications for the future. *Reviews of Infectious Diseases* 11, Suppl. 2, S353–9.

Fleischer, B. (1989) A conserved mechanism of T lymphocyte stimulation by microbial exotoxins. *Microbial Pathogenesis* 7, 79–83.

Fortier, A., Meltzer, M.S. and Nacy, C.A. (1984) Susceptibility of inbred mice to *Leishmania tropica* infection: genetic control of the development of cutaneous lesions in P/J mice. *Journal of Immunology* 133, 454–9.

French, D.L., Laskov, R. and Scharff, M.D. (1989) The role of somatic hypermutation in the generation of antibody diversity. *Science* 244, 1152–7.

Frommel, T.O. (1988) *Trypanosoma brucei rhodesiense*: effect of immunosuppression on the efficacy of melarsoprol treatment of infected mice. *Experimental Parasitology* 67, 364–6.

Gervais, F., Stevenson, M.M. and Skamene, E. (1984) Genetic control of resistance to *Listeria monocytogenes*: regulation of leukocytic inflammatory responses by the *Hc* locus. *Journal of Immunology* 132, 2078–83.

Gheorghiu, M., Mouton, D., Lecoeur, H., Lagranderie, M., Meuel, J.C. and Biozzi, G. (1985) Resistance of high and low antibody responder lines of mice to the growth of avirulent (BCG) and virulent (H37Rv) strains of mycobacteria. *Clinical and Experimental Immunology* 59, 177–84.

Goldstein, I.M. and Marder, S.R. (1983) Infections and hypocomplementaemia. *Annual Review of Medicine* 34, 47–53.

Grau, G.E., Fajardo, L.F., Piguet, P.-F., Allet, B., Lambert, P.-H. and Vassalli, P. (1987) Tumor necrosis factor (cachectin) as an essential mediator in murine cerebral malaria. *Science* 237, 1210–12.

Gray, D.F., Graham-Smith, H. and Noble, J.L. (1960) Variations in natural resistance to tuberculosis. *Journal of Hygiene* 58, 215–27.

Gray, G.D. (1987) Genetic resistance to haemonchosis in sheep. *Parasitology Today* 3, 253–5.

Grencis, R.K. and Pinkerton, S. (1990) Immunoregulation during infection with the nematode *Trichinella spiralis*: variation in immunity and cytokine production.

In: McDonald, T.T. (ed.), *Proceedings of International Congress of Mucosal Immunology,* Khuver Academic. (in press)

Gros, P., Skamene, E. and Forget, A. (1981) Genetic control of resistance to *Mycobacterium bovis* (BCG) in mice. *Journal of Immunology* 127, 2417–21.

Handlinger, J.H. and Rothwell, T.L.W. (1981) Studies on the response of basophil and eosinophil leucocytes and mast cells to the nematode *Trichostrongylus colubriformis*: comparison of cell populations in parasite resistant and susceptible guinea pigs. *International Journal for Parasitology* 11, 67–70.

Harrison, R.A., Bickle, Q. and Doenhoff, M.J. (1982) Factors affecting the acquisition of resistance against *Schistosoma mansoni* in the mouse. Evidence that the mechanisms which mediate resistance during early patent infections may lack immunological specificity. *Parasitology* 84, 93–110.

Heath, A.W., Devey, M.W., Brown, I.N., Richards, C.E. and Playfair, J.H.L. (1989) Interferon-gamma as an adjuvant in immunocompromised mice. *Immunology* 67, 520–4.

Heber-Katz, E. (1990) The autoimmune T cell receptor: epitopes, idiotopes and malatopes. *Clinical Immunology and Immunopathology* 55, 1–8.

Hirayama, K., Matsushita, S., Kikuchi, I., Iuchi, M., Ohta, N. and Sasazuki, T. (1987) HLA-DQ is epistatic to HLA-DR in controlling the immune response to schistosomal antigen in humans. *Nature* 327, 426–30.

Hitzig, W.H. (1968) The Swiss type of agammaglobulinaemia. In: Bergsma, D. and Good, R.A. (eds), *Immunologic Deficiency Diseases in Man.* Vol. 82, The National Foundation – March of Dimes, New York, p. 82.

Holland, G.P. and Steward, M.W. (1989) Antibody affinity maturation: the role of $CD8^+$ cells. *Clinical and Experimental Immunology* 78, 488–93.

Inverso, J.A. and Mansfield, J.M. (1983) Genetics of resistance to the African trypanosomes. II. Differences in virulence associated with VSSA expression amongst clones of *Trypanosoma rhodesiense*. *Journal of Immunology* 130, 412–17.

James, S., Correa-Oliveira, R. and Leonard, E. (1984) Defective vaccine-induced immunity to *Schistosoma mansoni* in P strain mice. II. Analysis of cellular responses. *Journal of Immunology* 133, 1587–93.

James, S., Skamene, E. and Meltzer, M.S. (1983) Macrophages as effectors of protective immunity in murine schistosomiasis. V. variation in macrophage, schistosomulicidal and tumoricidal activities among mouse strains and correlation with resistance to reinfection. *Journal of Immunology* 131, 948–53.

Janis, E.M., Kaufmann, S.H.E., Schwartz, R.H. and Pardoll, D.M. (1989) Activation of gamma/delta T cells in the primary immune response to *Mycobacterium tuberculosis*. *Science* 244, 713–16.

Jenkins, D.C. and Carrington, T.S. (1981) *Nematospiroides dubius*: the course of primary, secondary and tertiary infection in high and low responder Biozzi mice. *Parasitology* 82, 311–18.

Joysey, H.S. (1986) Experimental infection of high and low responder Biozzi mice with *Taenia crassiceps* (Cestoda). *International Journal of Parasitology* 16, 217–21.

Kee, K.C., Taylor, D.W., Cordingley, J.S., Butterworth, A.E. and Munro, A.G. (1986) Genetic influence on the antibody response to antigens of *Schistosoma mansoni*. *Parasite Immunology* 8, 565–74.

Kennedy, M.W. (1990) Book review. *Parasitology Today* 6, 168.

Kennedy, M.W., Tomlinson, L.A., Frasier, E.M. and Christie, J.F. (1990) The specificity of the antibody response to internal antigens of *Ascaris*: heterogeneity in humans, and MHC (H-2) control of the repertoire in mice. *Clinical and Experimental Immunology* 80, 219–24.

Kierszenbaum, F. and Howard, J.G. (1976) Mechanisms of resistance against experimental *Trypanosoma cruzi* infection: the importance of antibodies and antibody-forming capacity in the Biozzi high and low responder mice. *Journal of Immunology* 116, 1208–11.

Kindler, V., Sappino, A.-P., Grau, G.E., Piguet, P.-F. and Vassalli, P. (1989) The inducing role of tumor necrosis factor in the development of bactericidal granulomas during BCG infection. *Cell* 56, 731–40.

Kongshavn, P.A.L. (1986) Genetic control of resistance to *Listeria* infection. *Current Topics in Microbiology and Immunology* 124, 67–85.

Kotzin, B.L. and Palmer, E. (1987) The contribution of NZW genes to lupus-like disease in (NZB × NZQ)F$_1$ mice. *Journal of Experimental Medicine* 165, 1237–51.

Lachman, P.J. (1984) Inherited complement deficiencies. *Philosophical Transactions of the Royal Society, London, Series B* 306, 419–30.

Lammas, D.A., Mitchell, L.A. and Wakelin, D. (1989) Genetic control of eosinophilia. Analysis of production and response to eosinophil-differentiating factor in strains of mice infected with *Trichinella spiralis*. *Clinical and Experimental Immunology* 77, 137–43.

Levine, R.F. and Mansfield, J.M. (1984) Genetics of resistance to the African trypanosomes. III. Variant-specific antibody responses of H-2-compatible resistant and susceptible mice. *Journal of Immunology* 133, 1564–9.

Lewin, H.A. and Bernoco, D. (1986) Evidence for BoLA-linked resistance and susceptibility to subclinical progression of bovine leukemia virus infection. *Animal Genetics* 17, 197–207.

Lurie, M.B., Zappasodi, P. and Dannenberg, A.M. (1952) On the mechanism of genetic resistance to tuberculosis and its mode of inheritance. *American Journal of Human Genetics* 4, 302–14.

Lwin, M., Targett, G.A.T. and Doenhoff, M.J. (1987) Reduced efficacy of chemotherapy of *Plasmodium chabaudi* in T-cell deprived mice. *Transactions of the Royal Society of Tropical Hygiene and Medicine* 81, 899–902.

Lynch, C.J., Pierce-Chase, C.H. and Dubos, R. (1965) A genetic study of susceptibility to experimental tuberculosis in mice infected with mammalian tubercle bacilli. *Journal of Experimental Medicine* 121, 1051–70.

McGuire, T.C., Banks, K.L. and Poppie, M.J. (1975) Combined immunodeficiency in horses: characterization of the lymphocyte defect. *Clinical Immunology and Immunopathology* 3, 555.

Matsuda, H., Nakano, T., Kiso, Y. and Kitamura, Y. (1987) Normalization of anti-tick response of mast cell-deficient *W/W* mice by intracutaneous injection of cultured mast cells. *Journal of Parasitology* 73, 155–60.

Maudsley, D.J., Morris, A.G. and Tomkins, P.T. (1989) Regulation by interferon of the immune response to viruses via the major histocompatibility complex antigens. In: Dimmock, N.J. and Minor, P.D. (eds), *Immune Responses, Virus Infections and Disease*. IRL Press, Oxford, pp. 15–33.

Max, E. (1989) Immunoglobulins: molecular genetics. In: Paul, W.E. (ed.), *Funda-*

mental Immunology, 2nd edn., Raven Press, New York, pp. 235–90.

Miller, L.H. and Carter, R. (1976) Innate resistance in malaria. *Experimental Parasitology* 40, 132–46.

Mimms, C.A. (1987) *The Pathogenesis of Infectious Disease.* 3rd edn. Academic Press, London and New York.

Mitchell, G.F. (1982) Genetic variation in resistance of mice to *Taenia taeniaeformis*: analysis of host protective immunity and immune evasion. In: Flisser, A., Wilms, K., Lachette, J.P., Ridawa, C., Beltran, F. and Larralde, C. (eds), *Cysticercosis: Present State of Knowledge and Perspectives.* Academic Press, New York, pp. 575–84.

Mitchell, G.F. (1989) Glutathione S-transferases – potential components of anti-schistosome vaccines. *Parasitology Today* 5, 34–7.

Mitchell, G.F., Rajasekariah, G.R. and Rickard, M.D. (1980) A mechanism to account for mouse strain variation in resistance to the larval cestode, *Taenia taeniaeformis. Immunology* 39, 481–9.

Mock, B.A., Fortier, A.H., Potter, M., Blackwell, J. and Nacy, C.A. (1985) Genetic control of systemic *Leishmania major* infection: identification of subline differences for susceptibility to disease. *Current Topics in Microbiology and Immunology* 122, 115–21.

Mosmann, T.R. and Coffman, R.L. (1989) TH1 and TH2 cells: Different patterns of lymphokine secretion lead to different functional properties. *Annual Review of Immunology* 7, 145–73.

Muller, I., Pedrazzini, R., Farrell, J.P. and Louis, J. (1989) T-cell responses and immunity to experimental infection with *Leishmania major. Annual Review of Immunology* 7, 561–78.

Murray, M., Morrison, W.I. and Whitelaw, D.D. (1982) Host susceptibility to African trypanosomiasis: trypanotolerance. *Advances in Parasitology* 21, 1–68.

Nickoll, A.D. and Bonventre, P.F. (1977) Anomolous high native resistance of athymic mice to bacterial pathogens. *Infection and Immunity* 18, 636–45.

Noonan, D.J., Kofler, R., Singer, P.A., Cardenas, G., Dixon, F.J. and Theofilopolous, N.A. (1986) Delineation of a defect in T cell receptor genes of NZW mice predisposed to autoimmunity. *Journal of Experimental Medicine* 163, 644–53.

O'Brien, A.D., Rosenstreich, D.L., Metcalf, E.S. and Scher, I. (1980) Differential sensitivity of inbred mice to *Salmonella typhimurium*: a model for the genetic control of innate resistance to bacterial infection. In: Skamene, E., Kongshavn, P.A.L. and Landy, M. (eds), *Genetic Control of Natural Resistance to Infection and Malignancy.* Academic Press, New York, pp. 101–12.

O'Brien, A.D., Scher, I., Campbell, G.H., MacDermott, R.P. and Formal, S.B. (1979) Susceptibility of CBA/N mice to infection with *Salmonella typhimurium*: influence of the X-linked gene controlling B lymphocyte function. *Journal of Immunology* 123, 720–4.

Okazaki, K., Nishikawa, S.-I. and Sakano, H. (1988) Aberrant immunoglobulin gene rearrangement in SCID mouse bone marrow cells. *Journal of Immunology* 141, 1348–52.

Oksenberg, J.R., Stuart, S., Begovitch, A.B., Bell, R.B., Erlich, H.A., Steinman, L. and Bernard, C.C.A. (1990) Limited heterogeneity of rearranged T-cell V-alpha transcripts in brains of multiple sclerosis patients. *Nature* 345, 344–6.

Oldstone, M.B.A. (1989) Viral persistence. *Cell* 56, 517–20.

Oldstone, M.A., Dixon, F.J., Mitchell, G. and McDevitt, H. (1973) Histocompatibility linked genetic control of disease susceptibility. Murine lymphocytic choriomeningitis virus infection. *Journal of Experimental Medicine* 137, 1201–12.

O'Neill, H.C. and Blanden, R.V. (1983) Mechanisms determining innate resistance to ectromelia virus infection in C57BL mice. *Infection and Immunity* 41, 1391–4.

O'Neill, H.C., Blanden, R.V. and O'Neill, T.J. (1983) *H-2*-linked control of resistance to ectromelia virus infection in B10 congenic mice. *Immunogenetics* 18, 255–65.

Orme, I.M., Stokes, R.W. and Collins, F.M. (1985) Only two out of fifteen BCG strains follow the *Bcg* pattern. *Progress in Leukocyte Biology* 3, 285–9.

Outteridge, P.M., Windon, R.G. and Dineen, J.K. (1985) An association between a lymphocyte antigen in sheep and the response to vaccination against the parasite *Trychistrongylus colubriformis*. *International Journal for Parasitology* 15, 121–7.

Pantelouris, E.M. (1968) Absence of a thymus in a mouse mutant. *Nature* 217, 370–1.

Pazderka, F., Longenecker, B.M., Law, G.R.J., Stone, H.A. and Ruth, R.F. (1975) Histocompatibility of chicken populations selected for resistance to Marek's disease. *Immunogenetics* 2, 93.

Perlmutter, R.M., Nahm, M., Stein, K.E., Slack, J., Zitron, I., Paul, W.E. and Davie, J.M. (1979) Immunoglobulin subclass-specific immunodeficiency in mice with an X-linked B-lymphocyte defect. *Journal of Experimental Medicine* 149, 993–8.

Perrudet-Badoux, A., Binaghi, R.A. and Boussac-Aron, Y. (1978) *Trichinella spiralis* infection in mice: mechanism of the resistance in animals genetically selected for high and low antibody production. *Immunology* 35, 519–22.

Piguet, P.-F., Grau, G.E., Allet, B. and Vassalli, P. (1987) Tumor necrosis factor/cachectin is an effector of skin and gut lesions of the acute phase of graft-vs.-host disease. *Journal of Experimental Medicine* 166, 1280–9.

Plant, J.E. and Glynn, A.A. (1976) Genetics of resistance to infection with *Salmonella typhimurium* in mice. *Journal of Infectious Diseases* 133, 72–8.

Plant, J. and Glynn, A.A. (1982) Genetic control of resistance to *Salmonella typhimurium* infection in high and low antibody responder mice. *Clinical and Experimental Immunology* 50, 283–90.

Pollack, M.A. and Rich, R.R. (1985) The HLA complex and the pathogenesis of infectious disease. *Journal of Infectious Diseases* 151, 1–7.

Prehn, R.T. and Prehn, L.M. (1989) The flip side of tumor immunity. *Archives of Surgery* 124, 102–6.

Reed, N.D., Wakelin, D., Lammas, D.A. and Grencis, R.K. (1988) Genetic control of mast cell development in bone marrow cultures. Strain-dependent variation in cultures from inbred mice. *Clinical and Experimental Immunology* 73, 510–15.

Reidlinger, J., Grencis, R.K. and Wakelin, D. (1986) Antigen specific T cell lines transfer immunity against *Trichinella spiralis in vivo*. *Immunology* 58, 57–61.

Ross, J. and Sanders, M.F. (1987) Changes in the virulence of myxoma virus strains in Britain. *Epidemiology of Infections* 98, 113–17.

Roy, S., Scherer, M.T., Briner, T.J., Smith, J.A. and Gefter, M.L. (1989) Murine MHC polymorphism and T cell specificities. *Science* 244, 572–5.

Ruppel, A., Rother, U. and Diesfeld, H.J. (1982) *Schistosoma mansoni*: development of primary infections in mice genetically deficient or intact in the fifth component of complement. *Parasitology* 85, 315–23.

Schatz, D.G., Oettinger, M.A. and Baltimore, D. (1989) The V(D)J recombination activating gene, RAG-1. *Cell* 59, 1035–48.

Schell, K. (1960) Studies on the innate resistance of mice to infection with mousepox. I. Resistance and antibody production. *Australian Journal of Experimental Biology and Medicine* 38, 271–88.

Scott, P. (1989) The role of TH1 and TH2 cells in experimental cutaneous leishmaniasis. *Experimental Parasitology* 68, 369–72.

Schurr, E., Buschman, E., Malo, D., Gros, P. and Skamene, E. (1990) Immunogenetics of mycobacterial infections: Mouse–human homologies. *Journal of Infectious Diseases* 161, 634–9.

Schwartz, R.H. (1989) Acquisition of immunologic self-tolerance. *Cell* 57, 1073–81.

Skamene, E., Kongshavn, P.A.L. and Sachs, D.H. (1979) Resistance to *Listeria monocytogenes* in mice is genetically controlled by genes which are not linked to the *H*-2 complex. *Journal of Infectious Diseases* 139, 228–31.

Stevenson, M.M., Shenouda, G., Thomson, D.M.P. and Skamene, E. (1985) Genetically determined defect in chemotactic responsiveness of inflammatory macrophages from A/J mice. *Progress in Leukocyte Biology* 3, 577–84.

Steward, M.W. (1979) Chronic immune complex disease in mice: the role of antibody affinity. *Clinical and Experimental Immunology* 38, 414–23.

Steward, M.W., Reinhardt, M.C. and Staines, N.A. (1979) The genetic control of antibody affinity. *Immunology* 37, 697–703.

Titus, R.H., Lima, G.C., Engers, H.D. and Louis, J.A. (1984) Exacerbation of murine cutaneous leishmaniasis by adoptive transfer of parasite-specific helper T-cell populations capable of mediating *Leishmania major*-specific delayed-type hypersensitivity. *Journal of Immunology* 133, 1594–600.

Tonegawa, S. (1983) Somatic generation of antibody diversity. *Nature* 302, 575–81.

Vadas, M.A. (1982) Genetic control of eosinophilia in mice: gene(s) expressed in bone marrow-derived cells control high responsiveness. *Journal of Immunology* 128, 691–5.

Van Eden, W., Gonzalez, N.M., de Vries, R.R.P., Convit, J. and van Rood, J.J. (1985) HLA-linked control of predisposition to lepromatous leprosy. *Journal of Infectious Diseases* 151, 9–14.

Viens, P. and Targett, G.A.T. (1971) *Trypanosoma musculi* infection in intact and thymectomized CBA mice. *Transactions of the Royal Society of Tropical Medicine and Hygiene* 65, 424–9.

Viens, P., Targett, G.A.T., Wilson, V.C.I.C. and Edwards, C.I. (1972) The persistence of *Trypanosoma (Herpetesoma) musculi* in the kidneys of immune CBA mice. *Transactions of the Royal Society of Tropical Medicine and Hygiene* 66, 669–70.

Wakelin, D. (1985) Genetic control of immunity to helminth infections. *Parasitology Today* 1, 17–23.

Wakelin, D. (1989) Helminth infections. In: Wakelin, D.M. and Blackwell, J.M. (eds), *Genetics of Resistance to Bacterial and Parasitic Infection.* Taylor and Francis, London, pp. 153–224.

Wakelin, D.M. and Blackwell, J.M. (eds) (1989) *Genetics of Resistance to Bacterial and Parasitic Infection.* Taylor and Francis, London.

Wakelin, D. and Donachie, A.M. (1983) Genetic control of eosinophilia. Mouse strain variation in response to antigens of parasite origin. *Clinical and Experi-*

mental Immunology 51, 239–46.

Warren, K.S. (1968) Pathophysiology and pathogenesis of hepatosplenic schisto-somiasis mansoni. *Bulletin of the New York Academy of Medicine* 44, 280–94.

Wassom, D.L., Krco, C.J. and David, C.S. (1987) I-E expression and susceptibility to parasite infection. *Immunology Today* 8, 39–43.

Weatherall, D.J., Bell, J.I., Clegg, J.B., Flint, J., Higgs, D.R., Hill, A.V.S., Pasvol, G. and Thein, S.L. (1988) Genetic factors as determinants of infectious disease transmission in human communities. *Philosophical Transactions of the Royal Society (London) Series B*, 321, 1–22.

Westphal, H. (1989) Transgenic mammals and biotechnology. *FASEB Journal* 3, 117–20.

Wicker, L.S. and Scher, I. (1986) X-linked immune deficiency (*xid*) of CBA/N mice. *Contemporary Topics in Microbiology and Immunology* 124, 87–101.

Windon, R.G. and Dineen, J.K. (1980) The segregation of lambs into 'responders' and 'non-responders': response to vaccination with irradiated *Trychostrongylus colubriformis* larvae before weaning. *International Journal for Parasitology* 10, 65–73.

Windon, R.G. and Dineen, J.K. (1984) Parasitological and immunological com-petence of lambs selected for high and low responsiveness to vaccination with irradiated *Trichostrongylus colubriformis larvae*. In: Dineen, J.K. and Outter-idge, P.M. (eds) *Immunogenetic Approaches to the Control of Endoparasites*. CSIRO, Division of Animal Health, Melbourne, pp. 13–28.

Ytterberg, S.Y., Mahowald, M.L. and Messner, R.P. (1988) T cells are required for coxsackievirus B1 induced murine polymyositis. *Journal of Rheumatology* 15, 475–8.

Zschiesche, W. (ed.) (1987) *Immune Modulation by Infectious Agents*. VEB Gustav Fischer Verlag, Jena.

Chapter 4

Model Systems on the Genetic Basis of Disease Resistance

D. Wakelin

Department of Zoology, University of Nottingham, University Park, Nottingham NG7 2RD, UK

Summary

Resistance to disease in general, and to parasitic disease in particular, is known to be a genetically variable characteristic of domestic animals. Recognition of this fact makes it possible to think in terms of breeding selectively for enhanced resistance; knowledge of the mechanisms through which genetic variation is expressed makes it possible to think in terms of manipulating immune responsiveness selectively in order to improve overall breed resistance. Progress in both of these fields depends critically upon a detailed understanding of genetic and immunological influences on disease resistance, an understanding which must come initially from detailed studies in laboratory model systems, particularly those involving mice. Only in these relatively inexpensive, well-defined and manipulable hosts can parameters of infection, response and genotype be rigorously controlled and exploited.

This review will describe a range of host–parasite systems selected to cover the major groups of parasites responsible for disease in farm animals – Protozoa, Cestoda, Nematoda and Arthropoda. The diseases for which these systems are models include coccidiosis, trypanosomiasis, echinococcosis, gastrointestinal helminthiasis and tick infestation. For each system the genetic basis of resistance and susceptibility is defined and the roles of MHC-linked and background genes analysed. The immune and inflammatory responses controlled by these genes, and their functional activity in mediating resistance are discussed. The data arising from these laboratory studies are considered in relation to selective breeding programmes, to vaccination strategies, and to manipulation of host responsiveness designed to reduce immunopathology and prevent parasite-induced immunosuppression in domestic stock.

Introduction

Resistance to disease, and specifically infectious disease, operates at several levels. Innate resistance reflects those fixed properties of an animal (structural, behavioural, biochemical, physiological) which make it more-or-less insusceptible to the development of a particular disease organism. Acquired resistance describes those immune and immunologically mediated responses which adaptively protest an animal once it has experienced an infection. Both innate and acquired resistance reflect the phenotypic expression of genetically determined characteristics, and as such can be expected to vary among members of a genetically heterogenous population.

The empirical observation that individuals within flocks and herds of domestic animals do differ markedly in their resistance to infectious diseases has a long history. Descriptive studies of this phenomenon extend back for at least 60–70 years, and a number of important papers demonstrating the genetic basis of such differences were published in the 1950s (e.g. Champion, 1954; Whitlock, 1955). Despite the importance of genetic variation in disease resistance, detailed analyses of the ways in which this variation is controlled and expressed in domestic stock are difficult to undertake, and progress (with some important exceptions) has been relatively slow. There are many reasons for this situation; these include the logistic complexity and expense of large-animal research, and the long generation times and seasonal reproduction of many species. Equally important, however, has been the lack of genetically and immunologically well-defined lines of domestic animals, and the paucity of specific immunological reagents for investigating the components of their immune systems, although these limitations are no longer as severe as they once were. None of these difficulties apply to the use of laboratory rodents, however, and many models of infectious diseases have been explored in great detail using these experimental systems. Although in most cases the model has to make use of species related to the target infectious organism, rather than the organism itself, this is in itself not a serious limitation and such work has led to a detailed understanding of the general principles underlying the expression of genetic variation in resistance to infectious diseases (Wakelin and Blackwell, 1988). This understanding began with the classical studies by Gowen and Webster on bacterial and viral infections in mice (Gowen, 1951; Webster, 1937). It grew exponentially with the availability of genetically defined strains of mice and with the development of immunology and immunogenetics, and it continues with the application of the concepts and approaches of molecular biology and molecular genetics. This short review will focus on a few model systems involving parasitic infections where progress has been substantial and testable hypotheses developed. For a broader and more detailed coverage of the topic the reader is referred to

Table 4.1.   Useful rodent models in which genetic influences upon resistance to parasitic infections can be studied.

	Parasite	Host	Target
Protozoa	*Babesia* spp.	Mouse	Piroplasmosis
	Eimeria spp.	Mouse, Rat	Coccidiosis
	Trypanosoma spp.	Mouse	Trypanosomiasis
Platyhelminths	*Schistosoma* spp.	Mouse, Rat	Schistosomiasis
	Taenia taeniaeformis	Mouse	Cysticercosis, Hydatid
Nematoda	*Ascaris suum*	Mouse	Ascariasis
	Heligmosomoides polygyrus	Mouse	Intestinal nematodiasis
	Nippostrongylus brasiliensis	Mouse, Rat	Intestinal nematodiasis
	Trichinella spiralis	Mouse, Rat	Intestinal nematodiasis
	Trichostrongylus colubriformis	Guinea Pig	Trichostrongylidosis
	Trichuris muris	Mouse	Trichuriasis
Arthropoda	*Dermacentor variabilis*	Mouse	Tick infestation
	Haemaphysalis longicornis	Mouse	Tick infestation

the reviews by Rosenstreich *et al.* (1982), Skamene (1985), Wakelin and Blackwell (1988) and Wassom and Kelly (1990).

Rodent model systems

A wide spectrum of the parasitic infections affecting farm animals can be modelled in the laboratory using rodent hosts. Of these the most useful from the point of view of the immunogeneticist are those involving mice, rats and guinea pigs, where well-defined inbred strains are available (Table 4.1). Mice are the best defined species, both genetically and immunologically, and are the most manipulable experimentally. The range of strains available, and the ease of breeding, makes it possible to pinpoint genes responsible for observed differences in resistance (Tables 4.2 and 4.3) and to identify the ways in which these differences are inherited. The range of immunological reagents available for mice makes it possible to identify, remove or augment particular components of immune and inflammatory responses in order to determine their role in the expression of resistance.

The murine systems discussed here are selected to cover the major groups of parasites of farm animals – Protozoa, Nematoda and Arthropoda. They illustrate the usefulness of experimental approaches in defining the genetic basis of resistance, identifying the resistance mechanisms concerned and providing predictive models relevant to practical problems.

Table 4.2. Analysis of genetic differences in disease resistance: Range of mouse strains available for immunogenetic studies.

Strains	Characteristics
Random-bred	Individuals differ genetically ∴ model variation in populations of domestic stock
Inbred	Individuals genetically identical – different strains ∴ model variation seen between different individuals
MHC congenic	Different strains differ only at the MHC loci – ∴ allow influence of specific loci to be determined
Recombinant inbred	Inbred strains produced from crossing two standard inbred strains. Determination of strain distribution pattern of a given characteristic allows identification of nature of gene control involved and may identify chromosomal location of genes
Mutant	Different strains, carrying defined mutations affecting immunological or inflammatory responses

Table 4.3. Analysis of genetic differences in disease resistance: Expression of genetic differences.

Gene involved in control	
MHC-linked genes (HLA, BLA, OLA, H-2)	Non-MHC-linked genes (Background genes)
Effects expressed in infection	
Qualitative differences between individuals	Quantitative differences between individuals
(All-or-nothing responses)	(Level of response)
seen in:	*seen in*:
Resistance or susceptibility Antigen recognition Lymphocyte responsiveness	Degree of resistance Level of antibody response Level of inflammation

Protozoa

Coccidia

Coccidiosis has long been a major problem in the poultry industry and is increasingly important in the intensive rearing of calves, pigs and sheep. In chickens, breed variation in resistance to disease has been known for many years and is now receiving detailed study, the availability of inbred lines making it possible to define the importance of MHC-linked and non-MHC genes (Lillehoj *et al.*, 1989). A number of species of *Eimeria* infect rodents, and experimental studies have similarly shown genetic variation in resistance. *E. vermiformis* has proved to be a particularly good model for analysis of this variation in inbred mice (Rose and Wakelin, 1989). Primary infections are prolonged in susceptible strains, providing an easily measurable parameter of altered responsiveness (Fig. 4.1). MHC-linked genes play only a small part in determining response phenotype, although it is clear, from adoptive transfer and radiation chimaera studies, that CD4[+] T helper

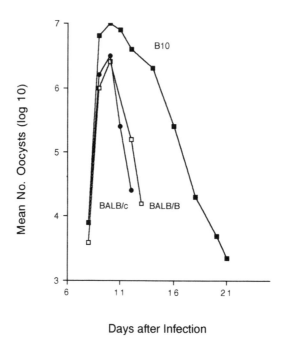

Days after Infection

Fig. 4.1. Time course of a primary infection with the coccidian *Eimeria vermiformis* in inbred mice. B10 mice (C57BL/10 – H-2^b) are susceptible and have a high, prolonged oocyst output. Both BALB/c [H-2^d] and BALB/B mice [H-2^b] are resistant, showing that MHC[H-2]-linked genes have little influence on responsiveness. (Redrawn from Joysey *et al.*, 1988.)

cells play a major role in the initiation and expression of resistance. It has recently been shown that effective resistance involves the cytokine IFNγ, and it can be proposed that response differences between resistant and susceptible strains may result from genetically determined variation in the time and degree of Th cell activation. This hypothesis is supported by experiments which show that T cell proliferation to *E. vermiformis* antigen occurs earlier in resistant than in susceptible strains, and that immune T cells capable of transferring resistance to naive animals likewise appear earlier in the mesenteric lymph nodes of resistant mice (Rose *et al.*, 1990). The reason for this differential responsiveness is not yet understood in terms of antigen recognition or Th cell kinetics. The demonstration that it occurs, and is correlated with resistance, implies that one way of enhancing prophylactic resistance to coccidial infection may be to design vaccines which selectively boost Th cell cytokine production, rather than attempt to elicit the (as yet unidentified) mechanisms responsible for the more complete form of challenge immunity seen after initial exposure. Even where T cell immunity appears not to control the kinetics of primary infections (as with many self-limiting coccidial species) it is clear that it contributes to limitation of pathology.

Trypanosomes

Trypanosomiasis continues to be a severe problem for cattle production in Africa, despite programmes aimed at controlling the tsetse fly vector. As will be discussed elsewhere, there is evidence of marked breed variation in ability to resist trypanosome infections, thus raising the possibility of selective breeding to enhance resistance in commercial herds. Murine systems using several species of trypanosome have contributed greatly both to our present understanding of trypanotolerance and of genetic variation in responsiveness to this group of parasites.

Studies with *T. congolense* showed that strains of mice varied considerably both in terms of parasitaemia and of survival (Jennings *et al.*, 1978). Genetic analysis of inheritance of resistance initially suggested involvement of two genes (Morrison and Murray, 1979), but later work produced data consistent with control by a single gene (Blackwell, 1988; Pinder, 1984). Although there is variation in survival times between H-2 congenic strains it does not appear that MHC-linked genes exert a significant influence (Morrison and Murray, 1979; Pinder *et al.*, 1985). Resistant strains of mice (e.g. C57BL/6) produce variant antigen-specific IgM antibodies earlier in infection than susceptible BALB/c, and this is true also of C57BL/6 nude mice, showing the T-independence of this response (Pinder *et al.*, 1985). In a similar study, Mitchell and Pearson (1985) showed that resistant C57BL/6 not only made high IgM responses but produced little specific IgG, whereas susceptible A/J mice showed a preferential switch to IgG isotypes. Work in mice with *T. rhodesiense* (de Gee and Mansfield, 1984) indicates that although isotype-specific responses may correlate with the ability to control the initial wave of parasitaemia, they do not correlate directly with ultimate survival, so that it is difficult to assess fully the functional significance of the *T. congolense* data. It may be necessary, as with *T. rhodesiense*, to look at differences in cytokine release (de Gee *et al.*, 1985), or to investigate, as with *T. brucei brucei*, the likelihood of parasite-induced immune suppression operating against B cell function (Newson *et al.*, 1990).

Cestoda

Larval tapeworms

Sheep and cattle act as intermediate hosts for the larval stages of several tapeworms. In addition to the problems associated with direct (*Taenia saginata*) or indirect (*Echinococcus granulosus*) infections of man, such parasites can cause economic losses because of carcasses being condemned. A convenient mouse model for studying the immunogenetics of these infections is *Taeniae taeniaeformis*, a tapeworm of the cat whose larvae develop in rodents. A common finding in early work was the unpredictability of establishing infections in mice. Detailed studies showed that this reflected non-MHC genetic differences between mouse strains, which were expressed in the rate of formation of specific antiparasite Ig antibodies (Mitchell *et al.*, 1977, 1980). Invading larvae can be killed in several ways, of which antibody-dependent complement-mediated damage is one. As the larvae develop in the liver they develop anticomplementary activity, which reduces or prevents the host's attack. If antibody can be made quickly enough the larvae can be killed, if antibody is made slowly the larvae become insusceptible and survive.

Although this is a simplistic model it allows a number of predictions to be made about immunity to *T. taeniaeformis* and, by extrapolation, to the economically important larval tapeworms. One prediction is that after an initial infection both resistant and susceptible hosts should express immunity to challenge, because both will have the capacity (through immunological memory) to develop antibodies rapidly. The other is that, if antibody levels can be raised by vaccination, both resistant and susceptible hosts should be protected against an initial infection. Both predictions are correct (see Conchedda and Ferretti, 1984 (challenge); Rajasekariah *et al.*, 1980 (vaccination)) and vaccination is achieved readily even in susceptible hosts (Fig. 4.2). More recent studies (Gibbens *et al.*, 1986) imply that the important element of the response necessary is not simply total antibody levels, but levels of specific antibodies against selected important antigens. That such antibodies can be raised in susceptible hosts is shown not only by

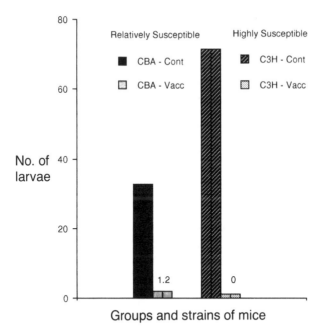

Fig. 4.2. Vaccination of susceptible strains of mice against infection with the larval cestode *Taenia taeniaeformis*. Both relatively and highly susceptible mice express almost complete resistance to infection after vaccination with parasite antigens. (Data from Rajasekariah *et al.*, 1980.)

the data referred to above, but also by the recent successful development of a recombinant vaccine against larval *Taenia* species in sheep (Johnson *et al.*, 1989).

Nematodes

The most important nematode parasites of farm animals are the gastro-intestinal worms. These are a complex of species whose composition and relative importance vary with the host and country concerned. Few of these species can be used directly in laboratory rodents, although guinea pigs are satisfactory hosts for *Trichostrongylus colubriformis*, but there are model systems that use related species, or other intestinal worms, which provide useful data on immunogenetic aspects of these infections.

Trichinella spiralis

The veterinary importance of *T. spiralis*, although locally significant, is relatively small. However, the low host specificity and ease of maintenance of this species make it an excellent model for studying the responses involved in controlling infections with intestinal worms. Infections generate strong protective responses that have been analysed in some detail (Wakelin *et al.*, 1990). The time course of infection and the degree of immunity elicited are variable between different strains of mice, as are almost all parameters of responsiveness (Table 4.4). Resistant strains expel adult worms rapidly from the intestine and in consequence acquire low muscle larval burdens (Fig. 4.3). Susceptible strains maintain reproducing worm populations in the intestine for significantly longer periods and develop heavy muscle burdens. Genetic variation is also seen in the timing and expression of immunity to reinfection.

The genes exerting the strongest influences on immunity lie outwith the MHC but, when H-2 congenic strains are studied, MHC-linked genes are seen to exert a clear effect. As with a majority of parasites, overall resistance appears to reflect the dominant influence of background genes, modulated within limits by MHC-linked genes. At least three genes within the MHC have been associated with response to *T. spiralis*: and these show complex interactions (Wassom and Kelly, 1990). Particular attention has been paid to the role of MHC class II-related genes in determining resistance. Whereas all mice express class II I-A gene products on their antigen-presenting cells only certain strains also express the corresponding class II I-E genes. Strains which express only IA are relatively resistant to *T. spiralis*; those that additionally express I-E are relatively susceptible. It was originally suggested that I-E expression was associated with a T cell-mediated immunosuppression that counteracted the enhanced resistance associated with IA gene expression (Wassom *et al.*, 1987). More recently, since the demonstration that there is deletion of certain gene sequences from the T cell receptor repertoire in IE positive mice, it has been proposed that, as a consequence, there may be a defective response in such mice to *T. spiralis* antigens (Wassom and Kelly, 1990). This may either be in

Table 4.4. The mouse–*Trichinella spiralis* model: Genetical variation parameters of the response to infection.

Parasitological	Immunological
Site of infection in gut	Antibody response
Survival of adult worms	Proliferative response
Fecundity of female worms	Inflammatory response
Number of muscle larvae	Degree of immune depression
Response to challenge infection	Response of vaccination

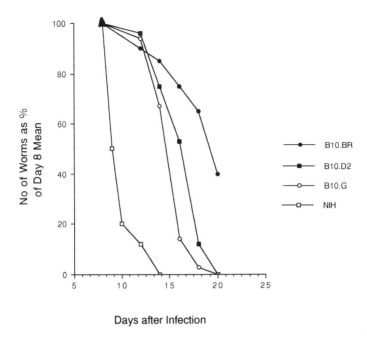

Fig. 4.3. Time course of a primary infection with the nematode *Trichinella spiralis* in inbred mice. NIH mice [H-2^q] are resistant and expel adult worms quickly from the intestine. The B10 congenic strains are susceptible but MHC [H-2]-linked genes influence the degree of susceptibility. B10.G [H-2^q] are more resistant than B10.D2 [H-2^d] or B10.BR [H-2^k].

absolute terms, i.e. in an absence of response, or in relative terms, i.e. in an imbalance of cytokine release from different T helper subsets. There are certainly differences between resistant and susceptible strains in the responsiveness of T helper cells, both qualitative and quantitative (Zhu and Bell, 1989) but it is not clear how significant these are in determining overall response phenotype. A series of experiments utilizing adoptive transfer, radiation chimaeras and reciprocal reconstitutions of histocompatible resistant and susceptible strains points to the important expression of genetic variation being in the bone marrow stem cells that provide the inflammatory cell populations necessary for the development of the intestinal changes which ultimately force worms out of the gut (Table 4.5) (Wakelin, 1988). *In vivo* and *in vitro* studies with two major inflammatory cell types, the mucosal mast cell and the eosinophil, support this contention. Both cell types are heavily T cell dependent for their development and differentiation, yet it can be shown that host strains with very different inflammatory responsiveness during infection have comparable abilities to produce the relevant T cell cytokines. Reciprocal transfers of immune T cells or of bone

Table 4.5.   The mouse–*Trichinella spiralis* model: Characteristics of the protective immune response to the intestinal stages.

---

1. Immunity is T cell-dependent
2. Antibody plays a major role in reducing worm growth and fecundity
3. Worm expulsion is mediated through inflammatory responses
4. Intestinal inflammation involves structural, functional and biochemical changes to the worms' environment
5. Mucosal mastocytosis and eosinophilia are important elements of the inflammatory response
6. Genetic variation in inflammatory responsiveness affecting mast cells and eosinophils operates at the level of the bone marrow precursor cells

---

marrow, and *in vitro* cultures of bone marrow, carried out under conditions of optimal cytokine stimulation, show that the difference lies in the capacity of the bone marrow to respond to T cell cytokines (Reed *et al.*, 1988; Lammas *et al.*, 1989).

It follows from these conclusions that vaccination is likely to be relatively less effective in boosting low responsiveness to *T. spiralis*, and other worms that are controlled by inflammation-dependent mechanisms, than it is in boosting low responsiveness to parasites controlled more directly by immunological responses (e.g. *Taenia taeniaeformis* and other larval tapeworms). This has been confirmed in the mouse–*T. spiralis* system (Wakelin *et al.*, 1986) and might explain the relative unresponsiveness of certain individual sheep to vaccination against *Trichostrongylus colubriformis*. Although the mechanisms of immunity operative against this species are not fully understood there is suggestive evidence that implicates inflammatory cell populations (Dawkins *et al.*, 1989; Dineen and Windon, 1980). In the guinea-pig model genetically determined resistance to infection with *T. colubriformis* is likewise correlated with enhanced inflammatory responsiveness (Handlinger and Rothwell, 1981). A further conclusion that may be drawn from the model studies is that the parameter of inflammatory responsiveness may be a better predictor of anti-worm responsive potential than other markers, including MHC markers (Cooper *et al.*, 1989).

Heligmosomoides (Nematospiroides) polygyrus

T. spiralis is unusual among many nematodes in the strength and speed of the intestinal response provoked. It is characteristic of most gastrointestinal nematodes that they give rise to long-lasting persistent infections. One possibility to explain this situation is that such worms have the ability actively to depress the protective responses of the host. The trichostrongyle

H. polygyrus produces chronic infections in most mouse strains, and exerts a powerfully suppressive effect on both homologous and heterologous immune responses in the host (Behnke, 1987). One manifestation of this is the suppression of inflammatory responses such as the mucosal masto-cytosis induced by *Trichinella spiralis* (Dehlawi and Wakelin, 1988), a suppression associated with a parasite factor-mediated effect upon host T cells. The chronicity of infection with *H. polygyrus* (and presumably, therefore, the degree of host suppression) is genetically variable between mouse strains, and influenced by both MHC-linked and non-MHC genes (Robinson *et al.*, 1989). Identification of the nature and expression of the genetic control involved, which may concern T cell cytokine profiles and T helper cell subset imbalance, may provide useful pointers to the ways in which the suppressive effects likely to occur in other chronic intestinal infections can be modulated in the host's favour.

H. polygyrus in mouse has also proved to be a valuable model for following the effects of strong selection for resistance to infection. Dobson and colleagues have carried out several series of selective breeding experiments (Brindley and Dobson, 1981, 1983; Sitepu and Dobson, 1982). They have not only established parameters associated with enhanced innate and acquired immunity to the parasite itself, but have also examined concurrent selection for resistance to other organisms, for changes in host character-istics (relevant to the question of 'productivity' vs 'resistance'), and changes in parasite biology. All of these aspects are important considerations to be borne in mind when selective breeding for parasite resistance is undertaken in domestic stock.

Arthropoda

Breed and individual variation in resistance of bovines to tick infestation is a well-established fact. European breeds (*Bos taurus*) as a whole are more susceptible than *Bos indicus* breeds, and this influences the impact of both tick disease *per se* and the diseases transmitted by these vectors. Recent progress in the development of anti-tick vaccines is likely to have a signifi-cant influence on these problems, but it is still important to understand (and utilize) the genetic and immunological variations that occur in tick resistance. A number of tick species will feed successfully on mice, providing useful models for such studies. Considerable mouse strain varia-tion in resistance has been recorded by a number of workers.

One important observation that has been made is that resistance, which affects the ability of ticks to feed, is associated with IgE antibodies and the development of inflammatory responses at the feeding site (Matsuda *et al.*, 1990). Mast cells play a key role in these responses. Mast cell-deficient mice (W/W^v strain) develop no resistance to infestations with *Haema-physalis longicornis*, yet do so if reconstituted with normal bone marrow

(Matsuda *et al.*, 1985). The local relevance of functional mast cells in resistance has been demonstrated using elegant grafting experiments with W/Wᵛ mice given skin from compatible mast cell-sufficient donors (Matsuda *et al.*, 1987).

Although inflammatory responses are likely to be involved in anti-tick immunity in cattle, it is not clear whether these responses are relevant to vaccine induced immunity. It appears that antibodies against gut-related antigens of the tick may play an important role in this immunity (Opdebeeck *et al.*, 1988). It can, therefore, be predicted that the ability to raise effective levels of the appropriate specific antibodies will also be subject to genetic variation, thus resulting in poorer protection in certain individuals.

Conclusions

The value of model systems can be judged in several ways, by their closeness to the situation modelled, by their predictive value, and by their capacity to provide fundamental insights. Models for studying the genetic basis of disease resistance fulfil some or all of these criteria to differing extents. In parasitological terms few are really close to the relationships that exist between ruminant hosts, and other farm animals, and their parasites. Nevertheless many model these host–parasite relationships in essence and thus provide data of direct relevance. The predictive value of models has been borne out in several ways. For example, almost all models show that clear-cut, genetically determined differences do exist in resistance and susceptibility, that resistance is heritable, and heritable as a dominant characteristic. This observation is critical for strategies designed to improve the disease resistance of stock by selective breeding. Similarly, despite assumptions to the contrary, model systems have shown that relatively few of the complex antigens presented by parasites are significant in terms of eliciting functional resistance to infection, and that immunological recognition of such antigens can be tightly regulated. This observation is critical for the rational design and use of defined vaccines.

Many models are providing fundamental insights into the mechanisms of resistance itself, an analysis that is simplified by direct comparisons of resistant and susceptible hosts, and into the ways in which genes influence the development and expression of these mechanisms. Progress towards the explanation of this control and expression in terms of molecular genetics and molecular immunology will provide the means of directly extrapolating functional explanations from the model to the target system. The close relationships that exist between, for example, the murine and other mammalian genomes, will make it possible to use murine disease-resistance genes to probe other genomes for their homologues. The essential similarities between all mammalian immune systems mean that data derived from rodents will have direct relevance to understanding the ways in which immune responses cope with parasite infection in hosts of economic importance. Progress in these areas has been substantial over recent years, and has established a firm basis for future exploitation of immunogenetics in the field of resistance to parasitic infections.

References

Behnke, J.M. (1987) Evasion of immunity by nematode parasites. *Advances in Parasitology* 26, 1–71.

Blackwell, J.M. (1988) Protozoan infections. In: Wakelin, D. and Blackwell, J.M. (eds), *Genetics of Resistance to Bacterial and Parasitic Infections.* Taylor and Francis, London, pp. 103–51.

Brindley, P.J. and Dobson, C. (1981) Genetic control of liability to infection with *Nematospiroides dubius* in mice: selection of refractory and liable populations of mice. *Parasitology* 83, 51–65.

Brindley, P.J. and Dobson, C. (1983) Genetic control of liability to infection with *Nematospiroides dubius* in mice; direct and correlated responses to selection of mice for faecal parasite egg count. *Parasitology* 87, 113–27.

Champion, L.R. (1954) The inheritance of resistance to cecal coccidiosis in the domestic fowl. *Poultry Science* 33, 670–81.

Conchedda, M. and Ferretti, G. (1984) Susceptibility of different strains of mice to various levels of infection with the eggs of *Taenia taeniaeformis. International Journal for Parasitology* 14, 541–6.

Cooper, D.W., van Oorschot, R.A.H., Piper, L.R. and Le Jambre, L.F. (1989) No association between the ovine leucocyte antigen (OLA) system in the Australian Merino and susceptibility to *Haemonchus contortus* infection. *International Journal for Parasitology* 19, 695–7.

Dawkins, H.J.S., Windon, R.G. and Eagleson, G.K. (1989) Eosinophil responses in sheep selected for high and low responsiveness to *Trichostrongylus colubriformis. International Journal for Parasitology* 19, 199–207.

De Gee, A.L.W. and Mansfield, J.M. (1984) Genetics of resistance to the African trypanosomes. IV. Resistance of radiation chimaeras to *Trypanosoma rhodesiense* infection. *Cellular Immunology* 87, 85–91.

De Gee, A.L.W., Sonnenfeld, G. and Mansfield, J.M. (1985) Genetics of resistance to the African trypanosomes. VI. Qualitative and quantitative differences in interferon production among susceptible and resistant mouse strains. *Journal of Immunology* 134, 2723–6.

Dehlawi, M.S. and Wakelin, D. (1988) Suppression of mucosal mastocytosis by *Nematospiroides dubius* results from an adult worm-mediated effect upon host lymphocytes. *Parasite Immunology* 10, 85–95.

Dineen, J.K. and Windon, R.G. (1980) The effect of sire selection on the response of lambs to vaccination with irradiated *Trichostrongylus colubriformis* larvae. *International Journal for Parasitology* 10, 189–96.

Gibbens, J.C., Harrison, L.J.S. and Parkhouse, R.M.E. (1986) Immunoglobin class responses to *Taenia taeniaeformis* in susceptible and resistant mice. *Parasite Immunology* 8, 491–502.

Gowen, J.W. (1951) Genetics and disease resistance. In: Dunn, L.C. (ed.), *Genetics in the 20th Century.* Macmillan, New York, pp. 401–29.

Handlinger, J.H. and Rothwell, T.L.W. (1981) Studies on the responses of basophil and eosinophil leucocytes and mast cells to the nematode *Trichostrongylus colubriformis*: comparison of cell populations in parasite resistant and susceptible guinea pigs. *International Journal for Parasitology* 11, 67–70.

Jennings, F.W., Whitelaw, D.D., Holmes, P.H. and Urquhart, G.M. (1978) The susceptibility of strains of mice to infection with *Trypanosoma congolense.* *Research in Veterinary Science* 25, 399–412.

Johnson, K.S., Harrison, G.B.L., Lightowlers, M.W., O'Hoy, K.L., Cougle, W.G., Dempster, R.P., Lawrence, S.B., Vinton, J.G., Heath, D.O. and Rickard, M.D. (1989) Vaccination against ovine cysticercosis using a defined recombinant antigen. *Nature* 338, 585–7.

Joysey, H.S., Wakelin, D. and Rose, M.E. (1988) Coccidiosis: resistance to infection with *Eimeria vermiformis* in mouse radiation chimaeras is determined by donor bone-marrow cells. *Infection and Immunity* 56, 1399–401.

Lammas, D.A., Mitchell, L.A. and Wakelin, D. (1989) Genetic control of eosinophilia. Analysis of production and response to eosinophil-differentiating factor in strains of mice infected with *Trichinella spiralis. Clinical and Experimental Immunology* 77, 137–43.

Lillehoj, H.S., Ruff, M.D., Bacon, L.D., Lamont, S.J. and Jeffers, T.K. (1989) Genetic control of immunity to *Eimeria tenella.* Interaction of MHC genes and non-MHC linked genes influences levels of disease susceptibility in chickens. *Veterinary Immunology and Immunopathology* 20, 135–48.

Matsuda, H., Fukui, K., Kiso, Y. and Kitamura, Y. (1985) Inability of genetically mast cell-deficient W/W^v to acquire resistance against larval *Haemaphysalis longicornis* ticks. *Journal of Parasitology* 71, 443–8.

Matsuda, H., Nakano, T., Kiso, Y. and Kitamura, Y. (1987) Normalization of anti-tick response to mast cell-deficient W/W^v mice by intracutaneous infection of cultured mast cells. *Journal of Parasitology* 73, 155–8.

Matsuda, H., Watanabe, N., Kiso, Y., Hirota, S., Kannan, Y., Azuma, M., Koyama, H. and Kitamura, Y. (1990) Necessity of IgE antibodies and mast cells for manifestation of resistance against larval *Haemophysalis longicornis* ticks in mice. *Journal of Immunology* 144, 259–62.

Mitchell, G.F., Goding, J.W. and Richard, M.D. (1977) Studies on immune responses to larval cestodes in mice: increased susceptibility of certain mouse strains and hypothymic mice to *Taenia taeniaeformis* and analysis of passive transfer with serum. *Australian Journal of Experimental Biology and Medical Science* 55, 165–86.

Mitchell, G.F., Rajasekariah, G.R. and Richard, M.D. (1980) A mechanism to account for mouse strain variation in resistance to the larval cestode, *Taenia taeniaeformis. Immunology* 39, 481–9.

Mitchell, L.A. and Pearson, T.W. (1985) Antibody responses during *Trypanosoma congolense* infection in resistant and susceptible inbred mice. *Progress in Leukocyte Biology* 3, 501–15.

Morrison, W.I. and Murray, M. (1979) *Trypanosoma congolense*: Inheritance of susceptibility to infection in inbred strains of mice. *Experimental Parasitology* 48, 364–74.

Newson, J., Mahan, S.M. and Black, S.J. (1990) Synthesis and secretion of immunoglobulin by spleen cells from resistant and susceptible mice infected with *Trypanosoma brucei brucei* GUTat 301. *Parasite Immunology* 12, 125–39.

Opdebeeck, J.P., Wong, J.Y.M., Jackson, L.A. and Dobson, C. (1988) Vaccines to protect Hereford cattle against the cattle tick, *Boophilus microplus. Immunology* 63, 363–7.

Pinder, M. (1984) *Trypanosoma congolense*: genetic control of resistance to infection in mice. *Infection and Immunity* 57, 185–94.

Pinder, M., Fumoux, F. and Roelants, G.E. (1985) Immune mechanisms and genetic control of natural resistance to *Trypanosoma congolense. Progress in Leukocyte Biology* 3, 495–500.

Rajasekariah, G.R., Mitchell, G.F. and Rickard, M.D. (1980) *Taenia taeniaeformis* in mice: protective immunization with oncospheres and their products. *International Journal for Parasitology* 10, 155–60.

Reed, D., Wakelin, D., Lammas, D.A. and Grencis, R.K. (1988) Genetic control of mast cell development in bone marrow cultures. Strain-dependent variation in cultures from inbred mice. *Clinical and Experimental Immunology* 73, 510–15.

Robinson, M., Wahid, F., Behnke, J.M. and Gilbert, F.S. (1989) Immunological relationships during primary infection with *Heligmosomoides polygyrus* (*Nematospiroides dubius*): dose-dependent expulsion of adult worms. *Parasitology* 98, 115–24.

Rose, M.E. and Wakelin, D. (1989) Mechanisms of immunity to coccidiosis. In: Yvore, P. (ed.), *Coccidia and Intestinal Coccidiomorphs. Proceedings of the 5th International Coccidiosis Conference* Les Coloques de l'INRA 49, pp. 527–40.

Rose, M.E., Wakelin, D. and Hesketh, D. (1990) *Eimeria vermiformis*: differences in the course of primary infection can be correlated with lymphocyte responsiveness in the BALB/c and C57BL/6 mouse, *Mus musculus. Experimental Parasitology* 71, 276–83.

Rosenstreich, D.L., Weinblatt, A.C. and O'Brien, A.D. (1982) Genetic control of resistance to infection in mice. *Critical Reviews in Immunology* 3, 263–330.

Sitepu, P. and Dobson, C. (1982) Genetic control of resistance to infection with *Nematospiroides dubius* in mice: selection of high and low immune responder populations of mice. *Parasitology* 85, 73–84.

Skamene, E. (ed.) (1985) Genetic control of host resistance to infection and malignancy. *Progress in Leukocyte Biology Vol. 3.* Alan R. Liss, New York.

Wakelin, D. (1988) Helminth infections. In: Wakelin, D. and Blackwell, J.M. (eds), *Genetics of Resistance to Bacterial and Parasitic Infection.* Taylor and Francis, London, pp. 153–224.

Wakelin, D. and Blackwell, J.M. (eds) (1988), *Genetics of Resistance to Bacterial and Parasitic Infection.* Taylor and Francis, London.

Wakelin, D., Harnett, W. and Parkhouse, R.M.E. (1991) Nematodes. In: Warren, K.S. and Agabian, N. (eds), *Immunology and Molecular Biology of Parasitic Infections*, 3rd ed, Blackwell Scientific Publications, Oxford (in press).

Wakelin, D., Mitchell, L.A., Donachie, A.M. and Grencis, R.K. (1986) Genetic control of immunity to *Trichinella spiralis* in mice. Response of rapid- and slow-responder strains to immunization with parasite antigens. *Parasite Immunology* 8, 159–70.

Wassom, D.L. and Kelly, E.A.B. (1990) The role of the major histocompatibility complex in resistance to parasite infections. *Critical Review of Immunology* 10, 31–52.

Wassom, D.L., Krco, C.J. and David, C.S. (1987) I-E expression and susceptibility to parasite infection. *Immunology Today* 8, 39–43.

Webster, L.T. (1937) Inheritance of resistance of mice to enteric bacterial and neurotropic virus infections. *Journal of Experimental Medicine* 65, 261–80.

Whitlock, J.H. (1955) A study of inheritance of resistance to trichostrongylidosis in sheep. *Cornell Veterinarian* 48, 127–33.

Zhu, D. and Bell, R.G. (1989) IL-2 production, IL-2 receptor expression, and IL-2 responsiveness of spleen and mesenteric lymph node cells from inbred mice infected with *Trichinella spiralis. Journal of Immunology* 142, 3262–7.

Section 2
Methodology of Breeding for Disease Resistance

Exposing vulnerable unprotected animals to the challenge of the diseases endemic in their own specific herd/flock environment could prove to be a costly and ineffective method of selecting for disease resistance in farm animals. A more targeted approach, where animals are selected for breeding on the basis of natural or artificial tests for resistance to specific diseases shows promise for a number of disease conditions. Several gene markers have been identified that can be used for the indirect selection for resistance genes. Studies have shown the involvement of two classes of MHC phenotype in disease resistance in cattle but, only for a few diseases such as mastitis, bovine leukosis virus and the protozoan parasite *Theilera parva*, is there evidence that the knowledge may have practical application. Biomolecular approaches to genotype identification can result from genome mapping work. The use of gene transfer to produce transgenic animals, incorporating specific genes for disease resistance, is now a practical feasibility in the laboratory. However, the success rate in producing transgenic farm animals has hitherto been low.

Finally consideration is given to the important problem of the associated effects of breeding for disease resistance. The relatively high level of heritability (circa 0.3) reported for a range of 15 diseases in Australia and New Zealand augurs well for the future. However, methods of efficiently incorporating disease resistance traits in selection indices for farm animals have yet to be established, and a high investment in research on the relevant genetic correlations and on the overall efficacy of the resultant breeding schemes will be required.

Chapter 5
Selection under Challenging Environments

Max F. Rothschild

Department of Animal Science, Iowa State University, Ames,
IA 50011, USA

Summary

Increased selection pressure applied to commercially important traits in production environments is often accompanied by increases in disease problems. At the same time selection for immune responsiveness and disease resistance has often been ignored by animal geneticists because of the difficulty of measuring these traits. Actual disease resistance to individual diseases would have to be measured under an environment that included disease challenge. Such testing could be prohibitively expensive. New opportunities to improve our understanding of the genetic nature of disease resistance now exist through the recent advances in molecular biology and immunology and make indirect selection for disease resistance possible. The major histocompatibility complex (MHC) genes exert a major role in control of disease resistance and all immune functions. Genetic considerations involved with testing and selection for disease resistance and improved immune responsiveness will require knowledge of the genetic correlations between disease resistance and immune responsiveness and production traits. Research suggests that antagonistic relationships between immune response, disease resistance and production traits might make simultaneous improvement of these traits difficult by conventional breeding and selection methods. Use of marker-assisted selection or gene-transfer methods with the genes of the MHC may offer an alternative approach for simultaneous improvement in all of these traits.

Introduction

Efficient production of animals for meat, milk and fibre is often disrupted by disease, which costs livestock producers and consumers billions of dollars each year. These considerable monetary losses result from mortality, subclinical losses due to poorer production efficiency, increased veterinary costs and product loss. Selection pressure on commercially important traits under stressful production settings also may increase the incidence of disease. Disease also impairs genetic improvement in production traits because efficiency of selection is reduced. Current methods to control disease include vaccination, medication, sanitation, isolation of animals from pathogens and eradication of certain diseases. These approaches, however, are not always effective. Lack of effectiveness of some vaccines and increased attempts by consumers to change drug-use regulations because of fears of contamination of human food sources underscore the need for more effective methods to combat disease.

A clear understanding of disease and the animal's defence systems is required for alternative approaches to disease control. The onset of disease is often the result of the interaction between an individual animal's genotype and the environment to which the animal is exposed. If an animal has a genetic predisposition for acquiring a disease, then environmental conditions, including standard disease-prevention methods may be only partly effective in preventing disease. An often-overlooked alternative approach to standard disease control methods would be selective breeding to increase disease resistance in livestock. Genetic resistance to disease involves many facets of the body's defence system and their interactions and is extremely complex.

In this chapter the current status of knowledge of methods to select for disease resistance and immune responsiveness in livestock in challenging environments is reviewed. Opportunities to improve disease resistance and immune responsiveness in livestock through marker-assisted selection and gene-transfer techniques are discussed.

Current status

Disease resistance

The presence of some diseases may result from strictly genetic control whereas others may be caused by a combination of genetic predisposition and exposure to pathogens. Disease resistance research has included measurement of genetic control of disease losses, estimation of heritabilities and characterization of breed or strain differences. Numerous examples of this research exist for leukosis and Marek's disease in poultry (Freeman and

Bumstead, 1987; Hutt, 1958), atrophic rhinitis and respiratory diseases in pigs (Kennedy and Moxley, 1980; Lundeheim, 1979), mastitis and ketosis in cattle (Philipsson *et al.*, 1980; Solbu, 1982) and parasite infestation and foot rot in sheep (Albers *et al.*, 1987; Bulgin *et al.*, 1988). Genetic control of certain diseases may be the result of the presence or absence of receptors that are inherited simply. Resistance to specific subgroups of leukosis virus in chickens seems to be inherited simply (Crittenden, 1975) and may be the result of not having the receptor for the virus. In swine, some individuals lack the receptor for K88, a cell-surface antigen on some *Escherichia coli*, and the *E. coli* cannot attach to the gut in these pigs (Gibbons *et al.*, 1977). Resistance to infection from the K88 strain of *E. coli* is recessive, but susceptible animals remain in the population because maternal antibodies allow some protection.

Selection for disease resistance under challenging environments

Depending on the species, breeders routinely select animals on the basis of from one to perhaps as many as 20 traits. This selection is practised under environments that challenge livestock greatly and may increase the incidence of disease if management is poor. Many researchers have examined approaches to selection for disease resistance (Gavora and Spencer, 1983; Rothschild, 1985, 1989; Warner *et al.*, 1987). These direct approaches, with some modifications, are found in Table 5.1. The simplest method would be to observe and select breeding stock for disease resistance under normal production conditions. This would have no negative effects on production of breeding stock but probably would not be informative because without disease (under good hygiene and management), expression of disease resistance would be questionable. Challenging breeding stock, progeny or sibs would be costly, depending on the severity of the disease challenge, and production could be adversely affected. This method would not be advised if death usually results from a particular disease. Accuracy would be limited unless sufficient numbers of progeny or sibs are tested. One alternative possibility to testing progeny or sibs would be to obtain a large number of clones of embryos from planned matings of breeding stock. Once raised, one set of these animals could be challenged with a specific disease or diseases. Selection of the other animals could then occur on the basis of results from the cloned animals. Because the animals tested were clones, accuracy would be equal to testing the individuals themselves.

Numerous other problems involved with disease challenge or selection under challenging environments exist. These include standardization of the level of challenge exposure to a particular disease and maintenance of isolation facilities for this type of selection. Consider selection under a challenging environment or a disease environment compared with a normal

Table 5.1. Direct approaches to selection for genetic resistance to disease.

Type of selection	Method	Effects on production of breeders	Expression of disease resistance	Cost
Direct	1. Observe breeding stock	0	Questionable	0
	2. Challenge breeding stock	Negative	Good	Low–high
	3. Challenge sibs or progeny of breeding stock	0	Good	Low–high
	4. Challenge clones	0	Excellent	Moderate–high

Source: Adapted from Gavora and Spencer (1983) and Rothschild (1985).

production setting. Although previous research suggests that animals selected under 'bad' environments perform as well in 'good' environments as those selected directly in 'good' environments (Falconer, 1960), if performance rankings of individuals or families seem to be different under the two environments, then genotype by environment interaction probably exists. Estimates of the genetic correlations between disease resistance, immune responsiveness and production traits under 'good' and 'bad' environments would be needed to assess the possibility of genetic improvement by this approach. Selection difficulties would arise if there are antagonistic correlations between the disease resistance and production traits. These problems become multiplicative if the number of diseases for which selection is practised is large. An index approach would be useful, but would require accurate estimates of all the genetic correlations between the disease resistance and performance traits. Such estimates are presently unavailable. Direct selection for disease resistance may also cause a reduction of genetic progress in other traits because of the increased total number of traits.

Selection for disease resistance using indirect approaches

Given the difficulties in selecting for disease resistance under challenging environments alternatives to those methods have been proposed (Table 5.2). Immune responsiveness has been suggested as an indirect indicator of disease resistance (Biozzi *et al.*, 1980; Buschmann *et al.*, 1985; Gavora and Spencer, 1983; Rothschild, 1989; Warner *et al.*, 1987). Early studies with mice (Biozzi *et al.*, 1980) have revealed that genetic control of antibody response to sheep red blood cells was moderately heritable and that selection for humoral immune response for one antigen may improve humoral immune response for other antigens. They also investigated genetic control of cell-mediated responses. Selection for increased humoral response to sheep red blood cells did not improve cell-mediated response, suggesting independence of these traits.

In livestock, the most extensive research with genetic control of immune response has been in poultry (Lamont and Dietert, 1990; van der Zijpp, 1983a). Genetic control of immune responsiveness to sheep red blood cells has been thoroughly investigated (Siegal and Gross, 1980; van der Zijpp and Leenstra, 1980). Heritability estimates have ranged from 0.28 to 0.38, suggesting that this trait is under moderate genetic control. Results suggested that immunization procedures, dosage and site of immunization all affect measurement and extent of genetic control (van der Zijpp, 1983b). Such details may make the use of immune response as an indicator of disease resistance more difficult. Other experiments have demonstrated that response to vaccination with Newcastle disease, *Salmonella pullorum* and *E. coli* are under moderate genetic control and that selection for high

Table 5.2. Indirect approaches to selection for genetic resistance to disease.

Type of selection	Method	Effects on production of breeders	Expression of disease resistance	Cost
Indirect	1. Vaccine challenge	0	Good	Low
	2. *In vitro* tests	0	Good	Low
	3. Genetic markers	0	Good	Low
Molecular genetics	Construct resistant genotypes	0	Good	High

Source: Adapted from Gavora and Spencer (1983) and Rothschild (1985).

and low antibody response following vaccination is effective (Lamont and Dietert, 1990).

Immune response experiments in swine using sheep red blood cells and DNP-hapten as antigens have revealed significant between-breed and within-breed differences (Buschmann *et al.*, 1974; Buschmann, 1980). In a separate series of experiments, the heritability of secondary immune response and peak response to bovine serum albumen were estimated to be 0.51 and 0.42, respectively (Huang, 1977). Using commercial vaccines for *Bordetella bronchiseptica* and pseudorabies virus, researchers demonstrated significant within-breed and between-breed differences, low to moderate heritabilities (0.05 to 0.52) but little or no non-additive (heterosis) genetic control of immune response (Rothschild *et al.*, 1984a,b; Meeker *et al.*, 1987a,b). Immune responses to vaccines that contained two *E. coli* antigens also were shown to be under moderate genetic control, with heritability estimates ranging from 0.29 to 0.45 (Edfors-Lilja *et al.*, 1985).

In cattle and sheep, experiments (Lie, 1979; Nguyen, 1984; Muggli *et al.*, 1987) have revealed genetic variation for immune response to a variety of antigens, including chicken red blood cells, human serum albumen and infectious bovine rhinotracheitis virus (IBRV). These results suggest that genetic control of immune response exists for all livestock species.

A second indirect approach would be to consider *in vitro* methods as indicators of disease resistance. These methods include, for instance, phagocytic and bactericidal actions of peripheral blood monocytes against disease agents such as *Salmonella typhimurium* and *Staphylococcus aureus* (Lacey *et al.*, 1990). Other such methods include neutrophil metabolic and phagocytic activity and lymphocyte blastogenesis in response to antigens. Use of mitogens as an indicator of cell-mediated response has revealed that genetic differences exist in poultry for the T cell mitogens phytohaemagglutinin and concanavalin (Lassila *et al.*, 1979). As with poultry, response to mitogen stimulation has also been demonstrated in swine for reactivity to phytohaemagglutinin and poke-weed mitogen. Efforts to create an immunocompetence index in swine have been only moderately successful (Buschmann *et al.*, 1985). *In vitro* procedures and their relationship to actual disease resistance is unknown in general.

A third method would be to locate marker genes associated with disease resistance. A new class of polymorphisms, restriction fragment length polymorphisms (RFLPs), has been developed. The RFLPs are created by using restriction endonucleases to cleave DNA molecules at specific sites. The fragments of DNA are then identified by using cloned DNA probes that detect specific homologous DNA fragments. RFLP analysis has been suggested as a means to identify varietal strains or parentage, to identify quantitative trait loci (QTL) and for use in genetic improvement programmes (Soller and Beckmann, 1983). RFLP analyses could employ a large number of random probes to attempt to cover the entire genome such that QTL could be identified or use loci-specific probes to determine if specific loci could serve as markers for disease resistance and immune response traits.

Given the complexity of immune response and its relationship with disease resistance, the search for marker genes associated with these traits is enormous. Modern immunology, however, has revealed that a group of genes, called the major histocompatibility complex (MHC) genes, seems to be intimately associated with both disease resistance and immune responsiveness. All higher life forms are known to possess a MHC that codes for the predominant cell-surface proteins on cells and tissues of each individual species. These antigens are markers of 'self' and are unique for animals other than identical twins or clones. Three classes of protein molecules, class I, class II and class III are encoded for by the MHC. The class I molecules are extremely polymorphic, whereas the class II antigens are less polymorphic. Class III molecules show only limited polymorphism. The function of the MHC genes is now generally known. The class I antigens act as restricting elements in T cell recognition of virally infected target cells and, thus, are necessary to generate an immune response. The class II genes (Ir genes) control the interaction of T cells, B cells and macrophages in the generation of the humoral immune response and participate, as well, in aspects of cellular immunity. The class III genes are intimately involved with the complement cascade, which ends with the lysis of the cell or virus particle to which antibody has bound. The structure and function of the MHC in livestock seems to be similar to that of humans and mice (Warner *et al.*, 1988).

The B complex, the MHC in chickens, has been extensively studied and shown to be involved with both immune response and disease resistance. More specifically, the B complex has been shown to be associated with immune response to synthetic antigens, bovine serum albumen, *Salmonella pullorum* bacterium, total IgG levels and cell-mediated responses (Lamont and Dietert, 1990; Nordskog, 1984). Resistance to Marek's disease, Rous Sarcoma virus, fowl cholera and lymphoid leukosis viruses has also been demonstrated to be associated with the chicken MHC (Lamont, 1989; Lamont and Dietert, 1983; van der Zijpp, 1983a). In retrospect, it seems that poultry-breeding companies have, in fact, indirectly selected for certain MHC genotypes by eliminating lines or strains that were susceptible to certain diseases.

In swine, perhaps the best-studied domestic species, several experiments have shown that the swine MHC (SLA complex) is associated with immune response following vaccination. Researchers, working with miniature inbred pigs and commercial strains, have demonstrated that the SLA complex is associated with immune response to hen egg-white lysozyme and *B. bronchiseptica* vaccine (Rothschild *et al.*, 1984a; Vaiman *et al.*, 1978b). The synthetic polypeptide antigen (T,G)-A-L has been shown to be under the control of the class II SLA genes in miniature pigs (Lunney *et al.*,

1986). Differences in quantitative levels of class III molecules have also been shown to be related to the SLA complex (Vaiman *et al.*, 1978a). Evidence of SLA association with parasite infection has been recently reported (Lunney and Murrell, 1988). The structure and function of the SLA complex has been reviewed by Warner and Rothschild (1990).

In other livestock species, reports of MHC involvement with immune response and disease resistance also exist. In cattle, the BoLA (bovine MHC) complex has been associated with intestinal parasites, tick suscepti- bility and mastitis (Spooner *et al.*, 1988). Bovine leukaemia virus has also been demonstrated to be associated with the BoLA complex (Lewin *et al.*, 1988). In the horse, the ELA (equine MHC) complex seems to be asso- ciated with equine sarcoid tumours (Meredith *et al.*, 1986). Resistance to scrapie (Millot *et al.*, 1988) was shown to be associated with the sheep MHC (OLA complex), as was response to vaccination against *Tricho- strongylus colubriformis* (Outeridge *et al.*, 1985). Opportunities exist for discovering additional relationships between the MHC and immune response and disease resistance in livestock.

The MHC genes, therefore, may serve as gene candidates for selection. In addition, other genes may be linked to disease resistance traits, and their eventual discovery will lead to some marker-assisted selection for increased disease resistance.

Interrelationships of disease resistance, immune responsiveness and production traits

When one considers the problems of selection under challenging environ- ments, the possibilities to make improvements in disease resistance and immune responsiveness seem limited. First, direct selection for disease resistance may be too costly. Second, results in a number of species suggest that humoral immune response to one antigen is not necessarily a good indicator of humoral response to other antigens or to other aspects of the immune response. Third, MHC genotypes known to be associated with susceptibility to one disease differ from those known to be associated with susceptibility to different diseases. The heterozygosity of genes that in- fluence immune response, especially those of the MHC, is an obvious advantage and polymorphism for MHC loci would allow species the opportunities to survive a variety of disease challenges. The interrelation- ships among disease resistance and immune responsiveness must also take into account their association with production traits. It might seem reason- able to believe that selection for improved production efficiency might also have indirectly selected for improved disease resistance and immune responsiveness. Given the use of medication and the frequent vaccination programmes that exist in livestock production, it seems that these husbandry practices may have masked genetic resistance to disease.

Experimental evidence of the relationships of disease resistance, immune response and production traits is scarce. In poultry, results are contradictory (Lamont, 1989; van der Zijpp, 1983b). In one experiment, selection for rapid growth rate was associated with increased susceptibility to Marek's disease. In another set of experiments, associations with Marek's disease and egg laying performance were contradictory. Immune response to sheep red blood cell seems to be antagonistically associated with growth rate in broilers. Immune response to Newcastle disease vaccine seems to be negatively correlated with hatchability in Japanese quail and with egg production in chickens.

In other livestock species the results are similar. Meeker *et al.* (1987a) found that growth rate was negatively correlated with immune response following vaccination with either a commercially prepared *B. bronchiseptica* or pseudorabies virus vaccine but that backfat at market weight was not correlated with either immune response. Pigs that have the intestinal receptor for K88 *E. coli* (and are therefore susceptible) grew faster and had better feed conversion (Edfors-Lilja *et al.*, 1986). Results with cattle suggest that immune response to IBRV vaccination was not associated with performance.

A final point to consider is that the MHC has been linked to a number of production traits (as reviewed in numerous articles in Warner *et al.*, 1988). These reports include an association of egg production with the chicken MHC complex, growth rate with the bovine MHC and growth and reproduction traits with the pig MHC. It seems reasonable to assume that, inasmuch as immune responsiveness and production traits both seem to be associated with the MHC, these genes should then be the ones to exploit in future selection programmes.

Future directions and opportunities

A number of limitations seem to exist that may impede progress towards breeding livestock, by conventional methods, that are more disease resistant and have improved immune responsiveness. The first of these is that limited funds are now committed to disease testing of different germplasm. This lack of funds has limited the development of our understanding of the extent of genetic resistance to disease. In addition, the general lack of associations among immune-response parameters and the negative correlations of some diseases and immune response with production traits make conventional selection methods extremely difficult.

The alternative approach would be to identify individual marker genes that are associated with disease resistance and improved immune response. The MHC genes seem to be likely candidates because they are already known to be associated with these traits. Once specific MHC alleles and other genes can be identified, they can be used in marker-assisted selection or transferred to create transgenic animals with improved disease resistance and immune responsiveness. The payoff for such research is likely to be extremely large. For livestock producers, the benefits of genetically engineered animals would most likely include increased protection from disease and improved production performance under challenging environments. In the end, the consumer also would benefit from these developments, which could greatly alter livestock production.

This article is Journal Paper No. J–14040 of the Iowa Agriculture and Home Economics Experiment Station, Ames, Projects 1901, 2594, 2609 and 2798.

References

Albers, G.A., Gray, G.D., Piper, L.R., Barker, J.S.F., Le Jambre, L.F. and Barger, I.A. (1987) The genetics of resistance and resilience to *Hamemonchus contortus* infection in young Merino sheep. *International Journal of Parasitology* 17, 1355–63.

Biozzi, G., Siqueira, M., Stiffel, C., Ibanez, O.M., Mouton, D. and Ferreira, V.C.A. (1980) Genetic selection for relevant immunological functions. In: *Progress in Immunology IV.* Academic Press, New York, pp. 432–57.

Bulgin, M.S., Lincoln, S.D., Parker, C.F., Sowth, P.J., Dahmen, J.J. and Lane, V.M. (1988) Genetic-associated resistance to foot rot in selected Targhee sheep. *Journal of the American Veterinary Medicine Association* 192, 512–15.

Buschmann, H. (1980) A selection experiment on the antibody forming capacity to DNP-hapten in pigs. *Animal Blood Groups and Biochemical Genetics* II (suppl. 1), 4.

Buschmann, H., Junge, V., Krausslich, H. and Radzikowski, A. (1974) A study of the immune response to sheep erythrocytes in several breeds of swine. *Medical Microbiology and Immunology* 159, 179–80.

Buschmann, H., Krausslich, H., Herrmann, H., Meyer, J. and Kleinschmidt, A. (1985) Quantitative immunological parameters in pigs – experience with the evaluation of an immunocompetence profile. *Zeitschrift Fur Tierzuchtung Züchtungsbiologie* 102, 189–99.

Crittenden, L.B. (1975) Two levels of genetic resistance to Marek's disease. *Avian Disease* 19, 281–92.

Edfors-Lilja, I., Gahne, B. and Petersson, H. (1985) Genetic influence on antibody response to two *Escherichia coli* antigens in pigs. II. Difference in response between paternal half sibs. *Zeitschrift Fur Tierzuchtung Züchtungsbiologie* 102, 308–17.

Edfors-Lilja, I., Petersson, H. and Gahne, B. (1986) Performance of pigs with and without the intestinal receptor for *Escherichia coli* K88. *Animal Production* 42, 381–8.

Falconer, D.S. (1960) Selection of mice for growth on high and low planes of nutrition. *Genetic Research* 1, 91–113.

Freeman, B.M. and Bumstead, N. (1987) Breeding for disease resistance – the prospective role of genetic manipulation. *Avian Pathology* 16, 353–65.

Gavora, J.S. and Spencer, J.L. (1983) Breeding for immune responsiveness and disease resistance. *Animal Blood Groups and Biochemical Genetics* 14, 159–80.

Gibbons, R.A., Selwood, R., Burrows, M. and Hunter, P.A. (1977) Inheritance of resistance to neonatal *Escherichia diarrhea* in the pig: Examination of the genetic system. *Theoretical and Applied Genetics* 55, 65–70.

Huang, J. (1977) *Quantitative Inheritance of Immunological Response in Swine.* Unpublished PhD dissertation. University of Hawaii, Honolulu, Hawaii.

Hutt, F.B. (1958) *Genetic Resistance to Disease in Domestic Animals.* Comstock, Ithaca, NY.

Kennedy, B.W. and Moxley, J.E. (1980) Genetic factors influencing atrophic rhinitis in the pig. *Animal Production* 30, 277–83.

Lacey, C., Wilkie, B.N., Kennedy, B.W. and Mallard, B.A. (1990) Genetic and other effects on bacterial phagocytosis and killing by cultured peripheral blood monocytes of SLA-defined miniature pigs. *Animal Genetics* 20, 371–82.

Lamont, S.J. (1989) The chicken major histocompatibility complex in disease resistance and poultry breeding. *Journal of Dairy Science* 72, 1328–33.

Lamont, S.J. and Dietert, R.R. (1990) Immunogenetics. In: Crawford, R.D. (ed.), *Poultry Breeding and Genetics.* Elsevier, Amsterdam.

Lassila, O., Nurmi, T. and Eskola, J. (1979) Genetic differences in the mitogenic response of peripheral blood lymphocytes in the chicken. *Journal of Immunogenetics* 6, 37–43.

Lewin, H.A., Wu, M.-C., Stewart, J.A. and Nolan, T.J. (1988) Association between BoLA and subclinical bovine leukemia virus infection in a herd of Holstein-Friesian cows. *Immunogenetics* 27, 338–44.

Lie, Ø. (1979) Genetic analysis of some immunological traits in young bulls. *Acta Veterinaria Scandinavica* 20, 372–86.

Lundeheim, N. (1979) Genetic analysis of respiratory diseases of pigs. *Acta Agriculturae Scandinavica* 29, 209–5.

Lunney, J. and Murrell, K.D. (1988) Immunogenetic analysis of *Trichinella spiralis* infections in swine. *Veterinary Parasitology* 29, 179–93.

Lunney, J.K., Pescovitz, M.D. and Sachs, D.H. (1986) The swine major histocompatibility complex: its structure and function. In: Tumbleson, M.E. (ed.), *Swine in Biomedical Research.* Plenum, New York, vol. 3, pp. 1821–36.

Meeker, D.L., Rothschild, M.F., Christian, L.L., Warner, C.M. and Hill, H.T. (1987a) Genetic control of immune response to pseudorabies and atrophic rhinitis vaccines I. Heterosis, general combining ability and relationship to growth and backfat. *Journal of Animal Science* 64, 407–13.

Meeker, D.L., Rothschild, M.F., Christian, L.L., Warner, C.M. and Hill, H.T. (1987b) Genetic control of immune response to pseudorabies and atrophic rhinitis vaccines II. Comparison of additive direct and maternal genetic effects. *Journal of Animal Science* 64, 414–19.

Meredith, D., Elser, A.H., Wolf, B., Soma, L.R., Donawick, W.J. and Lazary, S. (1986) Equine leukocyte antigens: relationships with sarcoid tumors and laminitis in two pure breeds. *Immunogenetics* 23, 221–5.

Millot, P., Chatelain, J., Dautheville, C., Salmon, D. and Cathala, F. (1988) Sheep major histocompatibility (OLA) complex: linkage between a scrapie suscepti-bility/resistance locus and the OLA complex in the Ile-de-France sheep progenies. *Immunogenetics* 27, 1–11.

Muggli, N.E., Hohenboken, W.D., Cundiff, L.V. and Mattson, D.E. (1987) Inherit-ance and interaction of immune traits in beef calves. *Journal of Animal Science* 64, 385–93.

Nguyen, T.C. (1984) The immune response in sheep. Analysis of age, sex and genetic effects on the quantitative antibody response to chicken red blood cells. *Veterinary Immunology and Immunopathology* 5, 237–45.

Nordskog, A.W. (1984) Selection for immune response as related to the major histocompatibility complex (MHC). *Annales Agriculturae Fenniae* 23, 255–9.

Outeridge, P.M., Windon, R.G. and Dinien, J.K. (1985) An association between a lymphocyte antigen in sheep and the response to vaccination against the parasite *Trichostrongylus colubriformis. International Journal of Parasitology* 15, 121–8.

Philipsson, J., Thafvelin, G. and HedebroVelander, I. (1980) Genetic studies on disease recordings in first lactation cows of Swedish dairy breeds. *Acta Agri-culturae Scandinavica* 30, 327–35.

Rothschild, M.F. (1985) Selection for disease resistance in the pig. *Pig News and Information* 6, 277–80.

Rothschild, M.F. (1989) Selective breeding for immune responsiveness and disease resistance in livestock. *AgBiotech News and Information* 3, 355–60.

Rothschild, M.F., Chen, H.L., Christian, L.L., Lie, W.R., Venier, L., Cooper, M., Briggs, C. and Warner, C.M. (1984a) Breed and swine lymphocyte antigen haplotype differences in agglutination titers following vaccination with *B. bronchiseptica. Journal of Animal Science* 59, 643–9.

Rothschild, M.F., Hill, H.T., Christian, L.L. and Warner, C.M. (1984b) Genetic differences in serum neutralization titers of pigs following vaccination with pseudorabies modified live virus. *American Journal of Veterinary Research* 45, 1216–18.

Siegal, P.B. and Gross, E.B. (1980) Production and persistence of antibodies in chickens to sheep red erythrocytes 1. Directional selection. *Poultry Science* 59, 1–5.

Solbu, H. (1982) Heritability estimates and progeny testing for mastitis, ketosis and 'all diseases'. *Zeitschrift Fur Tierzuchtung Züchtungsbiologie* 101, 210–19.

Soller, M. and Beckmann, J.S. (1983) Genetic polymorphism in varietal identifica-tion and genetic improvement. *Theoretical and Applied Genetics* 67, 25–33.

Spooner, R.L., Brown, P., Glass, E.J., Innes, E.A. and Williams, J.L. (1988) Characterization and function of the bovine MHC. In: Warner, C.M., Rothschild, M.F. and Lamont, S.J. (eds), *The Molecular Biology of the Major Histo-compatibility Complex of Domestic Animal Species.* Iowa State University Press, Ames, pp. 79–96.

Vaiman, M., Hauptman, G. and Mayer, S. (1978a) Influence of the major histo-compatibility complex in the pig (SLA) on haemolytic complement levels. *Journal of Immunogenetics* 5, 59–63.

Vaiman, M., Metzger, J.J., Renard, Ch. and Vila, J.P. (1978b) Immune response gene(s) controlling the humoral anti-lysozyme response (Ir-Lys) linked to the major histocompatibility complex SL-A in the pig. *Immunogenetics* 7, 231–8.

van der Zijpp, A.J. (1983a) Breeding for immune responsiveness and disease resistance. *World's Poultry Science Journal* 39, 118–31.

van der Zijpp, A.J. (1983b) The effect of the genetic origin, source of antigen, and dose of antigen in the immune response of cockerels. *Poultry Science* 62, 205–11.

van der Zijpp, A.J. and Leenstra, F.R. (1980) Genetic analysis of the humoral immune response of white leghorn chicks. *Poultry Science* 59, 1363–9.

Warner, C.M., Meeker, D.L. and Rothschild, M.F. (1987) Genetic control of immune responsiveness: A review of its use as a tool for selection for disease resistance. *Journal of Animal Science* 64, 159–80.

Warner, C.M. and Rothschild, M.F. (1990) The swine major histocompatibility complex (SLA). In: Srivastava, R. (ed.) *Immunogenetics of the MHC* (in press).

Warner, C.M., Rothschild, M.F. and Lamont, S.J. (1988) *The Molecular Biology of The Major Histocompatibility Complex of Domestic Animal Species.* Iowa State University Press, Ames, 193 pp.

Chapter 6

The Major Histocompatibility Complex and Disease Resistance in Cattle

Alan J. Teale, Stephen J. Kemp and W. Ivan Morrison
International Laboratory for Research on Animal Diseases
(ILRAD), P.O. Box 30709, Nairobi, Kenya

Summary

The process whereby foreign antigens are degraded within host cells and the resultant peptide fragments become associated with major histocompatibility complex (MHC) molecules on the cell surface is an essential step in the induction of immune responses. A remarkable feature of the class I and class II MHC molecules involved in this process is the high degree of polymorphism they display among individuals of a species. This creates the potential for MHC phenotype to exert an influence on the magnitude and quality of the immune response and hence on susceptibility to disease. It is, therefore, recognized that characterization of the MHC in farm animal species is essential for studies of both innate and acquired resistance to disease. In the last few years, the application of modern immunological and molecular biological techniques has resulted in considerable progress in our knowledge of the structure and function of the bovine MHC. The sizes of the class I and class II gene systems, their approximate genetic organization and their chromosomal location are known. Two loci encoding functionally important, polymorphic class I molecules have been identified, and class II molecules and genes equivalent to human DR and DQ have been characterized. Functional studies have shown that, as in other species, bovine class I and class II molecules act as restriction elements for $CD8^+$ and $CD4^+$ T lymphocytes, respectively. Despite considerable effort, relatively few significant associations of MHC with disease susceptibility or resistance have been detected in cattle. These include susceptibility to mastitis and to infection with bovine leukosis virus. Recent studies indicate that class I MHC phenotype influences which antigenic epitopes are recognized in cytotoxic T cell responses of cattle to the protozoan parasite *Theileria parva*, and thus affects the parasite strain specificity of the cytotoxic T cell response. Similar phenomena may also be of considerable significance in novel vaccines utilizing recombinant proteins or synthetic peptides which contain a limited number of antigenic epitopes.

Introduction

This chapter reviews what is currently known of the bovine major histo-compatibility complex (MHC), both class I and class II genes, their products and some aspects of their function. Linkage relationships of MHC genes, with each other and with non-MHC genes are also considered. Evidence for associations between MHC type in cattle and disease resistance/susceptibility is also reviewed.

Function of the MHC

The MHC is a highly polymorphic system of genes, the cell surface glyco-protein products of which function to present peptide fragments of proteins to T cells. For this reason the MHC has for some time been an important consideration in attempts to understand genetically controlled variation in disease resistance and susceptibility between individuals of a species.

The T cell is pivotal in classical immune responses. The ability of an individual T cell to respond to antigens derived from pathogens, is a consequence of recognition of a composite antigen comprising 'foreign' peptide together with 'self' MHC gene product on the surface of a presenting cell. The theoretical implications in a disease context are obvious. If an animal does not express an MHC gene product capable of presenting a particular pathogen component, a T cell dependent response cannot follow. The extreme polymorphism of MHC genes in cattle, as in mouse and man, suggests a mechanism for an MHC influence on immune response variation.

The bovine MHC includes class I and class II genes, both of which encode cell surface glycoproteins (Spooner *et al.*, 1988). The important distinctions between class I and class II products are their cell population distributions and the functional differences of the T cell subpopulations which respond to them. The class I glycoproteins occur on the surface of most nucleated cells (although in varying amounts on different cell types) as heterodimers in association with the relatively invariant β_2 micro-globulin.

T cells restricted by class I molecules are of the CD8$^+$ phenotype (Teale *et al.*, 1986) and include the majority of the cytolytic population of T cells (Morrison *et al.*, 1988). T cells restricted by class II molecules are of the

CD4$^+$ phenotype (Teale *et al.*, 1986) and include the majority of T cells generally classified as of 'helper' phenotype (Baldwin *et al.*, 1986; Morrison *et al.*, 1988).

The cytolytic T cells (CTL) constitute an immune effector mechanism and destroy cells displaying appropriate peptides in association with an appropriate class I heterodimer. The class II restricted helper T cell subset is generally regarded as an enabling element in the generation of specific immune responses. These cells produce cytokines essential for the maturation and maintenance of specific effector mechanisms, such as those mediated by antibodies and other T cells. In addition, they produce cytokines which are components of a complex network of regulatory molecules with influences on generation, maturation, multiplication and function of cells of other types, some of which are involved in resistance to pathogens in a non-specific manner. These include macrophages, granulocytes, NK cells and antigen-presenting cells.

Class I genes

Genes

The bovine class I genetic region has been provisionally assigned to chromosome 23 (Fries *et al.*, 1986, 1989). Evidence of multiple large class I-containing DNA fragments produced by digestion of the DNA of MHC homozygous animals with rare cutter restriction enzymes, suggests there are at least six class I genes in cattle (Bensaid *et al.*, 1989a). There could be many more. The class I region in those cattle studied occupies approximately 1.5 Mb of DNA.

With respect to the number of class I genes which have a functional product, Ennis *et al.* (1988) reported cloning of two distinct cDNAs from the bovine transformed cell line BL3. However, whether the clones are allelic variants of a single locus or products of two class I loci, is unclear. In our laboratory we have recently cloned two full-length cDNAs from the peripheral blood mononuclear (PBM) cells of an MHC-homozygous animal (Bensaid *et al.*, in press). This animal expresses the BoLA-Aw 10 antigen (Spooner *et al.*, 1979) and the KN104 antigen (Kemp *et al.*, 1990). Moreover, one of the cDNAs hybridizes with the mRNA of mouse L cells transfected with, and expressing, the BoLA-Aw 10 gene, whereas the other cDNA clone hybridizes with the mRNA of an equivalent transfectant expressing the KN104 gene. The same homozygous animal was used as the source of PBM for the construction of the cDNA library from which the two cDNAs were cloned and as the source of genomic DNA for transfection of the mouse L cells (Toye *et al.*, 1990). It therefore appears that there are at least two class I loci (A and B) expressed by bovine PBM.

Pulsed-field analysis of large restriction fragments of DNA suggests these loci are no more than 210 kb apart (Bensaid *et al.*, in press). Moreover, the w10 and KN104 genes are found to be associated in both *Bos indicus* and *Bos taurus* cattle in geographically distant locations (Kemp *et al.*, 1988) suggesting either very close linkage of the A and B loci in cattle or linkage disequilibrium.

Products

The typical association of a class I encoded glycoprotein with β_2-microglobulin was demonstrated for the bovine MHC by sodium dodecyl sulphate-polyacrylamide gel electrophoresis (SDS-PAGE) of class I gene products (Hoang-Xuan *et al.*, 1982a). Bensaid *et al.* (1988) reported a detailed biochemical study in which evidence was found of multiple products of single haplotypes. One of the haplotypes controlled expression of the w10 and KN104 allospecificities. The cloning of the cDNAs representing the Aw10 and BKN104 genes has subsequently validated the earlier biochemical analyses. Evidence of expression of activation-associated class I molecules on bovine cells has also been obtained (Bensaid *et al.*, 1989b).

Polymorphism in bovine class I gene products, as in other species, is normally characterized serologically. A degree of standardization of serological definitions, using complement-mediated cytotoxicity of alloantisera and monoclonal antibodies, has been achieved through a series of three comparison tests and subsequent international workshops (Anon., 1982; Bull *et al.*, 1989; Spooner *et al.*, 1979). The fourth comparison test has now been concluded and the results will be reported in the near future. The third workshop identified 33 class I antigens in 1298 cattle in 13 countries. The sample included individuals of 38 *B. taurus* and four *B. indicus* breeds as well as a small number of crossbred animals.

The standard microlymphocytoxicity assay for class I product definition has now been automated (Kemp *et al.*, 1987) and a software package specifically for data storage and analysis of cattle typing data has been reported (Kemp, 1988).

There is clear breed variation in class I gene frequencies (Caldwell *et al.*, 1979; Oliver *et al.*, 1981; Kemp *et al.*, 1988; Maillard *et al.*, 1989). In some locations it is possible to discriminate *B. indicus* and *B taurus* types on the basis of their class I polymorphisms; a situation which is very similar to that in man where there are clear population and racial differences in class I gene frequencies (Dick, 1978).

In addition to class I typing of cattle by serology, it is possible to refine definition of the serologically defined class I specificities on the basis of net charge of the molecules, by isoelectric focusing (IEF). In this way, gene products which are serologically indistinguishable can be differentiated

(Joosten *et al.*, 1990). Further, it is possible to 'type' animals on the basis of IEF band patterns for which serological definition is not possible because of lack of reagents (Joosten *et al.*, 1988).

Function

Since the demonstration that serological definitions of class I gene products correlate with the specificity of bovine alloreactive T cells (Teale *et al.*, 1985) and that T cells of the CD8$^+$ phenotype are restricted by class I gene products (Teale *et al.*, 1986), a great deal has been learned of cellular interactions involving class I recognition. These studies have focused on *Theileria parva*-specific T cell function; *T. parva* being an intracellular protozoan parasite of cattle capable of transforming host leucocytes and generating MHC-restricted T cell responses.

In cattle undergoing immunization or challenge with *T. parva*, cytolytic T cells capable of killing parasitized lymphocytes are detectable in the blood for a few days at the time of remission of the immunizing or challenge infection (Emery *et al.*, 1981). These effectors have been shown to belong to the CD8$^+$ subpopulation of T cells and to be restricted by class I MHC determinants (Morrison *et al.*, 1987). It has been possible to cultivate the specific cytolytic T cells *in vitro* and to derive cloned populations (Goddeeris *et al.*, 1986). The cytolytic activity of such clones can be inhibited by monoclonal antibodies (mAbs) specific for determinants on the restricting class I MHC molecule. By testing these clones on panels of target cells of different MHC phenotypes and by examining the capacity of mAbs specific for polymorphic determinants on class I MHC antigens to inhibit cytotoxicity, it has been possible, in some instances, to define the restricting class I MHC molecule (Goddeeris *et al.*, 1986, 1990). It is noted that in animals carrying the W10 and KN104 specificities, the cytolytic T cell responses are often restricted by KN104, i.e. the B locus product. However, in other instances, the response is restricted by A locus products (W.I. Morrison and B.M. Goddeeris, unpublished data). These findings demonstrate that the products of A and B class I loci both have an important functional role as restriction elements for CD8$^+$ cytolytic T cell responses.

Class II

Genes

Like the class I region, the class II region in cattle has been provisionally assigned to 23q13-23 by *in situ* hybridization of heterologous probes to bovine chromosome preparations (Fries *et al.*, 1986, 1989). A great deal of information has now been obtained on the constitution and structure of the bovine class II region. This has been possible because of the relatively greater structural diversity in class II genes by comparison with class I genes, which are less easily grouped into subfamilies. Andersson *et al.* (1986a,b) described restriction fragment length polymorphisms (RFLPs) in digested bovine DNA, detected with probes for DQ and DR alpha and beta genes. They also reported close linkage of the DQ and DR regions.

Subsequently, single bovine analogues of DQ beta and DZ alpha genes were reported (Andersson and Rask, 1988; Andersson *et al.*, 1988). Andersson *et al.* also showed that whereas there is a single DR alpha gene in cattle, there are at least three DR beta genes. At least one DR beta gene is a pseudogene (Muggli-Crockett and Stone, 1988). It also seems that (as in man and mouse) some bovine haplotypes have more DQ alpha and beta genes than others. Further, DY alpha and beta genes have been described in cattle, which have no established relationship with class II genes in other species (Andersson *et al.*, 1988).

With respect to linkage relationships, the class II genes fall into two groups. DQ and DR alpha and beta on the one hand, and DO and DY genes on the other. Moreover, there seems to be a relatively high recombination frequency between these class II subregions.

Class II gene products

Initial biochemical characterization of bovine molecules analogous to DR alpha and beta gene products in man was reported by Hoang-Xuan *et al.* (1982b). Polymorphism in DR beta products has been observed directly by IEF of molecules precipitated by anti-DR xenoantisera (Joosten *et al.*, 1989) and with a mouse anti-bovine monoclonal antibody (I. Joosten, personal communication). Importantly, Joosten *et al.* (1990) have been able to correlate IEF banding patterns with class II restriction fragment length polymorphisms (RFLPs) in two *B. taurus* cattle breeds.

It has so far proven difficult to develop a serological typing system for class II gene products which is as informative and simple to apply as the class I microlymphocytotoxicity assay. However, serological definition has received some attention (Newman *et al.*, 1982) and it is possible that interest in this area may be revived (Mackie and Stear, 1990). There is now also a developing system and degree of standardization of class II typing by IEF through inclusion of IEF analysis in the fourth BoLA comparison test (results to be reported).

Mixed lymphocyte reactivity (MLR) has been used in bovine systems to detect class II polymorphism (Usinger *et al.*, 1981) and a refined form of this, the use of CD4+ cloned alloreactive T lymphocytes in proliferation assays, has been described (Teale and Kemp, 1987). MLR has been particularly useful in characterizing human class II polymorphism, but has remained relatively underdeveloped in livestock species. In cattle, our own observations confirm the problematical nature of this approach which still persists when homozygous typing cells are used to characterize responder phenotypes.

Function

The specificity of CD4$^+$ alloreactive T cells (Baldwin *et al.*, 1986) for class II molecules has been known for some time (Teale *et al.*, 1986). Relatively little is known, however, of the significance of class II polymorphism in cattle for immune response variation. Lie *et al.* (1988) reported high and low antibody response to human serum albumin associated with inheritance of paternal MHC haplotypes in sire families of the Norwegian Red breed. They also found haplotypes associated with variation in response to (T,G)-A-L. Glass *et al.* (1990) reported studies of *in vitro* T cell responses to oval-bumin, which utilized a population of unrelated animals and two paternal half-sib families. As in the studies reported by Lie *et al.* (1988), haplotypes were identified on the basis of class I serology. Perhaps not surprisingly, there was no correlation of high and low proliferative responses with class I types in the unrelated animals, but evidence of influence of MHC haplotype on the level of response in the families was reported, with high and low response being inherited with different paternal haplotypes. There was no significant variation, however, in *in vitro* antibody levels to ovalbumin in any of the groups studied by Glass and colleagues. Confounding influences attributable to the inherent variability of bovine proliferation assays, the complexities due to the use of heterozygous donor cattle, the use of complex antigens and the possibility of T cell sequestration make interpretation of these results difficult.

Studies of responses to defined viral peptides (E. Glass, personal communication) and studies of responses to purified protozoan parasite molecules in homozygous cattle (H. Pereira and D.J. McKeever, personal communication) are presently being undertaken. The results of such studies will, it is hoped, give some indication of the potential influence of MHC class II type on antigen responsiveness, which may be reflected in disease susceptibility and vaccination responses. How important this may be in terms of livestock and vaccine development is, as yet, unclear.

Linkages in the bovine MHC region

As already indicated, there is close linkage of the bovine class I A and B genes. In this respect they are analogous to the human B and C genes. There is also close linkage of the DR and DQ genes on the one hand and other class II genes, on the other. There is also evidence of linkage disequilibrium in both class I and class II regions in cattle.

Linkage between the class I and II regions has been known for some time (Usinger *et al.*, 1981) and confirmed subsequently in various laboratories with a variety of techniques. As yet there has been no formal demonstration of physical linkage of class I and class II genes in cattle, even in studies utilizing pulsed-field analysis of large restriction fragments (Bensaid *et al.*, in press). It has, however, been possible using this technique to link the C4 complement gene with the class II region in cattle, confirming earlier results (Andersson *et al.*, 1988) which showed association of C4 with the DR and DQ genes. Andersson and colleagues also demonstrated linkage of TCP1 with the DO and DY genes.

The genes for steroid 21-hydroxylase (CYP21), prolactin (PRL) and glyoxalase 1 (GLO1) are known to be syntenic with the bovine MHC (Fries *et al.*, 1989). In our laboratory we have recently demonstrated cosegregation of microsatellite polymorphisms within the 21-hydroxylase gene with MHC haplotypes in a full-sib family of N'Dama (*Bos taurus*) cattle.

The M blood group locus has also been shown to be linked to the bovine MHC (Leveziel and Hines, 1984). Indeed, there is now evidence that this locus is within the class I region (Hines and Ross, 1987) and that M' may be BoLA-Aw16.

Disease associations

The first point to emerge from a critical analysis of the available information, is that as in man, there are few reported associations between MHC type and resistance or susceptibility to infectious diseases which approach being convincing. Solbu *et al.* (1982) reported an effect of the MHC on mastitis susceptibility with w2 being associated with resistance in Norwegian cattle and w16, with susceptibility.

Larsen *et al.* (1985) subsequently reported a statistical association between expression of the erythrocyte antigen M' and susceptibility to mastitis in Red Danish dairy cattle. As already indicated, there is now evidence that M' is a class I allele (Hines and Ross, 1987). However, in the studies undertaken by Larsen *et al.*, the effect was small.

A second example of a potentially important association is that reported by Lewin and Bernoco (1986). In this case there was association of BoLA haplotypes with progression of subclinical disease due to BLV infection at both family and herd levels. Subsequently it was reported (Lewin *et al.*, 1988; Palmer *et al.*, 1987; Stear *et al.*, 1988) that different haplotypes are associated in different herds, possibly indicating an effect of an MHC linked gene.

Weak associations of individual class I MHC antigens with high or low levels of infestation with *Boophilus microplus* and with low nematode

94 Chapter 6

faecal egg counts have been described (Stear *et al.*, 1989, 1990).

In a long-term study of performance and disease resistance in a herd of N'Dama cattle in Zaire (Kemp *et al.*, 1989; Trail *et al.*, 1989) an association between a serologically defined class I phenotype and susceptibility to the effects of trypanosome infection was observed. The significance of these observations is yet to be determined.

Perhaps, in view of what is understood of the MHC, the most obvious effect one might expect, is association with non-response to given antigens. Indeed, it is argued that the need to reduce the likelihood of this occurring, has driven the development of polymorphism in the MHC during evolution. As a consequence of this, and the fact that parasites are antigenically complex, it is perhaps naive to expect to find strong associations between class I or II types *per se* and infectious disease resistance, and possibly, even with susceptibility to important pathogens (at least in populations which have been under selection pressure for a reasonable period). However, in the case of T lymphocyte responses to complex antigens, it is clear that the MHC can influence which antigen epitopes are recognized by the host. This has important functional implications when considering pathogenic organisms which display antigenic polymorphism. Thus, depending on whether the T cell response is directed towards conserved or polymorphic antigenic epitopes, immunity between strains of an organism may either be cross-reactive or strain-specific. This phenomenon is well documented for T cell responses to influenza A viruses in mice and humans (Townsend and McMichael, 1985; Vitiello and Sherman, 1983). Recent studies indicate that such qualitative differences in response may also be important in determining the strain specificity of immune responses of cattle to *T. parva*. Antigenically different strains of *T. parva* show varying degrees of cross-protection. Of particular interest is the observation that among animals immunized with one strain of the parasite, some may be resistant to challenge with a second strain while others remain susceptible. Similar heterogeneity in the strain specificity of cytolytic T cell responses has been observed (Morrison *et al.*, 1987; Morrison and Goddeeris, 1990) and recent studies indicate that this correlates with cross-protection (E.L.N. Taracha, B.M. Goddeeris, S.P. Morzaria and W.I. Morrison, unpublished data). Experiments in which the MHC restriction and parasite strain specificities of a large number of cytolytic T cell clones from several cattle immunized with *T. parva* were analysed, indicated that class I MHC phenotype exerts a strong influence on the parasite strain-specificity of the cytolytic T cell response (Goddeeris *et al.*, 1990).

Another area in which MHC-determined differences in immune responses are of particular significance is in responses to vaccines consisting of antigenic subunits or synthetic peptides, which contain a limited number of T cell epitopes. Thus, in seeking to vaccinate outbred animals with defined antigens, it is necessary either to identify T cell epitopes which are universally recognized within the target animal population or to include multiple potential T cell epitopes to ensure that all vaccinated individuals will mount a strong T cell response.

The MHC region of cattle is large, and it is known that non-MHC genes of possible significance in disease, such as those encoding complement components and tumour necrosis factor are linked to it. Polymorphisms in such genes could have disease consequences, although at the present time there are no indications that this is the case.

Finally, it is clear that where associations between MHC type and disease resistance/susceptibility are sought, careful consideration must be given to data analysis (Ostergaard *et al.*, 1989). Without this, weak and spurious associations are not difficult to find.

References

Andersson, L. and Rask, L. (1988) Characterization of the MHC class II region in cattle. The number of DQ genes varies between haplotypes. *Immunogenetics* 27, 110–20.

Andersson, L., Bohme, J., Rask, L. and Peterson, P.A. (1986a) Genomic hybridisation of bovine class II major histocompatibility genes. I. Extensive polymorphism of DQ and DQB genes. *Animal Genetics* 17, 95–112.

Andersson, L., Bohme, J., Peterson, P.A. and Rask, L. (1986b) Genomic hybridisation of bovine class II major histocompatibility genes. 2. Polymorphism of DR genes and linkage disequilibrium in the DQ-DR region. *Animal Genetics* 17, 295–304.

Andersson, L., Lunden, A., Sigurdardottir, S., Davis, C.J. and Rask, L. (1988) Linkage relationships in the bovine MHC region. High recombination frequency between class II subregions. *Immunogenetics* 27, 273–80.

Anon (1982) Proceedings of the Second Bovine Lymphocyte Antigen (BoLA) Workshop (1982) *Animal Blood Groups and Biochemical Genetics* 13, 33–53.

Baldwin, C.L., Teale, A.J., Naessens, J.G., Goddeeris, B.M., MacHugh, N.D. and Morrison, W.I. (1986) Characterization of a subset of bovine T lymphocytes that express BoT4 by monoclonal antibodies and function similarly to lymphocytes defined by human T4 and Murine L3T4. *Journal of Immunology* 136, 4385–91.

Bensaid, A., Kaushal, A., Young, J.R., Kemp, S.J. and Teale, A.J. (1989a) Organisation génomique du complexe majeur d'histocompatibilité du bovin. *Annales de Recherches Vétérinaire* 20, 389–94.

Bensaid, A., Kaushal, A., MacHugh, N.D., Shapiro, S.Z. and Teale, A.J. (1989b) Biochemical characterization of activation associated bovine class I Major histocompatibility complex antigens. *Animal Genetics* 20, 241–55.

Bensaid, A., Naessens, J., Kemp, S.J., Black, S.J., Shapiro, S.Z. and Teale, A.J. (1988) An immunochemical analysis of class I (BoLA) molecules on the surface of bovine cells. *Immunogenetics* 27, 139–44.

Bensaid, A., Young, J.R., Kaushal, A. and Teale, A.J. Pulsed-field gel electrophoresis and its application in the physical analysis of the bovine MHC. In:

Schook, L.B., Lewin, J.A. and McLaren, D.G. (eds), *Gene Mapping: Strategies, Techniques and Applications*. Marcel Dekker, New York, in press.

Bull, R.W., Lewin, H.A., Wu, M.C. *et al.* (1989) Joint report of the Third International Bovine Lymphocyte Antigen (BoLA) Workshop, Helsinki, Finland. 27 July 1986. *Animals Genetics* 20, 109–32.

Caldwell, J., Cumberland, P.A., Weseli, D.F. and Williams, J.D. (1979) Breed differences in frequency of BoLA specificities. *Animal Blood Groups and Biochemical Genetics* 10, 93–8.

Dick, H.M. (1978) HLA-A-B and C serology and antigen reports. In: Bodmer, W.F. and Batchelor, H.R. (eds) *Histocompatibility Testing 1977.* Munksgaard, Copenhagen, pp. 157–204.

Emery, D.L., Eugui, E.M., Nelson, R.T. and Tenywa, T. (1981) Cell-mediated immune responses to *Theileria parva* (East Coast fever) during immunisation and lethal infections in cattle. *Immunology* 43, 323–35.

Ennis, P.D., Jackson, A.P. and Parham, P. (1988) Molecular cloning of bovine class I MHC cDNA. *Journal of Immunology* 141, 642–51.

Fries, R., Beckmann, J.S., Georges, M., Soller, M. and Womack, J. (1989) The bovine map. *Animal Genetics* 20, 2–29.

Fries, R., Hediger, R. and Stranzinger, G. (1986) Tentative chromosomal localization of the bovine histocompatibility complex by *in situ* hybridisation. *Animal Genetics* 17, 287–94.

Glass, E.J., Oliver, R.A. and Spooner, R.L. (1990) Variation in T-cell responses to ovalbumin in cattle: Evidence for Ir gene control. *Animal Genetics* 21, 15–28.

Goddeeris, B.M., Morrison, W.I., Teale, A.J., Bensaid, A. and Baldwin, C.L. (1986) Bovine cytotoxic T-cell clones specific for cells infected with the protozoan parasite *Theileria parva*: parasite strain specificity and class I major histocompatibility complex restriction. *Proceedings of the National Academy of Sciences USA* 83, 5238–42.

Goddeeris, B.M., Morrison, W.I., Toye, P.G. and Bishop, R. (1990) Strain specificity of bovine *Theileria parva*-specific cytotoxic T-cells is determined by the phenotype of the restriction class I MHC. *Immunology* 69, 38–44.

Hines, H.C. and Ross, M.J. (1987) Serological relationships among antigens of the BoLA and the bovine M blood group systems. *Animal Genetics* 18, 361–9.

Hoang-Zuang, M., Leveziel, H., Zilber, M-T., Parodim A-L. and Levy, D. (1982a) Immunochemical characterisation of major histocompatibility antigens in cattle. *Immunogenetics* 15, 207–11.

Hoang-Xuan, M., Charron, D., Zilber, M-T. and Levy, D. (1982b) Biochemical characterisation of class II bovine major histocompatibility complex antigens using cross-species reactive antibodies. *Immunogenetics* 15, 621–4.

Joosten, I., Hensen, E.J., Sanders, M.F. and Andersson, L. (1990) Bovine MHC class II restriction fragment length polymorphism linked to expressed polymorphism. *Immunogenetics* 31, 123–6.

Joosten, I., Oliver, R.A., Spooner, R.L., Williamson, J.L., Hepkema, B.G., Sanders, M.F. and Hensen, E.J. (1988) Characterisation of class I bovine lymphocyte antigens (BoLA) by one dimensional isoelectric focusing. *Animal Genetics* 19, 103–13.

Joosten, I., Sanders, M.F., Van der Poel, A., Williamson, J.L., Hepkema, B.G. and Hensen, E.J. (1989) Biochemically defined polymorphism of bovine MHC class

II antigens. *Immunogenetics* 29, 213–16.

Kemp, S.J. (1988) Automation of the lymphocytotoxicity test for the detection of anti class II antibodies. *Animal Genetics* 19 (Suppl. 1), 13–15.

Kemp, S.J., Gettinby, G., King, D. and Teale, A.J. (1987) BoLA–PC: a database package for the storage and analysis of animal tissue typing records on an IBM–PC microcomputer system. *Animal Genetics* 18, Suppl. 1, 16–17.

Kemp, S.J., Spooner, R.L. and Teale, A.J. (1988) A comparative study of major histocompatibility complex antigens in East African and European cattle breeds. *Animal Genetics* 19, 17–29.

Kemp, S.J., Trail, J.C.M., D'Ieteren, G.D.M. and Teale, A.J. (1989) Lymphocyte antigens as markers of trypanotolerance. *Immunobiology* 4 (Suppl.) 187.

Kemp. S.J., Tucker, E.M. and Teale, A.J. (1990) A bovine monoclonal antibody detecting a class I BoLA antigen. *Animal Genetics* (in press).

Larsen, B., Jensen, N.E., Madsen, P., Nielsen, S.M., Klastrup, O. and Madsen, P.S. (1985) Association of the M blood group with bovine mastitis. *Animal Blood Groups and Biochemical Genetics* 16, 165–73.

Leveziel, H. and Hines, H.C. (1984) Linkage in cattle between the major histo-compatibility complex (BoLA) and the M blood group system. *Génétique Sélec-tion Évolution* 16, 405–16.

Lewin, H.A. and Bernoco, D. (1986) Evidence for BoLA-linked resistance and susceptibility to subclinical progression of bovine leukaemia virus infection. *Animal Genetics* 17, 197–207.

Lewin, H.A., Wu, M.C., Stewart, J.A. and Nolan, T.J. (1988) Association between BoLA and subclinical progression of bovine leukaemia virus infection. *Animal Genetics* 17, 197.

Lie, O., Solbu, H., Larsen, H.J. and Spooner, R.L. (1988) Possible association of antibody responses to human serum albumin and (T,G)-A-L with the bovine major histocompatibility complex (BoLA). *Veterinary Immunology and Immunopathology* 11, 333.

Mackie, J.T. and Stear, M.J. (1990) The definition of five B lymphocyte alloantigens closely linked to BoLA class I antigens. *Animal Genetics* 21, 69–76.

Maillard, J.C., Kemp, S.J., Leveziel, H., Teale, A.J. and Queval, R. (1989) Le complexe majeur d'histocompatibilite de bovins ouest africains. Typage d'antigens lymphocytaires (BoLA) de taurins Baoule (*Bos taurus*) et de zebus Soudaniens (*Bos indicus*) du Burkina Faso (Afrique occidentale). *Revue d'Elevage et de Medicine Veterinaire des Pays Tropicaux* 42, 275–81.

Morrison, W.I., Baldwin, C.L., MacHugh, N.D., Teale, A.J., Goddeeris, B.M. and Ellis, J. (1988) Phenotypic and functional characterisation of bovine lympho-cytes. *Progress in Veterinary Microbiology and Immunology* Vol. 4, Karger, Basel, pp. 134–64.

Morrison, W.I. and Goddeeris, B.M. (1990) Cytotoxic T-cells in immunity to *Theileria parva* in cattle. In: Kaufman, S.H.E. (ed.), *T-cell Paradigms in Parasitic and Bacterial Infections Current Topics in Microbiology and Immunology* Vol. 155, Springer-Verlag, Berlin, pp. 77–93.

Morrison, W.I., Goddeeris, B.M., Teale, A.J., Croocock, C.M., Kemp, S.J. and Stagg, D.A. (1987) Cytotoxic T-cells elicited in cattle challenged with *Theileria parva* (Muguga): evidence for restriction by class I MHC determinants and parasite strain specificity. *Parasite Immunology* 9, 563–78.

Muggli-Crockett, N.E. and Stone, R.T. (1988) Identification of genetic variation in the bovine major histocompatibility complex DRB-like genes using sequenced bovine genomic probes. *Animal Genetics* 19, 213–25.

Newman, M.J., Adams, T.E. and Brandon, M.R. (1982) Serological and genetic identification of a bovine β lymphocyte alloantigen system. *Animal Blood Groups and Biochemical Genetics* 13, 123–39.

Oliver, R.A., McCoubrey, C.M., Millar, P., Morgan, A.L.G. and Spooner, R.L. (1981) A genetic study of bovine lymphocyte antigens (BoLA) and their frequency in several breeds. *Immunogenetics* 13, 127–32.

Ostergaard, J., Kristensen, B. and Andersen, S. (1989) Investigations in farm animal associations between the MHC system and disease resistance and fertility. *Livestock Production Science* 22, 49–67.

Palmer, C., Thurmond, M., Picanso, J., Brewer, A.W. and Bernoco, D. (1987) Susceptibility of cattle to Bovine Leukaemia Virus infection associated with BoLA type. *Proceedings of 91st Annual Meeting of US Animal Health Association*, Salt Lake City, 218.

Sigurdardottir, S., Lunden, A. and Andersson, L. (1988) Restriction fragment length polymorphism of DQ and DR class II genes of the bovine major histocompatibility complex. *Animal Genetics* 19, 133–50.

Solbu, N., Spooner, R.L. and Lie, O. (1982) A possible influence of the bovine major histocompatibility complex (BoLA) on mastitis. In: *Proceedings of the 2nd World Congress on Genetics Applied to Livestock Production*. Madrid, 4–8 October, VII, 368.

Spooner, R.L., Oliver, R.A., Sales, D.I. *et al.* (1979) Analysis of alloantisera against bovine lymphocytes. Joint report of the 1st International Bovine Lymphocyte Antigen (BoLA) Workshop. *Animal Blood Groups and Biochemical Genetics* 15, 63–86.

Spooner, R.L., Teale, A.J. and Cullen, P. (1988) The MHC of cattle and sheep. *Progress in Veterinary Microbiology and Immunology*, Vol. 4. Karger, Basel, pp. 88–107.

Stear, M.J., Dimmock, C.K., Newman, M.J. and Nicholas, F.W. (1988) BoLA antigens are associated with increased frequency of persistent lymphocytosis in bovine leukaemia virus infected cattle and with increased incidence of antibodies to bovine leukaemia virus. *Animal Genetics* 19, 151.

Stear, M.J., Nicholas, F.W., Brown, S.C. and Holroyd, R.G. (1989) Class I antigens of the bovine major histocompatibility system and resistance to the cattle tick (*Boophilus microplus*) assessed in three different seasons. *Veterinary Parasitology* 31, 303–15.

Stear, M.J., Hetzel, D.J.S., Brown, S.C., Gershwin, L.J., Mackinnon, M.J. and Nicholas, F.W. (1990) The relationships among ecto- and endoparasite levels, class I antigens of the bovine major histocompatibility system, immunoglobulin E levels and weight gain. *Veterinary Parasitology* 34, 303–21.

Teale, A.J. and Kemp, S.J. (1987) A study of BoLA class II antigens with BoT4[+] lymphocyte clones. *Animal Genetics* 18, 17–28.

Teale, A.J., Baldwin, C.L., Ellis, J.A., Newson, J., Goddeeris, B.M. and Morrison, W.I. (1986) Alloreactive bovine T lymphocyte clones: an analysis of function, phenotype and specificity. *Journal of Immunology* 136, 4392–8.

Teale, A.J., Morrison, W.I., Goddeeris, B.M., Groocock, C.M., Stagg, D.A. and

Spooner, R.L. (1985) Bovine alloreactive cytotoxic cells generated *in vitro*: target specificity in relation to BoLA phenotype. *Immunology* 55, 355–62.

Townsend, A.R.M. and McMichael, A.J. (1985) Specificity of cytotoxic T lymphocytes stimulated with influenza virus. Studies in mice and humans. In: *Immunobiology of HLA Class I and Class II Molecules, Progress in Allergy*, Vol. 36, S. Karger, Basel, pp. 10–43.

Toye, P., MacHugh, N.D., Bensaid, A.M., Alberti, S., Teale, A.J. and Morrison, W.I. (1990) Transfection into mouse L cells of genes encoding two serologically and functionally distinct bovine class I MHC molecules from a MHC homozygous animal: Evidence for a second class I locus in cattle. *Immunology* 70, 20–6.

Trail, J.C.M., D'Ieteren, G.D.M. and Teale, A.J. (1989) Trypanotolerance and the value of conserving livestock genetic resources. *Genome* 31, 805–12.

Usinger, W.R., Curie-Cohen, M., Benforado, K., Pringnitz, D., Rowe, R., Splitter, G.A. and Stone, W.H. (1981) Close linkage of the genes controlling serologically defined antigens and mixed lymphocyte reactivity. *Immunogenetics* 14, 423–8.

Vitiello, A. and Sherman, L.A. (1983) Recognition of influenza-infected cells by cytolytic T lymphocyte clones: determinant selection by class I restriction elements. *Journal of Immunology* 131, 1635–40.

Chapter 7

Molecular Biological Approaches and their Possible Applications

Alan L. Archibald

AFRC Institute of Animal Physiology and Genetics,
Edinburgh Research Station, Roslin, Midlothian EH25 9PS,
UK

Summary

Molecular biology can be expected to contribute to enhancing the disease resistance of farm animals in two complementary ways – molecular genotyping and gene transfer (transgenesis).

Molecular genotyping techniques allow the detection of the DNA polymorphism, which underlies the genetic variation between individuals. Such polymorphic marker loci can be used in 'marker assisted selection'. For example selection for a disease-resistance gene, for which there is no direct method of genotyping, can be effected by selection for the appropriate alleles at linked marker loci. The marker loci can also serve as starting points for the isolation of the disease (resistance) gene(s) by the process known as 'reverse genetics'. Maps of an animal's genome can be derived, such that any part of its genome is close to a polymorphic marker locus. The genetic (or linkage) maps of pigs, cattle, sheep and poultry being elaborated by a number of groups worldwide will facilitate the future mapping, manipulation and cloning of disease-resistance genes.

Genes can now be identified, isolated, manipulated in the laboratory to generate novel genes and then reintroduced into embryos of the same or different animals to yield transgenic animals. Although such transgenic pigs, sheep, cattle and poultry have all been produced, success rates are low. The development of more efficient and effective methods of gene transfer for livestock is underway. However, if this technology is to be exploited to enhance the disease resistance of farm animals, greater efforts will be needed to identify or design genes which have the potential to increase disease resistance.

Introduction

Disease in farm animals is the result of a complex interaction between pathogenic organisms, the environment and the host animal. As discussed in other contributions, the objective of this book is to consider breeding (i.e. modifying the host component) as a means of reducing disease incidence. Molecular genotyping and gene transfer offer complementary means of using molecular biology to modify the genetic make-up of the host. These two approaches will be described in greater detail here. The use of molecular biological techniques to enhance the disease resistance of animals through improved vaccines, monoclonal antibodies and other pharmaceutical products will not be discussed.

Molecular genotyping

Breeding for disease resistance by traditional means relies upon selecting the fittest individuals as parents of the next generation. However, disease resistance is a trait which can often be evaluated only in a challenge environment or in older animals. Furthermore, our understanding of the genetic architecture of variation in disease resistance is very poor. As with many other traits of economic importance, it is not known whether appropriate models should be based on a few genes or a few tens of genes; thus the relationship between disease resistance and single genes or their products remains uncertain.

In the absence of knowledge of gene products or genes with effects on disease resistance, one is left to look for genetic components to resistance by examining different breeds or large sibships, in which there appears to be variation in resistance. In order to try to localize loci which might contribute to this variation, it is useful to be able to mark regions of the genome. This marking can be effected by following the segregation of alleles at loci, for which genotyping systems exist (Geldermann, 1975). As the use of genetic markers underlies several of the approaches outlined by others in this book, it would be useful to review the methodology, with particular emphasis on the contribution of molecular biological methods.

The breakthrough which molecular biology/recombinant DNA techniques have brought over the past two decades to the search for marker loci is to make the entire genome accessible. Previously the accessibility of parts of the genome has been limited by the genotyping techniques available. Genetic differences between individuals arise from differences in the sequences found in their chromosomal DNA. Using molecular biological techniques, the molecular nature of these polymorphisms in the DNA strands can be examined directly.

Molecular genotyping methods

Landegren *et al.* (1988) and Cotton (1989) have reviewed some of the molecular biological methods, which can be used to detect DNA variation. However, the rate of technological advances in the field of molecular genetics is sufficiently rapid that these reviews are already quite dated. Here the discussion will be limited to methods which are finding applications in livestock.

Restriction fragment length polymorphisms (RFLPs)

Southern blot analysis, which is one of the basic techniques in molecular genetics, combines the ability of single-stranded nucleic acid molecules to hybridize to their complementary targets with the reproducible fractionation of DNA molecules with restriction enzymes. Specific restriction endonucleases cleave DNA at specific locations or recognition sites, including some composed of four base pair, five base pair, six base pair and eight base pair sequence motifs. In general terms the frequency with which particular restriction sites occur in the genome is related to the length of the recognition site. The size of any particular restriction fragment is determined by two recognition sites and is sensitive to mutations, which either disrupt the specificity of one of the sites, or which create a new internal recognition site. The DNA fragments generated by a restriction digest can be separated according to their size by electrophoresis in agarose gels and stained with a general DNA stain such as ethidium bromide. However, if the DNA was derived from a higher eukaryote, the fractionated DNA will appear as a continuous spectrum of molecules with a few discrete bands. These discrete bands arise from the repetitive DNA sequences in such genomes. The problem of detecting particular fragments within this myriad of bits of DNA was elegantly solved by Southern (1975). The fractionated DNA is denatured and transferred (or blotted) to an inert support medium such as nitrocellulose or nylon filter. The immobilized single-stranded DNA can be probed with a suitably tagged (usually radioactively) single-stranded nucleic acid. After completion of the hybridization step the unbound probe is washed away and the filter placed with an X-ray film for a few hours to several days in order to reveal the fragments, to which the probe has hybridized. A basic Southern blotting analysis allows the detection of two types of DNA variation or (RFLP) (Botstein *et al.*, 1980). First, changes to the recognition/cleavage sites of the enzyme used can be detected. These changes can be single base substitutions, insertions or deletions. Larger insertion or deletion events can also be revealed and are characterized by simultaneous changes to the restriction fragments generated by several different enzymes. Simple RFLPs are inherited in a codominant fashion and are generally diallelic. The probes used to detect

RFLPs can include cDNA or genomic clones of known genes or anonymous pieces of cloned DNA.

Variable number of tandem repeat loci (VNTRs)

One of the limitations of RFLP analyses for examining DNA variation is that the polymorphic information content (PIC) of most RFLP loci is low (Botstein *et al.*, 1980). However, there are some loci which are characterized by their hypervariability. The first of these hypervariable types of DNA are known as minisatellite or VNTR (variable number of tandem repeats) loci (Jeffreys *et al.*, 1985; Jeffreys, 1987; Nakamura *et al.*, 1987). The polymorphism arises from variation in the number of the repeat units at a particular locus, with the hypervariability apparently inherent to the repeat unit structure. To study VNTR loci, chromosomal DNA is digested with four or five base pair cutters, which have no recognition sites within the repeated units, in order to cleave the DNA close to the boundaries of the tandem repeat. In this manner the size of the resulting restriction fragments containing the repeats will be determined largely by the number of repeats in the array. Cloned copies of the repeats can be used as probes. If the post-hybridization washes are carried out at low stringency then probes of this nature will detect fragments from several related loci scattered throughout the genome. The complex multi-banded patterns produced under these conditions are known as DNA fingerprints (Jeffreys *et al.*, 1985). DNA fingerprints reveal many hypervariable loci simultaneously. However, as the alleles at any given VNTR locus may be very different in size, it is not always possible to identify the allelic relationships between bands in DNA fingerprints. If the post-hybridization washes are carried out at high stringency, or if unique DNA sequences flanking the tandem repeats are used as the probe, then the resulting autoradiograph will show only one or two fragments per individual corresponding to the alleles at the particular VNTR locus. The degree of heterozygosity at VNTR loci is generally very high and there may be 10 or more common allelic variants at each locus. Again alleles at VNTR loci are inherited in a straightforward Mendelian fashion. One of the disadvantages of VNTR loci is that in humans, at least, they are not evenly distributed throughout the genome, but rather tend to be found towards the telomeres (Royle *et al.*, 1988).

Simple tandem repeat and microsatellite loci/polymerase chain reaction

The STR (simple tandem repeat) loci constitute another class of hypervariable DNA. The simplest of the repeat motifs is a dinucleotide such as CpA, with the corresponding repeat loci known as microsatellites (Litt and Luty, 1989; Weber and May, 1989). Microsatellite loci appear to be inherently hypervariable probably for similar reasons to those postulated for the VNTR loci. In humans and mice it has been shown that micro-satellite loci are scattered throughout the genome unlike the VNTR loci. The number of repeats at microsatellite loci vary between 15 and 60 and alleles may differ in length by as little as one repeat or two nucleotides. Differences of this size are below the resolving power of agarose gel electro-phoresis and Southern blotting. However, these loci are well suited to analysis by the polymerase chain reaction (PCR) (Saiki *et al.*, 1988). Briefly, a pair of oligonucleotide primers of 20–25 nucleotides each are designed to anneal to unique DNA sequences either side of the target locus. The target DNA is denatured by heating, cooled to allow the primers to anneal to their complementary targets, and then the target DNA is copied using a thermostable DNA polymerase. The key to the PCR technique is that this sequence of denaturation, annealing and DNA synthesis is repeated several times over (25–30 times) thus making many copies of (or amplifying) the region of interest. The PCR products are then examined on a polyacrylamide gel electrophoresis system capable of resolving single base or dinucleotide differences in the fragment lengths of the allelic variants at such loci. The DNA fragments can be revealed by autoradiography having incorporated a small amount of radioactive nucleotide triphosphates in the PCR reaction. The attractions of the microsatellite loci are their apparent random (even) distribution, the minimal quantities of sample DNA required and the opportunities for semi-automation associated with the PCR element of the analysis. Ten or more allelic variants at individual microsatellite loci have been reported. These loci, like the VNTR loci, therefore have high PIC values, which make them valuable as genetic markers. All these polymorphisms detected directly at the DNA level are codominant; thus these loci can be genotyped completely and unambiguously.

Uses of genetic markers

Genetic marker loci may either be coincident with the locus controlling the trait of interest or may only be linked to the locus of interest. In the latter case, the identification (genotyping) of alleles at the marker locus can be used to predict which allele is present at a linked locus. Until the develop-ment of molecular approaches to the detection of genetic variation, the search for genetic markers was limited to the examination of gene products by electrophoretic, immunological or functional assays. Whilst many of these gene product markers have been used to look for associations with traits of interest, there have been very few successes with this approach. The number of 'marker loci' has been limited (probably less than fifty) and therefore the proportion of the genome which could be examined was small. Furthermore, association studies of this sort seldom used the complete panel of marker loci. One of the best examples of the use of biochemical genetic markers in livestock improvement is that of the *HAL* locus in pigs (Archibald and Imlah, 1985; Gahne and Juneja, 1985). This locus is also discussed in greater detail in a later chapter (Archibald, 1991).

The advantages of molecular genetic markers are that they allow access to a greater proportion of the genome, some classes of molecular markers are characterized by high PIC values, and the genotyping of molecular markers unlike many gene product markers is independent of developmental stage or sex. Molecular genotyping and linkage analysis have already been used extensively in prenatal diagnosis of a number of inherited disorders in man.

Molecular genetic markers are increasingly being used in the livestock species. For example, DNA fingerprints have been used to identify a genetic marker for muscle hypertrophy in cattle (Georges *et al.*, 1990a). RFLPs detected with a porcine GPI (glucose phosphate isomerase) clone can be used within families to predict genotypes at the *HAL* locus which controls susceptibility to halothane-induced malignant hyperthermia in pigs (Davies *et al.*, 1988). Further examples of studies of the relationship between disease resistance and a variety of types of genetic markers are described in other chapters in this book (e.g. Archibald, 1991; Hanset, 1991; Hunter and Hope, 1991; Raadsma, 1991).

The examples described above relate to specific restricted regions of the genome. Projects which have been initiated to produce genetic maps of the livestock species (Georges *et al.*, 1990b; Haley *et al.*, 1990), will exploit molecular genotyping to search whole genomes for genes influencing traits such as disease resistance.

Genetic marker maps of complete genomes

Complete genetic linkage maps made up of linked polymorphic marker loci evenly distributed through the genome can now be derived. A genetic map could be constructed by genotyping individuals in three generation pedigrees for each of the marker loci, with the information being used to detect linkage and estimate genetic recombination distances between the loci. The same and additional individuals would be measured for phenotypic traits such as disease resistance. These data are then used to identify loci which are responsible for variation in the measured traits (so called quantitative trait loci, or QTLs) and which are linked to the marker loci. With a complete RFLP map, all genes of moderate or large effect and of potential economic importance can be located and mapped. Thus these genetic maps together with statistical techniques currently being developed (Lander and Botstein, 1989) will show for the first time how many genes of moderate or large effect contribute to genetic variation in economically important traits, such as disease resistance.

A full RFLP based map has been produced for tomatoes and has been used to map genes controlling traits of commercial importance (Paterson *et al.*, 1988). A preliminary genetic map of man, also based on RFLPs (Donis-Keller *et al.*, 1987), has now been produced and more detailed genetic maps of individual human chromosomes are being developed. A high-resolution map of the mouse genome is being constructed based on PCR-analysed microsatellite loci (Love *et al.*, 1990). Maps of livestock species such as cattle and pigs are currently being elaborated (Georges *et al.*, 1990b; Haley *et al.*, 1990).

The identification and measurement of linkage relationships between QTLs and marker loci will allow the use of marker assisted selection (Beckman and Soller, 1987; Geldermann, 1990; Smith and Simpson, 1986; Soller and Beckman, 1990), such that animals are selected on their geno-types at marker loci as well as on their phenotypes. For example, a few genes for particular traits from an unimproved, yet disease-resistant, breed could be transferred into a modern, but disease-susceptible, breed without compromising the performance of the breed for the majority of traits for which it excels. Marker loci are used to select only those portions of the genome which are required from the unimproved breed.

In the longer term, it may be possible to isolate and clone QTLs, including those which contribute to disease resistance, known only by their map position. Such 'reverse genetics' has been successfully applied to clone disease loci in man, e.g. cystic fibrosis (Kerem *et al.*, 1989; Riordan *et al.*, 1989; Rommens *et al.*, 1989). Once cloned, the expression and function of the loci can be studied. This technique will reveal some of the estimated greater than 95% of genes which have yet to be located and provides potential material for other developments, such as gene transfer programmes.

Gene transfer

Selection, whether marker assisted or not, is constrained by the genetic variation for the trait of interest and by species boundaries. The traditional means of acquiring new and desirable genetic variants is by cross-breeding, perhaps to exotic stock lines. For example, the genes for prolificacy presumed to be present in Chinese pig breeds, such as the Meishan, have been introduced to European pig stocks by cross-breeding. However, the resulting cross-bred individuals also inherit a proportion of the undesirable features of their exotic parents – such as the high fat content of Meishan meat. Transgenesis (or gene transfer), which brings together the techno-logies of recombinant DNA, embryo manipulation, and embryo transfer offers the possibilities of introducing single new genes without any unwanted genetic baggage. The addition of extra and/or completely novel genes in this manner has the potential to increase the available genetic variation and therefore the scope for selection.

Methods for introducing cloned genes into mammals have been developed in mice (reviewed by Jaenisch, 1988). In 1980 Gordon *et al.* reported the first successful gene transfers and shortly after two critical features of the transferred genes (transgenes) – transmission to subsequent generations and expression – were demonstrated (Costantini and Lacy, 1981; Gordon and Ruddle, 1981; Stewart *et al.*, 1982; E. Wagner *et al.*, 1981; T. Wagner *et al.*, 1981). One of the early and most dramatic examples of transgene expression was the 'supermice' experiment of Palmiter *et al.* (1982), in which the transgene was a hybrid gene composed of regulatory DNA sequences from the mouse metallothionein-I gene and coding sequences from a rat growth hormone structural gene. Some of the transgenic mice carrying this hybrid gene grew much larger than non-transgenic control mice. The side-by-side photographic comparison of the transgenic and control mice, which appeared on the front cover of *Nature* in December 1982, fired the imaginations of animal breeders with the potential of gene transfer for genetic improvement of livestock.

Subsequently the techniques of gene transfer – microinjection, infection by retroviral vectors and embryo-stem cell transfer – have also been applied to livestock (for reviews see Clark, 1988; Clark *et al.*, in press; Pursel *et al.*, 1989; Wall *et al.*, 1990; Wilmut *et al.*, 1990). Microinjection of DNA into pronuclei of fertilized eggs has been used to produce transgenic rabbits, sheep, pigs and cattle (Brem *et al.*, 1985; Ebert *et al.*, 1988; Hammer *et al.*, 1985; Murray *et al.*, 1989; Roschlau *et al.*, 1989; Simons *et al.*, 1988; Vize *et al.*, 1988). Transgenic poultry have been generated by retroviral vector gene transfer (Bosselman *et al.*, 1989; Salter *et al.*, 1987) and experiments are in progress to transfer genes into the ovine and porcine genomes via putative pluripotent embryo-derived cells (Notarianni *et al.*, 1990).

Methods of gene transfer

Microinjection

Microinjection of naked DNA into pronuclei of fertilized eggs is the most commonly used method for the production of transgenic mammals. It is currently the method of choice for the transfer of genes into mice and the only proven method for gene transfer into large mammals (Brem *et al.*, 1985; Ebert *et al.*, 1988; Hammer *et al.*, 1985; Murray *et al.*, 1989; Roschlau *et al.*, 1989; Simons *et al.*, 1988; Vize *et al.*, 1988). A few hundred DNA molecules are injected into one of the two pronuclei of fertilized eggs which have been collected surgically from superovulated females. Localization of the pronuclei in mouse eggs is facilitated by the clarity of the cytoplasm and the prominence of the nucleolar organizers. In contrast, the cytoplasm of sheep, pig and cattle eggs is opaque and the pronuclei difficult to visualize. Careful microscopy with differential interference-contrast optics has been used to overcome this problem in sheep (Hammer *et al.*, 1985; Simons *et al.*, 1988). The pronuclei in porcine and bovine eggs can be seen more clearly if the eggs are centrifuged briefly (Biery *et al.*, 1988; Wall *et al.*, 1985).

The injected eggs are transferred surgically to recipient females. In mouse experiments the injected eggs are incubated overnight *in vitro* prior to transfer. Eggs which survive and develop during this period are transferred to pseudopregnant females. With sheep, three or four eggs per recipient are transferred almost immediately after injection to ewes, which have been hormonally synchronized with the donor ewes. For pigs, it is necessary to transfer large numbers of eggs into each recipient, as a minimum of about four surviving embryos are required to maintain the pregnancy. As 70–80% of the transferred embryos will be lost, it may be useful to place the embryos in a temporary surrogate mother prior to transfer to a final recipient, especially for cattle where the cost of recipient females is considerable (Biery *et al.*, 1988; Roschlau *et al.*, 1989).

Only a proportion of the injected and transferred eggs survive to yield a live born animal and only a fraction of these animals will have incorporated the injected DNA into their genomes. For mice and livestock approximately 5% and 1% respectively of such eggs give rise to transgenic individuals (Brinster *et al.*, 1985; Clark *et al.*, in press). The integrated DNA is generally found as a multiple copy array at a single chromosomal location. One of the advantages of the microinjection method of gene transfer is that the transgene is usually present in every cell, including the germ cells, and therefore is presumably integrated prior to the first cleavage. However, there is evidence from transgenic mouse experiments that 20–30% of founder transgenic individuals are mosaics, i.e. not all cells contain an integrated transgene (Whitelaw *et al.*, 1990; Wilkie *et al.*, 1986).

The low frequency of successful gene transfer by microinjection is not a major problem with mice. However, the extra costs, longer gestation times, longer generation intervals and smaller litter sizes associated with the mammalian livestock species, make widespread applications unlikely unless more efficient methods of gene transfer can be found. Many of these difficulties are also barriers to the optimizing of microinjection protocols in livestock. For example, there are no published estimates of the frequency of mosaicism in transgenic livestock. Nor is there a great deal of experimental data on the optimum for such parameters as the developmental stage of the eggs, DNA concentration and so on, in contrast to the detailed studies of the effects of such factors on the efficiency of generating transgenic mice (Brinster *et al.*, 1985).

Retroviral vectors

Genes have been transferred into mouse and chicken embryos using retroviral vectors (Bosselman *et al.*, 1989; Jähner *et al.*, 1985; Salter *et al.*, 1987; van der Putten *et al.*, 1985). Retroviral vectors have been created, which retain the ability to integrate copies of their genomes into host chromosomes, but which have lost the ability to produce subsequently further infective virus. The potential advantages of retroviral gene transfer are that non-surgically recovered multicellular embryos can be infected and that integration is usually a single copy event. In species, such as poultry, where the single cell embryo is particularly inaccessible and difficult to handle, retroviral vectors may be especially useful. Transgenic poultry have been generated by infecting early chick embryos with replication competent wild-type or recombinant retroviruses and although the resulting transgenic birds were viraemic germline transmission of the integrated virus was demonstrated (Salter *et al.*, 1987; Salter and Crittenden, 1988). More recently, Bosselman *et al.* (1989) described the integration of foreign DNA into the germline of chickens following injection of a replication-defective retroviral vector through the area pellucida into the subgerminal cavity of the blastoderm of unincubated eggs. Low levels of competent virus were found in some of the transgenic birds, but not in those selected for breeding. These lines of transgenic chickens have been created by retroviral vector-mediated gene transfer into the primordial germ cells present in the unincubated chick embryos. Retroviral vectors have also been used for gene transfer into ovine and porcine embryos (Hettle *et al.*, 1989; Petters *et al.*, 1988). However, as the analyses to determine whether the transfers had been successful were performed on 50-day and 42-day old fetuses respectively, it is not known whether the transgenes were present in the germ cells.

The disadvantage of using multicellular embryos is that the transgenic animals produced will be mosaics and the germline may be chimaeric. Transmission of the transgenes, therefore, may vary from the inefficient to the non-existent. For example, of 82 chicks shown to be transgenic by the presence of the transgene in blood cells only 33 had the transgene present in semen. The four males from this group selected for breeding transmitted the transgene to between 2 and 8% of their offspring (Bosselman *et al.*, 1989). Other disadvantages of the use of retroviral vectors are their limited capacity for foreign DNA sequences and the adverse effects of some retroviral sequences on expression of the transgene. Public acceptability may also prove a significant barrier to the use of retroviral vectors for gene transfer in livestock. For example, the remobilization of an apparently defective provirus reported by Crittenden and Salter (1990) tends to confirm the suspicion that validating the safety of retroviral vector systems may be difficult.

Embryo-derived stem cells

When grown under defined conditions cells isolated from the inner cell masses of blastocyst-stage embryos retain their pluripotency and do not differentiate (Evans and Kaufman, 1981). These embryo-derived stem cells (ES cells) can be used to recolonize the inner cell masses of blastula stage embryos (Bradley *et al.*, 1984; Zimmer and Gruss, 1989) and may contribute to both the somatic and germ cells of the developing animal (Gossler *et al.*, 1986). During the period in culture, genes can be transferred into the ES cells by calcium phosphate precipitation, microinjection, retroviral infection or electroporation and thus into the germ lines of a proportion of the subsequent chimaeric mice. The recent demonstrations of site-specific genetic changes by homologous recombination in ES cells means that gene transfer by this route can be used to modify existing genes, as well as to add new genes (Joyner *et al.*, 1989; Mansour *et al.*, 1988; Zimmer and Gruss, 1989; reviewed by Capecchi, 1989). The availability of ES cell lines for the livestock species would transform the prospects for gene transfer in farm animals. However, as yet mice are the only species from which functional ES cell lines have been derived. Preliminary results of attempts to produce ovine and porcine embryo-derived stem-cell lines have been reported (Handyside *et al.*, 1987; Notarianni *et al.*, 1990). It was anticipated that the identification of a growth factor which inhibits the differentiation of murine ES cells cultured *in vitro* would expedite efforts in several laboratories to establish ES cell lines for the livestock species (Moreau *et al.*, 1988; Smith *et al.*, 1988; Williams *et al.*, 1988). However, the most recent experiments to derive such ES cell lines have not overtly used these growth factors (Notarianni *et al.*, 1990). Notarianni and her colleagues have cultured cells isolated from the inner cell masses of both porcine and ovine blastocyst-stage embryos. When cultured on a feeder cell layer of mitotically inactivated STO fibroblasts under the appropriate conditions these cells display the morphological and growth characteristics of ES cells. It seems likely that these cells are pluripotent, as they can be induced to differentiate into a range of different cell types under the relevant conditions. The critical experiments to determine whether these cell lines have the potential to contribute to all the tissues of a chimaeric animal are in progress. Preliminary results with the porcine cells suggest that such ES-like cells marked with a transferred gene are still retained in the developing embryo (Notarianni, personal communication). Evidence of live-born chimaeric individuals and subsequent germ-line transmission of the transgenes are eagerly awaited.

As with the retroviral method of gene transfer the use of ES cells yields chimaeric founder transgenic individuals with the associated uncertainties concerning germline transmission to subsequent generations. The transfer of nuclei from transformed ES cells into enucleated oocytes or eggs would circumvent this problem of chimaerism. Nuclei from ovine and bovine inner cell masses, unlike their murine counterparts, are capable of supporting normal development after transfer into enucleated oocytes (McGrath and Solter, 1984; Smith and Wilmut, 1989). For sheep and cattle, therefore, ES cells in conjunction with nuclear transfer would represent an ideal gene transfer system. Such an ES cell based approach to gene transfer in cattle would avoid the expense and effort of collection and transfer of eggs as blastula stage embryos can be collected and returned non-surgically.

Can sperm be used as vector for gene transfer?

A recent report by Lavitrano *et al.* (1989) indicated that transgenic mice could be produced efficiently by *in vitro* fertilization with sperm which had simply been incubated with DNA. Furthermore, the same group reported successfully generating transgenic pigs with the same technique (Gandolfi *et al.*, 1989). Attempts by several of the world's most experienced gene transfer laboratories to repeat these experiments have yielded no transgenic individuals amongst over 1300 mice produced (Brinster *et al.*, 1989). If sperm could be used for gene transfer in the manner described by Lavitrano *et al.* (1989), gene transfer would become dramatically simpler. However, as this method would suffer from random integration of the transgene in the same manner as the microinjection route, I do not consider that it could make a useful contribution to gene transfer for improving performance, including disease resistance.

Which genes should be transferred?

The application of gene transfer technology to the genetic improvement of farm animals is constrained in two main ways. First, the efficiency of generating transgenic animals is too low. However, as described above, the development of better systems is underway. The second constraint on gene transfer for genetic improvement is the limited range of cloned genes which are known *a priori* to have an effect on the trait of interest, in this case disease resistance. The remainder of this chapter will address this question of 'which genes should be candidates for transfer when the objective is to enhance disease resistance?'.

It is important to remember when considering genes for transfer, that this technology is not limited to the use of natural genes. Not only can single genes be transferred, but prior to transfer the genes can be manipulated in the laboratory to generate entirely new genetic elements. For example, the regulatory sequences from one gene can be fused to the coding sequences of another gene. The expectation that, on introduction of such a hybrid gene into the genome of an animal, the manner and site of expression associated with the regulatory elements will be imposed upon the coding sequences, to which they had been fused, is often fulfilled.

The *Mx* gene

One host gene which is known to influence the pathogenesis of a viral infection is the murine *Mx1* gene (previously known as the *Mx* gene). The dominant $Mx1^+$ allele, which confers resistance to infection with influenza A and B viruses, is found in wild mice and in the laboratory strain A2G. The $Mx1^-$ allele present in other strains of laboratory mice fails to provide protection against infections with these viruses. Hybrid genes composed of a variety of different promoters and the murine $Mx1^+$ gene have been used to generate transgenic mice and pigs (Arnheiter *et al.* 1990; Brem *et al.*, 1988; Brenig *et al.*, 1990). The transgenic mouse data indicate that when the $Mx1^+$ gene is introduced with its own promoter it confers resistance to influenza infection (Arnheiter *et al.*, 1990). The use of the Mx promoter, which is responsive to both viral infection and interferon, was critical to the success of the experiment as the postulated deleterious effects of constitutive expression of Mx were avoided.

Three different promoter regions (human MTIIA-promoter, murine Mx promoter, Sv40 early promoter) were used for the gene transfers into pigs. No transgenics were found with an intact SV40-Mx construct, and no transgene expression was detected in the pigs with the human MTIIA-Mx construct. Only two of eight founder transgenic pigs with the murine Mx promoter expressed and transmitted the transgene. As yet the transgenic pigs themselves have not been subjected to the appropriate disease challenge. As homologues of the *Mx* gene have been found in other species including humans and pigs (Charleston *et al.*, 1990; Staeheli, 1990), the significance and efficacy of the experiment to transfer the A2G murine *Mx* gene into pigs remains to be proved. An analysis of molecular genetic variation at the porcine *Mx* locus may reveal possibilities to enhance resistance to influenza in pigs by marker assisted selection. As other *Mx*-related genes have been identified in mice and rats, it is possible that a family of proteins (genes) with differing antiviral specificities are awaiting identification for future gene transfer or selection applications.

Interferons

Although interferons might seem an obvious choice as general antiviral agents their constitutive expression can have adverse physiological consequences including toxicity and infertility. The perceived species specificity of interferons is not complete, with human alpha and beta interferon being able to act on mouse cells. Chen *et al.* (1988) reasoned that the adverse effects of constitutive expression might be avoided by using a human beta interferon gene in mice. Transgenic mice carrying a metallothionein-beta interferon hybrid gene exhibited enhanced resistance to challenge with pseudorabies virus (Chen *et al.*, 1988). Sera from these transgenic mice also protected cultured human cells from vesicular stomatitis virus infection.

Host expression of viral coat proteins

Whilst developing retroviral based gene transfer methods for use in poultry Crittenden and Salter (1990) have generated transgenic chickens with resistance to avian leukosis virus. For example, in one of the transgenic chickens lines produced by replication-competent retroviral gene transfer the retroviral insert does not encode infectious virus, but does express the subgroup A envelope glycoprotein. Transgenic chickens in this line (alv6) show resistance to infection by a field strain of subgroup A avian leukosis virus (ALV) (Crittenden and Salter, 1990; Salter and Crittenden, 1989). This form of resistance is attributable to the (transgenic) host expressing viral coat protein genes so as competitively to reduce the binding of infecting viruses to the host's receptor proteins, thus conferring resistance to infection. As some neoplasia were seen among the older alv6 birds, one would probably want to limit the transferred portion of the retrovirus genome to that encoding the coat protein. Other candidates for this approach to disease resistance would be the ovine lentiviruses discussed by Demartini *et al.* (1991) and the pathogen responsible for diarrhoea in pigs – *E. coli.* K88 – as described by Edfors-Lilja (1991). However, the molecules which act as receptors for viruses or bacteria may have other physiological functions which would be disrupted if the receptors were permanently blocked by excess coat protein.

Data from transgenic plant experiments suggest that it may also be possible to resist viral infections by interfering with the early stages of infection. The strategies employed include expression of viral coat protein genes (Angenent *et al.*, 1990; Powell *et al.*, 1986; Register and Beachy, 1988), and the expression of antisense versions of parts of the viral genome (Powell *et al.*, 1989). Antisense gene expression is effected by changing the relative orientations of the promoter and coding sequences of a gene *in vitro* prior to gene transfer. The antisense transgene transcript should be able to hybridize to its sense counterpart to form a double-stranded RNA. The double-stranded RNA should be refractory to the subsequent steps of RNA processing, export from the nucleus and translation or in the case of RNA viruses, RNA duplication.

Mastitis

A different approach is being explored to tackle the problem of bacterial pathogens such as *Staphylococcus aureus* which is one of the causes of mastitis. The enzyme lysostaphin which attacks the cell wall of many species of *Staphylococcus* can provide protection against *S. aureus* challenges when injected into the mammary gland (Sears *et al.*, 1988). The lysostaphin gene from *S. simulans* has been fused to the promoter of the ovine beta lactoglobulin gene and microinjected into fertilized mouse eggs (Williamson *et al.*, 1990). It is hoped that expression of lysostaphin *in situ* in the mammary gland and its secretion into milk will confer resistance to *S. aureus* mastitis. The beta-lactoglobulin promoter has already been shown to be effective at directing expression of foreign proteins in the mammary glands of transgenic mice (Archibald *et al.*, 1990). However, if lysostaphin is perceived to be an antibiotic, difficulties may be experienced with licensing such manipulations in dairy animals.

Other candidate genes for transfer

Among the other candidates for gene transfer where the objective is enhancing disease resistance are genes which encode immune functions. For example, it has been shown that the major histocompatibility complex (MHC) genes are expressed in a dose-dependent way in disomic, trisomic and tetrasomic chickens (Bloom *et al.*, 1988). Therefore, it may be possible to extend the immune repertoire of an animal by introducing additional MHC genes as suggested by Lamont (1989). Functionally rearranged immunoglobulin genes have been transferred into mice (Grosschedl *et al.*, 1984; Rusconi and Köhler, 1985; Storb *et al.*, 1986). Unfortunately, high-level expression of such immunoglobulin transgenes in B lymphocytes can act to suppress the rearrangement and expression of other immunoglobulin genes by the mechanism known as allelic exclusion (Iglesias *et al.*, 1987); Ritchie *et al.*, 1984; Weaver *et al.*, 1985). In this case gene transfer effectively reduces, rather than increases, the available genetic variation.

The use of gene transfer in livestock production

The pursuit of genetic improvements in disease resistance (or other traits) by gene transfer technology will be expensive. It is essential that the recipient genomes be representative of the best contemporary stock (Smith *et al.*, 1987). Thorough testing for the benefits and for even minor deleterious effects of the transgene will also be critical (Smith *et al.*, 1987). The genetic gain represented by a founder transgenic individual needs to be high in order to counter the cost and the loss of genetic variation associated with using a limited number of founder animals. Smith *et al.* (1987) have expressed reservations about the use of transgenesis in animal improvement. However, many of the arguments advanced by these authors are specific to the microinjection route of gene transfer, where there is no control over the copy number of the transgene, nor over its site of chromosomal integration. The development of embryo-derived stem cells for the livestock species especially when used in conjunction with homologous recombination and nuclear transfer will answer many of the criticisms. Both the number of copies of the transgene integrated into the chromosome and the site of integration will be controlled. By site-specific integration in clonal ES cells it will also be possible to generate individuals, in which the transgene loci will be identical. Furthermore if such transgenic nuclei can be transferred by nuclear transfer one will be creating clonal transgenics, which will be invaluable for the accurate and rapid assessment of the transgenic phenotypes. Clearly such an approach also has the potential to reduce genetic variation in the improved populations. However, if the creation of ES cell lines were to become routine then loss of variation need not be a problem. Although the ES cell approach seems attainable for the livestock mammals, poultry present a greater challenge. The legal, ethical and safety issues associated with the use of genetically transformed livestock will also need to be addressed (Berkowitz, 1990; Jones 1986).

Conclusions

Molecular genotyping is already making a significant impact on human genetics and medicine. Although the first applications of this technology to livestock are likely to be in the fields of pedigree verification and forensics, selection decisions based on molecular genetic marker genotypes are imminent for kappa-casein types, porcine stress, and BSE. In the next five to ten years, the results from the genetic mapping projects will determine whether marker-assisted selection has a future. The development of gene transfer methods for livestock based on the use of ES cells combined with nuclear transfer will provide an incentive for greater efforts to identify genes worth manipulating for agricultural purposes. The identification of genes with the ability to influence disease resistance is a common requirement for both these approaches and a task to which molecular biology will make significant contributions.

References

Angenent, G.C., van den Ouweland, J.M.W. and Bol, J.F. (1990) Susceptibility to virus infection of transgenic tobacco plants expressing structural and non-structural genes of tobacco rattle virus. *Virology* 175, 191–8.

Archibald, A.L. (1991) Inherited halothane induced malignant hypothermia in pigs. In: Owen, J.B. and Axford, R.F.E. (eds), *Breeding for Disease Resistance in Farm Animals.* CAB International, Wallingford, pp. 449–66.

Archibald, A.L. and Imlah, P. (1985) The halothane sensitivity locus and its linkage relationships. *Animal Blood Groups and Biochemical Genetics* 16, 253–63.

Archibald, A.L., McClenaghan, M., Hornsey, V., Simons, J.P. and Clark, A.J. (1990) High-level expression of biologically active human α1-antitrypsin in the milk of transgenic mice. *Proceedings of the National Academy of Sciences USA* 87, 5178–82.

Arnheiter, H., Skuntz, S., Noteborn, M., Chang, S. and Meier, E. (1990) Transgenic mice with intracellular immunity to influenza virus. *Cell* 62, 51–61.

Beckman, J.S. and Soller, M. (1987) Molecular markers in the genetic improvement of farm animals. *Bio/technology* 5, 573–6.

Berkowitz, D.B. (1990) The food safety of transgenic animals. *Bio/technology* 8, 819–25.

Biery, K.A., Bondioli, K.R. and De Mayo, F.J. (1988) Gene transfer by pronuclear injection in the bovine. *Theriogenology* 29, 224.

Bloom, S.E., Delaney, M.E., Muscarella, D.M., Dietret, R.R., Briles, W.E. and Briles, R.W. (1988) Gene expression in chickens aneuploid for the MHC-bearing chromosome. In: Warner, C.M., Rothschild, M.F. and Lamount, S.J. (eds), *Molecular Biology of the Major Histocompatibility Complex in Domestic Animal Species.* Iowa State University Press, Ames, pp. 3–21.

Bosselman, R.A., Hsu, R.-Y., Boggs, T., Hu, S., Bruszewski, J., Ou, S., Kozar, L., Martin, F., Green, C., Jacobsen, F., Nicolson, M., Schultz, J.A., Semon, K.M., Rishell, W. and Stewart, R.G. (1989) Germline transmission of exogenous genes in the chicken. *Science* 243, 533–5.

Botstein, D., White, R.L., Skolnick, M. and Davis, R.W. (1980) Construction of a genetic linkage map in man using restriction fragment length polymorphisms. *American Journal of Human Genetics* 32, 314–31.

Bradley, A., Evans, M., Kaufman, M.H. and Robertson, E. (1984) Formation of germ-line chimaeras from embryo-derived teratocarcinoma cell lines. *Nature* 309, 255–6.

Brem, G., Brenig, B., Goodman, H.M., Selden, R.C., Graf, F., Kruff, B., Springman, K., Hondele, J., Meyer, J., Winnacker, E.-L. and Kräußlich, H. (1985) Production of transgenic mice, rabbits and pigs by microinjection into pronuclei. *Zuchthygiene* 20, 251–2.

Brem, G., Brenig, B., Müller, M., Kräußlich, H. and Winnacker, E.-L. (1988) Production of transgenic pigs and possible application to pig breeding. *Occasional Publication of the British Society for Animal Production,* no. 12, 15–31.

Brenig, B., Müller, M. and Brem, G. (1990) Gene transfer in pigs. *Proceedings of the 4th World Congress on Genetics Applied to Livestock Production* XIII, 41–8.

Brinster, R.L., Chen, H.Y., Trumbauer, M.E., Yagle, M.K. and Palmiter, R.D. (1985) Factors affecting the efficiency of introducing foreign DNA into mice by microinjecting eggs. *Proceedings of the National Academy of Sciences USA* 82, 4438–42.

Brinster, R.L., Sandgren, E.P., Behringer, R.R. and Palmiter, R.D. (1989) No simple solution for making transgenic mice. *Cell* 59, 239–41.

Capecchi, M. (1989) The new mouse genetics: altering the genome by gene targeting. *Trends in Genetics* 5, 70–6.

Charleston, B., Lida, J. and Smith, I.K.M. (1990) Porcine Mx specific gene sequences and expression. *Association of Veterinary Teachers and Research Workers*, Scarborough, April, 1990, Abstract B44, p. 25.

Chen, X-Z., Yun, J.S. and Wagner, T.E. (1988) Enhanced viral resistance in transgenic mice expressing the human beta 1 interferon. *Journal of Virology* 62, 3883–7.

Clark, A.J. (1988) Gene transfer in animal production. *Occasional Publication of the British Society for Animal Production* no. 12, 1–14.

Clark, A.J., Archibald, A.L., McClenaghan, M., Simons, J.P., Whitelaw, C.B.A. and Wilmut, I. (1991) The germline manipulation of livestock: progress during the past five years. *Proceedings of the New Zealand Society of Animal Production* (in press).

Costantini, F. and Lacy, E. (1981) Introduction of a rabbit β-globin gene into the mouse germ line. *Nature* 294, 92–4.

Cotton, R.G.H. (1989) Detection of single base changes in nucleic acids. *Biochemical Journal* 263, 1–10.

Crittenden, L.B. and Salter, D.W. (1990) Transgenic chickens resistant to avian leukosis virus. *Proceedings of the 4th World Congress on Genetics Applied to Livestock Production* XVI, 453–6.

Davies, W., Harbitz, I., Fries, R., Stranzinger, G. and Hauge, J.G. (1988) Porcine malignant hyperthermia carrier detection and chromosomal assignment using a linked probe. *Animal Genetics* 19, 203–12.

DeMartini, J.C., Bowen, R.A., Carlson, J.O. and de la Concha-Bermejillo, A. (1991) Strategies for the genetic control of ovine lentivirus infections. In: Owen, J.B. and Axford, R.F.E. (eds), *Breeding for Disease Resistance in Farm Animals*. CAB International, Wallingford, pp. 293–314.

Donis-Keller, H., Green, P., Helms, C., Cartinhour, S., Weiffenbach, B., Stephens, K., Keith, T.P., Bowden, D.W., Smith, D.R., Lander, E.S., Botstein, D., Akots, G., Rediker, K.S., Gravius, T., Brown, V.A., Rising, M.B., Parker, C., Powers, J.A., Watt, D.E., Kauffman, E.R., Bricker, A., Phipps, P., Muller-Kahle, H., Fulton, T.R., Ng, S., Schumm, J.W., Braman, J.C., Knowlton, R.G., Barker, D.F., Crooks, S.M., Lincoln, S.E., Daly, M.J. and Abrahamson, J. (1987) A genetic linkage map of the human genome. *Cell* 51, 319–37.

Ebwert, K.M., Low, M.J., Overstrom, E.W., Buonomo, F.C., Baile, C.A., Roberts, T.M., Lee, A., Mandel, G. and Goodman, R.H. (1988) A Moloney MLV-rat somatotrophin fusion gene produces biologically active somatotrophin in a transgenic pig. *Molecular Endocrinology* 2, 277–83.

Edfors-Lilja, I. (1991) *E. coli* resistance in pigs. In: Owen, J.B. and Axford, R.F.E. (eds), *Breeding for Disease Resistance in Farm Animals*. CAB International, Wallingford, pp. 424–35.

Evans, M.J. and Kaufman, M.H. (1981) Establishment in culture of pluripotential cells from mouse embryos. *Nature* 292, 154–6.

Gahne, B. and Juneja, R.K. (1985) Prediction of the halothane (Hal) genotypes of pigs by deducing Hal, Phi, Po2, Pgd haplotypes of parents and offspring: results from a large-scale practice in Swedish breeds. *Animal Blood Groups and Biochemical Genetics* 16, 265–83.

Gandolfi, F., Lavitrano, M., Camaioni, I., Spadafora, C., Siracusa, G. and Lauria, A. (1989) The use of sperm-mediated gene transfer for the generation of transgenic pigs. *Journal of Reproduction and Fertility* Abstract Series no. 4, 10.

Geldermann, H. (1975) Investigations on inheritance of quantitative characters in animals by gene markers. 1. Methods. *Theoretical and Applied Genetics* 46, 319–30.

Geldermann, H. (1990) Mapping quantitative traits by means of genetic markers. *4th World Congress on Genetics Applied to Livestock Production* XIII, 97–106.

Georges, M., Lathrop, M., Hilbert, P., Marcotte, A., Schwers, A., Swillens, S., Vassart, G. and Hanset, R. (1990a) On the use of DNA fingerprints for linkage studies in cattle. *Genomics* 6, 461–74.

Georges, M., Mishra, A., Sargeant, L., Steele, M. and Zhao, X. (1990b) Progress towards a primary DNA marker map in cattle. *4th World Congress on Genetics Applied to Livestock Production* XIII, 107–112.

Gordon, J.W. and Ruddle, F.H. (1981) Integration and stable germ line transmission of genes injected into mouse pronuclei. *Science* 214, 1244–6.

Gordon, J.W., Scangos, G.A., Plotkin, D.J., Barbosa, J.A. and Ruddle, F.H. (1980) Genetic transformation of mouse embryos by microinjection of purified DNA. *Proceedings of the National Academy of Science USA* 77, 7380–4.

Gossler, A., Doetschman, T., Korn, R., Serfling, E. and Kemler, R. (1986) Transgenesis by means of blastocyst-derived embryonic stem cell lines. *Proceedings of the National Academy of Science USA* 83, 9065–9.

Grosschedl, R., Weaver, D., Baltimore, D. and Costantini, F. (1984) Introduction of a μ immunoglobulin gene into the mouse germ line: specific expression in lymphoid cells and synthesis of functional antibody. *Cell* 38, 647–58.

Haley, C.S., Archibald, A.L., Andersson, L., Bosma, A.A., Davies, W., Fredholm, M., Geldermann, H., Groenen, M., Gustavsson, I., Ollivier, L., Tucker, E.M. and Van de Weghe, A. (1990) The Pig Gene Mapping Project – PiGMaP. *4th World Congress on Genetics Applied to Livestock Production* XIII, 67–70.

Hammer, R.E., Pursel, V.G., Rexroad, C.E. Jr., Wall, R.J., Bolt, D.J., Ebert, K.M., Palmiter, R.D. and Brinster, R.L. (1985) Production of transgenic rabbits, sheep and pigs by microinjection. *Nature* 315, 680–3.

Handyside, A., Hooper, M.L., Kaufman, M.H. and Wilmut, I. (1987) Towards the isolation of embryonal stem cell lines from the sheep. *Roux's Archives of Developmental Biology* 196, 185–90.

Hanset, R. (1991) A major gene of muscular hypertrophy in the Belgian Blue cattle breed. In: Owen, J.B. and Axford, R.F.E. (eds), *Breeding for Disease Resistance in Farm Animals.* CAB International, Wallingford, pp. 467–78.

Hettle, S.J.H., Harvey, M.J.A., Cameron, E.R., Johnston, C.S. and Onions, D.E. (1989) Generation of transgenic sheep by sub-zonal injection of feline leukaemia virus. *Journal of Cell Biochemistry* Suppl. 13B, 180.

Hunter, N. and Hope, J. (1991) The genetics of scrapie susceptibility in sheep (and its implications for BSE). In: Owen, J.B. and Axford, R.F.E. (eds), *Breeding for Disease Resistance in Farm Animals.* CAB International, Wallingford, pp. 329–44.

Iglesias, A., Lamers, M. and Köhler, G. (1987) Expression of immunoglobulin delta chain causes allelic exclusion in transgenic mice. *Nature* 330, 482–4.

Jähner, D., Haase, K., Mulligan, R. and Jaenisch, R. (1985) Insertion of the bacterial gpt gene into the germ line of mice by retroviral infection. *Proceedings of the National Academy of Science, USA* 82, 6927–31.

Jaenisch, R. (1988) Transgenic animals. *Science* 240, 1468–74.

Jeffreys, A.J. (1987) Highly variable minisatellites and DNA fingerprints. *Biochemical Society Transactions* 15, 309–317.

Jeffreys, A.J., Wilson, V. and Thein, S.L. (1985) Hypervariable 'minisatellite' regions in human DNA. *Nature* 314, 67–73.

Jones, D.D. (1986) Legal and regulatory aspects of genetically engineered animals. *Basic Life Science* 37, 273–83.

Joyner, A.L., Skarnes, W.C. and Rossant, J. (1989) Production of a mutation in mouse En-2 gene by homologous recombination in embryonic stem cells. *Nature* 338, 153–6.

Kerem, B.-S., Rommens, J.M., Buchanan, J.A., Markiewicz, D., Cox, T.K., Chakravarti, A., Buchwald, M. and Tsui, L.-C. (1989) Identification of the cystic fibrosis gene: genetic analysis. *Science* 245, 1073–80.

Lamont, S.J. (1989) The chicken major histocompatibility complex in disease resistance and poultry breeding. *Journal of Dairy Science* 72, 1328–33.

Landegren, U., Kaiser, R., Caskey, C.T. and Hood, L. (1988) DNA diagnostics – molecular techniques and automation. *Science* 242, 229–37.

Lander, E.S. and Botstein, D. (1989) Mapping Mendelian factors underlying quantitative traits using RFLP linkage maps. *Genetics* 121, 185–99.

Lavitrano, M., Camaioni, A., Fazio, V.M., Dolci, S., Farace, M.G. and Spadafora, C. (1989) Sperm cells as vectors for introducing foreign DNA into eggs: genetic transformation of mice. *Cell* 57, 717–23.

Litt, M. and Luty, J.A. (1989) A hypervariable microsatellite revealed by in vitro amplification of a dinucleotide repeat within the cardiac muscle actin gene. *American Journal of Human Genetics* 44, 397–401.

Love, J.M., Knight, A.M., McAleer, M.A. and Todd, J.A. (1990) Towards construction of a high resolution map of the mouse genome using PCR-analysed microsatellites. *Nucleic Acids Research* 18, 4123–30.

McGrath, J. and Solter, D. (1984) Inability of mouse blastomere nuclei transferred to enucleated zygotes to support development in vitro. *Science* 226, 1317–19.

Mansour, S.L., Thomas, K.R. and Capecchi, M.R. (1988) Disruption of the proto-oncogene int-2 in mouse embryo-derived stem cells: a general strategy for targeting mutations to non-selectable genes. *Nature* 336, 348–52.

Moreau, J.-F., Donaldson, D.D., Bennett, F., Witek-Giannotti, J., Clark, S.C. and Wong, G.G. (1988) Leukaemia inhibitory factor is identical to the myeloid growth factor human interleukin for DA cells. *Nature* 336, 690–2.

Murray, J.D., Nancarrow, C.D., Marshall, J.T., Hazelton, I.G. and Ward, K.A. (1989) Production of transgenic Merino sheep by microinjection of ovine metallothionein-ovine growth hormone fusion genes. *Reproduction Fertility and Development* 1, 147–55.

Nakamura, Y., Leppert, M., O'Connell, P., Wolff, R., Holm, T., Culver, M., Martin, C., Fujimoto, E., Hoff, M., Kumlin, E. and White, R. (1987) Variable number of tandem repeat (VNTR) markers for human gene mapping. *Science* 235, 1616–22.

Notarianni, E., Galli, C., Laurie, S., Moor, R.M. and Evans, M.J. (1990) Derivation of pluripotent, embryonic cell lines from porcine and ovine blastocysts. *4th World*

Congress on Genetics Applied to Livestock Production XIII, 58–64.

Palmiter, R.D., Brinster, R.L., Hammer, R.E., Trumbauer, M.E., Rosenfeld, M.G., Birnberg, N.C. and Evans, R.M. (1982) Dramatic growth of mice that develop from eggs microinjected with metallothionein-growth hormone fusion genes. *Nature* 300, 611–15.

Paterson, A.H., Lander, E.S., Hewitt, J.D., Peterson, S., Lincoln, S.E. and Tanksley, S.D. (1988) Resolution of quantitative traits into Mendelian factors by using a complete linkage map of restriction fragment length polymorphisms. *Nature* 335, 721–26.

Petters, R.M., Johnson, B.H. and Shuman, R.M. (1988) Gene transfer to swine embryos using an avian retrovirus, *Genome* 30 (Suppl. 1), 448. (abstract 35.23.5).

Powell, P.A., Nelson, R.S., Barun, D., Hoffmann, N., Rogers, S.G., Fraley, R.T. and Beachy, R.N. (1986) Delay of disease development in transgenic plants that express the tobacco mosaic virus coat protein gene. *Science* 232, 738–43.

Powell, P.A., Stark, D.M., Sanders, P.R., and Beachy, R.N. (1989) Protection against tobacco mosaic virus in transgenic plants that express tobacco mosaic virus antisense RNA. *Proceedings of the National Academy of Science USA* 86, 6949–52.

Pursel, V.G., Pinkert, C.A., Miller, K.F., Bolt, D.J., Campbell, R.G., Palmiter, R.D., Brinster, R.L. and Hammer, R.E. (1989) Genetic engineering of livestock. *Science* 244, 1281–88.

Raadsma, H.W. (1991) Genetic variation in resistance to fleece rot and fly strike in sheep. In: Owen, J.B. and Axford, R.F.E. (eds), *Breeding for Disease Resistance in Farm Animals*. CAB International, Wallingford, pp. 263–90.

Register, J.C. and Beachy, R.N. (1988) Resistance to TMV in transgenic plants results from interference with an early event in infection. *Virology* 166, 524–32.

Riordan, J.R., Rommens, J.M., Kerem, B.-S., Alon, N., Rozmahel, R., Grzelczak, Z., Zielenski, J., Lok, S., Plasvsic, N., Chou, J.-L., Drumm, M.L., Iannuzzi, M.C., Collins, F.S. and Tsui, L.-C. (1989) Identification of the cystic fibrosis gene: cloning and characterization of complementary DNA. *Science* 245, 1066–73.

Ritchie, K.A., Brinster, R.L. and Storb, U. (1984) Allelic exclusion and control of endogenous immunoglobulin gene rearrangement in κ transgenic mice. *Nature* 312, 517–20.

Rommens, J.M., Iannuzzi, M.C., Kerem, B.-S., Drumm, M.L., Melmer, G., Dean, M., Rozmahel, R., Cole, J.L., Kennedy, D., Hidaka, N., Zsiga, M., Buchwald, M., Riordan, J.R., Tsui, L.-C. and Collins, F.S. (1989) Identification of the cystic fibrosis gene: chromosome walking and jumping. *Science* 245, 1059–65.

Roschlau, K., Rommel, P., Andreewa, L., Zackel, M., Roschlau, D., Zackel, B., Schwerin, M., Hühn, R. and Gazarjan, K.G. (1989) Gene transfer experiments in cattle. *Journal of Reproduction and Fertility* Suppl. 38, 153–60.

Royle, N.J., Clarkson, R.E., Wong, Z. and Jeffreys, A.J. (1988) Clustering of hypervariable minisatellites in the proterminal regions of human autosomes. *Genomics* 3, 352–60.

Rusconi, S. and Köhler, G. (1985) Transmission and expression of a specific pair of rearranged immunoglobulin μ and κ genes in a transgenic mouse line. *Nature* 314, 330–4.

Saiki, R.K., Gelfand, D.H., Stoffel, S., Scharf, S.J., Higuchi, R., Horn, G.T., Mullis, K.B. and Erlich, H.A. (1988) Primer-directed enzymatic amplification of DNA with a thermostable DNA polymerase. *Science* 239, 487–91.

Salter, D.W. and Crittenden, L.B. (1988) Gene insertion into the avian germ line. *Occasional Publication of the British Society for Animal Production* no. 12, 32–57.

Salter, D.W. and Crittenden, L.B. (1989) Artificial insertion of a dominant gene for resistance to avian leukosis virus into the germ line of the chicken. *Theoretical and Applied Genetics* 77, 457–61.

Salter, D.W., Smith, E.J., Hughes, S.H., Wright, S.E. and Crittenden, L.B. (1987) Transgenic chickens: insertion of retroviral genes into the chicken germ line. *Virology* 157, 236–40.

Sears, P.M., Smith, B.S., Polak, J., Gusik, S.N. and Blackburn, P. (1988) Lysostaphin efficacy for treatment of *Staphylococcus aureus* intramammary infection. *Journal of Dairy Science* 71, Suppl. 1, 244. (Abstract, P376).

Simons, J.P., Wilmut, I., Clark, A.J., Archibald, A.L., Bishop, J.O. and Lathe, R. (1988) Gene transfer into sheep. *Bio/technology* 6, 179–83.

Smith, A.G., Heath, J.K., Donaldson, D.D., Wong, G.G., Moreau, J., Stahl, M. and Rogers, D. (1988) Inhibition of pluripotential embryonic stem cell differentiation by purified polypeptides. *Nature* 336, 688–90.

Smith, C. and Simpson, S.P. (1986) The use of genetic polymorphisms in livestock improvement. *Journal of Animal Breeding and Genetics* 103, 205–17.

Smith, C., Meuwissen, T.H.E. and Gibson, J.P. (1987) On the use of transgenes in livestock improvement. *Animal Breeding Abstracts* 55, 1–10.

Smith, L.C. and Wilmut, I. (1989) Influence of nuclear and cytoplasmic activity on the development in vivo of sheep embryos after nuclear transplantation. *Biology of Reproduction* 40, 1027–35.

Soller, M. and Beckman, J.S. (1990) Molecular mapping of quantitative genes. *4th World Congress on Genetics Applied to Livestock Production* XIII, 93–6.

Southern, E.M. (1975) Detection of specific sequences among DNA fragments separated by gel electrophoresis. *Journal of Molecular Biology* 98, 503–17.

Staeheli, P. (1990) Interferon-induced proteins and the anti-viral state. *Advances in Virus Research* 38, 147–200.

Stewart, T.A., Wagner, E.F. and Mintz, B. (1982) Human β-globin gene sequences injected into mouse eggs, retained in adults and transmitted to progeny. *Science* 217, 1046–8.

Storb, U., Pinkert, C., Arp, B., Engler, P., Gollahon, K., Manz, J., Brady, W. and Brinster, R.L. (1986) Transgenic mice with μ and κ genes encoding antiphosphorylcholine antibodies. *Journal of Experimental Medicine* 164, 627–41.

van der Putten, H., Botteri, F.M., Miller, A.D., Rosenfeld, M.G., Fan, H., Evans, R.M. and Verma, I.M. (1985) Efficient insertion of genes into the mouse germ line via retroviral vectors. *Proceedings of the National Academy of Science USA* 82, 6148–52.

Vize, P.D., Michalska, A.E., Ashman, R., Lloyd, B., Stone, B.A., Quinn, P., Wells, J.R.E. and Seamark, R.F. (1988) Introduction of a porcine growth hormone fusion gene into transgenic pigs promotes growth. *Journal of Cell Science* 90, 295–300.

Wagner, E.F., Stewart, T.A. and Mintz, B. (1981) The human β-globin gene and a functional viral thymidine kinase gene in developing mice. *Proceedings of the National Academy of Science USA* 78, 5016–20.

Wagner, T.E., Hoppe, P.C., Jollick, J.D., Scholl, D.R., Hodinka, R.L. and Gault, J.B. (1981) Microinjection of a rabbit β-globin gene into zygotes and its subsequent expression in adult mice and their offspring. *Proceedings of the National Academy of Science USA* 78, 6376–80.

Wall, R.J., Bolt, D.J., Frels, W.I., Hawk, H.W., King, D., Pursel, V.G., Rexroad, C.E. and Rohan, R.M. (1990) Transgenic farm animals: current state of the art. *AgBiotech News and Information* 2, 391–5.

Wall, R.J., Pursel, V.G., Hammer, R.E. and Brinster, R.L. (1985) Development of porcine ova that were centrifuged to permit visualization of pronuclei and nuclei. *Biology of Reproduction* 32, 645–51.

Weaver, D., Costantini, F., Imanishi-Kari, T. and Baltimore, D. (1985) A transgenic immunoglobulin Mu gene prevents rearrangement of endogenous genes. *Cell* 42, 117–27.

Weber, J.L. and May, P.E. (1989) Abundant class of human DNA polymorphisms which can be typed using the polymerase chain reaction. *American Journal of Human Genetics* 44, 388–96.

Whitelaw, C.B.A., Archibald, A.L., Harris, S., McClenaghan, M., Simons, J.P., Springbett, A., Wallace, R. and Clark, A.J. (1990) Frequency of germline mosaicism in G0 transgenic mice. *Mouse Genome* 88, 114.

Wilkie, T.M., Brinster, R.L. and Palmiter, R.D. (1986) Germline and somatic mosaicism in transgenic mice. *Developmental Biology* 118, 9–18.

Williams, R.L., Hilton, D.J., Pease, S., Willson, T.A., Stewart, C.L., Gearing, D.P., Wagner, E.F., Metcalf, D., Nicola, N.A. and Gough, N.M. (1988) Myeloid leukaemia inhibitory factor maintains the developmental potential of embryonic stem cells. *Nature* 336, 684–7.

Williamson, C.M., Lax, A.J. and Bramley, A.J. (1990) A transgenic approach to the control of Staphylococcal mastitis. *Association of Veterinary Teachers and Research Workers*, Scarborough, April, 1990, Abstract C1, p. 29.

Wilmut, I., Archibald, A.L., Harris, S., McClenaghan, M., Simons, J.P., Whitelaw, C.B.A. and Clark, A.J. (1990) Methods of gene transfer and their potential use to modify milk composition. *Theriogenology* 33, 113–23.

Zimmer, A. and Gruss, P. (1989) Production of chimaeric mice containing embryonic stem (ES) cells carrying a homoeobox Hox 1.1 allele mutated by homologous recombination. *Nature* 338, 150–3.

Chapter 8

Screening and Selection for Disease Resistance – Repercussions for Genetic Improvement

C.A. Morris

Ministry of Agriculture and Fisheries, Ruakura Agricultural Centre, Private Bag, Hamilton, New Zealand

Summary

Resistance of animals to production diseases is often heritable. Twenty-three heritability estimates for 15 diseases of cattle and sheep in New Zealand and Australia have averaged 0.31, which is no lower than for say milk yield, body weight or fleece weight. Selection experiments for resistance or susceptibility to specific disease traits have been established, and significant responses have already been achieved. Correlated responses are now under study, with the hope of finding useful indicator traits. New DNA technologies may provide advances in rates of response to selection, or may possibly assist in finding a cure for some diseases.

Introduction

The challenge of production diseases to sheep and cattle is part of the usual farm environment. Meeting the challenge should be part of what makes an animal rank highly among the farmer's breeding objectives. In the 1980s and 1990s when drenches and sprays are considered less desirable by society, it has become important to provide an animal with the genes it needs to survive, and for it to produce with less and less artificial assistance. Animal production systems now need to function with less and less labour per 100 cows or per 1000 ewes. Just as important, the concept of reducing costs per animal through disease resistance (rather than only considering increasing gross income per animal) is indeed a valid objective for the bull breeder or ram breeder and his clients. Perhaps, with improved resistance, the farmer can keep most of the saved costs himself, whereas any increase in, say, percentage meat yield in the carcass may return some of the benefit to the farmer, but the rest will go to the processors, distributors, retailers and consumers.

This chapter provides a survey of some of the biological mechanisms being studied in sheep and cattle under extensive conditions, in developing resistance of farm animals to disease. Examples are restricted mainly to genetic variation and genetic changes in herds and flocks in New Zealand and Australia.

Selection for resistance: which philosophy?

There are two philosophies of selection for resistance that are prevalent in New Zealand. In the first (Type 1), the practice is to select mainly for production traits in the confidence that resistant animals will be among those selected. This is the philosophy used by the New Zealand Dairy Board (Ahlborn-Breier *et al.*, 1990), with about 20 proven bulls each year to be selected from a young bull team of 150. Dairy cows are expected to produce high yields when routinely challenged with production diseases such as mastitis, facial eczema (a fungal toxin on autumn pasture), and (to a lesser extent, because of routine precautions taken by farmers) pasture bloat and hypomagnesaemia.

In the second (Type 2), the practice is to treat the animal's response to a disease as another recorded trait and as part of a multi-trait breeding objective. This is the approach being adopted by some New Zealand sheep group breeding schemes, with selection objectives including the usual output traits (more lambs, meat and wool) and the recent inclusion of selection for resistance to such factors as facial eczema, internal parasites or footrot (Warren *et al.*, 1990).

It is too early to distinguish between the two types of selection procedure on merit. Type 1 assumes that animals with favourable genes for resistance and production will be selected, regardless of the size of the genetic correlation. For Type 2, as clearly explained by Piper and Barger (1988) and by Woolaston (1990), it is difficult to define a relative economic value for any disease trait. Animals with favourable genes for resistance would only be selected if the relative economic value for resistance and the genetic correlation with production traits were high enough. These genes could be combined with genes for production from the same flock/herd in later generations. A variant on the Type 2 philosophy might be to combine genes for production with genes from special selection flocks breeding for resistance, for example by using semen from these flocks.

Performance versus progeny testing

In choosing between performance and progeny testing, the usual compromise is between (a) a short generation interval and low accuracy with performance testing, *versus* (b) a longer generation interval but greater accuracy with progeny testing. In the case of disease traits, the candidate's performance may pose difficulties of measurement. The candidate can be tested if the disease is, say, bloat in cattle, but the situation needs to be considered further for a trait like resistance to internal parasites. If the 'disease' is quickly reversible in the animal after removing the challenge, a performance test can be run with few complications.

In the early generations of a project to select for resistance, some degree of protection from the disease will be necessary, because the candidate's self-protection may be limited. Exposing animals to an uncontrolled level of the natural challenge could be risky, as well as unethical, so that a controlled artificial challenge may have to be applied. One alternative (e.g. Baker *et al.*, 1990) is to employ best linear unbiased prediction (BLUP) (a statistical technique that allows the incorporation of data from all ancestors and half-sibs or other relatives) to supplement the performance test, and thus provide a more accurate breeding value (BV) estimate for the candidate. A milder challenge than that required for the straight performance test may be used for a BLUP assessment, especially when there are several paternal half-sibs tested. If an indicator trait has to be used, this also compromises accuracy for the goal itself (see below).

Screening

In the case of screening individuals for an elite herd/flock, some form of progeny testing (ideally a reference sire scheme) should be used. This solves the problem of different levels of challenge in different herds/flocks, but it can be expensive, depending on the species. As with most screening exercises, the potential genetic lift offered by screening for a polygenic trait declines after the initial foundation as the elite herd/flock improves in genetic merit. However, for a single gene trait, the screening opportunities do not diminish with time; identifying the outliers from a large population remains important.

In the case of some transmissible diseases it could be undesirable to introduce screened animals to the elite flock/herd; in this case, the use of artificial insemination or embryo transfer may have to be considered.

Indicator traits

When performance testing is used, some disease traits require the use of an indicator trait, for example resistance of sheep to internal parasites. The indicator most commonly used so far for resistance to internal parasites is the faecal egg count (Woolaston, 1990), on the assumption that ranking animals on egg counts will predict their ranking on future worm numbers. The biological assumptions include:

1. The rate of egg production by each parasite does not vary to an important degree.
2. Egg numbers and subsequent worm numbers closely reflect the level of parasitic challenge to the host. However, sheep that are tolerant of (rather than resistant to) worms will not be accurately identified by egg counting.
3. With some exceptions, worm species cannot be identified separately when egg counting.

Baker *et al.* (1990) have investigated some of these problems, with the parasite selection flocks of Romney sheep at Ruakura. They are selected for high (H) faecal egg count, control and low (L) faecal egg count, by subjection to two rounds of field challenge from about 3 to 8 months of age. For the 1989 lamb crop (year 6) the L flock means for faecal egg count (2 sample times combined) averaged 50.2% of the H flock means. This difference was 0.63 times the residual standard deviation on the original scale, or 0.84 times the residual standard deviation on a log scale (Baker, R.L., personal communication). Samples of the 1988 L and H flocks, additionally challenged indoors with a standard dose of worms of two species (*Haemonchus contortus* and *Trichostrongylus* spp.), were found to have subsequent faecal egg counts of 4,220 and 7,650 eggs/g ($P < 0.01$), and total burdens of 3,840 and 5,560 worms ($P < 0.01$), respectively. This demonstrates that assumptions 1 and 2 above were not seriously violated, at least when the objective is to change resistance rather than tolerance. In addition, growing out the faecal eggs into identifiable worm species showed that resistance was conferred to the host across the species under study (assumption 3); this was also observed in two experiments on sheep in Australia where selection was for or against resistance to worms (Windon *et al.*, 1987; Woolaston, 1990). Selecting for faecal egg count is likely to be a much less suitable criterion, however, for studies on *Ostertagia circumcincta*, because egg output during infection in that species is a poor indicator of parasitic burden in the host (Woolaston, 1990).

Similar testing must of course be carried out with the indicator(s) for each disease trait. This has been achieved, for example, with facial eczema in sheep, a disease causing liver damage and subsequent reductions in performance traits (Smeaton *et al.*, 1985). An enzyme, gamma-glutamyl-transferase (GGT), has been shown to be an indicator of liver damage; the relationship between log (GGT) and liver damage score is approximately linear (Towers and Stratton, 1978), with a correlation of +0.75.

Type of challenge

Using a natural challenge on pasture may expose animals (especially the susceptible ones) to an intolerable assault. However, to obtain a uniform natural challenge, animals need to have the same daily food intake and uniform diet composition.

If a natural challenge is used for selection, as at Ruakura with selection for worm resistance, then it is important to determine if the selection flocks also show corresponding differences under a standard artificial challenge (Baker *et al.*, 1990). Conversely if a standard artificial challenge is used for selection, such as with the facial eczema Resistant, Control and Susceptible selection flocks at Ruakura (Morris *et al.*, 1989), then the flock responses to a natural challenge must be confirmed (e.g. Towers and Wesselink, 1989).

In the facial eczema selection experiment, animals are normally grazed on fungicide-sprayed pasture during the facial eczema season, to ensure the survival of the Susceptible flock. A consequence, however, is that the genetic correlation of the disease trait with production is only relevant to that particular management environment. In the Ruakura flock (Morris *et al.*, 1989), the disease trait was not significantly correlated with production traits. It is possible, however, that the genetic correlations between the disease and production traits are negative (favourable) in commercial flocks where no fungicide is used. One way of testing this hypothesis would be to double the numbers of animals in the Resistant flock, and manage part of the Resistant flock on toxic pasture. A faster and cheaper solution may come from estimating genetic correlations in suitable data from a large commercial flock where there has been selection for production and for facial eczema resistance, without using fungicide.

Another example of different genetic correlations according to management system is given by Piper and Barger (1988). Their preliminary conclusions were that genetic correlations between resistance to worms and production traits were small (−0.05) when the parasitic challenge was low, but considerably higher (−0.40) when the challenge was moderate.

In the beef cattle selection experiment at Rockhampton, Queensland, where selection from 1966 to 1981 was for 18-month weight (Frisch, 1981), it was also demonstrated that the size of observed correlated responses depended on the environment in which the cattle were tested. Under one of the most difficult environments (the natural one, i.e. with heat stress, and also challenges from external parasites (cattle ticks) and worms), differences in production (i.e. food intake and growth rate) and in parasite burdens between the selection and control herds were large. In a less challenging environment, there was a reranking of animals for production. The implications are that the breeding objective (and the environment for testing) must be clearly defined, and that a client would lose out if he bought bulls or rams from a breeder selecting for other objectives.

Selection practices in some dairy industries

Solbu (1984) has reported heritabilities for mastitis, ketosis and 'all diseases' of 0.01, 0.02 and 0.02, respectively, in Norwegian dairy cattle. The Norwegian dairy industry is set up in such a way that index selection is applied to bulls for reduced disease incidence and increased milk yield. Genetic correlations of disease traits with milk yield, however, are very low in Norway (Solbu, 1984) and also in West Germany (Distl *et al.*, 1989). However, since a progeny test is being carried out in these countries anyway, including one more trait is relatively inexpensive. In contrast, as mentioned before, the New Zealand dairy industry does not include any disease traits in its selection objective. There is no simple answer as to whether the Norwegian or New Zealand approach is more efficient or more profitable, because disease traits such as facial eczema, bloat and hypo-magnesaemia are not routinely monitored in New Zealand sire proving scheme herds. Certainly, the New Zealand approach must be cheaper for the testing organization, but the value or the cost to the nation is not known.

Biological study of resistant and susceptible herds/flocks

One important advantage of having contemporary resistant and susceptible herds is that it is possible to study differences between herds that are repeatable and predictable.

In New Zealand and Australia, there are herds selected for greater or lesser susceptibility to bloat (Cockrem *et al.*, 1983), flocks selected for resistance or susceptibility to facial eczema (Morris *et al.*, 1989), to internal parasites (Woolaston, 1990) and to a neurotoxin in ryegrass pasture causing ryegrass staggers (Morris, C.A. and Towers, N.R., unpublished). In addition there are herds at Rockhampton (Queensland) selected for production traits in a stressful environment which are showing correlated changes in resistance to internal and external parasites (Frisch, 1981; Mackinnon *et al.*, 1990).

A list of heritability estimates for disease traits in cattle and sheep in New Zealand and Australia is given in Table 8.1. As expected, there was a wide range for the 23 estimates, but the overall average of 0.31 for 15 diseases suggests that there are good prospects of genetic change through selection for many disease traits. This average of heritability estimates is no lower than for traits such as milk yield, body weight or fleece weight.

A variety of traits, or biochemical and physiological traits may be measured in order to identify correlated factors for use as indicators. For example, the High and Low Susceptible bloat herds at Ruakura have been compared for salivary proteins (Cockrem *et al.*, 1983), body composition (Carruthers and Morris, 1988), food intake, rib girth and milk yield (McIntosh *et al.*, 1988), and differences in mineral concentrations such as sodium and potassium (Carruthers and Bryant, 1988). Some correlations of these traits have been found with bloat but they are small in size. The facial eczema selection flocks at Ruakura have also received intensive study; resistant and susceptible flocks differ in bile secretion (Fairclough *et al.*, 1982); hepatic copper levels (Munday *et al.*, 1984); plasma transferrins (Morris *et al.*, 1988) and in some properties of erythrocytic membranes (Upreti *et al.*, 1990).

Outside the southern hemisphere, magnesium and copper deficiencies provide relevant examples. A comparison of five cattle breeds and their crosses has been undertaken since the 1970s at Texas A&M University (McElhenney *et al.*, 1985). The breeds and crosses of beef cows were found to differ in sensitivity to hypomagnesaemia (Greene *et al.*, 1986) and these differences were partitioned by the Texas group into:

1. different intakes of magnesium among breeds/crosses (range, 18.3–25.0 g/day)
2. different apparent absorption rates of magnesium (range, 17–46% of intake)
3. different daily requirements for magnesium.

The degree of difference among breeds depended on such factors as lactation status and whether the intake of magnesium was limiting in the diet.

Long-term studies in Edinburgh on lamb mortality and copper deficiency have demonstrated a genetic basis. Wiener *et al.* (1985) described a selection experiment involving breeding for high or low plasma copper concentrations over a 5-year period. The difference due to selection widened to 0.5 mg/l (over 50% of the mean) after five years. Woolliams *et al.* (1986) demonstrated in the above flocks a difference of more than twofold between lamb mortality rates up to 24 weeks (12% versus 28% mortality), and concluded that 'decreased resistance to infection is a clinical consequence of ovine copper deficiency in the field, amenable to control by copper treatment and genetic selection'.

Detailed biological studies of selection herds/flocks may one day provide the information to identify a useful marker gene or protein for the trait. Alternatively, the preparation of herd crosses and then backcrosses to the parents (or interbreds among herd crosses) should generate the required

Table 8.1. Heritability (h^2) of some disease resistance[a] traits in cattle and sheep in New Zealand and Australia.

Trait	Type[b]	h^2(SE)	Reference	Comment
Dairy cattle				
Mastitis	N	0.04	Wickham (1979)	Mastitis score
Mastitis	N	0.04	Smit and Wickham (1986)	Somatic cell count
Pasture bloat	N	0.08	Wickham (1979)	
Pasture bloat	N	0.15(0.06)	Morris et al. (1990a)	
GGT level[c]	N	0.31(0.10)	Morris et al. (1990b)	
Plasma Mg	N	0.15(0.06)	Morris et al. (1990b)	Hypomagnesaemia?
Beef cattle				
Tick resistance[a]	N	0.39	Wharton et al. (1970)	
Tick resistance[a]	N	0.34(0.05)	Mackinnon et al. (1990)	
Worm resistance[a]	N	0.78	Seifert (1977)	
Worm resistance[a]	N	0.28(0.03)	Mackinnon et al. (1990)	
Buffalo fly resistance[a]	N	0.06	Mackinnon et al. (1990)	
Eye cancer	N	0.83(0.28)	French (1959)	Lid pigmentation
Sheep				
Worm resistance[a]	A	0.41(0.19)	Windon et al. (1987)	
Worm resistance[a]	A	0.27(0.13)	Piper (1987)	
Worm resistance[a]	A	0.30(0.10)	Albers et al. (1987)	
Worm resistance[a]	N	0.34(0.19)	Baker et al. (1990)	

Table 8.1. *Contd.*

Trait	Type[b]	h²(*SE*)	Reference	Comment
Worm tolerance[a]	A	0.09(0.07)	Albers *et al.* (1987)	Production during infection
Facial eczema	A	0.42(0.09)	Campbell *et al.* (1981)	Liver damage score
GGT level[c]	A	0.41	Johnson, D. L. (pers. comm.)	
GGT level[c]	A	0.33(0.17)	Morris *et al.* (1989)	
Ryegrass staggers	N	0.47(0.30)	Campbell (1987)	
Footrot resistance	N	0.17	Skerman *et al.* (1988)	
Fleece rot resistance	N	0.36(0.07)	McGuirk and Atkins (1984)	

[a]All resistance traits are cases where the host has become (more) resistant to the challenge.
[b]Type of challenge: A = artificial; N = natural.
[c]Gamma-glutamyltransferase: an enzyme indicating liver damage from a challenge such as facial eczema.

animals from which DNA may be obtained and used for new molecular biology studies (linkage studies and DNA testing). In general, DNA testing is progressing faster in humans than in farm animals, because of the better gene map and greater financial resources currently available for humans.

An example of a disease trait in farm animals to which DNA technology will soon be applied is the halothane gene in pigs (Archibald and Imlah, 1985). Very recently new genetic information has come to light on the related human disease, malignant hyperthermia, which appears to be controlled by the ryanodine receptor gene (RYR); this maps on human chromosome 19 and cDNA is now available for it (MacLennan *et al.*, 1990). The halothane gene has been mapped to the equivalent place on the pig genome and is also closely linked to the GPI gene. The human gene appears to be dominant whereas the halothane gene is probably recessive. Thus, linkage information and/or a gene probe could soon be used to select against the halothane gene in pigs.

In the case of facial eczema, it has been found that transferrin (an iron-carrying protein) is correlated with resistance (Morris *et al.*, 1988). One of the problems with genetic markers is that the size of the correlation is seldom high enough to justify sole reliance on the marker. An objective for the future would be to use a whole series of markers, thereby increasing the accuracy of final predictions (Blair *et al.*, 1990).

One other factor of potential importance is the major histocompatibility complex (MHC: Teale *et al.*, 1991). It has been reported that the MHC genotype can be changed by selection (Biozzi *et al.*, 1982). It remains to be seen how many of the disease traits mentioned so far will turn out to be linked to the MHC genes. In this context an important finding is that high immune responses are not always correlated with high resistance across the diseases so far tested (Biozzi *et al.*, 1982).

A better understanding of the genes and gene complexes controlling disease for farm animal species is likely to be crucial to advances in this area.

Conclusions

Progress is being made in selecting for disease resistance traits in farm animals; currently differences between resistant and susceptible herds or flocks are as much as twofold. In the future, new DNA technologies may have a great impact in breeding animals for disease resistance, by providing novel alleles. These technologies could increase rates of response to selection or possibly find a cure for some diseases. In this event, there will be an increasing need to test animals with the new supergenes in commercial environments.

References

Ahlborn-Breier, G., Bishop, S.C. and Wickham, B.W. (1990) Progress in New Zealand dairy cattle breeding – the implementation of a selection index for production and traits other than production. *Proceedings of the Australian Association of Animal Breeding and Genetics* 8, 339–43.

Albers, G.A.A., Gray, G.D., Piper, L.R., Barker, J.S.F., Le Jambre, L.F. and Barger, I.A. (1987) The genetics of resistance and resilience to *Haemonchus contortus* infection in young Merino sheep. *International Journal for Parasitology* 17, 1355–63.

Archibald, A.L. and Imlah, P. (1985) The halothane sensitivity locus and its linkage relationships. *Animal Blood Groups and Biochemical Genetics* 16, 253–63.

Baker, R.L., Watson, T.G., Bisset, S.A. and Vlassoff, A. (1990) Breeding Romney sheep which are resistant to gastro-intestinal parasites. *Proceedings of the Australian Association of Animal Breeding and Genetics* 8, 173–8.

Biozzi, G., Mouton, D., Heumann, A.M. and Bouthillier, Y. (1982) Genetic regulation of immunoresponsiveness in relation to resistance against infectious diseases. *Proceedings of the 2nd World Congress on Genetics Applied to Livestock Production* 5, 150–63.

Blair, H.T., McCutcheon, S.N. and Mackenzie, D.D.S. (1990) Physiological predictors of genetic merit. *Proceedings of the Australian Association of Animal Breeding and Genetics* 8, 133–42.

Campbell, A.G. (1987) Animal selection for disease resistance. *Proceedings of the Asian–Australasian Association of Animal Production* 4, 403 (Abstract).

Campbell, A.G., Meyer, H.H., Henderson, H.V. and Wesselink, C. (1981) Breeding for facial eczema resistance – a progress report. *Proceedings of the New Zealand Society of Animal Production* 41, 273–8.

Carruthers, V.R. and Bryant, A.M. (1988) Quantity and composition of digesta in the reticulo-rumen of cows offered two diets differing in K:Na ratio. *New Zealand Journal of Agricultural Research* 31, 121–7.

Carruthers, V.R. and Morris, C.A. (1988) Weights of some body organs from cattle selected for high and low susceptibility to bloat. *Proceedings of the New Zealand Society of Animal Production* 48, 147–9.

Cockrem, F.R.M., McIntosh, J.T. and McLaren, R.D. (1983) Selection for and against susceptibility to bloat in dairy cows – a review. *Proceedings of the New Zealand Society of Animal Production* 43, 101–6.

Distl, O., Wurm, A., Glibotic, A., Brem, G. and Krausslich, H. (1989) Analysis of relationships between veterinary recorded production diseases and milk production in dairy cows. *Livestock Production Science* 23, 67–78.

Fairclough, R.J., Smith, B.L. and Campbell, A.G. (1982) Biochemical studies on resistance to facial eczema. *Ministry of Agriculture and Fisheries Agricultural Research Division. Annual Report*, 1981/82, p. 57.

French, G.T. (1959) A clinical and genetic study of eye cancer in Hereford cattle. *Australian Veterinary Journal* 35, 474–81.

Frisch, J.E. (1981) Changes occurring in cattle as a consequence of selection for growth rate in a stressful environment. *Journal of Agricultural Science, Cambridge* 96, 23–38.

Greene, L.W., Solis, J.C., Byers, F.M. and Schelling, G.T. (1986) Apparent and true digestibility of magnesium in mature cows of five breeds and their crosses. *Journal of Animal Science* 63, 189–96.

McElhenney, W.H., Long, C.R., Baker, J.F. and Cartwright, T.C. (1985) Production characters of first-generation cows of a five-breed diallel: reproduction of young cows and preweaning performance of *inter se* calves. *Journal of Animal Science* 61, 55–65.

McGuirk, B.J. and Atkins, K.D. (1984) Fleece rot in Merino sheep. I. The heritability of fleece rot in unselected flocks of Medium-wool Peppin Merinos. *Australian Journal of Agricultural Research* 35, 423–34.

McIntosh, J.T., Morris, C.A., Cockrem, F.R.M., McLaren, R.D. and Gravett, I.M. (1988) Genetics of susceptibility to bloat in cattle. IV. Girth measurements, bloat scores, saliva proteins, blood components, and milk production, liveweight, and intake data. *New Zealand Journal of Agricultural Research* 31, 133–44.

Mackinnon, M.J., Meyer, K. and Hetzel, D.J.S. (1990) Genetic variation and covariation for growth, parasite resistance and heat tolerance in tropical cattle. *Livestock Production Science* (submitted).

MacLennan, D.H., Duff, C., Zorzato, F., Fujii, J., Phillips, M., Korneluk, R.G., Frodis, W., Britt, B.A. and Worton, R.G. (1990) Ryanodine receptor gene is a candidate for predisposition to malignant hyperthermia. *Nature* 343, 559–61.

Morris, C.A., Jordan, T.W., Loong, P.C., Lewis, M.H. and Towers, N.R. (1988) Associations between transferrin type and facial eczema susceptibility and some production traits in sheep. *New Zealand Journal of Agricultural Research* 31, 301–5.

Morris, C.A., Towers, N.R., Campbell, A.G., Meyer, H.H., Wesselink, C. and Wheeler, N. (1989) Responses achieved in Romney flocks selected for or against susceptibility to facial eczema, 1975–87. *New Zealand Journal of Agricultural Research* 32, 379–88.

Morris, C.A., Cockrem, F.R.M., Carruthers, V.R., McIntosh, J.T. and Cullen, N.G. (1990a) Response to divergent selection for bloat susceptibility in dairy cows. *New Zealand Journal of Agricultural Research* (in press).

Morris, C.A., Towers, N.R., Tempero, H.J., Cox, N.R. and Henderson, H.V. (1990b) Facial eczema in Jersey cattle: incidence, regional difference, heritability and correlation with production. *Proceedings of the New Zealand Society of Animal Production* 50, (in press).

Munday, R., Campbell, A.G. and Smith, B.L. (1984) Hepatic copper levels in sheep bred for susceptibility or resistance to facial eczema. *Ministry of Agriculture and Fisheries Agricultural Research Division Annual Report*, 1983/84, p. 61.

Piper, L.R. (1987) Genetic variation in resistance to internal parasites. In: McGuirk, B.J. (ed.), *Merino Improvement Programs in Australia*, Australian Wool Corporation, Melbourne, pp. 351–63.

Piper, L.R. and Barger, I.A. (1988) Resistance to gastro-intestinal strongyles: feasibility of a breeding programme. *Proceedings of the 3rd World Congress on Sheep and Beef Cattle Breeding* 1, 593–611.

Seifert, G.W. (1977) The genetics of helminth resistance in cattle. *Proceedings of the 3rd International Congress of the Society for Advancement of Breeding Researches in Asia and Oceania*, Canberra, Australia, February, pp. 7(4)–7(8).

Skerman, T.M., Johnson, D.L., Kane, D.W. and Clarke, J.N. (1988) Clinical footscald and footrot in a New Zealand Romney flock: phenotypic and genetic parameters. *Australian Journal of Agricultural Research* 39, 907–16.

Smeaton, D.C., Hockey, H-U.P. and Towers, N.R. (1985) Effects of facial eczema on ewe reproduction and ewe and lamb live weights. *Proceedings of the New Zealand Society of Animal Production* 45, 133–5.

Smit, H. and Wickham, B.W. (1986) Prediction of changes in somatic cell counts due to culling and selection. *Proceedings of the New Zealand Society of Animal Production* 46, 77–81.

Solbu, H. (1984) Disease recording in Norwegian dairy cattle. 2. Heritability estimates and progeny testing for mastitis, ketosis and "all diseases". *Zeitschrift fur Tierzuchtung und Zuchtungsbiologie* 101, 51–8.

Teale, A.J., Kemp, S.J. and Morrison, W.I. (1991) The major histocompatibility complex and disease resistance in cattle. In: Owen, J.B. and Axford, R.F.E. (eds), *Breeding for Disease Resistance in Farm Animals.* CAB International, Wallingford, pp. 86–99.

Towers, N.R. and Stratton, G.C. (1978) Serum gamma-glutamyltransferase as a measure of sporidesmin-induced liver damage in sheep. *New Zealand Veterinary Journal* 26, 109–12.

Towers, N.R. and Wesselink, C. (1989) Facial eczema in a 'resistant' flock. In: Southey, C.A. (ed.), *Facial Eczema: Breeding for Resistance,* Ministry of Agriculture and Fisheries, Pukekohe, New Zealand, pp. 14–15.

Upreti, G.C., Riches, P.C., Morris, C.A. and Gravett, I.M. (1990) Prediction of resistance/susceptibility to facial eczema, a hepatotoxic disease of sheep, by studies on erythrocyte membranes. *Proceedings of the International Conference on Biomembranes in Health and Disease,* India, November 1988 (in press).

Warren, H., Daniell, D. and Parker, A.G.H. (1990). Where we are at in sheep breeding. *Proceedings of the Australian Association of Animal Breeding and Genetics* 8, 201–3.

Wharton, R.H., Utech, K.B.W. and Turner, H.G. (1970) Resistance to the cattle tick, *Boophilus microplus* in a herd of Australian Illawarra Shorthorn cattle: its assessment and heritability. *Australian Journal of Agricultural Research* 21, 163–81.

Wickham, B.W. (1979) Genetic parameters and economic values of traits other than production for dairy cattle. *Proceedings of the New Zealand Society of Animal Production* 39, 180–93.

Wiener, G., Woolliams, J.A. and Woolliams, C. (1985) Genetic selection to produce lines of sheep differing in plasma copper concentrations. *Animal Production* 40, 465–73.

Windon, R.G., Dineen, J.K. and Wagland, B.M. (1987) Genetic control of immunological responsiveness against the intestinal nematode *Trichostrongylus colubriformis* in lambs. In: B.J. McGuirk (ed.), *Merino Improvement Programs in Australia,* Australian Wool Corporation, Melbourne, pp. 371–5.

Woolaston, R.R. (1990) Genetic improvement of resistance to internal parasites in sheep. *Proceedings of the Australian Association of Animal Breeding and Genetics* 8, 163–71.

Woolliams, C., Suttle, N.F., Woolliams, J.A., Jones, D.G. and Wiener, G. (1986) Studies on lambs from lines genetically selected for low and high copper status. 1. Differences in mortality. *Animal Production* 43, 293–301.

Section 3
Breeding for Resistance to Helminths

Helminth parasites have had a long association with their hosts – chiefly grazing animals under extensive conditions – but the interference of man in seeking to control and intensify animal systems has led to major problems of worm infestation. The set-back inflicted on helminths through the advent of anthelmintics is now seen, throughout the world, to have been only a temporary respite as drug resistance has been developed by the parasites and is now widespread. Fortunately, there is much evidence, as illustrated in this section, of genetic variation in the resistance of farm animals to helminths which can be harnessed for the future benefit of pastoralists.

Recent work, primarily in Australia, has seen the establishment of sheep flocks selected for resistance or susceptibility to various important helminths. These experiments have shown significant selection responses in the case of *Haemonchus, Ostertagia* and *Trichostrongylus* species and are also yielding invaluable information which will be needed for the development of cost-effective procedures for low anthelmintic input systems.

In Europe, active research into the parasites of cattle, sheep and goats has confirmed the presence of drug-resistant parasites and also the important possibilities for genetic improvement, particularly in relation to the development of acquired resistance. An encouraging feature is that selection of animals for low faecal egg counts bolsters innate resistance and simultaneously reduces pasture contamination. There is also welcome evidence that selection for resistance to one form of parasite can confer resistance to other forms.

Chapter 9
Breeding for Resistance to Trichostrongyle Nematodes in Sheep

G. Douglas Gray
Department of Animal Science, University of New England,
Armidale, NSW 2351, Australia

Summary

Sheep industries throughout the world are seeking alternative approaches to parasite control because of production losses due to parasitism, increasing frequency of drug-resistant parasites and consumer demand for minimal chemical use in sheep production. Increasing flock resistance by genetic improvement is one such alternative. The evidence for genetic variation in resistance both between and within breeds is extensive but in order to exploit this variation effectively there is a need for better methods of selecting superior breeding animals at an early age and to identify the effects of selection for production, resistance to other sheep diseases, and any adaptation of the parasite to increased levels of host resistance. An additional but very important aspect of research on genetically resistant flocks and breeds is that they are tools by which mechanisms of resistance can be investigated, leading to improved knowledge of how vaccines against trichostrongyles may be developed or their performance enhanced.

Within flocks it has been shown that selection for resistance, by selecting young lambs on the basis of faecal egg output, would result in improved productivity in parasitized environments, and would also predict improved resistance in older sheep, including the periparturient rise in ewes. Corresponding changes in resistance to other sheep diseases and possible adaptation by the parasite have not been fully investigated but that information is critical before parasite resistance can be included in industry breeding objectives.

Genetic improvement will not be the only method of reducing parasitism in any strategic control programme, but will be integrated with more effective use of chemicals, improved husbandry and pasture management, possibly vaccination and any other innovations in worm control.

Introduction

Trichostrongyle nematodes are major pathogens of sheep in most parts of the world. They cause diseases of the gastrointestinal tract which are expensive to prevent and treat, and infections can result in lost production and death (Barger, 1982; Beck *et al.*, 1985). Chemicals are, however, available which kill almost all the parasitic stages of all species of importance. Why then is it necessary to consider genetic improvement as an alternative method of control? Even with chemical control, substantial production losses can occur, current broad spectrum drenches are not effective in some important sheep-rearing areas, and there is increasing demand for sheep products which are free of potentially harmful chemicals. Further, no commercial vaccines for the prevention of gastrointestinal trichostrongyles have yet been developed.

Nowhere in the world are these problems more intensely focused than in Australia and New Zealand where parasites have developed resistance to modern anthelmintics. The long growing seasons and improved pastures which are so suitable for the production of high quality meat and wool also aid the development of the non-parasitic stages of trichostrongyles. The consequent need for regular, frequent drenching has resulted in a dramatic increase in the incidence of drench-resistant parasite populations (Waller, 1987, 1990). As a result, breeding for resistance to sheep nematodes is a high research priority for sheep producers. The major questions being asked by industry will be addressed in this chapter and are as follows:

1. How can resistant sheep be selected?
2. What degree of control can be obtained?
3. What are the consequences of increasing resistance for production and animal health?

Sources of genetic variation in resistance to parasites

Between breeds

There is substantial evidence for variation between breeds in resistance to *Haemonchus contortus* and *Ostertagia circumcincta* (reviewed by Barger, 1989; Gray, 1987; Gruner and Cabaret, 1988; Piper, 1987) and to a lesser extent *Trichostrongylus* spp. (Ross, 1970). The usefulness of resistance genes present in other breeds depends very much on the breed of sheep currently being used and the nature of the product. It makes sense for a meat producer to change breeds, for example, if Dorsets were shown to be more resistant than Border Leicesters. But the large differences between breeds have been found in comparisons between breeds with quite different production characteristics. For example, Red Masai (Preston and Allonby, 1978, 1979), Florida Native, St Croix and Barbados Blackbelly (Courney *et al.*, 1984) are breeds with coarse coloured wool and have all been shown to be highly resistant to *H. contortus* in comparison to European breeds and to Rambouillets or Merinos (Gray *et al.*, 1987; Table 9.1). Yet it is highly unlikely that a producer of high quality wool would undertake the lengthy process of interbreeding to acquire only the genetic characteristic of parasite resistance.

Table 9.1 summarizes many of the publications on breed variation in resistance to trichostrongyles. None of these studies take account of variation among sires within each of the breeds compared. The magnitude of between-sire differences is of the same order as the largest of the between-breed differences (Gray *et al.*, 1987) and in general, conclusions that the observed differences are due to breed rather than to the use of resistant sires should be treated with caution.

It is from such breed comparisons in sheep and cattle that it has been suggested that the adaptation of tropical breeds to parasitic stresses has resulted in breeds with low production potential (Frisch, 1981; Frisch and Vercoe, 1984) and that selective breeding for resistance within populations would result in reduced production. There is substantial evidence to support the contrary view (Piper, 1987), but it is certainly true that these tropical sheep breeds are not suitable for most non-tropical production systems. They are of course of intense interest: they are highly resistant and to find out why may help us to understand how sheep become resistant to parasites and how resistance genes are expressed in other breeds.

Between strains

There is little published information on variation between strains of single breeds mainly because of the difficulty in establishing flocks in which proper genetic comparisons can be made. One study in New Zealand (Watson, *et al.*, 1986) showed significant differences among five strains of Romneys, a Romney × Coopworth and a Romney × Border Leicester cross. An attempt to compare strains of Merinos (Gray *et al.*, 1990) failed due to the persistence of environmental effects before lambs were tested in a common environment. A further study of eight strains of Merinos, tested primarily to look for heterosis in resistance (Gray and Raadsma, 1986) failed to find any significant between-strain differences. Further genetic analyses of these data from 37 sires and 862 progeny over two years indicated that the heritability of resistance to artificial infection with

Table 9.1. Breeds of sheep with high levels of resistance in lambs to trichostrongyle nematodes.

Breed (n^a)	Comparison breeds (n^a)	Type of infection	Parasite species[d]	Age (months)	Reference
Romney	Rambouillet British Breeds	Field	Oc	6/20	Stewart et al. (1937)
Rambouillet (3)	Romney/Cheviot	Field	Hc	rams	Warwick et al. (1949)
Florida Native	Rambouillet	Field	Hc	'lambs'	Loggins et al. (1965)
Targhee[b] (8)	Suffolk (8)	Artificial	Oc + Hc	3–11	Scrivner (1967)
Merino	Targhee	Field	Hc	2–7	Colglazier et al. (1968)
Florida Native (120)	Rambouillet (60)	Field	Hc	ewes	Jilek and Bradley (1969)
Scottish Blackface	Dorset	Artificial	Tc	lambs	Ross (1970)
Florida Native (19)	Rambouillet (8)	Artificial	Hc	5	Rhadakrishnan et al. (1972)
Florida Native (33)	Rambouillet (8)	Artificial	Hc	5	Bradley et al. (1973)
Cigaja (10)	Merino (10)	Field	Strongylid	12	Cvetkovic et al. (1973)
Navajo (24)	Suffolk (11)				Knight et al. (1973)
	Rambouillet (23) Targhee (15) Corriedale (15)	Artificial	Hc	4–5	
Scottish Blackface (24)	Finn Dorset (22)	Artificial	Oc	7–10	Altaif and Dargie (1978)
Red Masai (16)	Merino (6)	Artificial	Hc	6–24	Preston and Allonby (1978)
Merino (5)	Awassi (15)	Artificial	Hc	5–6	Al-Khshali and Altaif (1979)
Columbia (50)	Suffolk (50)	Field	epg	ewes	Norman and Hohenboken (1979)
Red Masai (10)	Merino (120)	Field	Hc	ewes	Preston and Allonby (1979)
Barbados Blackbelly	British	Field/Art.	Cooperia Tsp. Osp		Yazwinski et al. (1979)
Dorset (87)	Crossbred (89)				
Barbados Blackbelly (26)	British + Crossbred (26)	Artificial	Hc	lambs	Yazwinski et al. (1981)

Table 9.1. *Contd.*

Breed (n^a)	Comparison breeds (n^a)	Type of infection	Parasite species[d]	Age (months)	Reference
Border Leicester X Merino (66)	Merino (66)	Artificial	*Osp*	ewes	Donald *et al.* (1982)
Florida Native (5)	Domestic (5)	Field	*Hc*	ewes	Courtney *et al.* (1985a)
St Croix (4)					
Florida Native (10)	Domestic[c] (14)	Artificial	*Hc*	5–6	Courtney *et al.* (1985b)
St Croix (9)					
Lacaune (50)	Romanov (50)	Field	*Oc, Nsp*	ewes	Gruner *et al.* (1986)
Florida Native	Domestic	Field	*Hc*	ewes	Zajac *et al.* (1988)

[a]Number of sheep of each breed.

[b]One Targhee + two Suffolk rams were compared.

[c]Rambouillet/Finn/Suffolk and Dorset-cross.

[d]*Hc*: *Haemonchus contortus*; *Tc*: *T. colubriformis*; *Oc*: *Ostertagia circumcincta*; *Osp*: *Ostertagia* species; *Tsp*: *Trichostrongylus* species; *Nsp*: *Nematodirus* species.

H. contortus was 0.02 ± 0.08, not significantly different from zero (Gray and Raadsma, unpublished). The environment in which this study was carried out (Western New South Wales, Australia) is dry and *H. contortus* is a rare parasite which was not found in any of the lambs before the test infection was imposed. This raises the question of what kind of stimulus is required to 'switch on' resistance genes but it leaves the question of strain variation in Merinos unanswered.

Within flocks

Estimates of the heritability of resistance to *H. contortus, T. colubriformis, Ostertagia circumcincta* and *Nematospiroides* spp. have ranged from 0.2 to 0.5. However, in view of the zero heritability estimate described above, it is pertinent to note that in all cases the sheep studied had previous exposure to the parasites with which they were later tested. Le Jambre (1978) and Watson *et al.* (1986) sampled faeces from ewes and lambs which were naturally infected with several nematode species and had been infected for several months before sampling. Windon and Dineen (1984) based their estimates of heritability of artificial infections with *T. colubriformis* in lambs which had recieved two prior vaccinations with irradiated *T. colubriformis* larvae. Albers *et al.* (1987) and Woolaston (1990) report heritabilities from artificially infected lambs which had been routinely drenched while they grazed lightly contaminated pastures. Thus there may be some debate as to whether lambs need to be 'primed' before genetic variation in resistance can be detected. This is relevant when considering the mechanisms by which resistance genes are expressed but since selection for parasite resistance is aimed at increasing resistance to on-farm disease, perhaps greatest weight should be given to estimates from naturally occurring infections.

Direct selection for resistance

Faecal egg counts

The ability to make genetic progress in resistance is totally dependent on the breeder's ability to identify superior animals from candidate breeding stock. Currently the most effective and the most studied method is to allow animals to become infected, or to infect them artificially, and estimate worm burden from the concentration of nematode eggs in the faeces several weeks after infection. Although egg output can be considered indirectly to measure the worm burden, it is also of importance *per se* (Albers and Gray, 1986), as it measures contamination of pasture which in turn determines the level of infection to which the grazing flock is exposed.

Faecal egg counts can be cheaply and easily performed at many laboratories and their use has formed the basis of most genetic studies of variation
both between and within breeds. The major disadvantages of this process
are that it is labour intensive (although it is hard to conceive a process that
uses less labour, at least in the field) and enables rankings only within each
flock or management group. Comparisons between flocks would require
extensive experiments at a single site or the use of a sire-referencing
scheme. Nevertheless, faecal egg counts are the standard against which all
potential techniques should be measured and their extensive use will
continue, at least while new techniques are being evaluated.

Indirect selection for resistance

Haemoglobin type

There has been considerable interest in haemoglobin type as a marker for
resistance to *H. contortus* with several reports indicating that sheep with
haemoglobin type AA (HbAA) are more resistant than HbAB sheep which
in turn are more resistant than HbBB sheep. Other reports have indicated
no relationship (reviewed by Windon, 1990). No general conclusion can be
made and it is possible that Hb type would only be useful in predicting
resistance to a bloodsucking parasite, such as *H. contortus*.

Ovine lymphocyte antigens

Some progress has been made in identifying single genes of the sheep major
histocompatibility complex (MHC) which may prove useful as selection
criteria. The combination of the ovine lymphocyte antigen (OLA) types,
SY1a and SY1b, in lambs vaccinated and challenged with *T. colubriformis*,
confers a higher level of resistance than in lambs which lack these alleles.
This effect has been observed both in experimental selection lines
(Outteridge *et al.*, 1985) and in the offspring of rams collected at random
from unrelated flocks (Outteridge *et al.*, 1988). A similar effect was found
in Romneys (Douch and Outteridge, 1989) which were naturally infected
with *T. colubriformis*. This combination of alleles occurs at a low frequency
in both Merinos and Romneys and therefore there is scope for substantially
increasing their frequency and therefore flock resistance. A further OLA
type, Sy6 was associated with high faecal egg count. On the other hand,
Cooper *et al.* (1989) have shown no association between a further 16 OLA
haplotypes and resistance to *H. contortus*. Riffkin and Yong (1984) found
no OLA association in sheep naturally infected with *Trichostrongylus* and
Ostertagia spp. Luffau *et al.* (1986) found no association with resistance to
H. contortus in Romanov sheep. Lack of significant associations may mean

that none exist or that the correct combination of antigen and infection was not examined. In any case much remains to be understood about the relationship between the sheep MHC and resistance.

Restriction fragment length polymorphisms

Linkage analysis of restriction fragment length polymorphisms may reveal associations with resistance to parasites and several research groups are currently engaged in this type of research. No detailed results have yet been published although Windon (1990), quoting results of D.J. Hulme of the CSIRO McMaster Laboratory in Sydney, is cautiously optimistic about one human MHC Class II probe which is associated with susceptibility to *T. colubriformis.*

Associated traits

Other polygenically determined traits which have been considered as indirect selection criteria include anaemia, blood biochemistry, productivity, presence of dags, blood eosinophilia, blood lymphocyte stimulation by parasite antigens and skin sensitivity to parasite antigen. This vast topic is being reviewed by Windon (1991). It is important to note here, however, that if any of these traits are to be used as indirect selection criteria, then the genetic correlations among resistance and all other traits in the breeding objective must be accurately estimated.

Consequences of selection for parasite resistance

Breeding objectives

Before considering the spectrum of characters that may have to be considered when breeding for resistance it is necessary to define what the successful outcome of a breeding programme might be, whether by conventional breeding for polygenic traits, the selection of single genes or markers, or by the future transfer of resistance genes alone from resistant to susceptible sheep. Only a small number of nematode species that inhabit the gastrointestinal tracts of sheep are of economic importance and the actual and relative importance of each species is dependent on the environmental influences on the free-living stages which undergo development on the pasture, and on the level of genetic and non-genetic susceptibility of the host. This last category would certainly include plane of nutrition, age and reproductive state and concurrent disease. For an individual sheep, the outcome of exposure to infective larvae is to acquire a burden of worms and it should be possible to rank sheep according to their worm burden from the most susceptible to the most resistant. This is an important point because it contrasts with other important sheep diseases (which may also be the targets for genetic improvement) such as footrot (Egerton, 1991) and blowfly strike (Raadsma, 1991) which can be categorized as 'all-or-nothing' traits: animals are either 'affected' or 'not affected'. The objective of a breeding programme therefore, may be to reduce progressively the worm burden until the parasites are eliminated from the environment or, more likely, to a level at which the cost of treatment, control and lost production is 'acceptable' or 'minimal' or perhaps too low to justify the continued cost of selection.

Piper and Barger (1988) have conceived an 'economic threshold' which is defined by the minimum numbers of worms likely to cause economic loss, for each of the three genera which are important in Australian sheep (Table 9.2). These are 'best guess' estimates and the relative importance of these parasites changes with climatic zone within Australia and throughout the world. Zones can be defined climatically but in this context it is surely more appropriate to define the parasite problem in terms of the drenching programme which has been designed to counter it (Waller, 1987). For each of these zones, drenching and other management practices, such as movement of susceptible sheep and the provision of clean pastures are designed to achieve the minimal or acceptable level of disease described above. For the animal breeder with an interest in internal parasitism, therefore, each of these zones will have its own breeding objective. The extent to which the objectives are similar depends totally on the sign and magnitude of the genetic correlations between all the traits, and on the economic weights given to the traits included in the objective.

Effects on productivity

It has been argued (Woolaston, 1990) that long-term profitability is the real objective for any production system, but that most sheep-breeding objectives include annual output of wool and meat as the major influences on

Table 9.2. Economic thresholds of three important trichostrongylid nematode species of sheep.

Parasite	Economic threshold (no. of adult worms)
Haemonchus contortus	2,000
Trichostrongylus colubriformis	10,000
Osteragia circumcincta	20,000–30,000

Source: Piper and Barger (1988).

breeding strategies. Piper and Barger (1988) used the best-available genetic correlations between parasite resistance and production traits to estimate the gains which could be made from including resistance in a current sheep-breeding objective (Atkins, 1987). The authors ignored the direct costs associated with drenching and assumed that all benefits, if any, would accrue from favourable correlations between resistance and productivity. In a typical Merino flock, exposed to average levels of infection, including faecal egg output as a selection criterion leads to a significant increase in overall genetic improvement (about 10% per year). In the absence of parasites there were very small gains. This highlights the need for the emphasis given to selecting for parasite resistance to be matched to the environment in which the sheep will produce, not just where they are bred. Missing from the study by Piper and Barger were the savings resulting from reduced frequency of drenching. But by far the greatest pressure to include parasite resistance in a breeding objective may well come, not from marginal economic gains, but to prevent the catastrophic losses which may occur when there is resistance to all anthelmintics in the parasite population. There is general agreement that, in sheep, genetic correlations between faecal egg output and production traits such as wool quality and growth and liveweight gain are zero when animals are uninfected and moderately favourable when animals are infected (Albers *et al.*, 1987; Woolaston, 1990). An additional feature of these data is the large standard errors associated with the genetic correlations. Ponzoni (1987) emphasizes the need for accurate estimates for inclusion in industry breeding schemes such as WOOLPLAN in Australia and this requires large sets of data, preferably from several independent flocks in different environments.

Effects on reproduction

Reproduction rate is weighted very heavily in current Merino breeding programmes (Ponzoni, 1987). Correlations between parasite resistance and reproduction rate in sheep are few and variable. Woolaston (1990) gives two estimates of r_g from two different flocks of -0.22 and $+0.64$, but remains optimistic because in a selection line which had been bred for high resistance to *H. contortus* (Woolaston, 1990) reproduction rate was unaffected. It was the line bred for susceptibility which lost fertility. Nevertheless, accurate estimates must be obtained if resistance is to be included in conventional breeding objectives.

Cattle bred for high fertility and high growth rates in tropical Australia have been shown to be less resistant to worm infections (McKinnon, 1990) but more resistant to tick infestation. This effect was more noticeable in *Bos indicus* than in *Bos taurus*, the former having higher levels of resistance to both worms and ticks. The author concludes that in parasitized environments, susceptible breeds (*Bos taurus*) gain most by acquiring resistance genes as the effect of these genes is to reduce the pathogenic effects of disease and therefore both growth and fertility may increase. In resistant breeds (*Bos indicus*), or as selection proceeds towards resistance, the gains from having fewer parasites are less and the true genetic relationship with reproduction is 'unmasked'.

Resistance to other diseases

Polymorphism in resistance to disease has been postulated to account for speciation, sexual dimorphism and social behaviour (Keymer and Read, 1990) and Clarke (1979) argues strongly that parasitism or disease is a major influence in generating and maintaining genetic diversity within species. Certainly one of the features of the genes which regulate the host response to disease is the high degree of heterogeneity in chromosome regions which control disease resistance, which includes the MHC (May, 1985). If disease resistance has been the driving force in generating diversity, what then are the consequences if we seek to restrict this diversity by selecting high, and perhaps more uniform, levels of production, and 'complete' resistance to disease, where animals are not affected pathologically. Studies on sheep have not yet provided an answer to this question. Two species have been extensively studied: mice and chickens. Mice selected for high and low levels of circulating antibody to foreign red cells (Biozzi, 1982) have been exposed to a wide range of infections and test of immune function. High Ab level mice are more resistant to malaria, trypanosome and worm infections but less resistant to infections with *Salmonella*, *Brucella*, *Mycobacterium*, *Yersinia* and *Leishmania*. In parallel studies, chickens selected in a similar way for high levels of Ab (van der Zijpp, 1989) have been shown to be more resistant to Marek's disease than those selected for low Ab levels. On the other hand, for example, Gross *et al.* (1988) have shown that those high Ab chickens are more susceptible to avian adenovirus 2. Buschman and Meyer (1989) argue for selecting pigs on the basis of a range of immune functions, to create an 'index of immune competence'. Among the 18 traits included in the index, there were 18 positive correlations, six negative correlations and 282 correlations were not significant. In the cattle study by McKinnon (1990) there was a negative correlation between resistance to ticks and resistance to worms. The overall lack of positive genetic correlations among immune traits and the possibility of negative correlations among resistance to important sheep diseases may indicate that conventional selection methods are not appropriate (Rothschild, 1989) for improving the genetic level of disease resistance.

The message from these studies is a cautionary one, that it should not necessarily be expected that sheep selected for resistance to one parasite, will be resistant to others and to other diseases and ailments which are controlled or caused by the immune system. The evidence is encouraging that the genetic correlations between resistance to gastrointestinal parasites are positive (Windon, 1990). There is, however, no published information on genetic correlations between worm resistance and the other important sheep diseases which, in Australia, are blowfly strike, fleece rot, footrot, salmonellosis and clostridial disease. Again it is emphasized that if disease resistance is to be included in a breeding objective, accurate estimates of the correlations among all the disease and important production traits must be known, and as we have seen, these may differ for each disease environment.

Age dependence of resistance

Attention has been focused on the response of lambs to acute infections of trichostrongyles. In lambs these parasites can cause acute disease and high mortality but older animals are also at risk. Immunity to *Haemonchus*, for example is weak and short lived, and may be lost after anthelmintic treatment (Barger, 1988) and adult ewes and rams can suffer from haemonchosis. So if the benefits of selection are to be fully realized they must persist throughout adult life. Extensive experiments have been carried out in Armidale on the UNE Resistance Flock in which there are highly resistant animals, all related to a founder ram, the Golden Ram (Albers *et al.*, 1987). It is pertinent to note here that although highly resistant, the offspring of this ram have not yet been shown to carry a single major resistance gene and on data collected so far the basis of resistance in this flock is likely to be polygenic (Woolaston *et al.*, 1990a). Table 9.3 summarizes experiments carried out on lambs and adult sheep which were infected with a single large dose of infective larvae. At all ages up to one, and at five

Table 9.3. Comparison of faecal egg count in resistant Merinos of the UNE Resistance Flock at different ages. Egg output was measured 5 weeks after a single infection of 10,000–25,000 infectious larvae of *Haemonchus contortus*.

Age of sheep	Resistant (n[a])	Unselected (n[a])	Reference
5– 8 months	2,600 (17)	12,700 (882)	Albers *et al.* (1987)
5–10 months	3,844 (49)	10,400 (49)	Gray *et al.* (unpublished)
10–12 months	4,200 (6)	9,900 (5)	Presson *et al.* (1988)
5 years	320 (15)	4,840 (20)	Gray and Thamsborg (unpublished)
7 years	3,840 (5)	3,972 (18)	Gray and Thamsborg (unpublished)

[a]No. of sheep infected.

years of age, the offspring of resistant parents have many fewer worms than unselected counterparts. At seven years of age, at the end of the ewes' useful life, the difference has disappeared. So resistance measured as a weaned lamb is an indicator to resistance levels for most of the productive life of the ewes. Woolaston (1990a) reports similar evidence for *Haemonchus*-resistant lambs from weaning to one year of age. This is important not only because the ewes are a source of wool and lambs, but also as an important source of infective larvae for lambs. Around parturition, egg output from the ewe increases. In Fig. 9.1 egg counts from 382 lambs artificially infected with *H. contortus* in 20 sire groups, have been correlated with the egg counts resulting from natural infection of 92 of these lambs two years later just after dropping their first lamb. Although the correlation is not strong, it is positive and again encourages the use of measurements of resistant lambs to predict resistance as adults. Florida Native ewes have been shown not to undergo a periparturient rise in egg count when compared with Rambouillet crosses (Courtney *et al.*, 1984). However, non-lambing ewes of both breeds of similar age, when artificially infected with *H. contortus*, were equally resistant (Courtney *et al.*, 1985b; Zajac *et al.*, 1988).

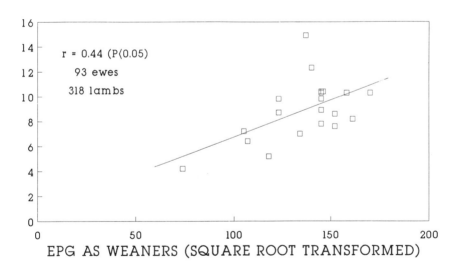

EPG AS WEANERS (SQUARE ROOT TRANSFORMED)

—□— Sire Group Means

Fig. 9.1. Sire group (UNE Resistance Flock) mean egg counts (20) of 318 lambs artificially infected with *Haemonchus contortus* one month after weaning and the mean natural worm egg counts, from the same sire groups, of 93 of these lambs 18 months later, just after their first lambing. (Data from Albers and Gray, unpublished)

Genetic resources for investigating resistance to parasites

Genetically resistant sheep have three important uses. First, when compared with normal or genetically susceptible animals, they provide tools for investigating the immune responses of sheep to parasite infections. This may lead to the development of new selection criteria for identifying resistant animals and assist in the development of new vaccines or improvements in their effectiveness. Second, they allow the estimation of the nature and magnitude of the correlated responses to other important production and disease traits. Third, comparison between resistant sources and the interbreeding of these resources may lead to better understanding of the genetic control of resistance and possibly to the more rapid developments of resistant bloodlines which can be released to industry.

In Australia and New Zealand there are six flocks which are being experimentally selected for resistance to trichostrongyles (Table 9.4). Progress in each flock varies but together they represent a substantial resource for research on the genetic basis and immune processes of resistance. At least three of these are sources of highly resistant lambs, perhaps sufficiently resistant to satisfy the demands of the 'minimal' breeding objective outline above. The levels of resistance of two *Haemonchus* resistant flocks after artificial and natural challenge are shown in Fig. 9.2. Red Masai and St Croix can be almost refractory to *H. contortus*. But there is little information on the nature of the resistance genes in each population or on the mechanisms by which they are expressed. Are they all the same?. At one extreme, different genes could be present in each population, leading to the intriguing possibility that their combination may further increase resistance. Alternatively there is the encouraging prospect that the genes may be identical and that these resistance genes are likely to be present in many sheep populations, even if at low frequencies. Although difficult, comparisons between these sources would, as comparisons between inbred populations of mice have shown (Wakelin, 1985), accelerate our understanding of parasite resistance.

Parasite control using resistance genes

There is no doubt that genes can control worm infections in sheep but when considering the scope for their practical use in industry it is useful to consider genetic control in the same way as control by a new anthelmintic. Although slightly different questions have to be answered, it would be poor marketing, if not illegal in most countries, to market a new drench whose range of efficacy, for example, had not been fully evaluated. The same is true for breeding for parasite resistance and Table 9.5 compares the problems associated with each approach.

Table 9.4. Trichostrongyle selection flocks in Australia and New Zealand.

Location	Date started	Selection criteria	Breed	Reference
Armidale, NSW	1976	Vaccination + challenge with *Tc*	Medium-wool Merino	Windon (1990)
Armidale, NSW	1977	Artificial challenge with *Hc*	Fine-wool Merino	Woolaston *et al.* (1990a)
Armidale, NSW	1980	Artificial challenge with *Hc*	Fine-medium wool Merino	Albers *et al.* (1987)
Hamilton, Vic.	1985	Artificial *Oc* infection + lymphocyte stimulation assay	Medium Merino	Woolaston (1990)
Ruakura/Wallaceville New Zealand	1987	Natural egg count (*Tc* + *Oc*)	Romney	Baker *et al.* (1990)
Rylington Park WA	1987	Index of natural egg count (*Tc* + *Oc*), liveweight gain + wool production	Mixed Merinos	Woolaston (1990)

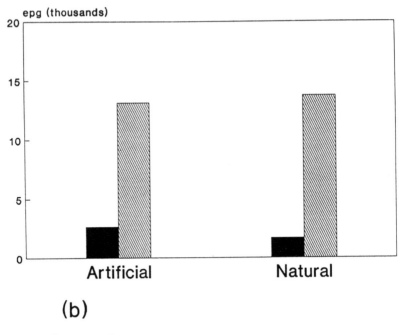

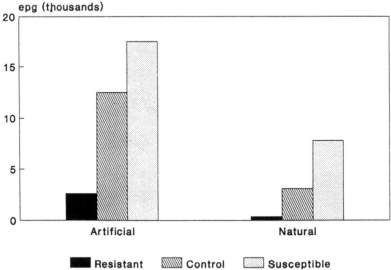

Fig. 9.2. Worm egg output 5–6 weeks after artificial infection with *Haemonchus contortus* or exposure to *H. contortus*-contaminated pasture in two genetically resistant flocks. (a) UNE resistance flock; (b) CSIRO resistance flock, (Data from Albers *et al.*, 1987; Gray unpublished; Woolaston *et al.*, 1990a).

Table 9.5. Comparison of some important considerations in the introduction of a new drug or of genetic control of parasites in sheep.

	Drench	Genetic improvement
Cost	Price of drench	Laboratory costs for measurement e.g. epg, OLA assay
Labour	Mustering/drenching	Mustering, sampling, laboratory analysis, data processing
Range of efficacy	Effects on related parasites – broad spectrum?	Genetic correlations with resistance related parasites and diseases
Toxicity	Effect on immune system – product quality/quantity – welfare	Genetic correlations with production, possible reduction in genetic diversity
Resistance	Adaption of parasite to drench	Adaption of parasite to increased host resistance

Conclusion

Improving levels of genetic resistance to trichostrongyle parasites should be regarded as another potential control method together with new drugs, vaccines and improved husbandry techniques. The long-term benefits of genetic improvement: low cost and reduced maintenance and absence of chemicals from the production system are offset to some extent by the long period required for selective breeding by conventional methods, and the uncertain response of the parasite population to increased host resistance. The use of single resistance genes or highly resistant alternative sources of breeding stock, may accelerate the process.

The problems created by the appearance of drench resistance has not only generated interest in genetic control of parasitism, but also in the integrated use of effective drenches, pasture management, nutrition and husbandry. The trend away from sheep monocultures which are supported by frequent use of chemicals could have many benefits, not only for parasite control, but also for the long-term sustainable use of pastures.

References

Albers, G.A.A. and Gray, G.D. (1986) Breeding for worm resistance: a perspective. *International Journal for Parasitology* 17, 559–66.

Albers, G.A.A., Gray G.D., Piper, L.R., Barker, J.S.F., Le Jambre. L.F. and Barger, I.A. (1987) The genetics of resistance and resilience to *Haemonchus contortus* infection in young Merino sheep. *International Journal for Parasitology* 17, 1355–63.

Al-Khshali, M.N. and Altaif, K.I. (1979) The response of Awassi and Merino sheep to primary infection with *Haemonchus contortus. Tropical Animal Health and Production* 11, 164–170.

Altaif, K.I. and Dargie, J.D. (1978) Genetic resistance to helminths. The influence of breed and haemoglobin type on the response of sheep to primary infections with *Haemonchus contortus. Parasitology* 77, 161–75.

Atkins, K.D. (1987) Potential responses to selection in Merino sheep given current industry structure and selection practices. In: McGuirk, B.J. (ed.), *Merino Improvement Programs in Australia* Australian Wool Corporation, Melbourne, pp. 299–312.

Baker, R.L., Watson, T.G., Bissett, S.A. and Vlassoff, A. (1990) Breeding Romney sheep which are resistant to gastrointestinal parasites. In: *Proceedings of the 8th Conference of the Australian Association of Animal Breeding and Genetics*, Ruakura, pp. 173–8.

Barger, I.A. (1982) Helminth parasites and animal production, In: Symons, L.E.A., Donald, A.D. and Dineen, J.K. (eds), *Biology and Control of Internal Parasites*, Academic Press, Sydney, pp. 133–55.

Barger, I.A. (1988) Resistance of young lambs to *Haemonchus contortus* infection, and its loss following anthelmintic treatment. *International Journal for Parasitology*, 18, 1107–9.

Barger, I.A. (1989) Genetic resistance of hosts and its influence on epidemiology. *Veterinary Parasitology* 32, 21–35.

Beck, A., Moir, B. and Meppem T. (1985) The cost of parasites to the Australian Sheep Industry. *Quarterly Review of the Rural Economy* 7, 336–43.

Biozzi, G. (1982) Correlation between genetic regulation of immune responsiveness and host defence against infections and tumours. *European Journal of Clinical Investigation* 12, 373–6.

Bradley, R.E., Rhadakrishnan, C.V., Patil-Kulkarni, V.G. and Loggins, P.E. (1973) Responses in Florida Native and Rambouillet lambs exposed to one and two oral doses of *Haemonchus contortus*. *American Journal of Veterinary Research* 34, 729–35.

Buschmann, H. and Meyer, J. (1989) An immune competence profile in swine. In: van der Zijpp, A.J. and Sybesma, W. (eds), *Improving Genetic Disease Resistance in Farm Animals*, Kluwer, Dordrecht, pp. 145–52.

Clarke, B.C. (1979) The evolution of genetic diversity. *Proceedings of the Royal Society of London B* 205, 453–74.

Colglazier, M.L., Lindahl, I.L., Turner, J.H., Wilson, G.I., Whitmore, G.E. and Wilson, R.L. (1968) Effect of management systems on the growth of lambs and development of internal parasitism. 2. Field trials involving medication with national formulary and purified grades of phenothiazine. *Journal of Parasitology* 54, 89–97.

Cooper, D.W., van Oorschot, R.A.H., Piper, L.R. and Le Jambre, L.F. (1989) No association between the ovine leucocyte antigen (OLA) system in the Australian Merino and susceptibility to *Haemonchus contortus* infection. *International Journal for Parasitology* 19, 695–7.

Courtney, C.H., Parker, C.F., McLure, K.E. and Herd, R.P. (1984) A comparison of the periparturient rise in fecal egg counts of exotic and domestic ewes. *International Journal for Parasitology* 14, 377–81.

Courtney, C.H., Parker, C.F., McLure, K.E. and Herd, R.P. (1985a) Resistance of exotic and domestic lambs to experimental infection with *Haemonchus contortus*. *International Journal for Parasitology* 15, 101–9.

Courtney, C.H., Parker, C.F., McLure, K.E. and Herd, R.P. (1985b) Resistance of nonlambing exotic and domestic ewes to naturally acquired gastrointestinal nematodes. *International Journal for Parasitology* 15, 239–43.

Cvetkovic, Li., Lepojez, O. and Vulic, I. (1973) Investigation of resistance of breeds of home produced sheep Cigaja, Merino Prekoce and Caucasus Merino to gastrointestinal strongyles under natural conditions of infection. *Veterinarski glasnik* 12, 867–72.

Donald, A.D., Morley, F.W.H., Waller, P.J., Axelson, A., Dobson, R.J. and Donnelly, J. (1982) Effects on reproduction, genotype and anthelmintic treatment of ewes on *Ostertagia* spp. populations. *International Journal for Parasitology* 12, 403–11.

Douch, P.G.C. and Outteridge, P.M. (1989) The relationship between ovine lymphocyte antigens and parasitological and production parameters in Romney sheep. *International Journal for Parasitology* 19, 35–41.

Egerton, J.R. and Raadsma, H.W. (1991) Breeding sheep for resistance to footrot. In: Owen, J.B. and Axford, R.F.E. (eds). *Breeding for Disease Resistance in Farm Animals*, CAB International, Wallingford, pp. 347–70.

Frisch, J.E. (1981) Changes occurring in cattle as a consequence of selection for growth rate in a stressful environment. *Journal of Agricultural Science, Cambridge* 96. 23–38.

Frisch, J.E. and Vercoe, J.E. (1984) An analysis of growth of different cattle genotypes reared in different environments *Journal of Agricultural Science, Cambridge* 96, 23–38.

Gray, G.D. (1987) Genetic resistance to haemonchosis in sheep. *Parasitology Today* 8, 253–5.

Gray, G.D., Albers, G.A.A. and Piper, L.R. (1990) Variation in resistance to internal parasite among sources of Australian Merinos. In: *Proceedings of the 4th World Congress of Genetics Applied to Livestock Production*, XV, 135–8.

Gray, G.D., Presson, B.L., Albers, G.A.A., Le Jambre, L.F., Piper, L.R. and Barker, J.S.F. (1987) Comparison of within- and between-breed variation in resistance to haemonchosis in sheep. In: McGuirk B.J. (ed.), *Merino Improvement Programs in Australia*, Australian Wool Corporation, Melbourne, pp. 365–9.

Gray, G.D. and Raadsma, H.W. (1986) Resilience to *Haemonchus contortus* infection in lambs crossbred between Merino strains and bloodlines. In: *Proceedings of the 3rd World Congress of Genetics Applied to Livestock Production*, Part XII, pp. 691–6.

Gross, W.B., Domermuth, C.H. and Siegel, P.B. (1988) Genetic and environmental effects on the response of chickens to avian adenovirus group 2 infection. *Avian Pathology* 17, 767–74.

Gruner, L. and Cabaret, J. (1988) Resistance of sheep and goats to helminth infections: a genetic basis. In: Thomson, E.F. and Thomson, F.S. (eds), *Improving Small Ruminant Productivity in Semi-Arid Areas*, Kluwer, Dordrecht pp. 257–65.

Gruner, L., Cabaret, J., Sauve, C. and Pailhories, R. (1986) Comparative susceptibility of Romanov and Lacaune sheep to gastrointestinal nematodes and small lungworms. *Veterinary Parasitology* 19, 85–93.

Jilek, A.F. and Bradley, R.E. (1969) Haemoglobin types and resistance to *Haemonchus contortus*. *American Journal of Veterinary Research* 30, 1773–8.

Keymer, A.E. and Read, A.F. (1990) The evolutionary biology of parasitism. *Parasitology Today* 6, 2–3.

Knight, R.A., Vegors, H.H. and Glimp, H.A. (1973) Effects of breed and date of birth of lambs on gastrointestinal nematode infections. *American Journal of Veterinary Research* 34, 323–7.

Le Jambre, L.F. (1978) Host genetic factors in helminth control. In: Donald, A.D., Southcott, W.H. and Dineen, J.K. (eds). *The Epidemiology and Control of Gastrointestinal Parasites of Sheep*. Division of Animal Health, CSIRO, Melbourne, Australia, pp. 137–41.

Loggins, P.E., Swanson, L.E. and Koger, M. (1965) Parasite levels in sheep as affected by heredity. *Journal of Animal Science* 24, 286–7.

Luffau, G., Nguyen, T.C., Cullen, P., Vu Tien Khang, J., Bouix, J. and Ricordeau, G. (1986) Genetic resistance to *Haemonchus contortus* in Romanov sheep. In: *Proceedings of the 3rd World Congress of Genetics Applied to Livestock Production*, Part XI, 683–90.

May, R.M. (1985) Host–parasite associations: their population biology and population genetics. In: Rollinson, D. and Anderson, R.M. (eds), *Ecology and Genetics of Host–Parasite Interactions*, Academic Press, New York, 243–62.

McKinnon, M.J. (1990) Genetic relationships between parasite resistance, growth and fertility in tropical beef cattle. In: *Proceedings of the 8th Conference of the Australian Association of Animal Breeding and Genetics*, Ruakura, pp. 163–71.

Norman, L.M. and Hohenboken, W. (1979) Genetic and environmental effects on internal parasites, foot soundness and attrition in crossbred ewes. *Journal of Animal Science* 48, 1329–37.

Outteridge, P.M., Windon, R.G. and Dineen, J.K. (1985) An association between a lymphocyte antigen in sheep and the response to vaccination against the parasite *Trichostrongylus colubriformis*. *International Journal for Parasitology* 15, 121–7.

Outteridge, P.M., Windon, R.G. and Dineen, J.K. (1988) An ovine lymphocyte marker for acquired resistance to *Trichostrongylus colubriformis*. *International Journal for Parasitology* 18, 853–8.

Piper, L.R. (1987) Genetic variation in resistance to internal parasites. In: McGuirk B.J. (ed.), *Merino Improvement Programs in Australia*, Australian Wool Corporation, Melbourne pp. 351–64.

Piper, L.R. and Barger, I.A. (1988) Resistance to gastrointestinal strongyles: feasibility of a breeding programme. In: *Proceedings of the 3rd World Congress of Sheep and Cattle Breeding*, Paris, Vol. 1, pp. 593–611.

Piper, L.R., Le Jambre, L.F., Southcott, W.H. and Ch'ang, T.S. (1978) Natural worm burdens in Dorset Horn, Merino and Corriedale Weaners and their crosses. *Proceedings of the Australian Society of Animal Production* 12, 276.

Ponzoni, R.W. (1987) Woolplan – design and implication for the Merino industry, In: McGuirk B.J. (ed.), *Merino Improvement Programs in Australia*, Australian Wool Corporation, Melbourne, pp. 25–40.

Presson, B.L., Gray, G.D. and Burgess, S.K. (1988) The effect of immunosuppressive therapy with dexamethasone on *Haemonchus contortus* infections in genetically resistant sheep. *Parasite Immunology* 10, 675–80.

Preston, J.M. and Allonby, E.W. (1978) The influence of breed on the susceptibility of sheep and goats to a single experimental infection with *Haemonchus contortus*. *Veterinary Record* 103, 509–12.

Preston, J.M. and Allonby, E.W. (1979) The influence of breed on the susceptibility of sheep to *Haemonchus contortus* infection in Kenya. *Research in Veterinary Science* 26, 134–9.

Raadsma, H.W. (1991) Genetic variation in resistance to fleece rot and fly strike in sheep. In: Owen, J.B. and Axford, R.F.E. (eds), *Breeding for Disease Resistance in Farm Animals*, CAB International, Wallingford, pp. 263–90.

Rhadakrishnan, C.V., Bradley, R.E. and Loggins, P.E. (1972) Host responses of worm-free Florida Native and Rambouillet lambs experimentally infected with *Haemonchus contortus*. *American Journal of Veterinary Research* 33, 817, 23.

Riffkin, G.G. and Yong, W.K. (1984) Recognition of sheep which have innate resistance to trichostrongylid nematode parasites. In: Dineen, J.K. and Outteridge, P.M. (eds), *Immunogenetic Approaches to the Control of Endoparasites*. CSIRO, Melbourne, pp. 30–40.

Ross, J.G. (1970) Genetic differences in the susceptibility of sheep to infection with *Trichostrongylus axei*. A comparison of Scottish Blackface and Dorset breeds. *Research in Veterinary Science* 11, 465–8.

Rothschild, M.F. (1989) Selective breeding for immune responsiveness and disease resistance in livestock. *Agbiotech News and Information* 1, 355–60.

Scrivner, L.H. (1967) Genetic resistance to ostertagiasis and haemonchosis in lambs. *Journal of the American Veterinary Medical Association* 151, 1443–6.

Stewart, M.A., Miller, R.F. and Douglas, J.R. (1937) Resistance of sheep of different breeds to infestation by *Ostertagia circumcincta. Journal of Agricultural Research* 55, 923–30.

van der Zijpp, A.J. (1983) Breeding for immune responsiveness and disease resistance. *World Poultry Science Journal* 39, 118–31.

Wakelin, D. (1985) Genetic control of immunity to helminth infections. *Parasitology Today* 1, 17–23.

Waller, P.J. (1987) Anthelmintic resistance and the future for roundworm control. *Veterinary Parasitology* 25, 177–91.

Waller, P.J. (1990) Resistance in nematode parasites of livestock to the benzimidazole anthelmintics. *Parasitology Today* 6, 127–9.

Warwick, B.L., Berry, R.O., Turk, R.D. and Morgan, C.O. (1949) Selection of sheep and goats for resistance to stomach worms, *Haemonchus contortus. Journal of Animal Science* 8, 609–10.

Watson, T.G., Baker, R.L. and Harvey, T.G. (1986) Genetic variation in resistance or tolerance to internal parasites in strains of sheep at Rotomahana. *Proceedings of the New Zealand Society of Animal Production* 46, 23–6.

Windon, R.G. (1990) Selective breeding for the control of nematodiasis in sheep. *Revues Scientifiques et Technicales de l'Office Internationale de Epizooties* 2, 555–76.

Windon, R.G. (1991) Genetic control of host responses involved in resistance to gastrointestinal nematodes of sheep. In: Owen, J.B. and Axford, R.F.E. (eds). *Breeding for Disease Resistance in Farm Animals.* CAB International, Wallingford, pp. 162–86.

Windon, R.G. and Dineen, J.K. (1984) Parasitological and immunological competence of lambs selected for high and low responsiveness to vaccination with irradiated *Trichostrongylus colubriformis* larvae. In: Dineen, J.K. and Outteridge, P.M. (eds), *Immunogenetic Approaches to the Control of Endoparasites,* CSIRO, Melbourne, pp. 11–18.

Woolaston, R.R. (1990) Genetic improvement of resistance to internal parasites of sheep. *Wool Technology and Sheep Breeding* 38, 1–6.

Woolaston, R.R, Barger, I.A. and Piper, L.R. (1990a) Response to helminth infection of sheep selected for resistance to *Haemonchus contortus. International Journal for Parasitology* 20, 1015–18.

Woolaston, R.R., Gray, G.D., Albers, G.A.A., Piper, L.R. and Barker, J.S.F. (1990b) Analysis for a major gene affecting parasite resistance in sheep. In: *Proceedings of the 4th World Congress of Genetics Applied to Livestock Production,* XV, 131–4.

Yazwinski, T.A., Goode, L., Moncul, D.J., Morgan, G.W. and Linnerud, D.J. (1979) Parasite resistance in straightbred and crossbred Barbados Blackbelly sheep. *Journal of Animal Science* 49, 919–26.

Yazwinski, T.A., Goode, L., Moncul, D.J., Morgan, G.W. and Linnerud, D.J. (1981) *Haemonchus contortus* resistance in straightbred and crossbred Barbados Blackbelly sheep. *Journal of Animal Science* 51, 279–84.

Zajac A.M., Herd, R.P. and McClure, K.E. (1988) Trichostrongylid parasite populations in pregnant or lactating and unmated Florida Native and Dorset Rambouillet ewes. *International Journal for Parasitology* 18, 981–5.

Chapter 10

Genetic Control of Host Responses Involved in Resistance to Gastrointestinal Nematodes of Sheep

R.G. Windon

CSIRO Division of Animal Health, Pastoral Research Laboratory, Armidale, NSW 2350, Australia

Summary

An alternative to the current reliance on anthelmintics for the control of gastrointestinal nematodes in sheep is the exploitation of host immunological responsiveness. These responses, involving both specific and non-specific components, play a central role in the regulation of parasite burdens. However, limitations occur because some categories of sheep are susceptible due to suppressed or immature immunological control. In this respect, most attention has focused on the relative unresponsiveness of lambs, and, within this category, it appears that genetic influences may determine the degree of susceptibility. Genetic constitution may, therefore, be a major constraint on the success of any vaccination strategy.

A number of studies in laboratory rodents and domestic animals have shown that various immunological activities are under genetic control. Responses to antigens are regulated by genes associated with the major histocompatibility complex (MHC), and various immune functions can be genetically manipulated by selection. Selection for immunological responsiveness to nematodes has also been successful in laboratory animals and sheep. For sheep, lines have been established in which lambs are either responsive or susceptible after vaccination and challenge with the intestinal nematode *Trichostrongylus colubriformis*. Animals from the high responder line have increased reactivity for a wide range of immunological responses. Other studies have shown that genetically determined resistance to the abomasal nematode *Haemonchus contortus* also has an immunological basis.

The existence of lambs having defined extremes of responsiveness to parasites facilitates studies into the mechanisms of resistance. Understanding these mechanisms is crucial for assessment of the specificity of selection in programmes directed against a single parasite species, identification of predictive markers with resistance, and development of suitable vaccines and vaccination strategies in unselected populations. The success of the genetic approach, however, may be limited by the ability of the parasite to adapt to withstand host responses. This aspect is currently receiving attention in the selected lines of sheep, and although preliminary results confirm that *T. colubriformis* can quickly adapt, further work is required to evaluate critically the significance of these findings.

Introduction

Current control procedures for gastrointestinal nematodes in sheep depend heavily on the use of anthelmintics in either a curative or preventive role (Donald and Waller, 1982). This reliance has imposed strong selection pressure on the parasite to withstand the single mode of action of these chemicals resulting in the emergence of anthelmintic-resistant strains (Waller, 1987). This, together with the massive total investment required to discover, develop and register new chemicals, and the long lead times before new drenches appear on the market (Hotson, 1985; Murray, 1989), stress the importance of preserving the effectiveness of the currently available anthelmintics. In addition, chemical usage is coming under increasing scrutiny as community awareness of environmental issues intensifies. With regard to anthelmintics, such concerns may be reflected in consumer demand for animal products and pastures free of chemical residues. For these reasons, alternative control strategies are being sought which will lessen the dependence on anthelmintics and reduce the spiralling cost of parasite control.

An option currently receiving considerable attention is utilization of host immune responses to control parasite infections. However, despite intensive efforts, particularly with recombinant DNA technology, commercially viable vaccines against gastrointestinal nematodes have yet to appear. A problem to be overcome in the successful application of vaccination procedures is the modulation of immune responsiveness by a number of intrinsic and extrinsic factors. Thus, some categories of animals are more susceptible to infection because immune control is suppressed (for example, in the lactating, nutritionally deprived or stressed) or not fully developed (as in the neonate) (Dineen, 1978). Within these categories, genetic influences appear to determine the level of susceptibility (see *et al.*, 1984; Dineen *et al.*, 1978). In particular, a number of studies have focused on the responses of young lambs because of the importance of protecting this age group until immune functions mature (Gregg *et al.*, 1978; Smith and Angus, 1980). Gray (1991) has shown that differences in worm burdens, as estimated by faecal worm-egg counts (FEC), between genetically resistant and susceptible sheep, are greatest in young animals, although still expressed up to seven years of age.

A number of programmes are underway in Australia and New Zealand to study genetic control of resistance to the gastrointestinal nematodes *Haemonchus contortus* and *Trichostrongylus colubriformis* in sheep (Gray, 1991; Windon, 1990; Woolaston, 1990). These programmes have demonstrated that host resistance in lambs is amenable to genetic manipulation by selection. An important consideration in this work is to utilize the animals having defined extremes of responsiveness to facilitate an understanding of the underlying basis of the selected trait. Although knowledge of the limiting mechanisms of resistance in sheep is at present only rudimentary, such an understanding is crucial to: (1) identify a marker which will identify resistant animals without the necessity for infection with the parasite, (2) determine the specificity of selection in terms of other related and unrelated parasite species and non-parasite pathogens, and (3) assess the potential for adaptation within the parasite population to withstand host resistance mechanisms. In addition, identification of limiting mechanisms responsible for resistance may allow the opportunity for various immunomanipulations designed to induce responsiveness in susceptible individuals of an outbred population. In this chapter genetic control of immunological responsiveness is examined with particular reference to those responses implicated in resistance to gastrointestinal nematodes.

Genetic control of immunological responses

Genetic regulation has been shown to be involved in the expression of most immunological functions (Wakelin, 1985). Evidence is primarily derived from two broad approaches: (1) the demonstration that immune response (Ir) genes determine reactivity to a variety of antigens, and (2) assortative matings in which predominantly laboratory animals have been selected for and against specific immune responses. In this second category, the parameters under selection include quantitative and qualitative aspects of humoral, cellular and phagocytic activity.

Immune response (Ir) genes

Ir genes are associated with the major histocompatibility complex (MHC) and encode for class II molecules on the membranes of a number of cells including macrophages, and T and B lymphocytes. These autosomal dominant genes have an important role in the recognition of antigen (Schwartz, 1985) and appear to modulate specific responses by way of factors (cytokines) produced by helper T cells (Benacerraf and Germain,

1978; Benacerraf, 1980; Miller, 1990). However, responses to some antigens are not MHC-linked (Dorf *et al.*, 1974), and rejection of parasites from mice involves both MHC and non-MHC genes (Else and Wakelin, 1988).

Selection for immunological functions

Antibody characteristics

The most intensively studied immunological function is that of antibody production. Biozzi *et al.* (1980, 1982) have carried out a number of selection experiments designed to examine the nature of genetic regulation of this response. In lines of mice selected for either high or low antibody levels after immunization with a number of natural immunogens, antibody production was found to be under polygenic control and had realized heritabilities of about 20%. In Selection 1, the most studied experiment, the basis of antibody responses to heterologous red blood cells was in fact determined by the rate of antigen catabolism in macrophages. Thus, an increased persistence of antigen in the high line compared to the low line stimulated greater antibody production. Furthermore, the quantity of antigen (threshold) required for inducing detectable antibody responses was greatly decreased in the high line compared to the low line. Equivalent cellular reactivity was observed in the high and low antibody lines suggesting that the genetic control of cellular and humoral immunity is independent.

Genetic control of antibody levels has since been confirmed in a number of other animal species. In experiments designed to parallel closely those of the Biozzi mice, guinea pigs (Ibanez *et al.*, 1980) and chickens (van der Zijpp and Nieuwland, 1989) have been selected for antibody responses to heterologous red cells. However, in contrast to the Biozzi mice, macrophage function was not the major cause of differences between the high and low lines of chickens (van der Zijpp *et al.*, 1988). Siegal and Gross (1980) have reported two separate bidirectional selection experiments in which chickens were bred for and against antibody production and antibody persistence to heterologous erythrocytes. Similarly, Buschmann *et al.* (1975; in Krausslich, 1984) consider that the porcine responses to DNP. BSA closely parallel those of the Biozzi mice. Antibody response of young bulls to human albumin was also found to be genetically determined (Lie, 1977).

Genetic regulation can also influence qualitative aspects of the antibody response. In this regard, Katz and Steward (1975) and Steward *et al.* (1979) have bred lines of mice having high or low affinity to human serum proteins. Because selection resulted in highly significant interline differences without influencing antibody levels, independent genetic control of antibody affinity and antibody production was proposed.

Cellular responses

Quantitative and qualitative aspects of cellular responsiveness under genetic control have been investigated by a number of workers. Chai (1966, 1975) established lines of mice having either high or low total leucocyte counts, the greatest proportion of leucocytes in the high line were lymphocytes, and in particular T lymphocytes. Mayerhofer *et al.* (1976; in Andresen, 1978) have estimated the heritability of blood lymphocyte counts in cattle to be 20%. Genetic control of cellular responsiveness has been demonstrated by Biozzi *et al.* (1980) who selected for and against the ability of mouse lymphocytes to incorporate tritiated thymidine after *in vitro* stimulation with the T cell mitogen, phytohaemagglutinin (PHA). Selection influenced the general reactivity of T cells as, in contrast to B cell mitogens where interline differences were small, responses similar to PHA occurred with another T cell mitogen, concanavalin A (con A). Gauthier-Rahman *et al.* (1983) also demonstrated that the Hi/PHA mice had increased delayed-type hypersensitivity and its *in vitro* correlate, cell migration inhibition.

Phagocytic responses

Genetic regulation of phagocytic activity of macrophages has been reported by Biozzi *et al.* (1980). It was possible to select for and against a phago-cytosis index, as measured by the clearance of colloidal carbon after triolein injection. Triolein was used to stimulate reticuloendothelial macrophages without influencing antibody responses, as it was concluded from earlier work, that unstimulated phagocytic function was a physiological homeostatic constant. However, Krausslich (1984) describes the results of successful bidirectional selection of mice for the ability to clear carbon. This selection increased liver and spleen weights and the numbers of phagocytosing cells in the high line, but the phagocytic activity of the single cell was unaffected.

General immunological responsiveness

The single character selection programmes described above demonstrate that many individual components of the immune repertoire are under independent polygenic control. This has prompted some workers to suggest the possibility of selection for a general improvement in overall immune responsiveness with the aim of achieving an increase in general disease resistance (Gavora and Spencer, 1978). In pigs, preliminary attempts have been made to identify criteria by which animals with increased immuno-competence can be selected (Buschmann and Meyer, 1989; Krausslich, 1984). An index of responsiveness has been constructed which includes traits assumed to influence disease resistance. These assays of quantitative and qualitative aspects of humoral and cellular immunity indicated significant individual variation. Furthermore, although some significant positive and negative correlations existed between traits, the vast majority of correlations were not significant (Buschmann and Meyer, 1989).

Host immune responses directed against parasites

In spite of intensive research over many years, the precise manner by which immune mechanisms act to cause parasite expulsion still remains poorly defined (Rothwell, 1989). It is clear that much remains to be learnt, especially of the mechanisms operating against intestinal parasites in domesticated animals and man. However, the study of laboratory rodent/parasite models has provided valuable information on the immune mechanisms involved in resistance. Thus, the concept of interactive components of the immune response involved in parasite rejection has been elucidated primarily in these models where mechanistic studies are aided by exaggerated reactions and the ability to adoptively transfer cells between individuals of inbred lines. In contrast, the relative lack of a consistently defined response and the unsuitability of the transfer techniques have limited these investigations in domestic ruminants. It is, however, possible that the responses observed in laboratory models may not be relevant in determining parasite regulation in the natural or field strains of host (Dineen and Wagland, 1982). In this respect, resistance mechanisms responsible for protection against parasites may vary between individuals of an outbred host population, in response to different stages of the parasite life cycle, and for different parasite species (Rothwell, 1989). Studies of immunological responses provoked by parasitic infection involve both specific and non-specific components (Dineen *et al.*, 1977). Humoral and lymphoid responses are determined by antigen recognition and therefore are specific in nature, whereas inflammatory responses once activated are non-specific in terms of their target. A brief description of the role each of these components plays in worm rejection follows.

Antibody responses

Ogilvie and Hockley (1968) demonstrated that *Nippostrongylus brasiliensis* in rats became morphologically damaged prior to expulsion, and Jones and Ogilvie (1971) found that this damage was mediated by antibodies. The transfer of mesenteric lymph node cells from immune, but not non-immune, donors were then able to initiate the expulsion of adult worms (Dineen *et al.*, 1973a, b). More recently, doubts concerning the necessity for antibody damage as a prerequisite for worm expulsion have been raised (Rothwell, 1989).

Lymphocyte responses

The nature of these cellular reactions probably reflects a regulatory function over other immunological and inflammatory components rather than a direct antiparasitic effect. Thus, activated T helper cells produce and secrete a number of cytokines which modulate the growth, differentiation or function of cells originating in bone marrow (see review by Miller, 1990). Miller (1990) considers that the equivalent of the mouse T helper class 2 subset will be more important to the regulation of responses directed against helminths because this cell type promotes cytokines responsible for mast cell and eosinophil differentiation, as well as the production of IgE and IgA.

Myeloid responses

The third component in immunologically mediated expulsion of *N. brasiliensis* involve the myeloid, or bone-marrow derived, cells and their products. Immune mesenteric lymph node cells or bone marrow cells alone could not initiate expulsion in irradiated hosts (Dineen and Kelly, 1973; Kelly *et al.*, 1973), suggesting that inflammatory responses are produced by the interaction of sensitized lymphocytes and myeloid cells. The cell types and cell products involved in this response have been discussed by Rothwell (1989) and Miller (1990). Mast cells and eosinophils are characteristically associated with gastrointestinal nematode infections, with both capable of generating and releasing a number of mediators which have antiparasitic activities when stimulated by membrane receptors for sensitizing antibodies (such as IgE and IgG) or the complement component C3. These mediators include the pharmacologically active amines (such as histamine and serotonin), prostaglandins, leukotrienes and eosinophilic major basic protein (Rothwell, 1989).

Genetic regulation of resistance to gastrointestinal nematodes

Evidence exists in both laboratory animals and domestic animals for genetic regulation of resistance to gastrointestinal nematodes. It is important to determine whether resistance to infection is acquired or innate. Whereas acquired responses are immunologically mediated and can occur during primary as well as secondary infection, innate resistance occurs because of biochemical or physiological properties of the host which are unsuitable for the parasite. Because of their nature, it is possible that innate responses are influenced to a greater degree by environmental factors. Genetic regulation and restriction of immunological reactivity play a major role in the existence of host variation in response to helminth infections (Wakelin, 1985). In addition, selection experiments involving both laboratory and domestic animals have indicated that the immune response plays a dominant part in expression of the selected trait.

Laboratory host/parasite models

Wakelin (1975a) was able to utilize between-animal variation for *Trichuris muris* worm burdens in an outbred population of mice to establish responder and non-responder lines. It was suggested that the genetic control of immunological responsiveness was a determining factor for the difference between lines. However, as mice from the non-responder line were capable of producing protective antibodies, it was proposed that the genetic control influencing susceptibility resided in another component of the expulsion mechanism (Wakelin, 1975b).

A programme in which mice were selected as being either refractory or liable to a single infection with *Nematospiroides dubius* has been described by Brindley and Dobson (1981). On the basis of partial protection transferred by sera from both lines, it was suggested that parasitological differences between lines was due to innate resistance (Brindley and Dobson, 1983a). However, mice from the refractory line exhibited greater levels of acquired immunity to subsequent infections, indicating either a pleiotrophic effect of genes determining innate resistance, or a partial linkage between genes controlling primary and secondary responses (Brindley and Dobson, 1982). In a parallel study, lines of mice were established with either high or low faecal egg counts after secondary infection with *N. dubius* (Sitepu and Dobson, 1982). The inheritance of this character was complex, but exhibited additive effects and partial dominance for high immune responsiveness (Brindley *et al.*, 1986).

Substantial interline differences occurred in guinea pigs selected for resistance and susceptibility after a single infection with *T. colubriformis* (Rothwell *et al.*, 1978). This difference was based on an earlier worm expulsion in the resistant line which required a functional lymphocyte component. Subsequently, Handlinger and Rothwell (1981) reported that animals from the resistant line had greater numbers of intestinal mast cells after infection, and circulating and tissue eosinophils before and after infection. Backcross analysis associated both resistance and susceptibility with regions of the MHC (Geczy and Rothwell, 1981).

Domestic animals

Empirical descriptions of genetic variation in resistance to gastrointestinal nematodes have been based on between and within breed comparisons of resistance. The evidence for this genetic control, particularly for *H. contortus*, has been well documented in recent times (Gray, 1991) and will not be detailed here. However, it should be mentioned that genetic variation within a breed may be as great as that occurring between breeds (Gray *et al.*, 1987).

Within flocks of sheep, nematode worm burdens exhibit a negative binomial distribution which empirically describes a situation of overdispersion (Barger, 1985). In overdispersion a relatively small proportion of the host population harbours a relatively large proportion of the parasite population. Manipulation of this within-breed variability by selection aims to reduce the numbers of susceptible animals such that the overall resistance of a flock can be increased. Within Australia and New Zealand, a number of programmes are underway to characterize the nature of the genetic regulation of resistance to nematodes and investigate the consequences of the single character selection procedures, which, in each case, are based on FEC after infection (Windon, 1990; Woolaston, 1990). Of these selection programmes, four have particular relevance to the present discussion.

The '*Trichostrongylus* selection lines' have been described by Windon and Dineen (1984), Windon *et al.* (1987) and Windon (1990). Selection and assortative matings have established lines of Merino sheep in which lambs are either responsive (high responders) or susceptible (low responders) after vaccination and challenge with *T. colubriformis* in pens. The selected trait, therefore, is overtly immunologically based and focuses on the responsiveness of young lambs to infection. The programme has now progressed to the fourth of arbitrarily designed generations, where the levels of protection (calculated by comparing vaccinated groups with those of sex-matched unvaccinated controls from an unselected line) had reached levels of 80–90% in high responders and −20 to −25% (that is, FEC higher than the relevant unvaccinates) in low responders (Windon, 1990). For each generation, high responder lambs had significantly lower FEC than low responders, and, within each line, female lambs were more responsive than males. In New Zealand, divergent lines of Romney sheep have been established on the basis of FEC resulting from naturally acquired field infection consisting mainly of *Trichostrongylus* and *Ostertagia* spp. (Baker *et al.*, 1990). Responses of lambs were monitored in two sampling periods separated by anthelmintic treatment. A greater response to selection in the second period probably reflects an acquired component in this genetic regulation.

Two programmes using Merino sheep have been established to investigate the genetic control of *H. contortus*. In the '*Haemonchus* selection lines', divergent lines have been established with either increased or decreased resistance to a single artificial infection when five to six months of age (Piper, 1987; Woolaston, 1990). Significant differences were observed between lines from the fifth year of the programme. The identification of a ram whose progeny were extremely resistant to *H. contortus* has been the basis of the '*Haemonchus* resistance flock' (Albers *et al.*, 1984, 1987). Backcross matings between relatives of this ram are being carried out to enable investigations into the nature of the genetic regulation. Individual responsiveness is based on FEC after an artificial challenge at four to give months of age while on pasture.

Immunocompetence in genetically resistant sheep

Utilization of lines of sheep having defined extremes of responsiveness of gastrointestinal nematodes makes it possible to undertake an examination of the mechanisms responsible for worm rejection, and opens the possibility of manipulating the genome of susceptible animals, identifying predictive markers particularly those genetically correlated with resistance, and identifying immunological procedures which aim to overcome the susceptibility of some animals. Studies into the immunocompetence of animals from each of the sheep selection experiments described above are still in the early stages. These have usually involved non-invasive assessment of the three broad arms of immune responsiveness, humoral, cellular and phagocytic activities, as the animals have been required for breeding. However, peripheral responses may not necessarily reflect the local protective immune reactions operating at the site of infection. It is anticipated that as the flocks expand and animals surplus to breeding requirements become available, an increasing effort will be devoted to understanding the local mechanisms underlying the selected traits in each programme.

Studies into the mechanisms of resistance in the *Trichostrongylus* selection flocks, suggest that selection for the broadly based character of response to vaccination and challenge with *T. colubriformis* has resulted in increased responsiveness across a wide range of immunological functions.

Antibody levels to parasite antigen, as measured by the complement fixation test (Windon and Dineen, 1981) and ELISA (Windon, unpublished) were higher in the high responder line compared to low responders. In these studies, antibody responsiveness at the time of challenge was inversely correlated to FEC after challenge (Windon and Dineen, 1984). *In vitro* blastogenic responses of antigen-reactive lymphocytes were also greater in high responders after vaccination and challenge (Dineen and Windon, 1980; Windon and Dineen, 1981). It is of interest to note that although high responders had greater antibody and blastogenic responses, responses were also detected in low responders. It has been proposed by Dineen (1963, 1978) and Dineen and Wagland (1982) that host responses can be directed against 'relevant' antigens (those which affect the survival of the parasite) and 'irrelevant' antigens (those which lead to a detectable response but does not result in protection). Relevant, or fitness, antigens are those against which immunoselection is directed as their detection determines the ultimate success or failure of a provoked response. These antigens may be difficult to identify and responses against them difficult to detect in a natural host/parasite relationship, whereas the lack of selective pressure on irrelevant antigens make them easily detected by the host. Results from the cellular and humoral assays may reflect qualitative and quantitative differences in high and low responders to relevant and irrelevant antigens. In addition, the cellular reactions may also indicate the production and release of lympho-kines, which exert an influence over the maturation and development of a number of effector cells which have been implicated as having a role in worm rejection (Miller, 1990). Although IgE has yet to be unequivocally identified in sheep, another isotype may have homocytotrophic antibody activity. This may be reflected in the higher antibody levels of high responders. Furthermore, the complement component C3 is capable of activating a number of cell types involved in phagocytosis and mediator release through membrane receptors (Ellner and Mahmoud, 1982). During infection, high responders had higher levels of C3 but not C4 (Groth *et al.*, 1987).

A number of parameters involved in the effector arm of the immune response have also been examined in the *Trichostrongylus* selection lines. Chemiluminescence activity, used in this instance as an indirect measure of *in vitro* phagocytosis in peripheral blood leucocytes, was significantly greater in high responder lambs (Windon and Dineen, 1984). This test is based on the production of oxygen radicals produced in response to membrane perturbation during phagocytosis and these may, in turn, have direct antiparasitic effects (Smith, 1989). Furthermore, oxygen species play a role in the production of a number of non-specific mediators such as prostaglandins, leukotrienes and thromboxanes and may stimulate histamine release from mast cells (Windon and Dineen, 1984). Jones *et al.* (1990) have shown that during infection high responders have greater levels of histamine in intestinal tissue, and higher concentrations of leukotrienes C4 and B4 in duodenal mucus. Of the cells capable of releasing these mediators, high responders had a greater abundance of globule leucocytes (thought to be degranulated mast cells) in the epithelium of the small intestine after challenge (Dineen and Windon, 1980) and elevated numbers of circulating eosinophils during the vaccination and challenge periods (Dawkins *et al.*, 1989). There was no difference, however, in tissue eosinophil numbers (Dineen and Windon, 1980). The precise role that mast cells and eosinophils play in the rejection of gastrointestinal nematodes has yet to be determined.

Immunological responses also appear to play a major role in the *Haemonchus* resistance flock. Presson *et al.* (1988) found that resistant animals had increased thymus weights and greater numbers of globule leucocytes in abomasal tissue than susceptible lambs. Treatment with the immunosuppressive agent dexamethasone abolished these parasitological and immunological differences. Further work (Gray, Barger, Le Jambre and Douch, unpublished) has shown that genetically resistant animals had elevated levels of LTC4 in intestinal mucus following field infection of predominantly *Trichostrongylus* spp.

Investigations into resistance mechanisms operating in the *Haemonchus* selection flocks have indicated that immunological functions may not be involved to the same extent as in the *Trichostrongylus* selection lines and the *Haemonchus* resistance flock (Gray, Gill and Woolaston, personal communication). After infection, animals from the increased and decreased resistance flocks had equivalent thymus weights and similar numbers of globule leucocytes in abosmasal tissue. In addition, dexamethasone treatment had only a minor influence on the parasitological differences between these lines. It was concluded that different resistance genes, expressed through different mechanisms, are operating in the *Haemonchus* resistance flock and the *Haemonchus* selection lines. Further investigations are planned to confirm and extend these studies into the nature of the genetic regulation in the *Haemonchus* selection programme. This instance serves to highlight the need for care in interpreting results from selection programmes based on a small number of founding animals. Individual variation in mechanistic responses to parasitic infection may exist, and those upon which selection programmes are based need not necessarily be representative of the majority of an outbred population.

Specificity of selection

The practical significance of selection programmes directed towards increasing resistance to gastrointestinal nematodes will, to a large extent, be dependent on an evaluation of correlated responses including those to other parasites and non-parasite pathogens. In an environment where other parasites and diseases pose a threat to the health and productivity of a flock, the value of selection for parasite resistance will be decreased if an increase in susceptibility occurs for other pathogens. However, in a manner similar to that described by Piper and Barger (1988), the importance of particular associations between resistance to nematodes and susceptibility to other diseases will not be relevant in all environments as it depends on the occurrence and significance of each. At present, little is known about these genetic correlations, although limited studies have been carried out in the Australian sheep selection programmes.

Windon (1990) has considered evidence that, in laboratory host/ parasite models, selection for responsiveness to non-parasite antigens and parasites can be non-specific. In this regard, the most significant reports have been those relating to the Biozzi mice (Biozzi *et al.*, 1982) selected for and against antibody response to heterologous red cells, and lines of mice selected for responsiveness to a single infection or multiple infections with *N. dubius* (Brindley and Dobson, 1983b; Sitepu *et al.*, 1984). This work has shown that selection for a particular immunological function (for example, antibody response) may exert non-specific effects where that function plays a dominant role in host defence or a determining role in immunological reactivity. On the other hand, if resistance operates through an immune function other than that being influenced by selection, suscept- ibility to these pathogens may be increased. Such results led Dineen (1985) to propose that selection for resistance to one organism may focus on particular limiting mechanisms and perturb an immunological equilibrium.

In the *Trichostrongylus* selection lines, high responder lambs exhibit significantly reduced FEC following artificial infection with other related and unrelated gastrointestinal species including anthelmintic-resistant *T. colubriformis*, *Trichostrongylus rugatus*, *Trichostrongylus axei* and *Ostertagia circumcincta* (Windon and Dineen, 1984; Windon *et al.*, 1987). Furthermore, high responders have lower FEC following natural field infection consisting predominantly of *Trichostrongylus* spp. and *Ostertagia* spp. Similarly, in the *Haemonchus* resistance flock, FEC were reduced in the resistance progeny after natural infection with *Trichostrongylus* spp. and *Ostertagia* spp. (Gray, Barger, Le Jambre and Douch, unpublished). Results from heterologous infections with *H. contortus* or *T. colubriformis* in the *Trichostrongylus* and *Haemonchus* selection lines, respectively, show that differences between resistant and susceptible lines are not as great as for the homologous species against which each selection is based (Windon, 1990; Woolaston, 1990). These observations may be attributable to the nature of the parasitism, site of infection, evasion strategies employed by each parasite, or the character and vigour of the mechanisms required for immunological control. Preliminary results from the *Trichostrongylus* selection lines suggests that immunological responses occurring during infection with *T. colubriformis* and *H. contortus* may differ. After infec- tion with *T. colubriformis*, high responder progeny had significantly increased circulating eosinophil numbers compared to low responders and these counts were inversely related to FEC (Dawkins *et al.*, 1989). Subse- quent vaccination and challenge of the same animals with *H. contortus* did not result in interline differences for this parameter, and there was no correlation with FEC after challenge (Windon, unpublished).

Further evaluation of the degree of specificity of selection in the sheep selection programmes is required in order to characterize responses to other gastrointestinal nematodes, other parasites (cestodes, trematodes and external parasites), and other disease pathogens. The underlying genetically controlled resistance mechanisms appear to play a major role in determining the nature of the specificity, and by understanding these, it should be possible to predict how selection for parasite resistance will influence susceptibility to other pathogens. Ultimately, this will allow an assessment of the merit of incorporating parasite resistance into a selection index from which superior animals are chosen for a particular environment.

Markers with resistance

Immunological or physiological parameters genetically linked to a trait of interest may provide a predictive marker by which that trait could be subjected to indirect selection. For parasite resistance, such a marker will identify resistant animals without the necessity for infection. Those parameters which have received most attention have been described by Windon (1990) and are discussed briefly here. It should be said, however, that the associations described to date are imprecise and the literature contains conflicting reports.

Haemoglobin (Hb) type

Initially, it was considered that sheep having type HbAA had lower *H. contortus* FEC than those with either HbAB or HbBB (Evans *et al.*, 1963). These observations have not been confirmed in animals selected for resistance to *H. contortus* (Albers *et al.*, 1984) or *T. colubriformis* (Windon and Dineen, 1984).

Lymphocyte blastogenic responses

Preinfection proliferative responses of lymphocytes stimulated with *H. contortus* or mixed *Ostertagia* and *Trichostrongylus* spp. antigen have been found to be inversely correlated with subsequent parasite resistance (Riffkin and Dobson, 1979; Riffkin and Yong, 1984). Such an association was not found in animals from the *Trichostrongylus* selection lines (Windon and Dineen, 1981).

Ovine lymphocyte antigens (OLA)

The association between MHC class I determinants on the surface of lymphocytes and acquired resistance to *T. colubriformis* has been studied by Outteridge *et al.* (1985, 1986). Using a panel of typing reagents consisting principally of pregnancy sera from the high responder, low responder and random (unselected) lines of the *Trichostrongylus* selection programme, two antigens were identified which had different frequencies in the high and low responder lines. Thus, SY1 and SY2 were more abundant in the high responder and low responder lines, respectively, and, in the random line, vaccinated animals possessing SY1 had significantly lower FEC. The association was confirmed in animals unrelated to the selected lines where matings were carried out on the basis of SY type (Outteridge *et al.*, 1988). Other workers, however, have not been able to confirm associations between OLA type and parasite resistance. Riffkin and Yong (1984) did not find an association between FEC resulting from field infection thought to be primarily *Trichostrongylus* and *Ostertagia* spp. in a small group of 39 ewes designated as being either resistant or susceptible, and no relationship was found for responsiveness to *H. contortus* in the UNE resistance flock (Gray, personal communication) or the *Haemonchus* selection flocks (Cooper *et al.*, 1989).

Restriction fragment length polymorphisms (RFLP)

Although no clear results have yet emerged, an area of considerable promise in the search for predictive markers is the linkage analysis between DNA markers (RFLPs) and segregating characters. Windon (1990) cited the unpublished results of D.J. Hulme at McMaster Laboratory in which a human MHC class II probe identified a particular DNA band which was associated with susceptibility within the *Trichostrongylus* selection lines. Unfortunately, this strong relationship did not prevail as, when a larger group of animals was examined, the difference in FEC between sheep in which this band was either present or absent became non-significant when the probability (P) value was corrected so as to guard against recognition of chance associations. This correction involves the standard practice of multiplication of the P value by the number of tests performed. Further work has shown that of 121 bands analysed so far, ten exerted an effect on FEC, but again significance was lost after correction of the probability value. Linkage analysis has also been carried out within sire groups in order to examine the influence of sire haplotype on FEC. In one family, FEC of progeny grouped according to the segregation of sire haplotypes was 720 epg and 1700 epg, respectively (P=0.0033; corrected P=0.043) (Hulme, unpublished). This work suggests that genes within or close to the MHC influence FEC, although these specific genes have yet to been identified. It is anticipated that if resistance mechanisms with major effects are identified, relevant 'candidate' genes for use as probes will accelerate this linkage analysis for identifying markers with resistance. Thus, in addition to genes associated with the MHC, candidate genes would include those having a functional basis.

Adaptation of the parasite to host responses

Like the host, the parasite is a complex organism in which genetic hetero-zygosity is maintained to ensure evolutionary fitness. It is, therefore, not unreasonable to expect the parasite population to respond to influences which threaten its survival. Such influences include environmental factors acting on free-living stages, protective host responses, and the use of anthelmintic treatments to control or remove parasite burdens. It is in the latter instance where the parasite's ability to adapt to external forces is most vividly seen. Thus, over a relatively short time interval, the parasite has, under provocation by severe selection pressure, adapted to the single mode of action of anthelmintics. In contrast, it has been considered that because host resistance operates through a variety of mechanisms and appears to be under polygenic control, comparable selection pressure would not be imposed (Albers *et al.*, 1984; Murray, 1989). However, if a selection programme for resistance to nematodes is based on a major gene, parasite adaptation could occur more rapidly. The ability of the parasite to adapt to host resistance mechanisms may limit the success of the genetic approach, and indeed, any control procedure based on immunological means, including vaccination.

Adaptation of the parasite has previously been thought of as occurring over evolutionary time. Various hypotheses have been proposed to account for this adaptation, including the evolutionary convergence in antigenicity (Damian, 1964; Dineen, 1963) immunological tolerance of particular parasite antigens (Sprent, 1962), or the masking of parasite antigens by host proteins (Smithers *et al.*, 1968). This suggests that a coevolutionary interaction between host and parasite exists, described as an 'incessant evolutionary dance' between virulence genes of parasites and resistance genes of their hosts (Haldane, 1949). Such an interaction is manifested in a dynamic equilibrium which counterbalances pathogenicity of tolerated parasite burdens (subthreshold levels) and reduced antigenic disparity between host and parasite (Dineen, 1963, 1978; Dineen and Wagland, 1982). As a corollary, Clarke (1976) considers that the interaction between host and pathogen populations maintains genetic diversity within each.

It is a common observation that within parasite cohorts some individuals are rejected earlier than others, and indeed, the remaining worms may survive for a considerable period (Behnke, 1987). Similar patterns have been described for the three common gastrointestinal species of sheep, *T. colubriformis* (Dobson *et al.*, 1990), *H. contortus* (Barger *et al.*, 1985) and *O. circumcincta* (Hong *et al.*, 1987). Although confounded by between-animal variation, such work provides empirical support for the view that existing within the parasite population are individuals capable of surviving and producing offspring in the face of strong host responses.

Direct evidence for parasite adaptation to host resistance mechanisms is at present limited and somewhat contradictory. Albers and Burgess (1988) carried out an experiment in which repeated infections of *H. contortus* were given to lambs either allowed to become resistant or remain susceptible to infection due to treatment with dexamethasone. Six serial passages were carried out in resistant lambs and nine through treated lambs. Groups of susceptible lambs were then infected with larvae from either the immune or susceptible donors. On the basis of FEC after challenge, the effect of serial passage through hosts of different immune status was small and non-significant. The authors argued that, in contrast to the development of anthelmintic resistance, the equivalent of three years of intense selection had not induced changes in infectivity within the *H. contortus* population. Adams (1988) also used serial passages of *H. contortus*. However, in this work, sheep were reinfected with a pool of larvae derived from the previous parasite generation ('suprapopulation reinfected'), or individuals were reinfected with larvae derived from the population they themselves carried in the previous infection ('infrapopulation reinfected'). Again, the conclusions drawn were that *H. contortus* could not adapt to withstand developing host immune responses, as the experiment ceased when no larvae could be recovered from the faeces of sheep during the sixth passage. Studies have also recently commenced to examine adaptive responses of *H. contortus* and *T. colubriformis* in animals from the increased and decreased resistance lines of the *H. contortus* selection experiment and aims to represent 30 generations of the parasite (Woolaston, 1990). Early results after five parasite generations indicate that there is no evidence for adaptation in *H. contortus* due to moderate selection pressure (Woolaston, personal communication).

A study using the *Trichostrongylus* selection lines has also addressed the question of parasite adaptation, with preliminary results being reported by Windon (1990). Fourth generation progeny from the high and low responder lines were vaccinated and challenged, using the usual testing procedures, with a strain of *T. colubriformis* routinely passaged in susceptible lambs at McMaster Laboratory. The faeces of 14 high responder rams (mean FEC after challenge 253 epg) and five low responders (mean FEC 4,453 epg) were collected during the challenge period, cultured and the resultant larvae recovered. Groups of six 8-month-old Border Leicester/Merino ewe lambs, chosen because of their strong acquired responses to *T. colubriformis*, were vaccinated with irradiated McMaster strain larvae or left unvaccinated. All animals were drenched and challenged with 20,000 larvae derived from either McMaster, low responder or high responder donors. FEC (± SE) in vaccinated animals challenged with McMaster, low responder and high responder larvae were 4 ± 3, 23 ± 18, and 181 ± 27, respectively. Examination of circulating eosinophil counts, serum antibody responses and intestinal histology showed no differences between vaccinated groups given the different larval types at challenge (Windon, Rothwell and Anderson, unpublished). These results indicate that, in contrast to *H. contortus*, one previous passage of *T. colubriformis* through resistant lambs led to animals otherwise solidly immune having FEC at a level capable of producing significant larval contamination on pasture. Further intensive selection of the parasite is planned so as to confirm and extend these findings. If the ability of the parasite to adapt is confirmed and selection can increase the proportion of these parasites surviving, the opportunity will exist to examine the antigenic, biochemical and genetic basis of this occurrence.

Fitness of adapted parasites will have an important bearing on the significance of this potential problem. In a recent report dealing with anthelmintic resistance in *H. contortus*, Maingi *et al.* (1990) proposed that there are three phases in the association between fitness and resistance: (a) where no anthelmintics are used, susceptible individual worms have a biological advantage over those having resistance genotypes, the latter therefore remain in low numbers, (b) after anthelmintics have been introduced and moderate levels of resistance occur in the worms, resistant individual worms show decreased biological fitness, possibly due to altered physiology associated with the loss of susceptibility mechanisms, and (c) due to resegregation of fitness with resistance, under severe selection pressure resistant worms had greater establishment, survival and reproduction. It will be important to determine whether similar events occur with parasites adapted to host resistance mechanisms.

Conclusions

As an alternative or complementary control option, the genetic approach offers a potentially valuable strategy to reduce the current heavy reliance on anthelmintics. In a computer simulation, genetically resistant lambs required treatment with anthelmintics at a frequency considerably less than susceptible lambs and resulted in a lower level of larval pasture contamination (Barger, 1989; Windon, 1990). Thus, the benefits derived from the genetic approach will be achieved from reduced costs of production and preservation of the effective life of currently available anthelmintics. An evaluation of the genetic gains that might be achieved and consideration of the problems involved in incorporating parasite resistance into commercial breeding enterprises have been discussed by Piper and Barger (1988). These workers estimate that income per ewe per year could be increased by 10% annually in a moderately parasitized environment after including parasite resistance in a selection index from which superior animals are chosen.

Crucial to the ultimate acceptance by industry of incorporating parasite resistance into commercial breeding programmes are: (1) the identification of predictive markers so that resistant animals can be identified without the need for infection; and (2) determination of the specificity of selection in terms of other parasite species (nematodes, cestodes, trematodes and external parasites) and non-parasite pathogens. Central to an understanding of these aspects will be the necessity to delineate the basis of genetic regulation in terms of limiting immunological mechanisms. It appears that the resistance mechanisms operating in sheep selection programmes for nematode resistance are principally immunologically based. Present knowledge, particularly for domestic ruminants, of invoked immune responses to nematodes are fragmentary and incomplete. However, it is anticipated that the use of genetically resistant animals possessing defined extremes of responsiveness will accelerate these investigations. Once mechanisms of importance have been identified, markers with resistance may be identified by gene probes or functional assays. Evidence from studies using laboratory rodents suggests that it will be possible to reduce predisposition of genetically resistant animals for susceptibility to other pathogens from the mechanisms found to be under genetic control. However, limited work to date with lambs selected for and against resistance to *T. colubriformis*, suggests that a wide range of immunological functions are under this genetic control.

An important determinant of the success of control measures based on immunological means will be the ability of the parasite to adapt quickly to resistance mechanisms under genetic control or induced by vaccination. Limited evidence suggests that *T. colubriformis* may be able to adapt, whereas, at this stage, this has not been demonstrated in *H. contortus*. Further work is required to evaluate critically the significance of these findings.

Acknowledgements

I wish to thank Mr D.J. Hulme for permission to report unpublished results of associations between RFLPs and resistance, and Drs G.D. Gray and R.R. Woolaston for making available unpublished observations from the *Haemonchus* resistance flock and the *Haemonchus* selection lines, respectively. Much of the Australian research into the genetic control of resistance to gastrointestinal nematodes in sheep, including the *Trichostrongylus* selection flocks, is currently receiving support from the Australian Wool Corporation on the recommendation of the Wool Research and Development Council.

References

Adams, D.B. (1988) Infection with *Haemonchus contortus* in sheep and the role of adaptive immunity in selection of the parasite. *International Journal for Parasitology* 18, 1071–5.

Albers, G.A.A. and Burgess, S.K. (1988) Serial passage of *Haemonchus contortus* in resistant and susceptible sheep. *Veterinary Parasitology* 28, 303–6.

Albers, G.A.A., Burgess, S.E., Adams, D.B., Barker, J.S.F., Le Jambre, L.F. and Piper, L.R. (1984) Breeding *Haemonchus contortus* resistant sheep – problems and prospects. In: Dineen, J.K. and Outteridge, P.M. (eds), *Immunogenetic Approaches to the Control of Endoparasites, with Particular Reference to Parasites of Sheep.* CSIRO Division of Animal Health, Melbourne, pp. 41–51.

Albers, G.A.A., Gray, G.D., Piper, L.R., Barker, J.S.F., Le Jambre, L.F. and Barger, I.A. (1987) The genetics of resistance and resilience to *Haemonchus contortus* infection in young Merino sheep. *International Journal for Parasitology* 17, 1355–63.

Andresen, E. (1978) On the possibility of breeding for genetic resistance to disease in cattle. European Association of Animal Production, 29th Annual Meeting, Stockholm, 8 pp.

Baker, R.L., Watson, T.G., Bisset, S.A. and Vlassoff, A. (1990) Breeding Romney sheep which are resistant to gastrointestinal parasites. In: *Proceedings of the 8th Conference of the Australian Association of Animal Breeding and Genetics, Ruakura, 5–9 February,* pp. 173–8.

Barger, I.A. (1985) The statistical distribution of trichostrongylid nematodes in grazing lambs. *International Journal for Parasitology* 15, 645–9.

Barger, I.A. (1989) Genetic resistance of hosts and its influence on epidemiology. *Veterinary Parasitology* 32, 21–35.

Barger, I.A., Le Jambre, L.F., Georgi, J.R. and Davies, H.I. (1985) Regulation of *Haemonchus contortus* populations in sheep exposed to continuous infections. *International Journal for Parasitology* 15, 529–33.

Behnke, J.M. (1987) Evasion of immunity by nematode parasites causing chronic infections. *Advances in Parasitology* 26, 1–71.

Benacerraf, B. (1980) Genetic control of the specificity of T lymphocytes and their regulatory products. In: Fougereau, M. and Dausset, J. (eds), *Immunology 80, Progress in Immunology IV.* Academic Press, London, pp. 419–31.

Benacerraf, B. and Germain, R.N. (1978) Immune response genes of the major histocompatibility complex. *Immunological Reviews* 38, 70–119.

Biozzi, G., Mouton, D., Heumann, A.M. and Bouthillier, Y. (1982) Genetic regulation of immunoresponsiveness in relation to resistance against infectious disease. In: *2nd World Congress on Genetics Applied to Livestock Production,* Madrid, 4–8 October, pp. 150–63.

Biozzi, G., Siqueira, M., Stiffel, C., Ibanez, O.M., Mouton, D. and Ferreira, V.C.A. (1980) Genetic selection for relevant immunological functions. In: Fougereau, M. and Daussett, J. (eds), *Immunology 80, Progress in Immunology IV.* Academic Press, London, pp. 432–57.

Brindley, P.J. and Dobson, C. (1981) Genetic control of liability to infection with *Nematospiroides dubius* in mice: selection of refractory and liable populations of mice. *Parasitology* 83, 51–65.

Brindley, P.J. and Dobson, C. (1982) Multiple infections with *Nematospiroides dubius* in mice selected for liability to a single infection. *Australian Journal of Experimental Biology and Medical Science* 60, 319–27.

Brindley, P.J. and Dobson, C. (1983a) Partitioning innate and acquired immunity in mice after infection with *Nematospiroides dubius. International Journal for Parasitology* 13, 503–7.

Brindley, P.J. and Dobson, C. (1983b) Host specificity in mice selected for innate immunity to *Nematospiroides dubius*: infection with *Nippostrongylus brasiliensis, Mesocestoides corti* and *Salmonella typhimurium. Zeitschrift fur Parasitenkunde* 69, 797–805.

Brindley, P.J., He, S., Sitepu, P., Pattie, W.A. and Dobson, C. (1986) Inheritance of immunity in mice to challenge infection with *Nematospiroides dubius. Heredity* 57, 53–8.

Buschmann, H. and Meyer, J. (1989) An immune competence profile in swine. In: van der Zijpp, A.J. and Sybesma, W. (eds), *Improving Genetic Disease Resistance in Farm Animals.* Kluwer, Dordrecht, pp. 145–52.

Chai, C.K. (1966) Selection for leukocyte counts in mice. *Genetical Research* 8, 125–42.

Chai, C.K. (1975) Genes associated with leukocyte production in mice. *Journal of Heredity* 66, 301–8.

Clarke, B. (1976) The ecological genetics of host–parasite relationships. *Symposium of the British Society of Parasitology* 14, 87–103.

Cooper, D.W., van Oorschot, A.H., Piper, L.R. and Le Jambre, L.F. (1989) No association between the ovine leucocyte antigen (OLA) system in the Australian Merino and susceptibility to *Haemonchus contortus* infection. *International Journal for Parasitology* 19, 695–7.

Courtney, C.H., Parker, C.F., McClure, K.E. and Herd, R.P. (1984) A comparison of the periparturient rise in faecal egg counts of exotic and domestic ewes. *International Journal for Parasitology* 14, 377–81.

Damian, R.T. (1964) Molecular mimicry: antigen sharing by parasite and host and its consequences. *American Naturalist* 98, 129–49.

Dawkins, H.J.S., Windon, R.G. and Eagleson, G.K. (1989) Eosinophil responses in sheep selected for high and low responsiveness to *Trichostrongylus colubriformis. International Journal for Parasitology* 19, 199–205.

Dineen, J.K. (1963) Immunological aspects of parasitism. *Nature (London)* 197, 268–9.

Dineen, J.K. (1978) The nature and role of immunological control in gastrointestinal helminthiasis. In: Donald, A.D., Southcott, W.H. and Dineen, J.K. (eds), *The Epidemiology and Control of Gastrointestinal Parasites of Sheep in Australia,* CSIRO Division of Animal Health, Melbourne, pp. 121–135.

Dineen, J.K. (1985) Host and host responses: alternative approaches to control parasites and parasitic disease. In: Anderson, N. and Waller, P.J. (eds), *Resistance in Nematodes to Anthelmintic Drugs,* CSIRO Division of Animal Health, Sydney, pp. 149–57.

Dineen, J.K., Gregg, P. and Lascelles, A.K. (1978) The response of lambs to vaccination at weaning with irradiated *Trichostrongylus colubriformis* larvae: segregation into 'responders' and 'non-responders'. *International Journal for Parasitology* 8, 59–63.

Dineen, J.K., Gregg, P., Windon, R.G. Donald, A.D. and Kelly, J.D. (1977) The role of immunologically specific and non-specific components of resistance in cross protection to intestinal nematodes. *International Journal for Parasitology* 7, 211–15.

Dineen, J.K. and Kelly, J.D. (1973) Expulsion of *Nippostrongylus brasiliensis* from the intestine of rats: the role of a cellular component derived from bone marrow. *International Archives of Allergy and Applied Immunology* 45, 759–66.

Dineen, J.K., Kelly, J.D, and Love, R.J. (1973a) The competence of lymphocytes obtained from immune and non-immune donors to cause expulsion of *Nippostrongylus brasiliensis* in the rat (DA strain). *International Archives of Allergy and Applied Immunology* 45, 504–12.

Dineen, J.K., Ogilvie, B.M. and Kelly, J.D. (1973b) Expulsion of *Nippostrongylus brasiliensis* from the intestine of rats: collaboration between humoral and cellular components of the immune response. *Immunology* 24, 467–76.

Dineen, J.K. and Wagland, B.M. (1982) Immunoregulation of parasites in natural host-parasite systems – with special reference to the gastrointestinal nematodes of sheep. In: Symons, L.E.A., Donald, A.D. and Dineen, J.K. (eds), *Biology and Control of Endoparasites*, Academic Press, Sydney, pp. 297–329.

Dineen, J.K. and Windon, R.G. (1980) The effect of sire selection on the response of lambs to vaccination with irradiated *Trichostrongylus colubriformis* larvae. *International Journal for Parasitology* 10, 189–96.

Dobson, R.J., Waller, P.J. and Donald, A.D. (1990) Population dynamics of *Trichostrongylus colubriformis* in sheep: the effect of infection rate on loss of adult parasites. *International Journal for Parasitology* 20, 359–63.

Donald, A.D. and Waller, P.J. (1982) Problems and prospects in the control of helminthiasis in sheep. In: Symons, L.E.A., Donald, A.D. and Dineen, J.K. (eds), *Biology and Control of Endoparasites*, Academic Press, Sydney, pp. 157–86.

Dorf, M.E., Dunham, E.K., Johnson, J.P. and Benacerraf, B. (1974) Genetic control of the immune response: the effect of non-H-2 linked genes on antibody production. *Journal of Immunology* 112, 1329–36.

Ellner, J.J. and Mahmoud, A.A.F. (1982) Phagocytes and worms: David and Goliath revisited. *Reviews of Infectious Diseases* 4, 698–714.

Else, K.J. and Wakelin, D. (1988) The effects of H-2 and non-H-2 genes on the expulsion of the nematode *Trichuris muris* from inbred and congenic mice. *Parasitology* 96, 543–50.

Evans, J.V., Blunt, M.H. and Southcott, W.H. (1963) The effect of infection with *Haemonchus contortus* on the sodium and potassium concentrations in the erythrocytes and plasma, in sheep of different haemoglobin type. *Australian Journal of Agricultural Research* 14, 549–58.

Gauthier-Rahman, S., Rouby, S.E., Liacopoulos-Briot, M., Stiffel, C., Decreusefond, C. and Liacopoulos, P. (1983) Delayed hypersensitivity and migration inhibition in two lines of mice genetically selected for high or low responsiveness to phytohaemagglutinin. *Cellular Immunology* 77, 249–65.

Gavora, J.S. and Spencer, J.L. (1978) Breeding for genetic resistance to disease: specific or general? *World's Poultry Science Journal* 34, 137–48.

Geczy, A.F. and Rothwell, T.L.W. (1981) Genes within the major histocompatibility complex of the guinea pig influence susceptibility to *Trichostrongylus colubriformis* infection. *Parasitology* 82, 281–6.

Gray, G.D., Presson, B.L., Albers, G.A.A., Le Jambre, L.F., Piper, L.R. and Barker, J.S.F. (1987) Comparison of within- and between-breed variation in resistance to haemonchosis in sheep. In: McGuirk, B.J. (ed.), *Merino Improvement Programs in Australia.* Australian Wool Corporation, Melbourne, pp. 365–9.

Gray, G.D. (1991) Breeding for resistance to trichostrongyle nematodes in sheep. In: Owen, J.B. and Axford, R.F.E. (eds), *Breeding for Disease Resistance in Farm Animals,* CAB International, Wallingford, pp. 139–61.

Gregg, P., Dineen, J.K., Rothwell, T.L.W. and Kelly, J.D. (1978) The effect of age on the response of sheep to vaccination with irradiated *Trichostrongylus colubriformis* larvae. *Veterinary Parasitology* 4, 35–48.

Groth, D.M., Wetherall, J.D., Outteridge, P.M., Windon, R.G., Richards, B. and Lee, I.R. (1987) Analysis of C3 and C4 in ovine plasma. *Complement* 4, 12–20.

Haldane, J.B.S. (1949) Disease and evolution. *La Ricerca Scientifica Supplement* 19, 68–76.

Handlinger, J.H. and Rothwell, T.L.W. (1981) Studies on the responses of basophil and eosinophil leucocytes and mast cells to the nematode *Trichostrongylus colubriformis*: comparison of cell populations in parasite resistant and susceptible guinea pigs. *International Journal for Parasitology* 11, 67–70.

Hong, C., Michel, J.F. and Lancaster, M.B. (1987) Observations on the dynamics of worm burdens in lambs infected daily with *Ostertagia circumcincta. International Journal for Parasitology* 17, 951–6.

Hotson, I.K. (1985) New developments in nematode control: the role of the animal health products industry. In: Anderson, N. and Waller, P.J. (eds), *Resistance in Nematodes to Anthelmintic Drugs.* CSIRO Division of Animal Health, Sydney, pp. 117–25.

Ibanez, O.M., Reis, M.S., Gennari, M., Ferreira, V.C.A., Sant'Anna, O.A., Siquerira, M. and Biozzi, G. (1980) Selective breeding for high and low antibody-responder lines of guinea pigs. *Immunogenetics* 10, 283–93.

Jones, V.E. and Ogilvie, B.M. (1971) Protective immunity to *Nippostrongylus brasiliensis*: the sequence of events which expels worms from the rat intestine. *Immunology* 20, 549–61.

Jones, W.O., Windon, R.G., Steel, J.W. and Outteridge, P.M. (1990) Changes in histamine and leukotriene concentrations in duodenal tissue and mucus after challenge in sheep which are high and low responders to vaccination with *Trichostrongylus colubriformis. International Journal for Parasitology* 20, 1075–9.

Katz, F.E. and Steward, M.W. (1975) The genetic control of antibody affinity in mice. *Immunology* 29, 543–8.

Kelly, J.D., Dineen, J.K. and Love, R.J. (1973) Expulsion of *Nippostrongylus brasiliensis* from the intestine of rats: evidence for a third component in the rejection mechanism. *International Archives of Allergy and Applied Immunology* 45, 767–79.

Krausslich, H. (1984) Possibilities and limits of breeding for immune responsiveness. *Journal of the South African Veterinary Association* 55, 11–17.

Lie, O. (1977) Genetic variation in the antibody response of young bulls. *Acta Veterinaria Scandinavica* 18, 572–4.

Maingi, N., Scott, M.E. and Prichard, R.K. (1990) Effect of selection pressure for thiabendazole resistance on fitness of *Haemonchus contortus* in sheep. *Parasitology* 100, 327–35.

Miller, H.R.P. (1990) Immunity to internal parasite. *Revue Scientifique et Technique, Office International des Epizooties* 9, 301–13.

Murray, P.K. (1989) Molecular vaccines against animal parasites. *Vaccine* 7, 291–9.

Ogilvie, B.M. and Hockley, D.J. (1968) Effects of immunity on *Nippostrongylus brasiliensis* adult worms: reversible and irreversible changes in infectivity, reproduction and morphology. *Journal of Parasitology* 54, 1073–84.

Outteridge, P.M., Windon, R.G. and Dineen, J.K. (1985) An association between a lymphocyte antigen in sheep and the response to vaccination against the parasite *Trichostrongylus colubriformis*. *International Journal for Parasitology* 15, 121–7.

Outteridge, P.M., Windon, R.G. and Dineen, J.K. (1988) An ovine lymphocyte antigen marker for acquired resistance to *Trichostrongylus colubriformis*. *International Journal for Parasitology* 18, 853–8.

Outteridge, P.M., Windon, R.G., Dineen, J.K. and Smith E.F. (1986) The relationship between ovine lymphocyte antigens and faecal egg counts of sheep selected for responsiveness to vaccination against *Trichostrongylus colubriformis*. *International Journal for Parasitology* 16, 369–74.

Piper, L.R. (1987) Genetic variation in resistance to internal parasites. In: McGuirk, B.J. (ed.), *Merino Improvement Programs in Australia*. Australian Wool Corporation, Melbourne, pp. 351–63.

Piper, L.R. and Barger, I.A. (1988) Resistance to gastrointestinal strongyles: feasibility of a breeding programme. In: *Proceedings of the 3rd World Congress of Sheep and Beef Cattle Breeding*, Paris, 19–23 June, Vol. 1, pp. 593–611.

Presson, B.L., Gray, G.D. and Burgess, S.K. (1988) The effect of immunosuppression with dexamethasone on *Haemonchus contortus* infections in genetically resistant Merino sheep. *Parasite Immunology* 10, 675–80.

Riffkin, G.G. and Dobson, C. (1979) Predicting resistance of sheep to *Haemonchus contortus* infections. *Veterinary Parasitology* 5, 365–78.

Riffkin, G.G. and Yong, W.K. (1984) Recognition of sheep which have innate resistance to Trichostrongylid nematode parasites. In: Dineen, J.K. and Outteridge, P.M. (eds), *Immunogenetic Approaches to the Control of Endoparasites, with Particular Reference to Parasites of Sheep*. CSIRO Division of Animal Health, Melbourne, pp. 30–40.

Rothwell, T.L.W. (1989) Immune expulsion of parasitic nematodes from the alimentary tract. *International Journal for Parasitology* 19, 139–68.

Rothwell, T.L.W., Le Jambre, L.F., Adams, D.B. and Love, R.J. (1978) *Trichostrongylus colubriformis* infection in guinea pigs: genetic basis for variation in susceptibility to infection among outbred animals. *Parasitology* 76, 201–9.

Schwartz, R.H. (1985) T lymphocyte recognition of antigen in association with gene products of the major histocompatibility complex. *Annual Reviews of Immunology* 3, 237–61.

Siegal, P.B. and Gross, W.B. (1980) Production and persistence of antibodies in chickens to sheep erythrocytes. 1. Directional selection. *Poultry Science* 59, 1–5.

Sitepu, P. and Dobson, C. (1982) Genetic control of resistance to infection with *Nematospiroides dubius* in mice: selection of high and low immune responder populations of mice. *Parasitology* 85, 73–84.

Sitepu, P., Dobson, C. and Brindley, P.J. (1984) Infection with *Salmonella typhimurium* and *Nippostrongylus brasiliensis* in mice selectively bred for high or low immune responsiveness to *Nematospiroides dubius*. *Australian Journal of*

Experimental Biology and Medical Science 62, 755–61.

Smith, N.C. (1989) The role of free oxygen radicals in the expulsion of primary infections of *Nippostrongylus brasiliensis. Parasitology Research* 75, 423–38.

Smith, W.D. and Angus, K.W. (1980) *Haemonchus contortus*: attempts to immunize lambs with irradiated larvae. *Research in Veterinary Science* 29, 45–50.

Smithers, S.R., Terry, R.J. and Hockley, D.J. (1968) Host antigens in schistosomiasis. *Proceedings of the Royal Society B* 171, 483–94.

Sprent, J.F.A. (1962) Parasitism, immunity, and evolution. In: Leeper, G.W. (ed.), *The Evolution of Living Organisms.* Melbourne University Press, Melbourne, pp. 149–65.

Steward, M.W., Reinhardt, M.C. and Staines, N.A. (1979) The genetic control of antibody affinity. Evidence from breeding studies with mice selectively bred for either high or low affinity antibody production. *Immunology* 37, 697–703.

van der Zijpp, A.J. and Nieuwland, M.G.B. (1989) The Biozzi model applied to the chicken. *Current Topics in Veterinary Medicine and Animal Science* 52, 160–8.

van der Zijpp, A.J., Scott, T.R., Glick, B. and Kreukniet, M.B. (1988) Interference with the humoral immune response in diverse genetic lines of chickens. I. The effect of carrageenan. *Veterinary Immunology and Immunopathology* 20, 53–60.

Wakelin, D. (1975a) Genetic control of immune responses to parasites: selection for responsiveness and non-responsiveness to *Trichuris muris* in random-bred mice. *Parasitology* 71, 377–84.

Wakelin, D. (1975b) Genetic control of immunity to parasites. *Parasitology* 71, xxv.

Wakelin, D. (1985) Genetic control of immunity to helminth infections. *Parasitology Today* 1, 17–23.

Waller, P.J. (1987) Anthelmintic resistance and the future for roundworm control. *Veterinary Parasitology* 25, 177–91.

Windon, R.G. (1990) Selective breeding for the control of nematodiasis in sheep. *Revue Scientifique et Technique, Office International des Epizooties* 9, 555–76.

Windon, R.G. and Dineen, J.K. (1981) The effect of selection of both sire and dam on the response of F1 generation lambs to vaccination with irradiated *Trichostrongylus colubriformis* larvae. *International Journal of Parasitology* 11, 11–18.

Windon R.G. and Dineen, J.K. (1984) Parasitological and immunological competence of lambs selected for high and low responsiveness to vaccination with irradiated *Trichostrongylus colubriformis* larvae. In: Dineen, J.K. and Outteridge, P.M. (eds), *Immunogenetic Approaches to the Control of Endoparasites, with Particular Reference to Parasites of Sheep.* CSIRO Division of Animal Health, Melbourne, pp. 13–28.

Windon, R.G., Dineen, J.K. and Wagland, B.M. (1987) Genetic control of immunological responsiveness against the intestinal nematode *Trichostrongylus colubriformis* in lambs. In: McGuirk, B.J. (ed.), *Merino Improvement Programs in Australia.* Australian Wool Corporation, Melbourne, pp. 371–5.

Woolaston, R.R. (1990) Genetic improvement of resistance to internal parasites in sheep. In: *Proceedings of the 8th Conference of the Australian Association of Animal Breeding and Genetics,* Ruakura, 5–9 February, pp. 163–71.

Chapter 11
Breeding for Helminth Resistance in Sheep and Goats

Lucas Gruner

INRA, Station de Pathologie Aviaire et de Parasitologie,
37380 Nouzilly, France

Summary

Is breeding for resistance useful for farmers or is it a tool for further studies on resistance mechanisms? The interest and limits of the different kinds of study are discussed, taking as examples some French results on the susceptibility of two breeds of sheep and two breeds of goat against *Teladorsagia circumcincta* and other helminth species including Proto-strongyles.

The distribution of helminth species depends on ecological and management factors and on the level of resistance. It is shown that resistance is not innate but is progressively acquired after a number of antigen stimulations. The regulation of the worm population occurs at different stages depending on the species of parasite. The rapidity of establishment and efficacy of the host resistance also depends on the parasite species and the susceptibility of sheep and goats to parasitic infection are not equal. The acquired resistance can be removed by anthelmintic or corticosteroid treatment and an important failure in the immune status occurs in lactating ewes. The specificity of the resistance is also limited.

As resistance is a complex and dynamic process of host–parasite relationships, no one simple parasitological parameter reflects these relations. Egg count is used in selection programmes and immunological markers are associated with resistant or susceptible animals. Selected strains are economically interesting in heavily infested areas; however it is as important to remove the most contaminative individuals from the flock.

Introduction

Sheep and goats show high variability in their susceptibility or resistance to helminth parasites. This is the result of a variety of factors. The principal effects examined are those related to sex, age, stage of reproductive cycle, plane of nutrition and genotype. The possibility of contact between parasite and host is regulated by ecological, behavioural and breeding management factors. Several examples are given to illustrate the relative importance of these factors on the occurrence of a helminth species and on the level of parasitism.

The terms resistance and susceptibility are notoriously difficult to define and they have been used in a variety of senses. How the resistance in a host is developed illustrates the difficulty of measurement. The assessment of resistance is of utmost importance in the setting up of the selection programmes.

Small ruminants harbour more than one species, from the same or from different families and from orders such as Nematodes, Cestodes and Trematodes. So it is important to verify the extent of the specificity of the resistance.

That certain breeds of sheep or goats are more resistant to helminth parasites than others was first documented by Stewart *et al.* (1937) for infection of sheep by a digestive-tract nematode. Since then numerous studies have been done and several reviews are available: gastrointestinal nematode and trypanosome infections (Dargie, 1982), gastrointestinal and pulmonary nematode infections (Gruner and Cabaret, 1988). It is also known that within breeds, the progeny of certain rams or ewes have increased resistance to gastrointestinal nematode infection (Albers *et al.*, 1984; Whitlock, 1958).

Two approaches have been followed in the development of ruminants that are genetically resistant to parasites: the exploitation of animal breeds having increased helminth resistance or within-breed selection for resistance. Gray (1991) and Windon (1991) present Australian sheep breeding programmes for resistance to *Haemonchus contortus* and *Trichostrongylus colubriformis*. The interest and limits of these different kinds of studies are discussed taking as examples some of our results on the susceptibility of sheep and goats to *Teladorsagia circumcincta* and other helminth species including Protostrongyles. Is breeding for resistance useful for farmers or is it a tool for further studies on the resistance mechanism?

The diversity of helminth parasites

There are mainly two types of development cycles for helminth parasites; in monoxen types (nematode species except Protostrongyles), adults in the host lay eggs that are deposited with faeces on the pasture. They develop into infective larvae that can continue their development only after ingestion by a host. The heteroxen cycle comprises one or two intermediate hosts (an acarid for tapeworms, an aquatic or terrestrial mollusc respectively for liver fluke and Protostrongyles, an ant and a terrestrial mollusc for the small liver fluke).

In a survey in sheep flocks from the Limousin region, the flock levels of parasitism were compared in relation to ecological and management factors. The parasites with intermediate hosts were primarily correlated with soil nature and secondarily with climatic factors; gastrointestinal nematodes were primarily correlated with spring rainfall and to a lesser extent to management factors. In another survey in 25 flocks of goats, management factors explained more than 60% of the variability of parasitic infection (Cabaret *et al.*, 1989). These considerations emphasize the importance of ecological factors in the distribution of the helminth species.

In France, the helminth fauna present in the Central and Western part with an atlantic climate can be distinguished from those in regions with a mediterranean climate, by using parasitological surveys in sheep and goat flocks (Table 11.1). In the first case, the most frequent gastrointestinal nematodes are: *Teladorsagia circumcincta, Trichostrongylus colubriformis, T. vitrinus* and, often at the end of the summer, *Haemonchus contortus. Nematodirus* species and *Moniezia* species are also present. Protostrongyles (*Muellerius* sp.) are predominantly found in goats. *Fasciola hepatica* is common. By contrast, in mediterranean regions, gastrointestinal nematodes (*T. circumcincta, T. vitrinus, Nematodirus* spp.) are as important as small lungworms (*Muellerius capillaris, Cystocaulus ocreatus* and *Neostrongylus linearis*). *Dicrocoelium lanceolatum* is very common, mainly in the East.

The pathogenicity of the different species is not the same and the effects on production, food intake, digestive upsets, and anaemia are relatively well known. There are very few data available for the protostrongyles: Pandey *et al.* (1984) demonstrated in Morocco their effects on ewe viability and fertility. Similar results were obtained in dry areas of Syria where gastrointestinal worms and lungworms occurred in chronic infection (Thomson *et al.*, 1990).

Resistance and its measurement

Resistance is commonly defined as the ability of the animal to suppress establishment and/or subsequent development of worm infection. The ability of the animal to maintain existing levels of productivity when infected is called resilience by the Australian researchers. The first aspect is illustrated with observations on sheep and on goats in France during recent years.

Table 11.1.   Diversity of helminth parasites in sheep (S) and goats (G) in France; 1, % of animals with positive egg or larval counts; 2, egg and larval counts; 3, worm burdens in mixed flocks; 4, % of animals with positive egg or larval counts; 5, sheep grazing irrigated pastures.

Parasite	Atlantic						Mediterranean		
	1		2 adult		3 young		4		5
	S	G	S	G	S	G	S	G	S
Digestive tract									
total strongyles			180	478			35	82	
Teladorsagia					3,100	1,900			++
Haemonchus					900	1,000			−
Trichostrongylus colubriformis				)	17,000	11,000			−
T. vitrinus				)					++
Nematodirus spp.	8	0.3			7	0	18	7	++
Moniezia spp.					+	−	16	44	+
Liver									
Dicrocoelium	9	1					23	17	++
Fasciola									+
Lungs									
Dictyocaulus									+
Protostrongyles	11	5	<1	40	−	−	18	72	++

Source: 1, Kerboeuf and Godu (1981); 2, 3, Mangeon and Cabaret (1987); 4, Brunet (1981); 5, Gruner and Cabaret (1985).

Innate or acquired resistance

In a flock of 300 ewes equally distributed between the Romanov and Merino d'Arles breeds and their crossbreeds, a significant breed effect was observed in the course of coproscopy surveys. To verify if this difference was innate, 60 three-month-old lambs, half Romanov and half Merino, were experimentally infected with one dose of 7,000 infective larvae (L3) of *T. circumcincta*. A month later, half the lambs were slaughtered and the worm burdens established. No breed difference was significant, either for egg counts or for worm burdens. The frequency of the worms in the host population was adjusted to a normal distribution (Fig. 11.1). The 30 remaining lambs were slaughtered after two months: the worm distribution was always normal, but worm mortality during this second month after

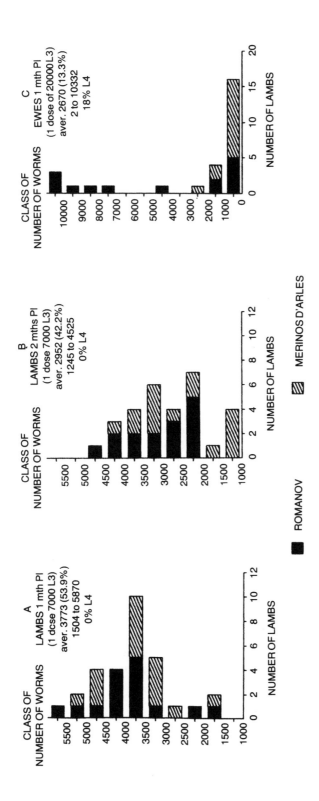

Fig. 11.1. Distribution of *Teladorsagia circumcincta* worms in experimentally infected Romanov and Merinos d'Arles sheep: three-month-old lambs slaughtered one (A) or two (B) months post-infection, and adult ewes (C) slaughtered one month post-infection (30 animals per group).

infection was higher in Merino lambs than in Romanov ones.

Similarly, adult ewes (grazing a pasture infected with different tricho-strongylid species during two to six years, then deparasitized), were infected with a dose of 20,000 L3 of the same *T. circumcincta* strain. The worm frequency after a month was adjusted to a negative binomial distribu-tion, with a very significant breed effect. This means that some animals remained very susceptible to this species, although most ewes had few worms (Gruner *et al.*, 1990). All the worms were adults in the lambs; in 90% of the ewes immature worms were present, and the average propor-tion remaining at the 4th stage was 18% of the burden.

Morphological study of the worms revealed no effect on the sex-ratio, but a reduction of the bodylength of the females and of the number of eggs *in utero* per female.

It was concluded that:

1. An innate resistance did not exist in young lambs when their immunocompetence was not yet established;

2. In contrast, adults having been in contact with the parasite for a long period resisted a challenge dose with two strategies: restraining the establishment of the worms and reducing the rapidity of worm develop-ment.

Similar results were observed for Saanen and Alpine milk goats: in mixed flocks, a breed effect was significant in trichostrongylid and proto-trongylid egg and larval outputs (Richard *et al.*, 1990). Two-month-old kids, experimentally infected with *T. circumcincta*, showed no breed differ-ence after a single infection (Richard and Cabaret, 1986). In naïve adults grazing pasture artificially infected with *T. circumcincta*, egg counts began to differ between the two breeds after four months (Richard, 1988). Resist-ance is acquired after a number of antigenic stimulations (ingestion of infective larvae); the speed of acquisition and its efficacy in terms of expul-sion of present worms and delayed development of new larvae depend on the individual host.

These examples lead to a better understanding of the different steps where resistance can act: (1) the rate of establishment of newly ingested infective stages; (2) the rate and the rapidity of development of the different stages into adults; (3) the survival of the adults; (4) the proli-ficacy of the females; and (5) the ability of the eggs to develop into free-living stages in the external environment (pastures).

Different types of population regulation

To obtain a greater understanding of the regulation of the worm population inside the host, Barger (1987) concluded that more useful information can be obtained from experiments using continuous infection with several estimates of worm population size. The mode of regulation is not the same for every species. For *T. circumcincta*, the number of larvae at the fourth stage in the abomasum lumen increases, the strategy of adult worms is cumulative with a density-dependent regulation of the number of eggs per female. Thus, very quickly, the egg output of the host becomes a poor reflection of the level of infection (Jackson and Christie, 1979). For *Trichostrongylus colubriformis*, *T. axei* and *T vitrinus*, different studies demonstrate a common strategy; there is a consistent sequence of accumulation of worms for a period inversely related to larval intake and age of the lamb, followed by resistance to establishment of new infection. The result is a plateau of total worm burden, as a consequence of an accumulation of larvae at the third stage and an expulsion of the older worm population. During the first two of these phases, egg production reflects the underlying worm burden, but the final expulsion of worms is preceded by a reduced faecal egg output and a decline in numbers of eggs *in utero* of female worms. For *Haemonchus contortus*, there is a good correlation between egg output and worm burden; with a new infection, there is a rapid expulsion of the adult worm population followed by a decrease of the egg output, then the development of the new population takes place.

For these main species, the death rate of adult worms is very dependent on the intake rate of infective larvae.

As a consequence of these different strategies of worm population regulation, the appreciation of the level of resistance is not alike for all species. The proportion of infective larvae that remain as L3 for *T. colubriformis* or L4 for *T. circumcincta* match with resistance; for *H. contortus*, the decrease in egg counts is a measure of the self-cure.

The rapidity of establishment and efficacy of the resistance

These depend on the parasite species. Young lambs acquired a good resistance against *Nematodirus* spp. after they had grazed for two or three months on infected pasture. Lambs are able to develop a good protection against *H. contortus* by the age of four months, whereas it is eight months against *T. circumcincta* and *T. colubriformis* (Barger, 1988). Six-month-old lambs, after eight weeks of daily dosing with 1,000 L3 of *T. vitrinus*, showed almost total resistance to establishment of incoming worms (Seaton *et al.*, 1989). For Protostrongyles, older animals are more heavily infested than younger ones (Cabaret *et al.*, 1980). However in experimental infection, the small lungworm *Muellerius capillaris* can develop into mature adults in 3–5-month-old lambs and kids (Sauerländer, 1988).

The susceptibility of sheep and goats to parasitic infection is not equal. Data are available from mixed flocks (Table 11.1). Goats are more susceptible to Trichostrongylid (with higher egg counts) and Protostrongylid infections (mainly *Muellerius* sp.) but are more resistant to *Nematodirus* sp. or *Moniezia* sp.. Nine-month-old Saanen goats had good protection against *T. colubriformis* 10 weeks after experimental infection at a dose of 10,000 L3/week (Pomroy and Charleston, 1989a). By the same method, goats failed to acquire resistance against *H. contortus* (Pomroy and Charleston, 1989b).

The stability of the acquired resistance

In *H. contortus* infection, the protection is more effective against a challenge infection if a moderate worm burden remains (Donald *et al.*, 1969). For this species, the protection is lost after an anthelmintic treatment (Benitez-Usher *et al.*, 1977; Douch, 1989; Luffau *et al.*, 1985). Immune animals without antigenic challenges kept good protection against a challenge infection after 42 days but they had entirely lost it after 84 days (Jackson *et al.*, 1988a). Corticosteroid treatment (with dexamethasone) totally abolished the protection and permitted the establishment of worms (Matthews *et al.*, 1979; Presson *et al.*, 1988). These last authors showed that selected resistant and susceptible sheep had the same worm establishment after dexamethasone treatment.

Another important failure in the immune status occurs in the parturient ewe. The increase of the egg counts is well known for Trichostrongylid infection from the last month of pregnancy until three to four months of lactation (periparturient rise). This was also observed in Protostrongylid infections (Cabaret *et al.*, 1980). In fact, the intensity of lactation is the main factor influencing this perturbation. In a comparison of Romanov and Merino breeds, four flocks of ewes of the two breeds and their cross were examined: one comprising dry ewes, the second with ewes weaned just after lambing, the third with ewes suckling one lamb, and the fourth with ewes suckling twins. Egg counts remained low and similar in the first two flocks with a significant breed effect (Romanov was greater than Merino and crossbred). Egg counts were much higher in the lactating flocks, and particularly for ewes with twins. Highest egg counts were observed in stressed ewes, i.e. Merino and young Romanov ewes with two lambs. From observations of the grazing behaviour, it appeared that these ewes ingested more herbage and infective larvae (Gruner *et al.*, 1989). Jackson *et al.* (1988b) demonstrated that the decrease of resistance in the ewes' last six weeks of pregnancy was expressed against *T. circumcincta* and not against *T. vitrinus*.

Criteria for genetic selection for resistance

As resistance is a complex and dynamic process of host–parasite relationships, no simple parasitological parameter seems to depict these relations. A selection on the limitation of worm establishment is interesting, as pathological effects depend on the worm number for a given species. However, no estimation of immature worm burden is possible without slaughtering the animal! The level of serum pepsinogen gives an appreciation of the worm burden of the abomasum. The limitation of egg excretion (egg counts and duration of the output) is of interest to limit pasture contamination and protect the entire flock. Egg and larval outputs are the easiest parameters that can be measured. In different situations, these counts represent the mature worm burdens.

Within breed comparison, egg or larval counts were often used, whereas few estimations of worm burdens or lung lesions were recorded (Gruner and Cabaret, 1988). Courtney *et al.* (1986) measured the periparturient rise and found that the Florida native ewes did not show this phenomenon, which is of the highest importance for pasture contamination.

The Australian selection for *Haemonchus*-resistant sheep is based on the result of one larval infective dose. How does this very peculiar criterion reflect the ability to limit the establishment of further infections? Some answer is given in the comparison of natural and experimental infection of the offspring of the most resistant ram compared with susceptible ones (Albers and Gray, 1989). The New Zealand programme is based on egg counts of naturally infected sheep on pastures where *Teladorsagia* and *Trichostrongylus* were present (Baker *et al.*, 1988; Watson *et al.*, 1986). This measure is a good reflection of the acquired resistance, and after three generations, 500 eggs/g of faeces separated the susceptible and resistant strains. This type of protocol cannot be used against a parasite such as *Haemonchus*, that is very dependent on an unpredictable rainfall. The heritability of a divergent selection for resistance to *T. circumcincta* and the genetic correlation between the natural and the experimental infections, run simultaneously on two similar flocks, are presently studied at INRA.

Other types of parameter have been tried as indicators of resistance. Performance parameters are rarely specific and the interpretation of their variation is unclear in field conditions. Albers *et al.* (1984) used PCV depletion as a reflection of the *Haemonchus* burden.

Immunological parameters are investigated as responsiveness to infection. Immunologically based resistance to helminths can be mediated through the inflammatory reaction or through mechanisms involving antibodies or complement-dependent cellular toxicity. Lymphoblastic and IgA-containing cells in the lymph played an important role in the protection of *Teladorsagia* and *Haemonchus* infections (Smith *et al.*, 1987; Smith, 1988). The number of globule leucocytes in the small intestine mucosa was higher in immunized-responder Romney lambs subject to a challenge infection with *T. colubriformis* (Douch, 1988). However, peripheral blood cellular and humoral responses did not differ in high and low responders (resistant or susceptible) to challenge infection with *T. colubriformis* after vaccination (Dawkins *et al.*, 1988). Progress is being made to find and to characterize host protective antigens that could be used as markers (O'Donnell *et al.*, 1989a, b; Outteridge *et al.*, 1988). The OLA combination SY1a+1b was found in more resistant sheep in Australia and also in New Zealand (Douch and Outteridge, 1989).

The specificity of resistance

Are animals selected on their resistance to a target species, resistant to other species? High responder lambs after vaccination against *T. colubriformis* were more resistant than low responders to challenge with *T. rugatus* and *T. circumcincta* but not with *H. contortus* (Windon and Dineen, 1984). Vaccination of lambs with irradiated larvae or parasite extracts of *T. colubriformis* or *H. contortus* gave only a specific protection (Adams, 1989; Adams *et al.*, 1989). Sheep immunized against *T. colubriformis* grazing a polyparasitized pasture before slaughter, showed a significant protection against the same species, but also to *T. axei*, *Nematodirus spathiger* and *T. circumcincta* (Douch, 1989). Greater numbers of immature stages of the last two species were present in the worm burdens, with a lower general establishment rate of the worms.

The last question is whether the parasite is able to cope with the host-protective immunity. Results with *H. contortus* showed that the helminth was unable to make this adjustment (Adams, 1988; Albers and Burgess, 1988).

Conclusion

The exploration of the genetic variability in helminth resistance of sheep and goats is a rich field of investigation because of the variety of host–parasite associations. The observations of repeatability permit the conclusion that the most susceptible or resistant individuals keep their status all their lives. Resistance seems to be as heritable as many performance traits, so selection is possible. Resistant lines of sheep and goats might be of economic interest in infected areas where production losses due to helminthosis are severe. However, it may be as important to remove the most contaminative individuals that perpetuate parasitism in the flock. To measure the selection progress, the resistant line is generally compared to a susceptible one or to a control flock without selection pressure. In fact, the two lines are of great interest for further studies on the mechanisms of resistance.

References

Adams, D.B. (1988) Infection with *Haemonchus contortus* in sheep and the role of adaptative immunity in selection of the parasite. *International Journal for Parasitology* 18, 1071–5.

Adams, D.B. (1989) A preliminary evaluation of factors affecting an experimental system for vaccination-and-challenge with *Haemonchus contortus* in sheep. *International Journal for Parasitology* 19, 169–75.

Adams, D.B., Anderson, B.H. and Windon, R.G. (1989) Cross-immunity between *Haemonchus contortus* and *Trichostrongylus colubriformis* in sheep. *International Journal for Parasitology* 19, 717–22.

Albers, G.A.A. and Burgess, S.K. (1988) Serial passage of *Haemonchus contortus* in resistant and susceptible sheep. *Veterinary Parasitology* 28, 303–6.

Albers, G.A.A. and Gray, G.D. (1989) The genetics of parasite resistance in sheep. In: Van Der Zijpp, A.J. and Sybesna, W. (eds), *Improving Genetic Resistance in Farm Animals*. Kluwer, Dordrecht, pp. 153–9.

Albers, G.A.A., Burgess, S.K., Adams, D.B., Barker, J.S.F., Le Jambre, L.F. and Piper, L.R. (1984) Breeding *Haemonchus contortus* resistant sheep. Problems and prospects. In: Dineen, J.K. and Outteridge, P.M. (eds), *Immunogenetic Approaches to the Control of Endoparasites*. Division of Animal Health, CSIRO, Melbourne, Australia, pp. 41–51.

Baker, R.L., Watson, T.G. and Harvey, T.G. (1988) Genetic variation in, and selection for, resistance or tolerance to internal nematode parasites in sheep. *Proceedings 3rd World Congress on Sheep and Beef Cattle Breeding*, Paris, t1, pp. 637–9.

Barger, I.A. (1987) Population regulation in Trichostrongylids of ruminants. *International Journal for Parasitology* 17, 531–40.

Barger, I.A. (1988) Resistance of young lambs to *Haemonchus contortus* infection, and its loss following anthelmintic treatment. *International Journal for Parasitology* 18, 1107–9.

Benitez-Usher, C., Armour, J., Duncan, J.L., Urquhart, G.M. and Gettinby, G. (1977) A study of some factors influencing the immunization of sheep against *Haemonchus contortus* using attenuated larvae. *Veterinary Parasitology* 3, 327–42.

Brunet, J. (1981) Le parasitisme des caprins dans l'Ardèche (1977–1978–1979). *Bulletin des Groupements Vétérinaires* 3, 58–66.

Cabaret, J., Dakkak, A. and Bahaida, B. (1980) On some factors influencing the output number of larvae of Protostrongylids of sheep in nature conditions. *Veterinary Quarterly* 2, 115–20.

Cabaret, J., Anjorand, N. and Leclerc, C. (1989) Parasitic risk factors on pastures of French dairy goat farms. *Small Ruminant Research* 2, 69–78.

Courtney, C.H., Gessner, R., Sholz, S.R. and Loggins, P.E. (1986) The periparturient rise in fecal egg counts in three strains of Florida Native ewes and its value in predicting resistance of lambs to *Haemonchus contortus*. *International Journal for Parasitology* 16, 185–9.

Dargie, J.D. (1982) The influence of genetic factors on the resistance of ruminants to gastrointestinal nematode and trypanosome infections. In: Owen, D.G. (ed.), *Animal Models in Parasitology*, Macmillan, London pp. 17–51.

Dawkins, H.J.S., Windon, R.G., Outteridge, P.M. and Dineen, J.K. (1988) Cellular and humoral responses of sheep with different levels of resistance to *Trichostrongylus colubriformis*. *International Journal for Parasitology* 18, 531–7.

Donald, A.D., Dineen, J.K. and Adams, D.B. (1969) The dynamics of the host-parasite relationship – VII. The effect of discontinuity of infection on resistance to *Haemonchus contortus* in sheep. *Parasitology* 59, 497–503.

Douch, P.G.C. (1988) The response of young Romney lambs to immunization with *Trichostrongylus colubriformis* larvae. *International Journal for Parasitology* 18, 1035–8.

Douch, P.G.C. (1989) The effects of immunization of sheep with *Trichostrongylus colubriformis* larvae on worm burdens acquired during grazing. *International Journal for Parasitology* 19, 177–81.

Douch, P.G.C. and Outteridge, P.M. (1989) The relationship between ovine lymophocyte antigens and parasitological and production parameters in Romney sheep. *International Journal for Parasitology* 19, 35–41.

Gray, G.D. (1991) Breeding for resistance to trichostrongyle nematodes in sheep. In: Owen, J.B. and Axford, R.F.E. (eds), *Breeding for Disease Resistance in Farm Animals*. CAB International, Wallingford, pp. 139–61.

Gruner, L., Bechet, G., Surhyahadi, S. and Cabaret, J. (1989) Relations between trichostrongylosis, grazing activity and larval distribution in lactating and non-lactating ewes of two breeds. *WAAVP 13th Conference*, 7–11 August, Berlin, GDR, abstract S2–9.

Gruner, L. and Cabaret, J. (1985) Utilisation des parcours méditerranéens et parasitisme interne des ovins. In: *Exploitation des milieux difficiles par les Ovins et les Caprins*, ITOVIC-SPEOC eds, pp. 307–35.

Gruner, L. and Cabaret, J. (1988) Resistance of sheep and goats to helminth infections: a genetic basis. In: Thomson, E.F. and Thomson, F.S. (eds), *Increasing Small Ruminants Productivity in Semi-arid Areas*, ICARDA, pp. 257–65.

Gruner, L., Cabaret, J., Cortet, J. and Sauvé, C. (1990) Distribution d'un *Trichostrongylidae* chez des ovins de deux races, jeunes et agés *VIIth International Congress of Parasitology*, 20–24 August, Paris, France.

Jackson, F. and Christie, M.G. (1979) Observation on the egg output resulting from continuous low level infections with *Ostertagia circumcincta* in lambs. *Research in Veterinary Science* 27, 244–5.

Jackson, F., Jackson, E. and Williams, J.T. (1988b) Susceptibility of the pre-parturient ewe to infection with *Trichostrongylus vitrinus* and *Ostertagia circumcincta*. *Research in Veterinary Science* 45, 213–18.

Jackson, F., Miller, H.R.P., Newlands, G.F.J., Wright, S.E. and Hay, L.A. (1988a) Immune exclusion of *Haemonchus contortus* larvae in sheep: dose dependency, steroid sensitivity and persistence of the response. *Research in Veterinary Science* 44, 320–3.

Kerboeuf, D. and Godu, J. (1981) Les strongyloses gastrointestinales. Données épidémiologiques et diagnostic chez les Caprins. *Bulletin des Groupements Vétérinaires* 3, 67–84.

Luffau, G., Pery, P. and Carrat, C. (1985) Interférence entre vermifugation et immunité dans les strongyloses gastrointestinales du mouton. *Annales de Recherches Vétèrinaires* 16, 17–23.

Mangeon, N. and Cabaret, J. (1987) Infestation comparée des ovins et des caprins en pâturages mixtes. *Bulletin des Groupements Techniques Vétérinaires* 4, 43–8.

Matthews, D., Brunsdon, R.V. and Vlassoff, A. (1979) Effect of dexamethasone on the ability of sheep to resist reinfection with nematodes. *Veterinary Parasitology* 5, 65–72.

O'Donnell, I.J., Dineen, J.K., Waglang, B.M., Letho, S., Werkmeister, J.A. and Ward, C.W. (1989a) A novel host-protective antigen from *Trichostrongylus colubriformis. International Journal for Parasitology* 19, 327–35.

O'Donnell, I.J., Dineen, J.K., Wagland, B.M., Letho, S., Dopheide, T.A.A., Grant, W.N. and Ward, C.W. (1989b) Characterization of the major immunogen in the excretory–secretory products of exsheathed third-stage larvae of *Trichostrongylus colubriformis. International Journal for Parasitology* 19, 793–802.

Outteridge, P.M., Windon, R.G. and Dineen, J.K. (1988) An ovine lymophocyte antigen marker for acquired resistance to *Trichostrongylus colubriformis. International Journal for Parasitology* 18, 853–8.

Pandey, V.S., Cabaret, J. and Fikri, A. (1984) The effect of strategic anthelmintic treatment on the breeding performance and survival of ewes naturally infected with gastrointestinal strongyles and protostrongyles. *Annales de Recherches Vétérinaires* 15, 491–6.

Pomroy, W.E. and Charleston, W.A.G. (1989a) Development of resistance to *Trichostrongylus colubriformis* in goats. *Veterinary Parasitology* 33, 283–8.

Pomroy, W.E. and Charleston, W.A.G. (1989b) Failure of young goats to acquire resistance to *Haemonchus contortus. New Zealand Veterinary Journal* 37, 23–6.

Presson, B.L., Gray, G.D. and Burgess, S.K. (1988) The effect of immunodepression with dexamethasone on *Haemonchus contortus* infection in genetically resistant Merino sheep. *Parasite Immunology* 10, 675–80.

Richard, S. (1988) Parasitisme helminthique des caprins: susceptibilité comparée des races Alpine et Saanen. Thèse Université de Tours, 104 pp.

Richard, S. and Cabaret, J. (1986) Caractéristiques de la réponse de jeunes chevreaux à une primo-infestation par le nématode *Teladorsagia circumcincta. Bulletin de la Société Française de Parasitologie* 4, 245–6.

Richard, S., Cabaret, J. and Cabourg, C. (1990) Genetic and environmental factors associated with nematode infection of dairy goats in Northwestern France. *Veterinary Parasitology* 36, 237–43.

Sauerländer, R. (1988) Experimental infection of sheep and goats with *Muellerius capillaris* (*Protostrongyldae, Nematoda*) *Journal of Veterinary Medicine, B* 35, 525–48.

Seaton, D.S., Jackson, F., Smith, W.D. and Angus, K.W. (1989) Development of immunity to incoming radiolabelled larvae in lambs continuously infected with *Trichostrongylus colubriformis. Research in Veterinary Science* 46, 22–6.

Smith, W.D. (1988) Mechanisms of immunity to gastrointestinal nematodes of sheep. In: Thomson, E.F. and Thomson, F.S. (eds), *Increasing Small Ruminant Productivity in Semi-arid Areas*, ICARDA, pp. 275–86.

Smith, W.D., Jackson, F., Graham, R., Jackson, E. and Williams, J. (1987) Mucosal IgA production and lymph cell traffic following prolonged low level infections of *Ostertagia circumcinta* in sheep. *Research in Veterinary Science*, 43, 320–6.

Stewart, M.A., Miller, R.F. and Douglas, J.R. (1937) Resistance of sheep of different breeds to infestation by *Ostertagia circumcincta. Journal of Agricultural Research* 55, 923–30.

Thomson, E.F., Orita, G., Bahhady, F.A., Rhodes, C. and Giangasperro, M. (1990) Effect of treating ewes with an anthelmintic on mortality and lambing rate in farm flocks. *CIHEAM/EAAP meeting*, 10–17 October, Rabat, Morocco.

Watson, T.G., Baker, R.L. and Harvey, T.G. (1986) Genetic variation in resistance or tolerance to internal nematode parasites in strains of sheep at Rotomahana. *Proceedings of the New Zealand Society of Animal Production* 46, 23–6.

Whitlock, J.H. (1958) The inheritance of resistance to trichostrongylosis in sheep-I. Demonstration of the validity of the phenomena. *Cornell Veterinary* 48, 127–33.

Windon, R.G. (1991) Genetic control of host responses involved in resistance to gastrointestinal nematodes of sheep. In: Owen, J.B. and Axford, R.F.E. (eds), *Breeding for Disease Resistance in Farm Animals*. CAB International, Wallingford, pp. 162–86.

Windon, R.G. and Dineen, J.K. (1984) Parasitological and immunological competences of lambs selected for high and low responsiveness to vaccination with irradiated *Trichostrongylus colubriformis*. In: Dineen, J.K. and Outteridge, P.M. (eds), *Immunological Approaches in the Control of Endoparasites*, Division of Animal Health, CSIRO, Melbourne, Australia, pp. 13–29.

Section 4
Breeding for Resistance to Diseases Involving Flies and Ticks

Flies and ticks are involved in a number of major problems for farm animals both in terms of direct effects, as in the case of the blowfly and, even more important, indirect effects as vectors of disease. One of the most striking and promising developments reported in this book is the progress made in combating tsetse fly transmitted trypanosomiasis in cattle and sheep. The work reported on the identification of trypanotolerant animals, particularly among the N'Dama cattle in Africa, promises to be of major importance to agricultural progress, potentially releasing a large area of Africa for cattle production. Ticks have also posed problems throughout the world, through their effect in causing anaemia and by their involvement in transmitting a wide range of tick-borne diseases. Like the tsetse fly, they have made certain locations, admittedly more geographically confined than in the case of the tsetse fly, 'no-go' areas for cattle and sheep. Tick resistance in cattle breeds is now a well-documented phenomenon and the development of resistance in cattle is a means of reducing the expense of acaricides and the dangers of selecting for ticks resistant to them.

Finally the work on developing disease resistance in sheep (particularly Merinos) to fly strike and fleece rot shows an encouraging picture, with heritability estimates of 0.3–0.4 for susceptibility to these diseases. It is now possible to conceive of effective practical breeding schemes with the inclusion of resistance as an objective.

Chapter 12

Trypanosomiasis in Cattle:
Prospects for Control

Max Murray[1], M.J. Stear[1], J.C.M. Trail[2], G.D.M. d'Ieteren[2],
K. Agyemang[3] and R.H. Dwinger[3]
[1]*Department of Veterinary Medicine, University of Glasgow,
Veterinary School, Bearsden, Glasgow G61 1QH, UK,*
[2]*International Livestock Centre for Africa (ILCA), PO Box
46847, Nairobi, Kenya and* [3]*International Trypanotolerance
Centre (ITC), PMB 14, Banjul, The Gambia*

Summary

Tsetse-transmitted trypanosomiasis is possibly the major constraint on
livestock and agricultural production in Africa. The disease affects both
humans and their livestock with the sociological, medical and veterinary
costs estimated at over 50 billion dollars annually. No vaccine is available
because of the phenomenon of antigenic variation. Currently, control of the
disease is largely dependent on reduction of tsetse fly populations by
insecticides or on the use of trypanocidal drugs. However, recent advances
in our knowledge of the biology of tsetse, of trypanosomes and of the host,
are offering hope for alternative methods of control. In particular, there is
an increasing awareness that the use of African cattle could contribute to
reducing the ravages of this disease. Over a period of several thousand
years, certain breeds of cattle, such as the taurine N'Dama and West
African Shorthorn have developed the ability to thrive in tsetse-infested
areas where there is a high risk of trypanosome infection. This trait has
been termed trypanotolerance. These trypanotolerant breeds of cattle are
environmentally extremely well adapted and are now recognized as having
considerable production potential. Trypanotolerance is associated with the
ability to control parasitaemia and to resist the development of anaemia in
the face of infection. Both these criteria, are highly heritable and genetically
correlated with production. Thus, the means now exists to identify trypanotolerant animals and begin rational breeding programmes to maximize
cattle production in the vast tsetse-infested areas of Africa.

Introduction

Pathogenic species of salivarian trypanosomes are present throughout vast areas of Africa, Asia, Latin America and the Middle East and cause disease in cattle, sheep, goats, water buffalo, pigs, horses, camels, wildlife and man. In Africa, the major pathogenic trypanosome species are transmitted by the tsetse fly (genus *Glossina*) and include *Trypanosoma congolense, T. vivax, T. brucei* and *T. simiae.* Two closely related subspecies of *T. brucei, T. rhodesiense* and *T. gambiense* cause sleeping sickness in man. Non-tsetse transmitted forms of trypanosomiasis also occur in Africa, as well as in the Middle East, Asia and Latin America. The most important pathogen, under these circumstances, is *T. evansi.* This parasite can cause severe disease in horses and camels and lead to significant losses in production and performance in cattle and water buffalo.

Currently, tsetse infest 11 million km^2 of Africa, about 37% of the continent, affecting 40 countries (FAO–WHO–OIE, 1982). It is considered that 7 million km^2 of this area would otherwise be suitable for livestock and mixed agriculture without stress to the environment, if trypanosomiasis could be controlled (MacLennan, 1980). No other continent is dominated by one disease as Africa is dominated by tsetse-transmitted trypanosomiasis. Tsetse-transmitted trypanosomiasis not only results in severe losses in production in domestic livestock due to poor growth, weight loss, low milk yield, reduced capacity for work, infertility and abortion, but excludes domestic livestock from large areas of Africa. Currently, of a total population of approximately 173 million cattle, only about 44 million are located in the tsetse-infested zone (IBAR, 1989). The situation with regard to sheep, goats, pigs, horses, donkeys and camels is probably as serious but it is poorly documented.

Domestic animals and wildlife act as reservoir hosts for the human pathogens *T. rhodesiense* and *T. gambiense.* Human trypanosomiasis is an important constraint on rural development in Africa. It causes disruption of communities with the resultant depletion of human resources on which agricultural communities depend. At present, it is estimated that some 50 million people in some 36 countries are at risk (Molyneux, 1986); there are several serious active foci of the disease in Africa, e.g., in Uganda, Sudan, Zambia, Ivory Coast, Zaire, Angola and Mozambique and the number of new cases reported every year has increased to around 20 thousand (Molyneux, 1986).

Thus, the presence of tsetse results in widespread rural instability, as well as causing severe losses in production in a massive area of Africa. In 1963, the annual loss in meat production alone was estimated at US$ five billion (Murray and Gray, 1984). This figure excludes the value of milk, hides and the importance of cattle in mixed agriculture. In Africa, 80% of traction power is non-mechanized. It has been calculated that the availability of a draught ox to a family unit can increase agricultural output sixfold (McDowell, 1977). Furthermore, the manure provided by livestock is essential for the production of food and cash crops and is a potential source of energy in the form of biogas. If all these factors are considered, it has been estimated that livestock and agricultural development of tsetse-infested Africa could generate a further US$50 billion annually. Because of the paucity of reliable data, even this figure may be a gross under-estimate.

Many factors contribute to the magnitude of the problem. One is the complexity of the disease. In cattle, three species of trypanosome, *T. congolense*, *T. vivax* and *T. brucei*, cause disease, either individually or jointly. These trypanosomes are transmitted cyclically by tsetse, of which there are some 36 species and subspecies, each adapted to different climatic and ecological conditions (Ford, 1971). While tsetse are not the only vectors of African trypanosomes, cyclical transmission of infection re-presents the most important problem because, once the tsetse fly becomes infected, it remains infective for a long period, in contrast to the ephemeral nature of non-cyclical transmission. At the same time, trypanosomes infect a wide range of hosts including wild and domestic animals. The former do not suffer severe clinical disease but become carriers and constitute an important reservoir of infection. The success of the trypanosome as a parasite is to a large extent due to the ability to undergo antigenic variation, i.e., change a single glycoprotein (Variant Surface Glycoprotein (VSG): Cross, 1975) which covers the pellicular surface, thereby evading host immune responses and establishing a persistent infection. Added to the complexity of multiple variable antigen types expressed during a single infection, each trypanosome species comprises an unknown number of different strains or serodemes, all capable of elaborating a different repertoire of variable antigen types (Van Meirvenne *et al.*, 1977). As a result, no effective vaccine has been produced for use in the field.

Background to control

In addition to the complexity of the disease and its epidemiology, other factors contributing to the failure to contain and reduce the problem include the enormous area affected and the limitations of the methods currently available for control.

The main approaches to control have been directed towards eradicating or reducing the number of tsetse or have involved the use of trypanocidal drugs. Both can be effective if properly applied, but all too frequently they cannot be sustained and the net effect to date at the continental level has been limited. However, recent developments in the understanding of tsetse biology, in the strategic use of drugs, in our knowledge of trypanosome biology, and, in the realization of the importance of host genetic resistance have identified promising new approaches.

Control of tsetse populations

Attempts to control tsetse have been made for over 60 years. Initially, they included eradication of wildlife, clearing of fly barriers to prevent the advance of the vector, and widespread clearing of bush to destroy breeding habitats. Subsequently, the principal method employed to control tsetse has been the use of insecticides. The insecticides used fall into two categories, residual and non-residual. Residual insecticides (DDT and more recently dieldrin) are usually applied as a single application by hand-operated sprays that deliver the insecticide to sites where resting tsetse are known to alight. Non-residual insecticides (endosulphan) require several applications, and are applied mainly by fixed wing aircraft or helicopters. Where insecticide control measures have been properly implemented significant success was achieved, e.g., Nigeria, Zimbabwe, Bostwana and Zambia (MacLennan, 1980, 1981). Despite the potential efficacy of tsetse control by insecticides, there are severe limitations to this approach. The costs are high, and in Africa, there is a lack of trained personnel to implement insecticide control programmes. Furthermore, natural or man-made barriers are required to defend sprayed areas and prevent reinvasion, and constant surveillance for early detection of reinvasion is essential. Finally, there are increasing demands for restricted use of insecticides because of possible environmental impact on fauna and flora. In this respect, it should be emphasized, however, that although non-target organisms can be affected by antitsetse spraying and quantities of insecticide remain in the environment, these effects appear to be transitory, rarely lasting for more than a year (Jordan, 1986). Current research involves the development of new potent insecticides with low toxic environmental effect, e.g., synthetic pyrethroids.

More recently, a completely new approach to tsetse control has been conceived. Traps and screens have been used for many years as a means of sampling tsetse populations. However, with advances in the design, and the identification of colours and odours which attract tsetse, increasing attention is being given to the use of traps and targets as a method of tsetse control. Elegant studies carried out in Zimbabwe (Vale, 1987) have culminated in the development of simple insecticide-impregnated visual targets incorporating chemicals such as carbon dioxide, acetone and 1-octen-3-ol which attract tsetse. These developments are of considerable importance with respect to introducing a simple, and environmentally safe method of control. However, this approach has not been successful with all species of tsetse and costs can be high. There is hope that future research may remove these constraints. Another development in odour-baited control has been extended to the concept of a live odour-producing target, by the treatment of cattle with synthetic pyrethroids applied as a dipwash, a spray or a pour-on formulation. Preliminary results are promising (Connor, 1989).

The use of trypanocidal drugs

Escalating costs and other constraints, discussed above on initiating and maintaining tsetse control campaigns, have led to the livestock industries in the vast tsetse-infected areas of Africa being almost completely reliant on the use of trypanocidal drugs to prevent or treat the disease. Without these drugs, the situation would be even more disastrous. Despite the need and demand for effective trypanocides, no new drug has been produced for commercial use in the last 30 years and there would appear to be no immediate prospects of new drugs becoming available. The cost of registering a new drug for use in animal trypanosomiasis is regarded by pharmaceutical companies as too high in relation to the likely financial return. This is despite the fact that the potential market includes some 120 million cattle, sheep and goats which are exposed to infection. Even if these animals were treated only twice per year, potentially some 240 million doses would be required.

There is an increasing number of reports of the successful use of trypanocidal drugs in cattle under ranch or village management. Thus, some 12,000 Boran are maintained in Mkwaja Ranch in Tanzania in an area where Boran cattle rapidly succumb to trypanosomiasis if left untreated. As a result of the strategic use of isometamidium chloride (a prophylactic trypanocidal drug) in combination with diminazene aceturate (a therapeutic trypanocidal drug), the level of productivity achieved was close to that of Boran reared in tsetse-free conditions on ranches in Kenya considered among the best in the world (Trail *et al.*, 1985). At the same time, a similar drug strategy was implemented in East African Zebu cattle (700 head) under village management in Kenya and resulted in a 20% increase in performance (Maloo *et al.*, 1988). In both these situations, the level of tsetse challenge was considered high. On the other hand, where disease risk is low and where it is possible to examine individual animals at regular intervals, therapeutic trypanocidal drug control strategies have been successfully employed, e.g., at Kilifi Plantations on the coast of Kenya. This dairy ranch is one of the biggest in Africa, supporting 800 breeding females based on Sahiwal × Ayrshire. The owner was virtually out of business because of trypanosome-induced abortion storms until he successfully introduced a systematic therapeutic drug strategy which he has maintained for over 20 years (Murray *et al.*, 1982; Wissocq *et al.*, 1983).

The fact that these control programmes were carried out in villages and ranches, on large numbers of cattle, over a long period of time, with financially successful results, and with no evidence of significant drug resistance, offers some hope in short-term for livestock and consequently socioeconomic development programmes in tsetse-infested areas of Africa. Nevertheless, it must be emphasized that the implementation of drug control programmes does require a degree of competent management and the constant availability of trypanocidal drugs. At the same time, if no new drugs are developed, there must be concern that the repeated use of the drugs currently available could lead to serious drug resistance problems in the long-term.

Prospects for vaccination

The major constraint to the development of a vaccine against trypanosomiasis is the phenomenon of antigenic variation (Murray and Urquhart, 1977). However, while the repertoire of these antigens generated by bloodstream forms of the parasite is large (greater than 1,000), the repertoire of antigens produced by metacyclic parasites following transmission through the tsetse is much more limited and would appear relatively constant (Crowe *et al.*, 1983). Thus, it has been possible to immunize cattle and goats against tsetse-transmitted homologous (but not heterologous) strains of *T. congolense* and *T. brucei* (Morrison *et al.*, 1985), but not *T. vivax* (Emery *et al.*, 1987), by prior exposures to metacyclic parasites. Nevertheless, the feasibility of production and the efficacy of a vaccine against metacyclic trypanosomes will depend on the relative stability of the metacyclic antigen repertoire for each species of trypanosome and on the number of strains which occur in the field. Currently, research is directed towards these objectives. It is thought, however, that the number of different strains is likely to be prohibitively large for the production of a cocktail vaccine containing the appropriate metacyclic antigens. As a result, the development of a vaccine against African trypanosomiasis has been considered unlikely.

Most research has concentrated on the VSGs of the trypanosome and data on the subcellular distribution and properties of other antigens are surprisingly scarce. Recently, a flagellar pocket membrane fraction has been identified in *T. rhodesiense* on the parasite surface at the emergence of the flagellum from the flagellar pocket. This would appear to be non-variable between different serodemes and to have protective potential (Olenick *et al.*, 1988). At the same time, receptor-mediated endocytosis of low-density lipoprotein (LDL) and transferrin has been demonstrated with *T. brucei* (Coppens *et al.*, 1987, 1988). While cholesterol is the major sterol in the membrane of the trypanosome, there is no evidence that it can be synthesized by the trypanosome. Cholesterol is not freely available in the mammalian bloodstream but is buried within LDL particles. As a result, it was hypothesized and confirmed by studies *in vitro* that the ability to endocytose LDL is essential for optimal trypanosome growth: removal of LDL or addition of antibodies against the purified LDL receptors inhibits growth. The receptor appears to be highly conserved and to be localized to membrane of the flagellar pocket and to be completely absent from the rest of the pellicular membrane.

It has also been demonstrated that *T. brucei* parasites bind epidermal growth factor (EGF) and that binding modifies protein kinase activity and the growth rate of the parasites *in vitro* (Hide *et al.*, 1989). Furthermore, antibodies to mammalian EGF receptor were found to precipitate a surface polypeptide of the parasite which in turn bound EGF. Thus, it would appear that trypanosomes possess a surface growth factor receptor with considerable homology to the EGF receptor, although its location in the parasite remains to be identified.

Thus, possible target(s) for either chemotherapy or immunotherapy do exist and renewed consideration can be given to the feasibility of developing new drugs or a vaccine.

Trypanotolerance: genetic resistance

Because of the limitations of the current methods for control and the likelihood that a vaccine will not become available in the immediate future, increasing consideration is now being given to the use of trypanotolerant breeds of domestic animals as a sustainable approach to livestock development in tsetse-infested areas. It has long been recognized that certain breeds of cattle, as well as many species of wild Bovidae and Suidae, possess the ability to survive and to be productive in tsetse-infested areas without the aid of treatment where other breeds rapidly succumb to the disease (Pierre, 1906; Murray *et al.*, 1982). This trait is termed trypanotolerance and is generally attributed to the *Bos taurus* breeds of cattle in West and Central Africa, namely the N'Dama (Fig. 12.1; Roberts and Gray, 1973) and the West African Shorthorn (Roelants, 1986). While there is also evidence that significant differences in resistance to trypanosomiasis occur among *Bos indicus* breeds (Cunningham, 1966; Njogu *et al.*, 1985), most *Bos indicus* cattle in tsetse-infested areas require regular treatment or are found only on the fringes of fly belts. Imported breeds cannot be maintained even in areas of low tsetse risk without intensive drug therapy.

It is thought, on the basis of rock paintings and engravings, that the Hamitic Longhorn breed, from which the N'Dama is descended, arrived in the Nile Delta from the Near East at about 5000 BC, and the Shorthorn cattle were introduced into the same area between 2750 and 2500 BC

Fig. 12.1. N'Dama cow in The Gambia.

(Epstein, 1971; Payne, 1964). On the other hand, *Bos indicus* cattle, which are the most prevalent cattle type in Africa, did not become numerous in sub-Saharan Africa until after the Arab invasions of 669 AD, although they were recognized in Egypt between 2000 and 1500 BC. It is worth recalling that wild Bovidae, which are extremely resistant to trypanosomiasis (Murray *et al.*, 1981), emerged in Africa some 20–40 million years ago (Leakey and Lewin, 1977) and that tsetse probably originated even earlier (Ford, 1971).

While trypanotolerant breeds of cattle are a well-recognized component of livestock production in West and Central Africa, they represent only a small proportion of Africa's cattle population. We estimate that of the 44 million cattle believed to be located in tsetse-infested areas (IBAR, 1989), only 10.7 million or 24% can be regarded as being related to the trypanotolerant breeds (Hoste *et al.*, 1989); 4.86 million are N'Dama, 2.95 million are West African Shorthorn and 2.89 million are crosses between zebu and trypanotolerant breeds, i.e., only about 6% of Africa's 173 million cattle are of the trypanotolerant type. Failure to exploit these breeds can be attributed to the belief that they are not productive because of their small size (Stephen, 1966) and to the view that their trypanotolerance was limited to resistance to local trypanosome populations and that, as a result, their 'tolerance' would break down if they were moved to distant tsetse-infested locations where different trypanosome strains were present.

However, these views have not been substantiated by more recent detailed investigations. In a survey of the status of trypanotolerant cattle in 18 countries in West and Central Africa, indices of productivity were computed using all the production data available for different breeds (ILCA, 1979). It was found in areas where the tsetse fly risk was low or zero that the productivity of the N'Dama and West African shorthorn was equal to that of the physically larger trypanosusceptible Zebu. Directly comparable data between breeds were not available in many areas because the level of tsetse challenge was such that only trypanotolerant breeds were present. Work carried out at the International Trypanotolerance Centre (ITC) in The Gambia is now providing remarkable new data on the productive potential of the N'Dama. In the past, little attention was paid to milk production in N'Dama and it was generally assumed to be too low even to record. However, recent analysis of 668 lactation records of N'Dama cattle maintained under village management in the Gambia showed a milk offtake of between 400 and 600 kg over a lactation period of approximately 14 months (Agyemang *et al.*, in press). Such a yield is impressive for animals which had an average bodyweight of 225 kg and which had to survive a seven-month long dry season under constant tsetse challenge. The overall productivity index incorporating milk offtake, calf weaning weight, calving rate and viability gave an annual value (Agyemang *et al.*, in press), much higher than that recorded for Zebu, under similar management elsewhere in Africa (Otchere, 1983). Furthermore, supplementary feeding of N'Dama with local by-products, which are normally discarded, resulted in growth rates as much as 1 kg per day, producing two-year-old animals weighing over 300 kg (D.A. Little, personal communication). Thus, the N'Dama, hitherto largely ignored as a provider of milk and meat, possess considerable potential as a dual purpose animal and appear to have on a metabolic weight basis an overall performance capacity superior to that of certain other larger breeds.

With respect to the basis of resistance to trypanosomiasis, evidence that trypanotolerance is not due only to resistance acquired to local trypanosome populations has been provided by the successful establishment of cattle from West Africa in distant tsetse-infested areas of West and Central Africa (Fig. 12.2). An example is the introduction of Lagune (a West African Shorthorn) in 1904 and N'Dama in 1920 into Zaire (Mortelmans and Kageruka, 1976) and more recently N'Dama into the Central African Republic, Gabon and Congo (ILCA, 1979).

Criteria for trypanotolerance

Field and experimental studies carried out on several different breeds of cattle, including Ayrshire, Friesian, Holstein, Hereford and their crosses, as well as indigenous African breeds such as Zebu, Boran, West African shorthorn and N'Dama have confirmed the superior resistance, or trypano-

Fig. 12.2. N'Dama heifers being exported from The Gambia.

tolerance, of the latter two breeds, as judged by the ability to resist the effects of infection, i.e., not only to survive, but to gain weight and re-produce. This was shown for *T. congolense*, *T. vivax* and *T. brucei*, the three species of trypanosome which are pathogenic in cattle (Murray *et al.*, 1982; Murray and Dexter, 1988). It was found that the intensity, pre-valence and duration of the accompanying parasitaemia were reduced in the trypanotolerant breeds compared to the other breeds (Murray and Morrison, 1979). At the same time, anaemia, which is an inevitable and probably the single most important pathogenic consequence of a trypano-some-infection in cattle (Hornby, 1921; Murray, 1979), was consistently less severe in trypanotolerant breeds (Dargie *et al.*, 1979). Thus, the ability to control parasitaemia and to resist anaemia appears as key indicators of the trypanotolerance trait.

The validity of the foregoing experiments and the conclusions drawn have been and must be called into question for several reasons. In many studies, the disease history, in particular whether or not a previous infection with trypanosomes had been experienced, was not known. In some cases, groups of animals were exposed to field challenge or wild caught tsetse, a situation in which it is not possible to guarantee a uniform infective challenge. In other experiments, cattle were infected by syringe inoculation with bloodstream forms of the trypanosome; some workers, but not ourselves, believe that this method of infection might give misleading results as the skin in which the tsetse deposits the infective organisms, is bypassed by syringe inoculation. Furthermore, in the majority of studies, the organisms used for the challenge infection had not been characterized in terms of virulence, antigenicity and are likely to have consisted of a mixture of different strains.

However, all these variables were eliminated in an experiment carried out at the International Laboratory for Research on Animal Diseases, Kenya (Paling *et al.*, 1988). The N'Dama in this study were obtained as embryos from donors in The Gambia and implanted into surrogate Boran mothers in Kenya (Jordt *et al.*, 1986). All cattle were born and reared in an area of Kenya, free from trypanosomes. Starting at one year of age, eight of these N'Dama were each infected on four consecutive occasions with one of four clones of *T. congolense* transmitted by the tsetse *Glossina morsitans centralis*. While each clone was known to belong to a different antigenic serodeme of *T. congolense*, all had been shown to be virulent. The infectivity of every tsetse used was confirmed prior to use on cattle. Boran (*Bos indicus*) cattle of corresponding age were challenged at the same time. All animals were maintained on a high *ad libitum* plane of nutrition. To prevent death any infected animal with a packed red cell volume (PCV) value equal to 15% or less was treated with the trypanocidal drug, diminazene aceturate; previous experience had shown it was at this PCV value or less that death might occur.

This investigation completely confirmed the superior resistance of N'Dama cattle when compared to Boran. N'Dama with no previous experience of trypanosomiasis born to Boran dams that had never encountered the parasite, nevertheless resisted the effects of infection by trypanosomes. During the course of the four different challenge experiments, no N'Dama required trypanocidal drug treatment; in contrast, all infected Boran did. Moreover, while severe weight losses were experienced by all the trypanosome-infected Boran, compared to uninfected controls, trypanosome infection did not affect liveweight gains in the N'Dama, when compared to uninfected N'Dama controls. In contrast to the Boran, infected female N'Dama continued to show normal oestrous cycle activity and were even successfully superovulated when infected (Lorenzini *et al.*, 1988).

As in previous studies, the superior capacity of the N'Dama in resisting the effects of infection appeared to lie in the ability to control parasitaemia and resist anaemia. Dargie *et al.* (1979) have suggested that the N'Dama's ability to resist anaemia was in fact a direct reflection of its capacity to control parasitaemia. However, in Paling's (1988) studies when mean parasitaemia and PCV values of N'Dama were computed for individual animals over four infection periods, no direct correlation could be established, i.e. certain N'Dama demonstrated high PCV values during all four infections whereas others showed better parasite control; individual animals showed the ability to resist anaemia or to control parasitaemia in a consistent and repeatable fashion for all four experiments. It was, therefore, concluded that while both processes are under genetic control, they are not the same trait.

Another criterion for trypanotolerance may be related to the immune response. As parasitaemia wave remission is effected by antibodies directed against the surface coat antigens of the trypanosome (Murray and Urquhart, 1977), it has generally been assumed that the superior capacity of trypanotolerant animals to control parasitaemia is associated with a better immune response. However at present, there are only a few preliminary studies in N'Dama infected with *T. vivax* (Desowitz, 1959) and Bauole (West African shorthorn) infected with *T. congolense* (Akol *et al.*, 1986; Pinder *et al.*, 1988) which indicate that this might be the case; this is an important area requiring further research. Differences in cellular responses between breeds of cattle resulting in differential generation of key cytokines, such as tumour necrosis factor (Hotez *et al.*, 1984), might also play a role in parasite growth regulation and or in erythropoietic responses.

Thus, at least three criteria can be associated with trypanotolerance: the ability to resist anaemia, the ability to control parasitaemia and possibly the ability to mount a more effective immune response. It is important to know how each of these processes relate to the capacity of the animal to perform and be productive, and if any of them could be used as a marker to select for the trypanotolerance trait.

In the past, it has not always been clear whether the ability of certain breeds to survive in a tsetse-infested area was due to the fact that such animals were less liable to tsetse attack or to the fact that they were more tolerant to the effects of infection. Although it is not known if trypanotolerant breeds of cattle are less liable to tsetse attack, it has been shown that where tsetse challenge is high trypanotolerant breeds such as the N'Dama are just as likely to become infected as trypanosusceptible Zebu (Murray *et al.*, 1981). At the same time, N'Dama and Zebu are equally susceptible to the establishment of an experimental infection transmitted by tsetse (Paling *et al.*, 1988).

Thus, it has been clearly shown that trypanotolerance must be regarded as an innate capacity of the host to resist the effects of infection or as defined by Pagot (1974) the 'racial aptitude (of cattle) to maintain themselves in good condition and to reproduce while harbouring trypanosomes without showing clinical signs of the disease'.

It is now essential that critical experiments similar to that of Paling *et al.* (1988) are carried out on other breeds of cattle believed to be trypanotolerant, such as the West African shorthorn and the Orma Boran, as well as candidate breeds of sheep, e.g. the West African Djallonke, Red Maasai, and goats, e.g. the West African Dwarf and East African, to determine if the foregoing conclusions relating to N'Dama cattle can be applied to other breeds of cattle or to other species.

Environmental factors affecting trypanotolerance

While it has now been definitively established that the trypanotolerance characteristic is a genetically determined trait, there is evidence to indicate that its stability can be affected by several environmental factors. Over-work, intercurrent disease, repeated bleeding, as well as the stress of pregnancy, parturition, suckling and lactation have been incriminated as factors reducing trypanotolerance (Murray *et al.*, 1981, 1982). However, in this respect, probably the most important single factor is nutritional status. It has been found that the lack of adequate nutrition which can occur under field conditions can result in increased severity of anaemia and weight loss in trypanosome-infected N'Dama (Agyemang *et al.*, 1990). However, when such cattle were supplemented even with only small amounts of local by-products, including groundnut cake, rice bran and milled *Andropogon* hay, not only were growth rates improved but the anaemia which developed was less severe. It was of interest that the plane of nutrition did not alter the initial rate of development of anaemia, indicating that nutrition had no effect on the initial trypanosome assault on the erythropoietic system. However, subsequently cattle on the improved diet were apparently able to respond more effectively as they recovered from the trypanosome-induced anaemia more rapidly.

There are several reports which indicate that trypanotolerance can be enhanced by previous exposure (reviewed by Murray *et al.*, 1982). An explanation for these findings has been provided by the critical experiments carried out by Paling *et al.* (1988). It was found over the course of four infection periods in N'Dama cattle infected with *T. congolense* that the overall severity of the anaemia produced became progressively and significantly less (Fig. 12.3), despite the fact that there was no evidence of acquisition of immunity to the parasite, as assessed by the intensity, prevalence and duration of parasitaemia in each of the four infection periods. It was concluded that N'Dama cattle possessed an innate ability to acquire resistance to the disease, as assessed by 'increasing resistance to anaemia'. Corresponding erythropoietic responses were not observed in Boran cattle (Fig. 12.3).

The genetic control of trypanotolerance

Stewart (1951) reported cross-breeding studies involving N'Dama, Zebu and Ghanaian Shorthorn (a trypanotolerant genetic mix) in which he produced a larger, more productive animal that retained its resistance to trypanosomiasis. Similarly, it was found that N'Dama/Zebu crossbreeds retained a significant degree of trypanotolerance when exposed to natural tsetse challenge (Chandler, 1952). In breeding experiments in the Ivory Coast involving large numbers of N'Dama and Jersey, it was observed that

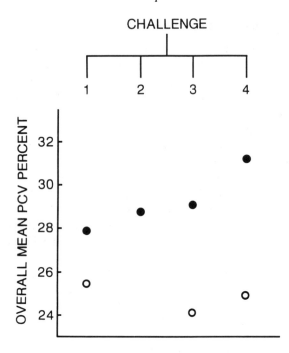

Fig. 12.3. The mean PCV over each infection period in N'Dama (●) and Boran cattle (○) subjected on four consecutive occasions to challenge with tsetse infected with one of four different clones of *Trypanosoma congolense*. Whereas with each challenge the overall mean PCV of the N'Dama progressively increased, no change occurred with the Boran. There were no data for the second challenge experiment on the Boran. (Data reproduced with permission of Paling *et al.*, 1988.)

the F1 cross produced an excellent animal as regards growth and milk production (Letteneur, 1978). It was stated that such crosses retained their tolerance, although no information was given on the level of tsetse challenge or on the prevalence of trypanosomes. However, crossbreeds with greater than 50% Jersey background appeared to be less hardy.

A major constraint to establishing estimates of heritability of trypanotolerance in cattle has been related to the difficulty of defining and measuring the trait, i.e., what precisely is trypanotolerance (Dolan, 1987). As described, it has now been shown that at least two factors are associated with the trait, the ability to control parasitaemia and the ability to resist anaemia. Moreover, evidence has been accumulating that the severity of anaemia, as measured by PVC value, in trypanosome-infected cattle is correlated with production traits, such as reproductive performance and growth, suggesting the PCV values, and possibly parasitaemia profiles, during the course of a trypanosome infection might serve as selection criteria for trypanotolerance (Trail *et al.*, 1988).

Recently, these possibilities, were considered in field research programmes involving N'Dama cattle in Zaire and Gabon (Trail *et al.*, 1990a, b; 1991). These programmes were designed to measure the effect of control of anaemia on animal productivity relative to that of other aspects of trypanotolerance, e.g., control of parasitaemia; to look at the practicality of its assessment early in an animal's life; and to evaluate the possibility of its improvement through a genetic selection programme, or by use of marker genes, or both.

In brief, it was found that the ability to resist anaemia and control parasitaemia was significantly associated with N'Dama performance in both adults and calves, the capacity to control PCV had a fourfold greater effect on performance than the capacity to control parasitaemia. The repeatability of PCV values between calving intervals was shown to be reasonably high and almost equal to that of calf weaning weight. Subsequently, when all environmental and parasitaemia information was taken into account, the heritability of growth and of average PCV following infection reached 0.39 $\pm$ 0.31 and 0.64 $\pm$ 0.33 (estimate $\pm$ standard error) respectively, with a genetic correlation of 0.70 $\pm$ 0.42 between average PCV and growth. As both the heritability and the genetic correlation increased when para-sitaemia data were included in the analysis, clearly the inclusion of para-sitaemia evaluation in the assessment of productive potential was advantageous and important.

As major differences in the capacity to control anaemia could be deter-mined within six weeks of infection and the effect of calf PCV values on performance was at least as important as dam PCV values, the possibility now exists of carrying out selection tests for trypanotolerance on young animals.

Other genetic attributes of the N'Dama

In addition to the trypanotolerance trait, trypanotolerant breeds of cattle and in particular the N'Dama would appear to have other genetic ad-vantages that must contribute to their potential for use in livestock develop-ment programmes in the tropics. They are reported to be resistant to several other important infectious diseases, including a number of tick-borne infections such as dermatophilosis (Coleman, 1967; Stewart, 1937), heart-water (*Cowdria ruminantium*), anaplasmosis and babesiosis (Epstein, 1971). These findings might indicate a greater resistance to ticks *per se*. N'Dama may also possess some degree of resistance to helminthiasis (H. Kaufmann, personal communication). In this respect, the Red Maasai sheep of Kenya have been shown to be significantly more resistant than the Merino not only to trypanosomiasis (Griffin and Allonby, 1979) but also to haemonchosis (Preston and Allonby, 1979). There is also evidence to indicate that indigenous African cattle such as the N'Dama as well as wild

Bovidae are more resistant to environmental constraints because of superior physiological adaptation in terms of food utilization, heat tolerance and water conservation (Murray *et al.*, 1982).

Discussion

It is now clear that Africa possesses unique genetic resources. These resources can now provide sustainable environmentally sound solutions for the vast disease problems currently confronting the continent. The resistance to trypanosomiasis and to several other important infectious diseases possessed by breeds of cattle such as the N'Dama can constitute an important additional approach to national and regional disease control programmes. The fact that these breeds also possess considerable production potential offers an unparalleled opportunity to improve livestock production in the vast areas of Africa dominated by tsetse, ticks and helminths.

Genetic resistance does not mean refractoriness to disease under all conditions. High levels of resistance and production need to be supported by adequate management and nutrition. Nevertheless, it is likely that the more-resistant breeds will be less demanding from a management point of view. For example, one might speculate that even where vaccines are or do become available, they will be more effective in cattle which also possess an innate natural resistance to the disease in question, as has been demonstrated in mouse model systems (Mitchell *et al.*, 1982). It could also be that in trypanotolerant cattle chemotherapy might be more effectively and economically applied both in terms of numbers of doses needed and also on the level of dose required. In this respect, it has been shown, at least in mice, that trypanocidal drugs are more efficacious in mice which are innately more resistant and whose immune system is intact (Bitonti *et al.*, 1986; De Gee *et al.*, 1983). It is also of interest that in trypanosome-infected N'Dama trypanocidal drug treatment resulted in a rapid complete recovery period for PCV value of one month, with a 70% recovery in 9 days (Trail *et al.*, 1990b). While there were no trypanosusceptible controls, in the authors' experience equivalent infections in Boran cattle would have taken several weeks to recover normal PVC values after trypanocidal drug therapy.

It is now known that the ability of N'Dama cattle to control parasitaemia and to resist the development of anaemia is correlated with the capacity to be productive, and that these criteria are under strong genetic control. These are extremely exciting findings. A means now exists for selecting trypanotolerant animals and for implementing rational breeding programmes with N'dama cattle.

An understanding of the mechanisms responsible for the regulation of parasite growth, for the control of development of anaemia, for the development of protective humoral and cellular responses, and for high productivity under conditions of trypanosome challenge is now required. Such knowledge might permit the identification of the genes which regulate resistance to disease thereby allowing the selection of stock without having to infect animals. It might also lead to the development of novel chemo-therapeutic and immunotherapeutic strategies targeted at the regulation of parasite growth or the induction of resistance to development of the anaemia. Finally, the identification of the loci responsible for trypano-tolerance might permit their isolation and subsequent transfection into trypanosusceptible breeds of cattle, possibly leading to the production of new breeds custom built for specific purposes.

Twenty years ago the N'Dama were regarded by experts as an historic relic and their trypanotolerance a doubtful biological oddity. It is now increasingly accepted that the introduction or expansion of N'Dama cattle into tsetse-infested areas may be the only practical sustainable way to increase livestock production and develop mixed farming in those areas.

References

Agyemang, K., Dwinger, R.H., Grieve, A.S. and Bah, M.L. (19$$) Milk production characteristics and productivity of N'Dama cattle kept under village management in The Gambia. *Journal of Dairy Science* (in press).

Agyemang, K., Dwinger, R.H., Touray, B.N., Jeannin, P., Fofana, D. and Grieve, A.S. (1990) Effects of nutrition on degree of anaemia and liveweight changes in N'Dama cattle infected with trypanosomes. *Livestock Production Science* 26, 39–51.

Akol, G.W.O., Authie, E., Pinder, M., Moloo, S.K., Roelants, G.E. and Murray, M. (1986) Susceptibility and immune responses of Zebu and taurine cattle of West Africa to infection with *Trypanosoma congolense* transmitted by *Glossina morsitans centralis*. *Veterinary Immunology and Immunopathology* 11, 361–73.

Bitonti, A.J., McCann, P.P. and Sjoerdsma, A. (1986) Necessity of antibody response in the treatment of African trypanosomiasis with α-difluoro-methylornithine. *Biochemical Pharmacology* 30, 331–4.

Chandler, R.L. (1952) Comparative tolerance of West African N'Dama cattle to trypanosomiasis. *Annals of Tropical Medicine and Parasitology* 46, 127–34.

Coleman, C.H. (1967) Cutaneous streptothricosis of cattle in West Africa. *Veterinary Record* 81, 251–4.

Connor, R.J. (1989) *Final Report of the Regional Trypanosomiasis Expert*. Regional Tsetse and Trypanosomiasis Control Programme Malawi, Mozambique, Zambia and Zimbabwe. European Communities, FGU-Kronberg Consulting Engineering GmbH.

Coppens, I., Baudhuin, P., Opperdoes, F.R. and Courtoy, P.J. (1988) Receptors for the host low density lipoproteins on the hemoflagellate *Trypanosoma brucei*: purification and involvement in the growth of the parasite. *Proceedings of the*

National Academy of Sciences of the United States of America 85, 6753–7.

Coppens, I., Opperdoes, F.R., Courtoy, P.J. and Baudhuin, P. (1987) Receptor-mediated endocytosis in the bloodstream form of *Trypanosoma brucei. Journal of Protozoology* 34, 465–73.

Cross, G.A.M. (1975) Identification, purification and properties of clone-specific glycoprotein antigens constituting the surface coat of *Trypanosoma brucei. Parasitology* 71, 393–417.

Crowe, J.S., Barry, J.D., Luckins, A.G., Ross, C.A. and Vickerman, K. (1983) All metacyclic variable antigen types of *Trypanosoma congolense* identified using monoclonal antibodies. *Nature* 306, 389–91.

Cunningham, M.P. (1966) Immunity in bovine trypanosmiasis. *East African Medical Journal* 43, 394–7.

Dargie, J.D., Murray, P.K., Murray, M., Grimshaw, W.T.R. and McIntyre, W.I.M. (1979) Bovine trypanosomiasis: the red cell kinetics of N'Dama and Zebu cattle infected with *Trypanosoma congolense. Parasitology* 78, 271–86.

De Gee, A.L.W., McCann, P.P. and Mansfield, J.M. (1983) Role of antibody in the elimination of trypanosomes after DL-difluoromethylornithine chemotherapy. *Journal of Parasitology* 69, 818–22.

Desowitz, R.S. (1959) Studies on immunity and host parasite relationships. I. The immunological response of resistant and susceptible breeds of cattle to trypanosomal challenge. *Annals of Tropical Medicine and Parasitology* 53, 293–313.

Dolan, R.B. (1987) Genetics and trypanotolerance. *Parasitology Today* 3, 137–43.

Emery, D.L., Moloo, S.K. and Murray, M. (1987) Failure of *Trypanosoma vivax* to generate protective immunity in goats against transmission by *Glossina morsitans morsitans. Transactions of the Royal Society of Tropical Medicine and Hygiene* 81, 611.

Epstein, H. (1971) *The Origin of the Domestic Animals of Africa.* Vols 1 and 2. Africana, New York.

FAO-WHO-OIE (1982) In: Kouba, V. (ed.), *Animal Health Yearbook 1981*, no. 18. Food and Agriculture Organization of the United Nations, Rome.

Ford, J. (1971) *The Role of the African Trypanosomiases in African Ecology: a Study of the Tsetse Fly Problem.* Clarendon Press, Oxford.

Griffin, L. and Allonby, E.W. (1979) Trypanotolerance in breeds of sheep and goats with an experimental infection *Trypanosoma congolense. Veterinary Parasitology* 5, 97–105.

Hide, G., Gray, A., Harrison, C.M. and Trait, A. (1989) Identification of an epidermal growth factor receptor homologue in trypanosomes. *Molecular Biochemistry and Parasitology*, 36, 51–60.

Hornby, H.E. (1921) Trypanosomes and trypanosomiasis of cattle. *Journal of Comparative Pathology* 34, 211–40.

Hoste, C.H., Chalon, E., d'Ieteren, G.D.M. and Trail, J.C.M. (1989) *Trypanotolerant Livestock in West and Central Africa*, vol. III. *Situation after a decade.* FAO Animal Production and Health Paper 20/3. Food and Agriculture Organization of the United Nations, Rome.

Hotez, P.J., Le Trang, N., Fairlamb, A.H. and Cerami, A. (1984) Lipoprotein lipase suppression in 3T3-L1 cells by a haemoprotozoan-induced mediator from peritoneal exudate cells. *Parasite Immunology* 6, 203–9.

IBAR (1989) Cattle Distribution Maps. Interafrican Bureau for Animal Resources.

Organisation of African Unity, Nairobi, Kenya.

ILCA (1979) *Trypanotolerant Livestock in West and Central Africa.* Monograph No. 2. International Livestock Centre for Africa, Addis Ababa, Ethiopia.

Jordan, A.M. (1986) *Trypanosomiasis Control and African Rural Development.* Longman, London and New York.

Jordt, T., Mahon, G.D., Touray, G.N., Ngulo, W.K., Morrison, W.I., Rawle, J. and Murray, M. (1986) Successful transfer of frozen N'Dama embryos from The Gambia to Kenya. *Tropical Animal Health and Production* 18, 65–75.

Leakey, R.E. and Lewin, R. (1977) *Origins.* MacDonald and Jane's London, London, pp. 1–17.

Letteneur, L. (1978) Crossbreeding N'Dama and Jersey cattle in Ivory Coast. *World Animal Review* 27, 36–42.

Lorenzini, E., Scott, J.R., Paling, R.W. and Jordt, T. (1988) The effects of *Trypanosoma congolense* infection on the reproductive cycle of N'Dama and Boran heifers. In: *Livestock Production in Tsetse Affected Areas of Africa.* International Livestock Centre for Africa and the International Laboratory for Research on Animal Diseases, Nairobi, Kenya, pp. 168–73.

McDowell, R.E. (1977) *Ruminant Products: More Meat than Milk.* Winrock International Livestock and Training Centre, Marilton, Arkansas.

MacLennan, K.J.R. (1980) Tsetse-transmitted trypanosomiasis in relation to the rural economy in Africa: Part 1. Tsetse infection. *World Animal Review* 36, 2–17.

MacLennan, K.J.R. (1981) Tsetse-transmitted trypanosomiasis in relation to the rural economy in Africa: Part 2. Techniques in use for the control or eradication of tsetse infestations. *World Animal Review* 37, 9–19.

Maloo, S.H., Chema, S., Connor, R., Durkin, J., Kimotho, P., Maehl, J.H.H., Mukendi, F., Murray, M., Rarienga, J.M. and Trail, J.C.M. (1988) The use of chemoprophylaxis in East African Zebu village cattle exposed to trypanosomiasis in Muhaka area, Kenya. In: *Livestock Production in Tsetse Affected Areas of Africa.* International Livestock Centre for Africa, and the International Laboratory for Research on Animal Diseases, Nairobi, Kenya, pp. 283–8.

Mitchell, G.F., Anders, R.F., Brown, G.V., Handman, E., Roberts-Thomson, F.C., Chapman, C.B., Forsyth, K.P., Kahl L.P. and Cruise, K.M. (1982) Analysis of infection characteristics and antiparasite immune responses in resistant compared to susceptible hosts. *Immunological Reviews* 61, 137–88.

Molyneux, D.H. (1986) African trypanosomiasis. *Clinics in Tropical Medicine and Communicable Diseases* 1, 535–55.

Morrison, W.I., Murray, M. and Akol, G.W.O. (1985) Immune responses of cattle to African trypanosomes. In: Tizard, I. (ed.), *Immunology and Pathogenesis of Trypanosomiasis.* CRC Press, Boca Raton, Florida, pp. 103–31.

Mortelmans, J. and Kageruka, P. (1976) Trypanotolerant cattle breeds in Zaire. *World Animal Review* 19, 14–7.

Murray, M. (1979) Anaemia of bovine African trypanosomiasis: an overview. In: Losos, G. and Chouinard, A. (eds), *Pathogenicity of Trypanosomes.* IDRC 132e. pp. 121–7.

Murray, M., Clifford, D.J., Gettinby, G., Snow, W.F. and McIntyre, W.I.M. (1981) A study of the susceptibility to African trypanosomiasis of N'Dama and Zebu cattle in an area of *Glossina morsitans submorsitans* challenge. *Veterinary Record* 109, 503–10.

Murray, M. and Dexter, T.M. (1988) Anaemia of bovine African trypanosomiasis. *Acta Tropica* (Basel) 45, 389–432.

Murray, M. and Gray, A.R. (1984) The current situation on animal trypanosomiasis in Africa. *Preventive Veterinary Medicine* 2, 23–30.

Murray, M., Grootenhuis, J.G., Akol, G.W.O., Emery, D.L., Shapiro, S.Z., Dar Faiqa, Bovell, D.L. and Paris, J. (1981) Potential application of research on African trypanosomiasis in wildlife and preliminary studies on animals exposed to tsetse infected with *Trypanosoma congolense.* In: Garstad, L., Nestel, B. and Graham, M. (eds), *Wildlife Diseases Research and Economic Development.* IDRC-179e, pp. 40–5.

Murray, M. and Morrison, W.I. (1979) Parasitaemia and host susceptibility to African trypanosomiasis. In: Losos, G. and Chouinard, A. (eds), *Pathogenicity of Trypanosomes*, IDRC No. 132e, pp. 71–81.

Murray, M., Morrison, W.I. and Whitelaw, D.D. (1982) Host susceptibility to African trypanosomiasis: Trypanotolerance. In: Baker, J.R. and Muller, R. (eds), *Advances in Parasitology,* vol. 21, Academic Press, London, pp. 1–68.

Murray, M. and Urquhart, G.M. (1977) Immunoprophylaxis against African trypanosomiasis. In: Miller, L.H., Pino, J.A. and McKelvey, Jr. J.J. (eds), *Immunity to Blood Parasites of Animals and Man.* Plenum, New York. pp. 209–41.

Njogu, A.R., Dolan, R.B., Wilson, A.J. and Sayer, P.D. (1985) Trypanotolerance in East African Orma Boran cattle. *Veterinary Record* 117, 632–6.

Olenick, J.G., Wolff, R., Nauman, R.K. and McLaughlin, J. (1988) A flagellar pocket membrane fraction from *Trypanosoma brucei rhodesiense*: immunogold localization and nonvariant protection. *Infection and Immunity* 56, 92–8.

Otchere, E. (1983) The productivity of white Fulani (Bunaji) cattle in pastoralist herds on the Kaduna plains of Nigeria. *ILCA Programme, Kaduna, Nigeria.* International Livestock Centre for Africa, Addis Ababa, Ethiopia.

Pagot, J. (1974) Les races trypanotolerantes. In: *Les Moyens de Lutte Contre les Trypanosomes et eurs Vecteurs: Actes du Colloque, Paris 1974,* Institut d'Elevage et de Medicine Veterinaire des Pays Tropicaux, Paris, pp. 235–48.

Paling, R.W., Moloo, S.K. and Scott, J.R. (1988) The relationship between parasitaemia and anaemia in N'Dama and Zebu cattle following four sequential challenges with *Glossina morsitans centralis* infected with *Trypanosoma congolense. International Scientific Council for Trypanosomiasis Research and Control.* 19th Meeting. Lome, Togo. OAU/STRC, Publication No. 114, pp. 256–64.

Payne, W.J.A. (1964) The origin of domestic cattle in Africa. *Empire Journal of Experimental Agriculture* 32, 97–113.

Pierre, C. (1906) L'elevage dans l'Afrique Occidentale Francaise. Gouvernement General de l'Afrique Occidentale Francaise, Paris.

Pinder, M., Bauer, J., Van Melick, A. and Fumoux, F. (1988) Immune responses of trypanoresistant and trypanosusceptible cattle after cyclic infection with *Trypanosoma congolense. Veterinary Immunology and Immunopathology* 18, 245–57.

Preston, J.M. and Allonby, E.W. (1979) The influence of breed on the susceptibility of sheep to *Haemonchus contortus* infection in Kenya. *Research in Veterinary Science* 26, 134–9.

Roberts, D.J. and Gray, A.R. (1973) Studies on trypanosome-resistant cattle. II. The effect of trypanosomiasis on N'Dama, Muturu and Zebu cattle. *Tropical*

Animal Health and Production 5, 220–33.

Roelants, G.E. (1986) Natural resistance to African trypanosomiasis. *Parasite Immunology* 8, 1–10.

Stephen, L.E. (1966) Observations on the resistance of West African N'Dama and Zebu to trypanosomiasis following challenge by wild *Glossina morsitans* from an early age. *Annals of Tropical Medicine and Parasitology* 60, 230–46.

Stewart, J.L. (1937) The cattle of the Gold Coast. *Veterinary Record* 49, 1289–97.

Stewart, J.L. (1951) The West African Shorthorn cattle. Their value to Africa as trypanosomiasis resistant animals. *Veterinary Record* 63, 454–7.

Trail, J.C.M., Colardelle, C., D'Ieteren, G.D.M., Dumont, P., Itty, P., Jeannin, P., Maehl, J.H.H., Nagda, S.M., Ordner, G., Paling, R.W., Rarieya, J.M., Thorpe, W. and Yangari, G. (1988) Evaluation of criteria of trypanotolerance. In: *Livestock Production in Tsetse Affected Areas of Africa.* International Livestock Centre for Africa and the International Laboratory for Research on Animal Diseases, Nairobi, Kenya, pp. 425–9.

Trail, J.C.M., d'Ieteren, G.D.M., Feron, A., Kakiese, O., Mulungo, M. and Pelo, M. (1990a) Effect of trypanosome infection, control of parasitaemia and control of anaemia development on productivity of N'Dama cattle. *Acta Tropica* 48, 37–45.

Trail, J.C.M., d'Ieteren, G.D.M., Colardelle C., Maille, J.C., Ordner, G., Sauveroche, B. and Yangari, G. (1990b) Evaluation of a field test for trypanotolerance in young N'Dama cattle. *Acta Tropica* 48, 47–57.

Trail, J.C.M., d'Ieteren, G.D.M. and Murray, M. (1991) Practical aspects of developing genetic resistance to trypanosomiasis. In: Owen, J.B. and Axford, R.F.E. (eds), *Breeding for Disease Resistance in Farm Animals.* CAB International, Wallingford, pp. 224–34.

Trail, J.C.M., Sones, K., Jibbo, J.M.C., Durkin, J., Light, D.E. and Murray, M. (1985) *Productivity of Boran Cattle Maintained by Chemoprophylaxis under Trypanosomiasis Risk.* ILCA Research Report no. 9. International Livestock Centre for Africa. Addis Ababa, Ethiopia.

Vale, G.A. (1987) Prospects for tsetse control. *International Journal for Parasitology* 17, 665–70.

Van Meirvenne, N., Magnus, E. and Vervoort, T. (1977) Comparison of variable antigen types produced by trypanosome strains of the sub-genus *Trypanozoon.* *Annales de la Societe Belge de Medecine Tropicale* 57, 409–23.

Wissocq, Y.J., Trail, J.C.M., Wilson, A.D. and Murray, M. (1983) Genetic resistance to trypanosomiasis and its potential economic benefits in cattle in East Africa. *International Scientific Council for Trypanosomiasis Research and Control.* 17th Meeting. Arusha, Tanzania, 1981. OAU/STRC. Publication No. 112, pp. 361–4.

Chapter 13

Practical Aspects of Developing Genetic Resistance to Trypanosomiasis

J.C.M. Trail[1], G.D.M. d'Ieteren[1] and Max Murray[2]

[1]International Livestock Centre for Africa (ILCA), P O Box 46847, Nairobi, Kenya and [2]Department of Veterinary Medicine, University of Glasgow Veterinary School, Bearsden, Glasgow G61 1QH, UK

Summary

Even within the trypanotolerant N'Dama breed some cattle are more resistant to trypanosomiasis than others. The packed red cell volume (PCV) under conditions of exposure to trypanosomes is a simple, rapid and easily estimated indicator of the extent of anaemia. The repeatability of PCV is reasonably high and similar to calf weaning weight. The first heritability estimates are encouragingly high and there are positive genetic correlations with production traits. These results suggest that cattle with improved resistance to the effects of trypanosomiasis can be identified and selection based on PCV could be used to improve both productivity and disease resistance.

Introduction

Major constraints in putting genetic resistance to trypanosomiasis to practical use have been difficulties in the definition and measurement of criteria of trypanotolerance. Work reviewed by Murray *et al.* (1991) shows that trypanotolerance is associated with at least two quantifiable factors, the ability to control parasitaemia and the ability to control the development of anaemia. This has encouraged field studies with N'Dama cattle in sites in a number of African countries collaborating within an African Trypanotolerant Livestock Network. In this network, sites have been established to provide baseline data for livestock development in tsetse-infested areas and to evaluate different methods for controlling trypanosomiasis (ILCA/ILRAD, 1988).

224

The studies covered in this chapter were designed to assess the relationships between criteria of trypanotolerance and livestock performance, to evaluate whether these criteria can be assessed early in an animal's lifetime, and to determine the degree to which they are under genetic control. Results from these studies are used to discuss the practical possibilities for increasing trypanotolerance levels through selection programmes in N'Dama cattle.

Relationships between criteria of trypanotolerance and livestock performance

At Mushie Ranch, Zaire, 146 calving interval records were built up from 64 N'Dama cows maintained for 3.5 years under a high natural tsetse-trypanosomiasis challenge (Trail *et al.*, 1991a). Monthly blood samples were examined by the dark-ground phase-contrast buffy coat method to detect the presence of trypanosomes (Murray *et al.*, 1977). The species of trypanosome was identified and the intensity of infection quantified as a parasitaemia score (Paris *et al.*, 1982). The degree of anaemia was quantified by measuring packed red cell volume percent (PCV). Attempts were made to control other possible causes of anaemia; ticks by weekly dipping and internal parasites through a pasture management system involving extensive grazing conditions, no night paddocks, and regular burning of pastures. Periodic faecal sampling and examination of blood smears for blood parasites other than trypanosomes were carried out when PCV% dropped to 20%.

Trypanotolerance criteria and cow productivity

The mean monthly trypanosome prevalence (the percentage of animals determined as being parasitaemic at a monthly examination) was 9.5% per month. Only three cows were never detected as parasitaemic during the period. *T. congolense* accounted for 79% of infections and *T. vivax* 21%. No animals were recorded with abnormal faecal egg burdens and no blood parasites other than *Theileria mutans* were detected, those rarely. The comparative influences of time detected parasitaemic, parasitaemia intensity representing control of development of parasitaemia, and PCV value representing control of development of anaemia were measured on calving interval, calf weaning weight and cow productivity (weight of weaner calf per cow per annum) using least squares mixed model procedures (Harvey, 1977).

Results are summarized in Table 13.1 and significant findings included cows detected parasitaemic for a low length of time having a 14% shorter calving interval and a 15% higher productivity than their contemporaries

Table 13.1. Comparative sizes of influences of trypanotolerance criteria on productivity.

Criteria	Calving interval		Calf weaning weight		Cow productivity[a]	
	days	%	kg	%	kg	%
Low versus high time detected parasitaemic	−68	−14.2	2.8	2.1	17.1	15.5
Low versus high parasitaemia score (within high time detected parasitaemic)	−20	−4.1	3.2	2.4	5.6	5.2
High versus low PCV (within low time detected parasitaemic)	−27	−6.3	7.8	5.8	12.6	10.4
High versus low PCV (within high time detected parasitaemic)	−59	−11.5	12.2	9.4	23.7	24.1

[a]Weight of weaner calf per cow per annum.

that were parasitaemic for a high length of time. The effects of parasitaemia intensity were not significant. In contrast, animals maintaining a high PCV value had an 11% shorter calving interval; a 9% heavier calf weaning weight; and a 24% superior cow productivity over those maintaining a low PCV value. So, control of development of anaemia, as measured by average PCV value, appeared to be the criterion of trypanotolerance most closely linked to overall cow productivity, in this production system where attempts had been made to control systematically other possible causes of anaemia.

Repeatabilities of trypanotolerance and performance traits

Repeatabilities, between calving intervals, for the three trypanotolerance criteria and the three performance measures, all expressed as traits of the cow, were computed. Traits with significant repeatabilities were calf weaning weight (0.35), average PCV over the calving interval (0.33) and time detected parasitaemic during the calving interval (0.23). The repeatabilities of the various traits set upper limits to their degrees of genetic determination and heritabilities (Falconer, 1981). The repeatability of PCV value was reasonably high and almost equal to that of calf weaning weight. Thus the ability to control development of anaemia, as indicated by PCV value, might well be a useful criterion of trypanotolerance with which to identify more trypanotolerant individual animals.

Criteria assessment in young animals

Simultaneous evaluation of the relative effects of criteria of trypano-tolerance, in both the preweaner calf and its dam, on calf performance (weaning weight) showed that calf PCV values were at least as important as dam PCV values. With similar monthly trypanosome prevalences (9.2% and 9.5%, respectively) in the calf a high PCV value within a high time parasitaemic increased weaning weight by 9.9 kg or 7.7% and in the dam increased the calf weaning weight by 8.8 kg or 6.1%. Thus evaluation of criteria of trypanotolerance in an animal might be feasible before it reached maturity, but would need to be sufficiently long after weaning for the preweaning influence of the dam to have disappeared.

A field test for trypanotolerance in young N'Dama cattle

The Zaire study clearly indicated that investigations into practical field tests for trypanotolerance should focus on the use of post-weaners, maintained for varying lengths of time in as high natural tsetse-trypanosomiasis challenge situations as possible. Trail *et al.* (1991b) report on three such tests, in which a total of 436 one-year-old N'Dama cattle were maintained for 12, 18 and 24 weeks under a medium tsetse-trypanosomiasis challenge at the Government Ranch of the Office Gabonais d'Amelioration et de Produc-tion de Viande (OGAPROV) in Gabon.

Every 4 weeks in the first and every 2 weeks in the second and third tests, blood samples were examined by the dark-ground phase-contrast buffy coat method to detect the presence of trypanosomes (Murray *et al.*, 1977) and the intensity of infection quantified as a parasitaemia score (Paris *et al.*, 1982). The degree of anaemia was estimated by measuring the PCV. Attempts were made to control other possible causes of anaemia; ticks by weekly dipping and internal parasites through 3-monthly dosing with anthelmintics. Periodic faecal sampling and examination of blood smears for blood parasites other than trypanosomes when PCV% dropped to 20%, was carried out. On the last day of the first and second tests all animals were treated with Samorin (isometamidium chloride) at the rate of 1.0 mg/kg bodyweight by intramuscular injection, to evaluate the recovery of PCV values.

Trypanotolerance criteria and animal growth

The comparative sizes of the influence of parasitaemia control and anaemia control on N'Dama daily liveweight gain are summarized in Table 13.2. Under trypanosome prevalences of 25%, 31% and 9%, respectively during the three tests, ability to control the development of anaemia had a major effect on daily weight gain, four times that of the ability to control parasitaemia. Above-average PCV values, as a measure of anaemia control, resulted in a 44% superior daily weight gain over below-average PCV values.

Three separate measures of anaemia control were compared, average PCV over the test period, lowest PCV reached, and average PCV when detected as parasitaemic. The first two measures were more closely associated with animal performance than the last. The first two could both serve as indices of the control of development of anaemia, although the effect of average PCV over the test period was significant within more of the parasitaemic time periods than was that of lowest PCV reached.

From the point of view of detecting major differences in anaemia

Table 13.2. Comparative sizes of influences of parasitaemia control and anaemia control on growth.

Test number	Statistic	Parasitaemia control		Anaemia control	
		High score	Low score	High PCV	Low PCV
1	Animal numbers	57	57	58	56
	Growth (g/day)	248	273	283	238
	se	23.5	23.9	21.0	20.6
2	Animal numbers	51	51	46	56
	Growth (g/day)	97	106	135	68
	se	14.4	11.0	27.2	26.3
3	Animal numbers	27	49	40	36
	Growth (g/day)	81	89	111	59
	se	15.6	19.5	21.1	32.0
Overall	Animal numbers	135	157	144	148
	Growth (g/day)	142	156	176	122
	se	19.0	19.1	23.2	25.9
Mean difference in growth (%)			+10	+44	

Constructed from least squares means for classes tested within all periods when detected as parasitaemic.

control, the minimal test period for an animal to be detected as para-sitaemic was shown to be approximately 6 weeks. When this situation was simulated, using only animals that became parasitaemic in the first one-third of each test period, the effects of anaemia control on weight gain were very similar to those obtained using the complete test data. Thus, in practice, a suitable test would be where natural infection could be effected as early as possible and anaemia control measurements carried out over a period of 6 weeks of detected parasitaemia. Alternatively, if a satisfactory correlation could be obtained between natural infection effects and those of an experimental alternative, a practical test could become still more feasible.

Post-test recovery of PCV values after trypanocidal drug treatment

At the completion of the first test, trypanosome prevalence was 30%, and the PCV values of groups that had been detected parasitaemic for varying percentages of the 12 week test period ranged from 33.1% for those never detected as parasitaemic, to 18.8% for those parasitaemic for 80% of the period. Coefficients of variation, indicating the amount of variation in PCV values within each group, ranged from 15.9% in animals never detected as parasitaemic to 26.5% in animals parasitaemic for 80% of the test period. On the last day of the test all animals were treated with Samorin at the rate of 1 mg/kg. When PCVs were measured 30 days later, major recovery of PCV had taken place, even those parasitaemic for 80% of the test period had a PCV value of 34%, and the coefficients of variation within all groups were reduced to 8.7%. In the second test, 73% of the recovery achieved one month after the trypanocidal drug treatment, had been reached in 9 days.

Genetic aspects of criteria of trypanotolerance

Results from research on genetic aspects of criteria of trypanotolerance are currently available in two main areas. Information on the inheritance of anaemia development and its genetic correlation with animal performance, will allow the possibilities of improvement through conventional genetic selection programmes to be assessed. Determination of any linkages between criteria of trypanotolerance and major histocompatibility complex (MHC) and common leucocyte antigens (CLA) will allow alternative/additional genetic progress to be made.

Genetic aspects of control of anaemia development

For the first time in a field test situation, the genetic parameters of measures of control of anaemia have been evaluated (Trail *et al.*, in press). At the OGAPROV ranch in Gabon, 148 N'Dama cattle, progeny of 29 different sires, were exposed to a short-term natural trypanosome challenge at one year of age. The animals had been born over a 3 month period in five separate multi-sire herds, an average of six sires in each herd produced a minimum of two progeny each. The calves remained in these five herds until weaning at 8 months of age, when they were separated into a male and female weaner herd, where they were maintained until the 3 month test had been completed. Parentage was determined through blood grouping techniques.

Eleven times, i.e., once every 8 or 9 days, blood samples were examined by the dark-ground phase-contrast buffy coat method to detect the presence of trypanosomes (Murray *et al.*, 1977) and the intensity of the infection quantified as a parasitaemia score (Paris *et al.*, 1982). The degree of anaemia was estimated by measuring the PCV. Attempts were made to control other possible causes of anaemia, ticks by weekly dipping and internal parasites through 3-monthly dosing with anthelmintics. Periodic faecal sampling and examination of blood smears for blood parasites other than trypanosomes was carried out when PCV dropped to 20%.

In animals detected as parasitaemic, those with above average 'average PCV' values or above average 'lowest PCV reached', had 34% and 35%, respectively, higher daily weight gains than those with values below average. Even when not detected as parasitaemic, those with above average 'average PCV values' or above average 'lowest PCV reached', had 14% and 12%, respectively, higher gain, suggesting that a proportion of these animals may have been parasitaemic.

The heritabilities of, and genetic and phenotypic correlations between, growth, average PCV and lowest PCV reached on test are shown in Table 13.3. The heritability of body weight at the start of the test when animals averaged 50 weeks of age was 0.49 ± 0.32. This is within the normally reported range for this trait, while the large standard error could be a reflection of the small number of five progeny available per sire. When all environmental and parasitaemia information was taken into account, the heritability of growth over the test period was 0.39 ± 0.31, again within the expected range for growth over a 3 month period. The heritabilities of both PCV measures were higher than the corresponding heritability of growth in all analyses and were 0.64 ± 0.33 and 0.50 ± 0.32 when all environmental and parasitaemia information was utilized.

When all environmental and parasitaemia information was taken into account, the genetic correlation between average PCV and growth was 0.70 ± 0.42 and between lowest PCV reached and growth, 0.28 ± 0.55. These values, coupled with the higher heritabilities of the PCV measures, indicate some possibility of selection on PCV values for control of anaemia development.

Table 13.3. Heritabilities of and genetic and phenotypic correlations between growth, average PCV, and lowest PCV reached, on test.

	Growth	Average PCV	Lowest PCV reached
(a) Parasitaemia detection and parasitaemia score not included in analysis			
Growth	0.22±0.28	0.41±0.73	−0.13±0.74
Average PCV	0.35	0.35±0.30	0.96±0.20
Lowest PCV reached	0.29	0.72	0.48±0.31
(b) Parasitaemia detection included in analysis			
Growth	0.38±0.30	0.71±0.42	0.28±0.55
Average PCV	0.32	0.63±0.33	0.99±0.17
Lowest PCV reached	0.25	0.66	0.51±0.32
(c) Parasitaemia detection and parasitaemia score included in analysis			
Growth	0.39±0.31	0.70±0.42	0.28±0.55
Average PCV	0.32	0.64±0.33	1.00±0.17
Lowest PCV reached	0.25	0.67	0.50±0.32

Heritability is value on diagonal.
Genetic correlation is value above diagonal.
Phenotypic correlation is value below diagonal.

Both heritabilities and genetic correlations increased in size when parasitaemia information was included in the analysis, although only 28% of the animals were positively detected as parasitaemic. The more sensitive field tests being developed for trypanosome detection, e.g., Majiwa (1989) and Nantulya and Lindquist (1989), will allow more accurate classification of parasitaemia status, which could in turn further raise the heritability estimates of PCV measures and their genetic correlations with performance.

Eventually, if all animals could become parasitaemic during a test, either through very high natural infection or through an acceptable experimental alternative, the heritability estimates of PCV measures could be higher still. In this situation, the test period could be reduced to the minimum time needed to evaluate accurately the PCV values of the animals when parasitaemic. This would allow the economic impact of the deleterious effects of trypanosome infection on animal performance to be minimized. Over a shortened test period PCV values can be much more accurately measured than can animal performance. Thus, selection on the most appropriate

PCV measure could well allow rapid progress to be made in genetic improvement of trypanotolerant N'Dama cattle.

Association of MHC and CLA phenotypes with performance and health parameters

The objective is to attempt to determine whether there are indications of associations between phenotype and various aspects of health and productivity in N'Dama cattle under trypanosome challenge. Trail *et al.* (1989) examined the 64 breeding cows and 146 progeny at Mushie Ranch, Zaire, referred to above.

A least squares analysis of the effects of MHC and CLA phenotypes on three performance parameters and on detected parasitaemia and average PCV values was made. Reactivities with one MHC-defining reagent and two CLA-defining mAb were found to have significant and independent associations (Table 13.4). No other significant associations were observed. There is an indication that the MHC phenotype defined by the allo-antiserum ECA 121 is negatively associated with maintenance of PCV in

Table 13.4. Least-squares means for effects of IL-A37 defined CLA phenotype on performance, IL-A39 defined CLA phenotype on parasitaemias detected, and ECA 121 defined MHC phenotype on average PCV.

Effect	Positive	Negative
IL-A37 defined CLA phenotype on performance		
Calving interval (days)	420±16.1	473±14.5
Calf weaning weight (kg)	138± 3.0	133± 2.7
Cow productivity (kg)[a]	125± 3.6	109± 3.3
IL-A39 defined CLA phenotype on parasitaemias detected (no./annum)		
Parturition to conception	1.01±0.601	1.84±0.300
Conception to next parturition	0.30±0.220	0.72±0.110
Parturition to parturition	0.54±0.253	1.05±0.126
ECA 121-defined MHC phenotype on average PCV (%)		
Parturition to conception	31.3±0.58	33.6±0.56
Conception to next parturition	31.9±0.63	34.1±0.61
Parturition to parturition	31.9±0.58	34.1±0.56

Note: The number of records were for IL-A37 positive and negative 67 and 79; for IL-A39 positive and negative 30 and 116; and for ECA 121 positive and negative 70 and 76, respectively.
[a]Weight of weaner calf per cow per year.

N'Dama cows under trypanosome challenge. There are also clear indications that reactivities with the CLA-defining mAb IL-A37 and IL-A39 are positively associated with superior performance and with fewer parasitaemic episodes, respectively.

Identification of genes responsible for trypanotolerance traits clearly should not be restricted to the MHC. This particular genetic region accounts for approximately one-thousandth of the genome. In addition to this statistical consideration, even if emphasis is placed only on genes involved with immune function, such genes are not restricted to the MHC. Significantly, these results show that there are genes encoding common leucocyte antigens, not linked with the MHC, which may be of importance for the genetic control of trypanotolerance traits.

While phenotypic associations between certain genetic markers and some aspects of trypanotolerance are being detected, fortuitous statistically significant associations can arise unless genetic structure of the animal sample is known. A very rigorous genetic analysis of associations between markers and resistance aspects is now possible with animal samples such as those recorded at Mushie Ranch, Zaire and OGAPROV, Gabon.

Conclusions

While there is an increasing number of examples of innate resistance to disease being identified in domestic livestock, trypanotolerance is one of the best recognized and one that is being thoroughly investigated. Recent field studies reported in this chapter are starting to provide the basic tools with which the trypanotolerance trait can be identified and exploited.

Thus the control of the development of anaemia as measured by PCV value during the course of a trypanosome infection has a major effect on production. The repeatability of PCV values has been shown to be reasonably high and almost equal to that of calf weaning weight. The effect of calf PCV values on performance indicates that animals can be assessed at the post-weaning stage. As major differences in the capacity to control anaemia can be determined within six weeks of infection, the possibility now exists of carrying out selection tests for trypanotolerance on young animals, using either natural infection or an acceptable experimental alternative. The rapid post-test recovery of PCV values following a single trypanocidal drug treatment indicates that such tests will have minimal deleterious effects on animal performance.

Major possibilities therefore already exist for such tests to be used in identifying replacement stock likely to provide optimal inputs to herd productivity. The first encouraging heritability estimates of PCV values and their positive genetic correlations with animal performance on test, hold promise for the implementation of practical genetic selection programmes.

The more sensitive field tests being developed for trypanosome detection will allow more accurate classification of parasitaemia status, which could further raise the heritability estimates of PCV measures and their genetic correlations with animal performance. Finally, the possibility of marker gene identification is an additional potential tool for the identification and use of more trypanotolerant individuals.

References

Falconer, D.S. (1981) *Introduction to Quantitative Genetics*, Longman, London.

Harvey, W.R. (1977) *User's Guide for Least-Squares and Maximum Likelihood Computer Programs*, Ohio State University, Columbus.

ILCA/ILRAD (1988) Proceedings of a meeting on livestock production in tsetse affected areas of Africa. ILCA/ILRAD, Nairobi, Kenya.

Majiwa, F.A.O. (1989) Recombinant DNA probes as tools for epidemiological studies of parasitic diseases. *Discovery and Innovation* 1, 35–40.

Murray, C., Murray, M., Murray, P.K., Morrison, W.I., Pyne, C. and McIntyre, W.I.M. (1977) Diagnosis of African trypanosomiasis in cattle. Improved parasitological and serological techniques, In: *International Scientific Council for Trypanosomiasis Research and Control*. 15th Meeting, The Gambia, OAU/STRC. Publication no. 110, pp. 247–54.

Murray, M., Stear, M.J., Trail, J.C.M., d'Ieteren, G.D.M., Agyemang, K. and Dwinger, R.H. (1991) Trypanosomiasis in cattle: prospects for control. In: Axford, R.F.E. and Owen, J.B. (eds), In: *Breeding for Disease Resistance in Farm Animals*. CAB International, Wallingford, pp. 203–23.

Nantulya, V.M. and Lindquist, K.J. (1989) Antigen-detection enzyme immunoassay for the diagnosis of *Trypanosoma vivax*, *T. congolense* and *T. brucei* infections. *Tropical Medicine and Parasitology*, 40, 267–72.

Paris, J., Murray, M. and McOdimba, F. (1982) An evaluation of the sensitivity of current parasitological techniques for the diagnosis of bovine African trypanosomiasis. *Acta Tropica* 39, 307–16.

Trail, J.C.M., d'Ieteren, G.D.M. and Teale, A.J. (1989) Trypanotolerance and the value of conserving livestock genetic resources. *Genome* 31, 805–12.

Trail, J.C.M., d'Ieteren, G.D.M., Feron, A., Kakiese, O., Mulungo, M. and Pelo, M. (1991a) Effect of trypanosome infection, control of parasitaemia and control of anaemia development on productivity of N'Dama cattle. *Acta Tropica* 48, 37–45.

Trail, J.C.M., d'Ieteren, G.D.M., Colardelle, C., Maille, J.C., Ordner, G., Sauveroche, B. and Yangari, G. (1991b). Evaluation of a field test for trypanotolerance in young N'Dama cattle. *Acta Tropica* 48, 47–57.

Trail, J.C.M., d'Ieteren, G.D.M., Maille, J.C. and Yangari, G. Genetic aspects of control of anaemia development in trypanotolerant N'Dama cattle. *Acta Tropica* (in press).

Chapter 14

Theileriosis and Evidence for Genetic Resistance

Roger L. Spooner and Duncan Brown
AFRC Institute of Animal Physiology and Genetic Research
Roslin, Midlothian EH25 9PS, UK

Summary

Tropical theileriosis caused by *Theileria annulata* occurs widely from Morocco to China. East Coast fever caused by *Theileria parva* is restricted to East and Central Africa. They both kill susceptible cattle in around 20 days. Evidence that local breeds are more resistant to the diseases than imported breeds exists but not all reports point in the same direction, some show no difference and in one large study that the imported breed is more resistant. There is no genetic evidence for variation in susceptibility and there is no good test for susceptibility or resistance. The diseases are both tick transmitted and the differences in resistance to the diseases may result from differences in tick resistance. There are a number of other components of the pathogenesis where genetic resistance might play a role. Differences in immune recognition of parasite antigens is one. Another is anaemia, which appears to correlate with susceptibility to tropical theileriosis as it does with susceptibility to trypanosomiasis. There is a need for detailed studies of the genetics of theilerial resistance/susceptibility in cattle families.

Introduction

Tropical theileriosis is a protozoan disease of major importance in the tropics and subtropics. It is caused by the protozoan *Theileria annulata* and is transmitted by ticks of the genus *Hyalomma*. It is very widespread and affects cattle from Morocco to China. More than 200 million cattle are at risk and the losses to cattle production are estimated at over £600 million annually. Another closely related disease, East Coast fever (ECF), occurs only in East Africa and is caused by the related protozoan *Theileria parva* and is transmitted by a different tick *Rhipicephalus appendiculatus*.

Both diseases involve the transmission of sporozoites from tick to bovine host. A unique feature common to both parasites is the ability of these sporozoites to infect cells of the immune system and to transform them into rapidly growing cells with the parasite dividing in synchrony with the host cell (Hulliger *et al.*, 1964). These macroschizont infected cells then develop initially in the draining lymph node prior to dissemination throughout the body in this form. Merozoites are later released which enter red cells to produce a piroplasm stage. These piroplasms are then taken up by ticks and following further development in the tick another cycle of the parasites takes place. There is severe leucopenia in both diseases though more marked with East Coast fever. Anaemia is also a pathognomonic feature of tropical theileriosis. With both diseases there is marked pyrexia and death in around 20 days in fully susceptible cattle. There are, however, a number of very interesting and important differences between the two diseases. As noted above they are geographically quite distinct, since they are transmitted by different species of ticks. In fact only in one country, Sudan, do both diseases occur with tropical theileriosis being endemic in the north and ECF in the very south of the country. There are also important differences in the ease of vaccination; with tropical theileriosis it is possible to use infected cells as a vaccine whereas with ECF the only effective method is by infection and treatment.

It was not known whether the difference between the two diseases depends on the inherent ability of the two parasites to parasitize host cells or for some other reason such as differences in the cell that is permissive to infection. It was already known that *T. parva* cell lines lacked surface immunoglobulin (Duffus *et al.*, 1978) and were positive for various T cell markers (Lalor *et al.*, 1986) but lacked bovine monocyte markers (Spooner *et al.*, 1988). They were all positive for major histocompatibility complex (MHC) class I and class II antigens (Morrison *et al.*, 1986; Spooner and Brown, 1980). *T. annulata*-infected cell lines on the other hand although positive for MHC class I and class II antigens, were negative for monocyte markers and, particularly significantly, were totally negative for T cell markers (Spooner and Brown, 1980; Spooner *et al.*, 1988). It was then shown that *T. annulata* infected and transformed macrophages and B cells but not T cells whereas *T. parva* infected T cells very efficiently and would not infect macrophages (Spooner *et al.*, 1989).

There were thus major differences between the two parasites but it was not clear how they gave rise to the differences seen in the field in the two diseases.

The type of vaccination operational at present uses a live parasite and so there is always the possibility of infection being passed on from immunized to non-immunized animals if the appropriate ticks are present. There would be obvious advantages in having a dead vaccine. It would be even better if one could breed for resistance to the diseases and not need to vaccinate.

Evidence for resistance

Unfortunately the hard evidence for genetic control of resistance to either tropical theileriosis or East Coast fever are lacking. Host susceptibility to theileriosis is perhaps best exemplified by the water or swamp buffalo, *Bubalus bubalis* which is evidently almost completely refractory to tropical theileriosis caused by *T. annulata* yet highly susceptible to East Coast fever caused by either *T. parva parva* (Bradshaw, 1924; Lambelin *et al.*, 1960) or *T. parva lawrencei* (Burridge *et al.*, 1973).

There is evidence, often anecdotal, that there are breed differences in resistance with European cattle exported to the tropics being particularly susceptible. In India, indigenous Zebu cattle (*Bos indicus*) are claimed to be resistant to *T. annulata* infection, whereas in North Africa the indigenous cattle, *Bos taurus*, are also thought to be almost equally resistant to the same organism. In both these regions tropical theileriosis is almost exclusively a disease of imported cattle, predominantly, but not exclusively of European taurine origin (Edwards, 1927). However, not all these data point to the local breeds being resistant and in one experiment 373 Algerian cattle, reared in disease-free conditions were challenged alongside 842 Aubrac cattle recently imported from France and the mortality rate was 23% in the indigenous stock and 12% in the exotic animal (Sergent *et al.*, 1945).

In Morocco the main casualties are imported animals soon after their importation (Ouhelli, personal communication). What is much less clear is whether pure European breeds that have been resident in a country for several years are more susceptible than local breeds. Kachani (1990) has challenged native Friesian and local Moroccan cattle and shown that the mortality was much earlier in the Friesians than in the local breed. However, the locals succumbed after several months during which time there was a substantial loss in weight. Recent studies in Morocco (E.J. Flach and H. Ouhelli, unpublished observations) indicate that in an intensive farming region in Morocco there is a high incidence of infection with *T. annulata* and the main work of the field veterinarians during the months of May to August is in relation to cases of theileriosis. These animals are uniformly of European breeds. Evidence from Morocco of field disease in local breeds is very difficult to obtain and one must conclude that it is not a significant problem.

Breed or species resistance is also seen with ECF and is highlighted by the apparent insusceptibility of local cattle to the disease prevalent in their region. Similar breed variation in the 'tick-resistance' of zebu cattle to the one-host tick *Boophilus microplus* is seen in Australia. In East Africa, where ECF is endemic the indigenous East African shorthorned Zebu (*Bos indicus*) was shown to be more resistant than Guernseys and Nganda/ Jersey crosses (Guilbride and Opwata, 1963). The Guernseys were dying on average in 22 days, the Jersey/Nganda crosses in 41 days and the pure Ngandas in 65 days; but they were all dying. It has also been reported by Paling *et al.* (1990) that Ankole × Jersey crosses are more susceptible to ECF (mortality 32%) than pure Ankole (mortality 64%) to experimental challenge with *T. parva.*

It is evident, however, that in many cases Zebu cattle are not necessarily resistant to theileriosis. Imported Santa Gertrudis (and thus 'exotic') Zebu types and breeds introduced to *T. annulata* endemic areas of North Africa have proved highly susceptible to tropical theileriosis (Kachani, 1990). Improved East African shorthorned Zebu, as Boran cattle, originating in Northern Kenya where neither ECF nor its vector tick *Rhipicephalus appendiculatus* are present, have been shown to be as susceptible as imported European taurine breeds in the field situation (Stobbs, 1966). However, it must be noted that in these field studies the diagnosis of the cause of death was not always possible and in the study in Uganda described by Stobbs (1966) several other diseases, including streptothricosis were prevalent. Thus the interpretations of the results with regard to ECF have to be treated with caution but it does underline that there is little difference in general fitness between Boran and European breeds in this environment. When such Boran cattle are selected for productivity and kept free of challenge, morbidity and mortality from ECF is very high indeed. Susceptibility is indistinguishable from that of *Bos taurus* 'grade' cattle where morbidity and mortality from ECF may be in the region of 90% (Brocklesby *et al.*, 1961). It is hardly surprising that they are considered by ILRAD as susceptible controls.

Indigenous cattle derived from ECF-endemic areas do, however, show evidence of ECF resistance even though they may have been maintained free from tick or disease challenge for a number of generations. In their study on the resistance of Jersey × Nganda calves to ECF mentioned above, Guilbride and Opwata (1963) showed that the pure Nganda (*Bos indicus*) controls survived three times longer than susceptible Guernsey (*Bos taurus*) controls and longer than cross-bred calves either immunized against ECF or derived from immune dams. A component of this survival was undoubtedly the fact that the Guernseys picked up ticks in the field four times as quickly as the Nganda calves (Guilbride and Opwata, 1963, Table 1). Thus the rate of exposure to disease challenge and the severity of that challenge was likely to be much greater for the exotic, taurine cattle than their indigenous, Zebu counterparts. The quantum of infection in theileriosis is a major factor in determining the outcome of infection (Cunningham *et al.*, 1974), thus the more ticks that feed the greater the morbidity and mortality from the disease (Wilson, 1950; Yeoman, 1956). However, at the susceptible end of the scale it is not possible, simply by controlling tick numbers, to determine differences in susceptibility. Even a single tick, infected at the lowest possible level – 1 acinus – may inoculate up to 40,000 sporozoites when it feeds (Fawcett *et al.*, 1982). This is enough to kill most taurine and 'exotic' Zebu cattle. To determine relative susceptibility in such cattle it is necessary to titrate defined quanta of infective sporozoites. When this was done it was shown that even Northern Kenyan Boran cattle are more resistant than European breeds (Radley, 1978), but only when sporozoite doses much less than that inoculated by a single tick were administered.

What is evident, however, is that resistance to ixodid ticks, and particularly resistance to the two- and three-host vectors of bovine theileriosis – *Hyalomma* spp. and *Rhipicephalus appendiculatus* – will have a major part to play in determining death or survival of cattle in an endemic area (Young *et al.*, 1988). Even though one infected tick may provide a lethal challenge, reduction in the numbers of ticks overall and of infected ticks in the population, will reduce the chances of an infested tick feeding on a susceptible animal. Thus, to date, the chief and, until recently, only means of control exerted over theileriosis, and ECF in particular, has been the short-interval application of acaricides to cattle. When such a programme breaks down the consequences may be disastrous (Norval, 1979). An integrated scheme is thus proposed for control of theileriosis and other tick-borne diseases in Africa (Young *et al.*, 1988), of which a major component is the use of cattle genetically resistant to ticks (Spickett *et al.*, 1989), which might reduce the need for acaricides (Tatchell *et al.*, 1986), and in which the acquisition of immunological 'resistance' to ticks would further reduce the overall tick burden (de Castro *et al.*, 1985).

What may be much more difficult is the integration of this tick resistance with the sort of productivity required by the intensive land use forced on the ever-increasing population in areas of the world where theileriosis is a threat. It is apparent that, even highly tick-resistant game animals suffer severely from tick predation when confined and tick numbers build up (Norval and Lightfoot, 1982). Moreover, attempts to improve productivity of East African shorthorned Zebu reared in an ECF endemic area by crossing with ECF-immune Borans failed due to the very high mortality rate in F_1 and F_2 generations (Stobbs, 1966), almost solely due to ECF.

Potential components of resistance

There are several stages at which genetic resistance might play a role. These include:

1. Resistance to ticks. As noted above tick resistance may play an important role.

2. Resistance to sporozoite entry into cells of the immune system. There are clearly differences in virulence between different strains of parasite, particularly with *T. parva*. This could have a significant effect on apparent differences between breeds in apparent susceptibility.

3. Ability to control the pathology.

(a) Leucopenia: there is substantial but variable leucopenia in theileriosis. Whether the ability to maintain a reasonable white cell concentration is genetically controlled is not known, but it could be.

(b) Anaemia: a major finding in both trypanosomiasis and tropical theileriosis is a severe drop in the packed cell volume (PCV) in those cattle that are likely to die. There appears to be a high correlation between the parasitaemia, the drop in haematocrit and the mortality. As has been shown by Trail *et al.* (1991) strong evidence is accumulating that there is a genetic component in the ability of cattle to maintain their PCV at normal levels under trypanosomiasis challenge. It would be very interesting to see whether the same trait is important in relation to theileriosis.

4. Immune response: there is evidence that the most important protective mechanism in both diseases is a cytotoxic T cell mediated killing of parasite infected cells (Innes *et al.*, 1989; Morrison *et al.*, 1986; Pearson *et al.*,1979; Preston *et al.*, 1983). Immune animals generate an MHC-restricted CTL response at the time that the disease is resolving. There are differences in the frequencies of different MHC class I antigens in different breeds of cattle but it is not known whether this is related in any way to the resistance. Recent studies by Goddeeris *et al.* (1990) have shown that with East Coast fever, where different strains of the parasite do not cross-immunize perfectly, there is evidence that the CTL generated against one strain of the parasite are restricted by different class I alleles from those directed against another strain. Whether this is related to innate resistance is not known.

In the field of disease genetics there are extensive, poorly defined data on breed differences in resistance to a variety of diseases. This is not the same thing as hard genetic evidence. Thus with theileriosis as is starting to be done with trypanosomiasis it is essential to study the segregation of resistance within families. To do this some form of test for resistance is required: good tests do not exist. One can attempt to infect with extremely low doses of parasite. But although high doses override individual differences low doses produce effects that are difficult to quantify. Very low doses of parasite although not producing disease in all animals can produce lethal infections in those animals that do become ill.

In summary, there is evidence for breed differences in susceptibility to both tropical theileriosis and East Coast fever. They do not always agree and there is no direct genetic evidence. Studies within families are required and to go with these a clearly definable test to estimate the level of resistance must be developed. Only then will data be obtained which will enable the identification of the genes controlling resistance.

References

Bradshaw, J.T.C. (1924) *Report of the Proceedings of the Vth Pan African Veterinary Conference*, Nairobi, 1923, p. 30.

Brocklesby, D.W., Barnett, S.F. and Scott, G.R. (1961) Morbidity and mortality rates in East coast fever (*Theileria parva* infection) and their application to drug screening procedures. *British Veterinary Journal* 117, 529–31.

Burridge, M.J. and Odeke, G.M. (1973) *Theileria lawrencei* infection in the Indian water buffaloe *Bubalus bubalis. Experimental Parasitology* 34, 257–61.

Cunningham, M.P., Brown, C.G.D., Burridge, M.J., Musoke, A.J., Purnell, R.E., Radley, E.E. and Sempebwa, C. (1974). East Coast fever: titration in cattle of suspensions of *Theileria parva* derived from ticks. *British Veterinary Journal* 130, 336–45.

de Castro, J.J., Cunningham, M.P., Dolan, T.T., Dransfield, R.D., Newson, R.M. and Young, A.S. (1985) Effects on the artificial infestations with the tick *Rhipicephalus appendiculatus. Parasitology* 90, 21–3.

Dolan, T.T., Linyoni, A., Mbogo, S.K. and Young, A.S. (1984) Comparison of long-acting oxytetracycline and parvaquone in immunisation against East coast fever by infection and treatment. *Research in Veterinary Science* 37, 175–8.

Duffus, W.P.H., Wagner, G.G. and Preston, J.M. (1978) Initial studies on the properties of a bovine lymphoid cell culture line infected with *Theileria parva. Clinical and Experimental Immunology* 34, 347–53.

Edwards, J.T. (1927) The acclimatisation of imported stock. *Agricultural Journal of India* 22 411–24.

Fawcett, D.W., Buscher, G. and Doxsey, S. (1982) Salivary gland of the tick vector of East Coast fever. III The ultrastructure of sporogony in *Theileria parva. Tissue Cell* 14, 183–206.

Goddeeris, B.M., Morrison, W.I., Toye, P.G. and Bishop, R. (1990) Strain specificity of bovine *Theileria parva* specific cytotoxic T cells is determined by the phenotype of the restricting class I MHC. *Immunology* 69 38–44.

Guilbride, P.D.L. and Opwata, B. (1963) Observations on the resistance of Jersey/ Nganda calves to East Coast Fever, (*Theileria parva*). *Bulletin of Epizootilogical Diseases of Africa.*

Hulliger, L., Wilde, J.K.H., Brown, C.G.D. and Turner, L. (1964) Mode of multiplication of *Theileria* in cultures of bovine lymphocytic cells. *Nature* 203, 728–30.

Innes, E.A., Millar, P., Brown, C.G.D. and Spooner, R.L. (1989) The development and specificity of cytotoxic cells in cattle immunized with autologous or allogeneic *Theileria annulata*-infected lymphoblastoid cell lines. *Parasite Immuno-logy* 11, 1–12.

Kachani, M. (1990) Ph.D. Thesis, Brunel University.

Lalor, P.A., Morrison, W.I., Goddeeris, B.M., Jack, R.M. and Black, S.J. (1986) Monoclonal antibodies identify phenotypically and functionally distinct cell types in the bovine lymphoid system. *Veterinary Immunology and Immunopathology* 13, 121–40.

Lambelin, G., Estors, F., van Vaerenbergh, R. and Mammerick, M. (1960) Sensibilite du buffle d'Asie aux principales maladies a protozoaires du betail aux Congo belge. *Annales de la Societe Belge de Medicine Tropicale* 40 189–97.

Morrison, W.I., Lalor, P.A., Goddeeris, B.M. and Teale, A.J. (1986) Theileriosis: antigens and host–parasite interactions. In: Pearson, T.W. (ed.), *Parasite Antigens: Towards New Strategies for Vaccines.* Marcel Dekker, New York, pp. 167–213.

Norval, R.A.I. (1979) Tick infestations and tick borne diseases in Zimbabwe/ Rhodesia. *Journal of the South African Veterinary Association* 50, 289–92.

Norval, R.A.I. and Lightfoot, C.J. (1982) Tick problems in wildlife in Zimbabwe. Factors influencing the occurrence and abundance of *Rhipicephalus appendiculatus. Zimbabwe Veterinary Journal* 13, 11–20.

Paling, R.W., Mpangala, C., Luttikhuizen, B. and Sibomana, G. (1990) The exposure of Ankole cross-Bred cattle to theileriosis in Rwanda. *Tropical Animal Health and Production* (in press).

Pearson, T.W., Lundin, L.B., Dolan, T.T. and Stagg, D.A. (1979) Cell mediated immunity to *Theileria*-transformed cell lines. *Nature* 281, 678–80.

Pipano, E. and Fish, L. (1982) Cultivation of erythrocyte stages of *Babesia bovis* and *Theileria annulata in vitro. Journal of Protozoology* 29, 543–5.

Preston, P.M., Brown, C.G.D. and Spooner, R.L. (1983) Cell-mediated cytotoxicity in *Theileria annulata* infection in cattle with evidence for BoLA restriction. *Clinical and Experimental Immunology* 53, 88–100.

Radley, D.E. (1978) *Immunisation against East Coast Fever by Chemoprophylaxis.* Food and Agriculture Organisation, Rome.

Sergent, E., Donatienl, A., Parrot, L. and Lestoquard, F. (1945) *Etudes sur les piroplasmoses bovines.* Institut Pasteur d'Algerie.

Spickett, A.M., de Klerk, D., Enslin, C.B. and Scholtz, M.M. (1989) Resistance of Nguni, Bonsmara and Hereford cattle to ticks in a Bushveld region of South Africa. *Onderstepoort, Journal of Veterinary Research* 56, 245–50.

Spooner, R.L. and Brown, C.G.D. (1980) Bovine lymphocyte antigens (BoLA) of bovine lymphocytes and derived lymphoblastoid lines transformed by *Theileria parva* and *Theileria annulata. Parasite Immunology* 2, 163–74.

Spooner, R.L., Innes, E.A., Glass, E.J., Millar, P. and Brown, C.G.D. (1988) Bovine mononuclear cell lines transformed by *Theileria annulata* express different subpopulation markers. *Parasite Immunology,* 10, 619–29.

Spooner, R.L., Innes, E.A., Glass, E.J. and Brown, C.G.D. (1989) *Theileria annulata* and *T. parva* infect and transform different bovine mononuclear cells. *Immunology* 66, 284–8.

Stobbs, T.H. (1966) The introduction of Boran cattle into an ECF endemic area. *East African Agriculture and Forestry Journal* 31, 298–304.

Tatchell, R.J., Chimwani, D., Chirchir, S.J., Ong'are, J.O., Mwangi, E., Rinkanya, F.

and Whittington, D. (1986) A study of the justification of intensive tick control in Kenyan Rangelands. *Veterinary Record* 119 401–3.

Teale, A.J. (1983) The major histocompatibility complex of cattle with particular reference to some aspects of East Coast fever. PhD thesis, University of Edinburgh.

Trail, J.C.M., d'Ieteren, G.D.M. and Murray, M. (1991) Practical aspects of developing geneticresistance to trypanosomiasis. In: Owen, J.B. and Axford, R.F.E. (eds), *Breeding for Diseases Resistance in Farm Animals.* CAB International, Wallingford, pp. 224–34.

Wilson, S.G. (1950) An experimental study of East Coast fever in Uganda. 1) A study of the type of East Coast fever reactions produced when the number of ticks is controlled. *Parasitology* 40, 195–209.

Yeoman, G.H. (1956) *The Occurrence of East Coast Fever (bovine Theileria parva infection) in Tanganyika and its Large-scale Control by Dipping.* Government Printers, Dar-es-Salaam, p. 6.

Young, A.S., Groocock, C.M. and Kariuki, D.P. (1988) Integrated control of ticks and tick-borne diseases of cattle in Africa. *Parasitology* 96, 403–32.

Chapter 15

Resistance to Ixodid Ticks in Cattle with an Assessment of its Role in Tick Control in Africa

J.J. de Castro

*GCP/ZAM/044/DEN 'Strategic Tick Control' Project, c/o
FAO Office, PO Box 30563, Lusaka, Zambia*

Summary

Host resistance to ticks is a complex and still poorly understood phenomenon. Different defence mechanisms including tick avoidance, grooming,
skin characteristics and more specific immunological responses are involved
in reducing the numbers of ticks parasitizing cattle. Host resistance has
successfully been used in Australia for the control of *Boophilus microplus*.
Culling of heavily infested animals is practised and cattle have been
selected and crossed to create new tick-resistant breeds.

In Africa, where several important tick species occur, trials to examine
tick resistance in different breeds of cattle and to assess cross-resistance
between different tick species as well as the repeatability of tick counts have
been started. At the same time, work is in progress to simplify the assessment of tick burdens on cattle. Preliminary results indicate the feasibility of
selecting cattle for tick resistance and improving the present levels by
culling of tick-infested animals.

The availability of field immunization methods against the major tick-
borne diseases together with a better understanding of the effects of ticks
on cattle have opened the way to integrated tick management in Africa.
Excellent cattle breeds exist and the majority are Zebu animals with high
tick resistance levels. However, the need to increase production forces
many African governments to introduce exotic breeds of low tick resistance
while at the same time spending large sums of money in tick control. There
is a need to plan cattle improvement to avoid losing host resistance and to
evaluate the productivity and tick resistance of the indigenous breeds with
the objective of reducing acaricide use for economic reasons and to post-
pone the development of resistance to acaricides.

Introduction

Control of ticks on cattle is generally directed to eliminating the risk of transmission of tick-borne diseases but also to avoid other direct losses known as 'tick worry'. These include the combined effects of the introduction of toxins, irritation, anaemia, and general unthriftiness. The possible consequences are losses in productivity, such as reduced growth rate and milk production (including udder damage mainly from *Amblyomma* species), hide damage, reduced fertility and even death.

Although alternative non-chemical tick control methods such as pasture spelling, predators and parasites, sterile male release, tick-resistant cattle and vaccination with tick antigens exist (Cunningham, 1981), they are at an experimental stage and tick control measures are still based on the use of acaricides.

Tick resistance in cattle is chiefly linked to Zebu (*Bos indicus*) animals although it is also present in some *Bos taurus* breeds particularly the Jersey breed. It is a widespread, though invisible tick control method present throughout the world and particularly effective in areas where pure Zebu animals are present such as in pastoralist and small cattle holder systems in Africa.

Host resistance to ticks is present in other domestic, wild and laboratory animals and the reviews by Willadsen (1980) and Brown (1985) cover all aspects of host resistance to ticks.

This chapter deals with host resistance to ticks in cattle and its potential use as a tick control method. Host resistance to ticks includes all interacting factors – anatomical, behavioural, physiological and immunological – reducing the number of ticks successfully attaching and engorging on the host.

Host resistance: a complex phenomenon

Tick avoidance

Contact is the first step in the tick–host relationship and different strategies are used by ticks in order to find a host. They have been described as either 'ambushers' or 'hunters' (Waladde and Rice, 1982) to define whether they wait for the host as in the case of the *Boophilus* and *Rhipicephalus* species or actively seek a host as in the case of the adults of some *Amblyomma* and *Hyalomma* species.

Cattle can detect the presence of ticks on the grass and avoid areas where ticks are present. This was demonstrated by Sutherst *et al.* (1986) for *Boophilus microplus* in Australia. Norval *et al.* (1988b) in Zimbabwe also observed avoidance behaviour with adult *Rhipicephalus appendiculatus* –

another questing tick – and avoidance was attributed to the sighting of the ticks.

In a trial in Kenya (de Castro, unpublished work), Masai (*B. indicus*) cattle were observed on three successive occasions for seven 24-hour intervals. They collected many fewer ticks than Hereford-cross (*B. indicus* × *B. taurus*) cattle when grazed together. Avoidance of ticks may help to explain the difference in the number of adult ticks between the two breeds (Table 15.1). However, it is as likely that the indigenous Zebu animals may have been better adapted to the conditions than the introduced animals. Sighting could have accounted for reducing the differences of *R. appendiculatus* but not for *Amblyomma variegatum* since this tick – hidden in the matt – becomes active only when cattle are in close proximity. It is, therefore, possible that other factors such as grooming and the early effect of the immune response may have been responsible for the differences in these observations.

Grooming and skin thickness

Grooming has been reported in *B. microplus* as a response to the release of histamine and other mediators involved in the hypersensitivity response. However, the dislodging of ticks – particularly in the case of 2- and 3-host ticks – by grooming may simply respond to irritation of ticks walking on the skin whilst seeking their predilection sites, before attachment.

The importance of grooming in controlling tick numbers has been assessed by several workers. Snowball (1956) and Bennett (1969) carried out experiments to evaluate the importance of grooming. They compared groups of cattle which could groom freely with others fitted with anti-grooming devices. Significantly less ticks were found on those animals which were able to groom. Later, however, tick numbers decreased on restrained animals (Bennett, 1969) and a proportion of larvae lost (18–39%) could not be accounted for by grooming alone (Koudstaal *et al.*,

Table 15.1. Daily means ± SE of tick burdens of five Hereford and five Masai cattle during three weeks of observations.

Week	Female *R. appendiculatus*		Female *Amblyomma* spp.	
	Hereford	Masai	Hereford	Masai
1	72.6±5.4	29.3±1.7	5.1±0.5	2.3±0.3
2	54.8±3.2	24.1±0.9	6.3±0.7	1.7±0.3
3	49.6±2.0	19.3±1.1	13.8±1.2	2.5±0.4

Source: de Castro, J.J. (unpublished work).

1978). Tatchell and Bennett (1969) demonstrated the involvement of histamine in host grooming and further studies (Schleger *et al.*, 1981b) have shown that histamine, liberated by mast cell degranulation, by inflicting cutaneous pain, is important in the initiation of the grooming process.

African *B. indicus* cattle naturally self-groom and groom each other frequently and thoroughly (Fig. 15.1). Under natural tick challenge in the field, tick-infested cattle spent more time grooming than tick-free animals at the expense of other normal behavioural events, suggesting that it also plays an important role in reducing the numbers of multi-host ticks (de Castro *et al.*, 1985b). Grooming was also important in reducing artificial infestations with *R. appendiculatus* (de Castro *et al.*, 1985a) and Norval *et al.* (1988a) regarded it as the main factor in the reduction in the number of *Amblyomma hebraeum* in cattle.

Skin thickness appears to play an important role in host resistance to ticks. However, there is still contradiction on whether thick (Ali, 1989; Bonsma, 1944) or thin skins (Brown, 1985; de Castro, unpublished work) are associated with resistance. However, these different observations have been made on different cattle breeds in different locations and other skin characteristics such as coat type, hair density and skin secretions may have influenced the results.

Only a proportion of ticks are not avoided or 'groomed-off' by the

Fig. 15.1. African cattle engaged in mutual grooming as frequently observed in the field.

animals and go on successfully to attach, becoming the target of other, more specific, mechanisms of the host's immune system.

Involvement of the immune system

In 1912 Munro-Hull observed resistance to the tick *B. microplus* in his herd of cattle in Australia, and reported that some, but not all of his animals rejected the ticks and remained in good condition. Johnston and Bancroft (1919) suggested that the condition was hereditary.

Several workers have since studied the ways in which the feeding of *B. microplus* on resistant cattle is interfered with. Although most of the losses of *B. microplus* occur during the first 24 hours (Roberts, 1968), few of those which die do so on the host (Koudstaal *et al.*, 1978). In spite of the observations by Riek (1962) who recorded the trapping of numerous larvae in the serous exudate of resistant animals, this event does not seem to occur very often (Roberts, 1971). It appears that larvae spend more time than other instars unattached on the host (Koudstaal *et al.*, 1978) and make significantly more attachments of short duration on resistant hosts (Kemp *et al.*, 1976; Roberts 1971), rendering them vulnerable to host grooming. This hypersensitivity reaction does not damage larvae since they are able to move in attempts to find new attachments (Kemp *et al.*, 1976). Furthermore, Roberts (1971) collected larvae from immune and non-immune hosts after they had been 19 hours on the host and found no differences in their viability. All this information suggests that the host reaction interferes with the ticks indirectly, by impairing their feeding which eventually causes their death or removal.

Schleger *et al.* (1976) studied the cellular responses of cattle of different *B. microplus* resistance. Highly resistant animals showed mast cell degranulation and eosinophil accumulation in the lesion and the level of protective resistance to ticks was related to the degree of hyperaemia which followed larval infestation (Hales *et al.*, 1981), to the number of arterio-venous anastomoses in the hair-follicular layer above the sebaceous-gland level of cattle skin (Schleger *et al.*, 1981a), and to the concentration and degranulation of eosinophils at the attachment site of the larvae (Schleger *et al.*, 1981b).

Antibody responses to tick infestation in cattle have been described by Brossard (1976) who showed a significant increase in serum gamma-globulin concentration following infestation with *B. microplus*. This was further confirmed when Roberts and Kerr (1976) transferred *B. microplus* resistance with sera.

Innate or acquired resistance?

Several workers have shown the existence of acquired resistance to ticks in cattle. Wagland (1975) compared tick-naive Brahman (*B. indicus*) and Shorthorn (*B. taurus*) cattle during four successive infestations with *B. microplus* larvae and obtained similar numbers of engorged females after the first infestation in both breeds. By the fourth infestation, however, the *B. indicus* animals yielded less than one-quarter of the Shorthorn cattle. He concluded that resistance to *B. microplus* was acquired, though some innate factors seemed to be involved in relation to duration of feed and weight of engorged females.

O'Kelly and Spiers (1976) compared the development of resistance to *B. microplus* in very young calves, showing that Zebu cattle (*B. indicus*) yielded significantly less engorged females than the Hereford (*B. taurus*) and Shorthorn crosses when infested for the first time. They also concluded that an innate component was involved in complementing the acquired factor.

Roberts (1968) designed an experiment to differentiate between innate and acquired resistance by comparing Droughtmaster (*B. indicus*) and Shorthorn steers. He showed that the degree of natural resistance varied as did the ability to acquire resistance.

Wagland (1978) demonstrated that Brahman (*B. indicus*) cattle are able to develop some measurable degree of resistance during the first three days of tick infestation whereas Shorthorn needed 20 days, perhaps explaining the earlier results of Roberts (1968) and O'Kelly and Spiers (1976).

It is now generally accepted that host resistance to ticks occurs in all breeds to a greater or lesser degree. However, in a population of *B. indicus* cattle, the proportion of animals with the ability to express resistance to ticks is greater than in *B. taurus* cattle and this ability wanes as the percentage of *B. indicus* blood decreases.

Importance of cattle breed

Several workers have compared *B. taurus* with *B. indicus* cattle and their relative capacity to develop resistance to *B. microplus*. Seifert (1971) analysed 3,000 tick counts made on over 1,000 animals of several breeds and described no differences between Afrikander (*B. indicus*) and Brahman crosses, though these carried significantly less ticks than the British (*B. taurus*) breeds studied. These results were supported by the findings of Utech *et al.* (1978) and Sutherst *et al.* (1988).

A good account of the meaning of individual variation in tick resistance is given by Wilkinson (1955) who described a situation where no adult *B.*

microplus developed on one animal whereas 1,458 adults completed their life cycle on a similar one.

The phenomenon of varying resistance to tick infestation in different cattle breeds has been much less studied in parts of the world other than Australia. Garris *et al.* (1979) in the USA found that Brahman cattle carried significantly more adults and nymphs of *Amblyomma americanum* than Hereford cattle when the two breeds were compared in field studies.

Bonsma (1944) in South Africa demonstrated that Afrikander cattle were more resistant to *A. hebraeum* and heartwater than British breeds. Of the total number of ticks naturally infesting the cattle, the former carried a mean of 9.6% of the ticks compared with 90.4% of the latter. The increased tick resistance of Criollo (*B. taurus*) cattle has also been stressed (Ulloa and de Alba, 1957) as well as differences observed between *B. indicus* and *B. taurus* and Kenana and Butana (*B. indicus*) breeds in the Sudan (Latif, 1984).

De Castro (unpublished work) also observed that Masai (*B. indicus*) cattle had much lower burdens than Boran (*B. indicus*) and Ayrshire, Friesian and Hereford (*B. taurus*) cross *B. indicus*. Similar results were obtained by Ali (1989) comparing Horro and Boran (*B. indicus*) with their crosses with Jersey, Simmental and Friesian (*B. taurus*) in western Ethiopia. Five animals of each breed, which previously had the same tick challenge, were herded together for four months under natural tick challenge composed mainly by *Boophilus decoloratus*, *Amblyomma cohaerens*, *Rhipicephalus evertsi evertsi* and *Rhipicephalus praetextatus*. Tables 15.2 and 15.3 present the comparison between means of the burdens of different tick species between breeds. Horro cattle invariably carried fewer ticks of all species than the other breeds, including Boran. Jersey crosses showed relatively high levels of resistance and Friesian crosses, very low resistance levels.

In search for host resistance markers

No easily recognizable characteristics have yet been found associated with host resistance to ticks. The abundance of arteriovenous anastomoses in the skin has been associated with resistance to *B. microplus* by Schleger *et al.* (1981a). The number of arteriovenous anastomoses in animals of low tick resistance was less than in those animals of high resistance. This observation may help to explain the positive relationship between hyperaemia and host resistance to *B. microplus* reported by Hales *et al.* (1981). Thick hide and high vascularity was one of the criteria adopted in the selection of cattle (Bonsma, 1944). This author stressed that ticks avoided animals with sleek coats and high sebum secretion as well as areas of the body where subcutaneous muscles were present. Sleek coats were later associated with

Table 15.2. Two-way analyses of variance (with repeated measurements) of the burdens of different tick species present on the cattle.

Source	DF	SS	MS	F
B. decoloratus				
Month	3	317.9	106.0	7.4***
Breed	7	885.8	126.5	8.8***
Month × Breed	21	114.7	5.5	0.4 NS
A. cohaerens				
Month	3	49.8	16.6	9.6***
Breed	7	68.2	9.7	5.7***
Month × Breed	21	30.6	1.5	0.9 NS
R.e. evertsi				
Month	3	47.4	15.8	16.4***
Breed	7	66.4	9.5	9.8***
Month × Breed	21	36.5	1.7	1.8 NS
R. praetextatus				
Month	3	40.2	13.4	16.1***
Breed	7	15.1	2.2	2.6*
Month × Breed	21	22.0	1.1	1.3 NS
Total tick burden				
Month	3	372.7	124.3	10.2***
Breed	7	945.6	135.1	11.1***
Month × Breed	21	96.8	4.6	0.4 NS

* $P < 0.05$.
*** $P < 0.001$.
NS, Not significant.
Source: Ali (1989).

animals in good condition and hence cattle of unimpaired tick resistance due to stress (Turner and Schegler, 1960).

O'Kelly and Spiers (1983) demonstrated that rectal and skin temperatures were lower in those animals, within breeds, that showed the highest tick resistance.

A relationship between phenotypes of the bovine major histocompatibility system and resistance to *B. microplus* has been described (Stear *et al.*, 1984). However, although significant, its role on resistance to *B. microplus*

Chapter 15

Table 15.3. Comparison of mean tick burdens in the different cattle breeds using least significant differences.

Tick species	Breed							
	HH	BJ	HS	HJ	BB	BF	BS	HF
B. decoloratus	16.3	49.3	59.6	47.7	55.9	58.6	101.3	162.5
	HH	BB	BJ	HS	HJ	HS	BS	BF
A. cohaerens	1.8	4.6	5.1	5.9	8.2	7.8	11.8	14.5
	HS	HH	BS	HJ	BJ	BB	BF	HF
R. e. evertsi	1.6	3.1	4.4	5.9	6.4	7.5	10.4	12.9
	HH	HS	BB	HJ	BJ	BS	HF	BF
R. praetextatus	0.3	1.0	1.2	2.3	2.5	2.7	4.8	4.8
	HH	HS	BJ	HJ	BB	BF	BS	HF
Total burden	21.5	68.1	63.3	61.2	69.2	88.3	120.2	188.0

Means underlined, not significantly different from one another.
Differences at $P < 0.05$.
Breeds of cattle: Boran (BB), Boran × Friesian (BF), Boran × Jersey (BJ), Boran × Simmental (BS), Horro (HH), Horro × Friesian (HF), Horro × Jersey (HJ) and Horro × Simmental (HS).
Source: Ali (1989).

is small. A correlation between the distribution of amylase genes (Am^B) and tick burden in Droughtmaster cattle was recorded by Francis and Ashton (1967).

Work in Kenya (de Castro, unpublished work) failed to show any relationship between tick burdens and coat colour or number of skin folds in the neck of Boran cattle.

The selection of cattle for tick resistance

Since Hull's observations, resistance to ticks has been considered a heritable character. This characteristic was studied subsequently with the idea of selecting cattle for tick resistance and significant progress has been made in Australia with the development of breeds of cattle which are resistant to ticks and at the same time show good productivity.

Tests to assess host resistance

Cattle resistance to the one-host tick *B. microplus* has been measured by either rating them after a known larval challenge or by ranking the animals while undergoing equal natural tick infestation (Sutherst *et al.*, 1979).

Rating of cattle to three-host tick *R. appendiculatus* was described by de Castro *et al.* (1985a) and this was the technique used by Latif (1984), Norval *et al.* (1988b) and de Castro *et al.* (1989). Ranking of cattle for two- and three-host tick infestations is also possible by assessing the number of adult ticks attached on the cattle (Kaiser *et al.*, 1982).

Since these techniques are often time consuming, the use of simpler tests has been proposed (Hewetson, 1978). Visual assessments of tick numbers have been used in Africa (de Castro, unpublished work). In this method, three independent observers assessed tick loads on cattle into five classes namely very high, high, medium, low and very low tick burdens. When the results were compared with the true tick counts on the animals, they were found to be highly correlated (Table 15.4) because larger African ticks are easily seen on animals with sleek coats. These results encourage further study towards its future use.

Table 15.4. Degree of concordance between three observers performing simultaneous and independent visual assessments of tick burden on 75 bulls.

	2	3
1	0.88[a]	0.80
	0.93[b]	0.88
2		0.88
		0.93

Note: all values significant at 0.001% level.
Source: de Castro, J.J. (unpublished work).
[a]Kendall
[b]Spearman

Selecting for host resistance

Hewetson (1972) showed that heritability ranged between 40 and 50%, and Utech *et al.* (1978) concluded that resistance in Brahman cattle appeared to be dominant. This is further stressed by Utech and Wharton (1982) who obtained similar results following 15 years of selection of *B. taurus* cattle with just one crossing of *B. taurus* with a Brahman bull. They concluded that tick-resistant bulls should be introduced into the artificial insemination programmes in Australia as a way of rapidly improving tick resistance. Tick-resistant breeds of cattle have now been developed in an effort to find animals which are able to produce, particularly milk, under tick challenge and in a tropical environment (Hayman, 1974; Reason, 1983). The development of these breeds started with the parallel development of cows with acceptable levels of milk production and bulls with high tick resistance and from there excellent dairy breeds have been created (Reason, 1983).

Despite the progress made in Australia, where they are concerned with only one tick species of economic importance (*B. microplus*) in cattle, there is less optimism from other parts of the world, especially Africa, where mixed species infestations are prevalent. The pioneering work of Bonsma (1944) set the criteria for breeding of 'tick repellent' cattle which included good heat tolerance, thick hides, high vascularity and some coat colours. He also recommended the culling of those cattle which carry more ticks, in agreement with Wilkinson (1955) and Kaiser *et al.* (1982), but this has not been widely applied in the field.

Despite the importance of ticks for the African cattle industry, very little work followed on the selection of cattle for host resistance in Africa.

A trial with the long-term objective of improving the level of host resistance to ticks in cattle was started in 1987 at Mutara Ranch, Laikipia, Kenya. The ranch is the home of the National Boran stud and the animals have been selected for growth rate and productivity since 1927. Since culling of those animals with undesirable characteristics takes place at three years, the work was started with the complete bull herd of 18–24 months, approximately 100 animals, with the intention of continuing every year.

For six months half-body tick collections were carried out while at the same time data on other parameters such as visual estimates of tick burdens, skin thickness, number of skin folds and coat colour were also obtained in order to investigate possible relationships between them and tick burdens.

The most common tick species observed was *B. decoloratus* followed in decreasing order by *Rhipicephalus evertsi evertsi, Hyalomma truncatum, R. pulchellus, R. appendiculatus* and *R. hurti.*

Good correlations between burdens of *B. decoloratus, H. truncatum, R. e. eversti* and *R. pulchellus* were recorded (Table 15.5) suggesting that

Table 15.5.   Correlation between collections of different tick species and total tick burden. Only two months shown.

	H.t.	R.a.	R.e.	R.h.	R.p.	Total
January						
B.d.	0.43**	0.08	0.10	−0.01	0.03	0.93**
H.t.		0.08	0.20	0.30**	0.13	0.65**
R.a.			0.03	−0.12	−0.13	0.21*
R.e.				−0.07	0.08	0.32**
R.h.					0.29**	0.12
R.p.						0.11
June						
B.d.	0.28**	0.08	0.07	−0.05	0.31**	0.95**
H.t.		0.25**	0.34**	−0.05	0.42**	0.47**
R.a.			0.13	−0.07	0.21*	0.16
R.e.				−0.05	0.23*	0.31**
R.h.					−0.08	−0.04
R.p.						0.49**

$*P < 0.05, **P < 0.01.$
Source: de Castro, J.J. (unpublished work).

animals are able to develop resistance to several tick species. However, poor correlations were observed for *R. appendiculatus*. This supports findings of low resistance to *R. appendiculatus* by Kaiser *et al.* (1982) and the poor cross-resistance between this tick species and *R. pulchellus* described by de Castro *et al.* (1989) under experimental conditions and comparisons between *R. appendiculatus*, *R. pravus* and *R. pulchellus* (Tatchell, 1986). However, numbers of *R. appendiculatus* were low and this may have interfered with the analyses of the results.

Repeatability of tick burdens on the different animals was studied by ranking the animals into different classes according to the degree of infestation and the results are encouraging (Table 15.6) as were those of Ali (1989) (Table 15.7).

Since *B. decoloratus* was the predominant tick species in these two trials, only limited information was obtained on other important tick species such as *R. appendiculatus* and further work is needed in this respect.

The future of tick-resistant cattle in tick control

As a result of many years of research, Australia has successfully tackled the problem of acaricide resistance in *B. microplus* by increasing Zebu blood in

Table 15.6.   Rank correlation for *B. decoloratus* and total tick burden for the six months of the trial.

	Months				
	F	M	A	M	J
B. decoloratus					
J	0.112[a]	0.262	0.367	0.300	0.321
	(0.136)	(0.328)	(0.435)	(0.365)	(0.386)
F		0.305	0.246	0.234	0.266
		(0.378)	(0.284)	(0.264)	(0.317)
M			0.328	0.355	0.323
			(0.395)	(0.421)	(0.403)
A				0.540	0.420
				(0.611)	(0.508)
M					0.474
					(0.558)
Total tick burden					
J	0.622	0.594	0.626	0.630	0.619
F		0.391	0.617	0.657	0.605
M			0.408	0.423	0.480
A				0.769	0.748
M					0.779

[a] Kendall (with Spearman rank in parentheses) correlation coefficients. With the exception of the correlation between *B. decoloratus* ranks between January and February, all other correlations highly significant ($P > 0.01$). *Source*: de Castro, J.J. (unpublished work).

cattle and regularly culling those animals with heavy tick burdens. Additionally, new breeds of highly productive tick-resistant cattle, better adapted to the environment have been developed, strengthening the role of host resistance in tick control even further and reducing the need for chemical methods. Further, vaccination of cattle using tick antigens has also shown encouraging results (Willadsen, 1986).

The situation in Africa in regard to ticks and cattle is an enigma. Although a number of cross-bred animals exist, the great majority of cattle in Africa today are of *B. indicus* stock, kept under traditional management with low productivity (in some circumstances because of these management systems) but with high levels of host resistance to ticks. However, regular and intensive acaricide tick control is still practised due to the implementation of schemes based on policies dictated many years ago directed to the eradication of tick-borne diseases. Further, the ever-

Table 15.7. Rank (Spearman) correlations of different tick species and total tick burden (TTB) for the four months of the trial.

	November	December	January
B. decoloratus			
December	0.64***		
January	0.62***	0.71***	
February	0.55***	0.83***	0.83***
A. cohaerens			
December	0.36*		
January	0.42**	0.64***	
February	0.56***	0.46**	0.46**
R. e. evertsi			
December	0.46**		
January	0.27*	0.13	
February	0.06	0.06	0.13
R. praetextatus			
December	−0.16		
January	0.10	0.04	
February	0.04	0.12	0.40**
Total tick burden			
December	0.70***		
January	0.67***	0.80***	
February	0.61***	0.80***	0.88***

$*P < 0.05, **P < 0.01, ***P < 0.001.$
Source: Ali (1989).

increasing needs for human food – particularly milk – due to population growth exerts great pressure on African governments to improve their livestock. This is done through artificial insemination schemes by crossing indigenous animals with European breeds, mainly Friesian, which rapidly improves milk production with the unwanted consequences of lowering resistance to ticks and tick-borne diseases, making acaricide use still more frequent and continuous. Most governments, therefore, have to spend increasing amounts of their scarce foreign exchange in measures to protect these 'improved' cattle. This situation has created a total dependency on acaricides, and the grave consequences of a break in this practice have been clearly described (Norval, 1978). Further, the increasing costs of these chemicals, although not as dramatic, contributes to the losses.

In areas where no tick control is carried out, endemic stability between ticks, tick-borne diseases and cattle exists (Young *et al.*, 1981). Recent research has showed that immunization against *Theileria parva* is possible (Young, 1985) and that cattle can tolerate relatively high tick burdens (de Castro *et al.*, 1985a, b; Norval *et al.*, 1988b; Pegram *et al.*, 1989) opening the way for the reduction of acaricide application and the re-establishment of endemic stability in many areas of Africa

Several FAO projects and other national and international research bodies have started to apply large-scale immunization procedures against tick-borne diseases and this has enabled new strategic tick control methods with reduced use of acaricides to be put into practice. In Burundi, Zambia and Zimbabwe such schemes have been successfully introduced with considerable economic advantages.

In a situation of endemic stability tick control can be greatly influenced by the level of host resistance of the cattle. Therefore, the reduction of tick resistance by the steady introduction of cross-breds could be detrimental in the long run. However, since the introduction of cross-breds in Africa will continue, perhaps crossing with other breeds such as Jersey should be considered as well as limiting the proportion of exotic blood introduced, coupled with continuous and compulsory culling of heavily infested animals. The selection for host resistance to ticks of highly productive breeds such as Boran or Sahiwal (*B. indicus*) offers another alternative.

Finally, most African cattle breeds have been neglected. It may be appropriate for one or more of the existing livestock research centres in Africa – or a newly established one – to start a programme in order to evaluate the African Zebu for both productivity and disease resistance. It is possible that these sturdy little animals when compared fairly with other cattle types may provide a few answers for the African livestock industry in general and the tick and tick-borne disease problem in particular.

Acknowledgements

Work described here was carried out while the author was at the International Centre of Insect Physiology and Ecology in Kenya and in Ethiopia with the UNDP/FAO ETH/83/023 'Tick Survey' Project.

The author is grateful to Dr R.J. Tatchell for his helpful criticism of the manuscript.

References

Ali, M. (1989) Study of host resistance to ticks in different breeds of cattle at Bako, Ethiopia. Unpublished D.V.M. Paper. Faculty of Veterinary Medicine, Addis Ababa University.

Bennett, G.F. (1969) *Boophilus microplus* (Acarina: Ixodidae): Experimental infestations on cattle restrained from grooming. *Experimental Parasitology* 26, 323–8.

Bonsma, J.C. (1944) Hereditary heartwater-resistant characters in cattle. *Farming in South Africa* 19, 71–96.

Brossard, M. (1976) Relations immunologiques entre bovins et tiques, plus particulierement entre bovins et *Boophilus microplus. Acta Tropica* 33, 15–36.

Brown, S.J. (1985) Immunology of acquired resistance to ticks. *Parasitology Today* 1, 166–71.

Cunningham, M.P. (1981) Biological control of ticks with particular reference to *Rhipicephalus appendiculatus.* In: Irvin, A.D., Cunningham, M.P. and Young, A.S. (eds), *Advances in the Control of Theileriosis.* Nijhoff, The Hague, pp. 160–4.

De Castro, J.J., Cunningham, M.P., Dolan, T.T., Deansfield, R.D. and Young, A.S. (1985a) Effects on cattle of artificial infestations with the tick *Rhipicephalus appendiculatus. Parasitology* 90, 21–33.

De Castro, J.J., Newson, R.M. and Herbert, I.H. (1989) Resistance in cattle against *Rhipicephalus appendiculatus* Neumann, 1901 with an assessment of crossresistance to *Rhipicephalus pulchellus* Gerstacker, 1873 (Acari: Ixodidae). *Experimental and Applied Acarology* 6, 237–44.

De Castro, J.J., Young, A.S., Dransfield, R.D., Cunningham, M.P. and Dolan, T.T. (1985b) Effects of tick infestation on Boran (*Bos indicus*) cattle immunised against theileriosis in an endemic area of Kenya. *Research in Veterinary Science* 39, 279–88.

Francis, J. and Ashton, G.C. (1967) Tick resistance in cattle: its stability and correlation with various genetic characteristics. *Australian Journal of Experimental Biology and Medical Science* 45, 131–40.

Garris, G.I., Stacey, B.R., Hair, J.A. and McNew, R.W. (1979) A comparison of Lone Star Ticks on Brahman and Hereford cattle. *Journal of Economic Entomology* 72, 869–72.

Hales, J.R.S., Schleger, A.V., Kemp, D.H. and Fawcett, A.A. (1981) Cutaneous hyperaemia elicited by larvae of the Cattle Tick, *Boophilus microplus. Australian Journal of Biological Sciences* 34, 37–46.

Hayman, R.H. (1974) The development of the Australian Milking Zebu. *World Animal Review* 11, 31–5.

Hewetson, R.W. (1972) The inheritance of resistance by cattle to Cattle Tick. *Australian Veterinary Journal* 48, 299–303.

Hewetson, R.W. (1978) Selection of cattle for resistance against *Boophilus microplus* In: Wilde, J.K.H. (ed.), *Tick-borne Diseases and their Vectors.* University of Edinburgh, Centre for Tropical Veterinary Medicine, Edinburgh, pp. 258–61.

Johnston, T.H. and Bancroft, M.J. (1919) Report on Mr. Munro Hull's claims regarding tick-resistant cattle. *Queensland Agricultural Journal* 11, 31–40.

Kaiser, M.N., Sutherst, R.W. and Bourne, A.S. (1982) Relationship between ticks and Zebu cattle in southern Uganda. *Tropical Animal Health and Production* 14, 63–74.

Kemp, D.H. Koudstaal, D., Roberts, J.A. and Kerr, J.D. (1976) *Boophilus microplus*: the effect of host resistance on larval attachments and growth. *Parasitology* 73, 123–36.

Koudstaal, D., Kemp, D.H. and Kerr, J.D. (1978) *Boophilus microplus*: rejection of larvae from British breed cattle. *Parasitology* 76, 379–86.

Latif, A.A. (1984) Resistance to natural tick infestations in different breeds of cattle in the Sudan. *Insect Science and its Application* 5, 95–7.

Munro-Hull, G.W. (1912) The cattle tick – A remedy. *Queensland Agricultural Journal* 29, 294–6.

Norval, R.A.I. (1978) The effects of partial breakdown of dipping in African areas in Rhodesia. *Rhodesian Veterinary Journal* 9, 9–16.

Norval, R.A.I., Floyd, R.B. and Kerr, J.D. (1988a) Ability of adults of *Amblyomma hebraeum* (Acarina: Ixodidae) to feed repeatedly on sheep and cattle. *Veterinary Parasitology* 29, 351–5.

Norval, R.A.I., Sutherst, R.W., Kurki, J., Gibson, J.D. and Kerr, J.D. (1988b) The effect of the Brown Ear tick (*Rhipicephalus appendiculatus*) on the growth of Sanga and European breed cattle. *Veterinary Parasitology* 30, 149–64.

O'Kelly, J.C. and Spiers, W.G. (1976) Resistance to *Boophilus microplus* (Canestrini) in genetically different types of calves in early life. *Journal of Parasitology* 62, 312–17.

O'Kelly, J.C. and Spiers, W.G. (1983) Observations on body temperature of the host and resistance to the tick *Boophilus microplus* (Acari: Ixodidae). *Journal of Medical Entomology* 20, 498–505.

Pegram, R.G., Lemche, J., Chizyuka, H.G.B., Sutherst, R.W., Kerr, J.D. and McCosker, P.J. (1989) Effect of tick control on liveweight gain of cattle in Central Zambia. *Medical and Veterinary Entomology* 3, 313–20.

Reason, G.K. (1983) Dairy cows with tick resistance: twenty years of the Australian Friesian Sahiwal. *Queensland Agricultural Journal* 109, 135–8.

Riek, R.F. (1962) Studies on the reactions of animals to infestation with ticks. VI. Resistance of cattle to infestation with the tick *Boophilus microplus* (Canestrini). *Australian Journal of Agricultural Research* 13, 532–50.

Roberts, J.A. (1968) Acquisition by the host of resistance to the Cattle Tick, *Boophilus microplus* (Canestrini). *Journal of Parasitology* 54, 657–62.

Roberts, J.A. (1971) Behaviour of larvae of the Cattle Tick, *Boophilus microplus* (Canestrini), on cattle of differing degrees of resistance. *Journal of Parasitology* 57, 651–6.

Roberts, J.A. and Kerr, J.D. (1976) *Boophilus microplus*: Passive transfer of resistance in cattle. *Journal of Parasitology* 62, 485–9.

Schleger, A.V., Lincoln, D.T. and Bourne, A.S. (1981a) Arteriovenous anastomoses in the dermal vasculature of the skin of *Bos taurus* cattle, and their relationship with resistance to the tick, *Boophilus microplus*. *Australian Journal of Biological Sciences* 34, 27–35.

Schleger, A.V., Lincoln, D.T. and Kemp, D.H. (1981b) A putative role for eosinophils in tick rejection. *Experientia* 37, 49–50.

Schleger, A.V., Lincoln, D.T., McKenna, R.V., Kemp, D.H. and Roberts, J.A. (1976) *Boophilus microplus*: cellular responses to larval attachment and their relationship to host resistance. *Australian Journal of Biological Sciences* 29, 499–512.

Seifert, G.W. (1971) Variations between and within breeds of cattle in resistance to field infestations of the Cattle Tick (*Boophilus microplus*). *Australian Journal of Agricultural Research* 22, 159–68.

Snowball, G.J. (1956) The effect of self-licking by cattle on infestations of Cattle Tick *Boophilus microplus* Canestrini. *Australian Journal of Agricultural Research* 7, 227–32.

Stear, M.J., Newman, M.J., Nicholas, F.W., Brown, S.C. and Holroyd, R.G. (1984) Tick resistance and the major histocompatibility system. *Australian Journal of Experimental Biology and Medical Science* 62, 47–52.

Sutherst, R.W., Floyd, R.B., Bourne, A.S. and Dallwitz, M.J. (1986) Cattle grazing behaviour regulates tick populations. *Experientia* 42, 194–6.

Sutherst, R.W., Maywald, G.F., Bourne, A.S., Sutherland, I.D. and Stegeman, D.A. (1988) Ecology of the cattle tick (*Boophilus microplus*) in Subtropical Australia. II Resistance of different breeds of cattle. *Australian Journal of Agricultural Research* 39, 299–308.

Sutherst, R.W., Wharton, R.H., Cook, I.M., Sutherland, I.D. and Bourne, A.S. (1979) Long-term population studies on the Cattle Tick (*Boophilus microplus*) on untreated cattle selected for different levels of tick resistance. *Australian Journal of Agricultural Research* 30, 353–68.

Tatchell, R.J. (1986) Interactions between ticks and their hosts. In: Howell, M.J. (ed.), *Parasitology – Quo Vadit?*. Australian Academy of Science, Canberra, pp. 597–606.

Tatchell, R.J. and Bennett, G.F. (1969) *Boophilus microplus*: antihistaminic and tranquilising drugs and cattle resistance. *Experimental Parasitology* 26, 369–77.

Turner, H.G. and Schleger, A.V. (1960). The significance of coat type in cattle. *Australian Journal of Agricultural Research* 11, 645–63.

Ulloa, G. and de Alba, J. (1957) Resistance to ectoparasites in some races of cattle. *Turrialba* 7, 8–12.

Utech, K.B.W. and Wharton, R.H. (1982) Breeding for resistance to *Boophilus microplus* in Australian Illawarra Shorthorn and Brahman × Australian Illawarra Shorthorn cattle. *Australian Veterinary Journal* 58, 41–6.

Utech, K.B.W., Wharton, R.H. and Kerr, J.D. (1978) Resistance to *Boophilus microplus* (Canestrini) in different breeds of cattle. *Australian Journal of Agricultural Research* 29, 885–95.

Wagland, B.M. (1975) Host resistance to Cattle Tick (*Boophilus microplus*) in Brahman (*Bos indicus*) cattle. I. Responses of previously unexposed cattle to four infestations with 20,000 larvae. *Australian Journal of Agricultural Research* 26, 1073–80.

Wagland, B.M. (1978) Host resistance to Cattle Tick (*Boophilus microplus*) in Brahman (*Bos indicus*) cattle. II. The dynamics of resistance in previously unexposed and exposed cattle. *Australian Journal of Agricultural Research* 29, 395–400.

Waladde, S.M. and Rice, M.J. (1982) The sensory basis of tick feeding behaviour. In: Obenchain, F.D. and Galun, R. (eds), *Physiology of Ticks*. Pergamon, Oxford, pp. 71–118.

Wilkinson, P.R. (1955) Observations on infestations of undipped cattle of British breeds with the Cattle Tick, *Boophilus microplus* (Canestrini). *Australian*

Journal of Agricultural Research 6, 655–65.

Willadsen, P. (1980) Immunity to ticks. *Advances in Parasitology* 18, 293–313.

Willadsen, P. (1986) Immunological approaches to the control of ticks. In: Howell, M.J. (ed.), *Parasitology – Quo Vadit?*. Australian Academy of Science, Brisbane, pp. 671–7.

Young, A.S. (1985) Immunization of cattle against theileriosis in the Trans-Mara Division of Kenya. A comparison of trials under traditional Maasai management with trials on a ranch development. In: Irvin, A.D. (ed.), *Immunization against Theileriosis in Africa*. ILRAD, Nairobi, pp. 64–8.

Young, A.S., Leitch, B.L. and Newson, R.M. (1981) The occurrence of a *Theileria parva* carrier state in cattle from an East Coast fever endemic area of Kenya. In: Irvin, A.D., Cunningham, M.P. and Young, A.S. (eds), *Advances in the Control of Theileriosis*. Nijhoff, The Hague, pp. 60–2.

Chapter 16

Genetic Variation in Resistance to Fleece Rot and Flystrike in Sheep

H.W. Raadsma

Department of Animal Health, University of Sydney,
Private Mailbag 3, Camden, NSW 2570, Australia

Summary

Fleece rot and body strike are economically significant production diseases in Merino sheep. Genetic variation in the predisposition of sheep to these diseases is reflected in a moderate to high (0.3–0.4) heritability of liability and substantial differences between major strains and bloodlines within the Merino breed. Low selection differentials, as a function of low prevalence, and the inclusion of other important production objectives in well-designed breeding plans, shows that reduction in susceptibility is likely to be very slow (1–1.5% per generation). Although fleece rot can be considered as a breeding objective in its own right, it is also a suitable indicator trait for resistance to body strike. More efficient indirect selection criteria are needed when sheep are selected in low risk environments. A number of fleece traits were promising as indicator traits for fleece rot, but have proved to be very inefficient when included with other selection criteria for an index seeking improvement in a number of production objectives.

A better understanding of mechanisms responsible for resistance, combined with optimum use of specialized genetic resources for experimentation should lead to the identification of new indicator traits. The immunological response of resistant and susceptible sheep to fleece rot and the use of restriction fragment length polymorphism of genomic DNA probed with Class II MHC gene probes, have shown promise.

The design of breeding programmes should focus on projected advantages of long-term reductions in disease status of the flock, and the exploitation of differences between flocks in susceptibility to these diseases.

Introduction

Fleece rot (a superficial bacterial dermatitis) and flystrike (cutaneous myiasis, invasion of the fleece and skin with maggots from primary blowflies) are two skin diseases of sheep which are of economic significance to the Australian sheep industry. A detailed review on the scope for genetic improvement and some of the management options available to minimize the impact of these diseases has been presented by Raadsma and Rogan (1987) and Raadsma (1987). This chapter will, therefore, only summarize the information presented up to 1987 and highlight relevant developments since then.

Major forms of flystrike, their prevalence and control options

Lucilia cuprina (the Australian Sheep Blowfly) is the species responsible for over 90% of the primary attacks. *Calliphora* spp. and *Lucilia sericata* are responsible for the remainder. Infestation of sheep with larvae from dipterous flies only occurs when the host has been predisposed for one or several reasons.

The site of predisposition determines the nature and prevalence of strike normally seen in the field. For this reason it is important to classify strikes and evaluate breeding in light of management options available to control each of the forms of strike separately.

Breech strike

This is a collective term for all strikes occurring on the crutch and tail region of sheep and is considered the most common form of strike in Australia (Belschner, 1937; Watts *et al.*, 1979). Sheep are predisposed by soiling of the breech area with urine and faeces and the consequent development of a dermatitis (Bull, 1931; Seddon, 1967). The major factors which influence the prevalence of breech strike are sex (with ewes more frequently affected than males), wool and tail length, and particularly the number and location of caudal folds in the crutch region (Bellschner, 1937; Seddon *et al.*, 1931; Watts *et al.*, 1979). The two major forms of prevention are surgical removal of skin folds in the crutch region (Mules operation) and docking of the tail to an optimum length. This procedure is permanent and reduces the prevalence of breech strike from 60–80% in ewes to less than 1% when combined with crutching. These management procedures are so effective in reducing the problem, that programmes investigating the scope for selective breeding were suspended in the 1930s. There are animal welfare arguments for less mulesing so the evaluation of alternative control options, including selective breeding, are being re-examined (James, personal communication).

Poll strike

Poll or head strike predominantly occurs in young rams. Although it can develop in wounds sustained during fighting, the majority of strikes seem to develop at the base of the horns, presumably due to bacterial activity induced by sweating and excessive skin debris at this site (Seddon, 1967). Raadsma (1987) observed that this form of strike is the major type of strike in flocks of mulesed rams. Chemical control is practised.

Pizzle strike

This strike occurs mainly in young rams and wethers when the long wool around the preputial opening becomes soiled with urine or exudate from balanitis 'pizzle rot' (Belschner, 1937). The prevalence of overt strikes appears to be low in most cases (< 2% of all male sheep affected) (Watts *et al.*, 1979), except in special circumstances where the belly wool is matted with burrs, or matted when sheep graze long wet pasture. Removal of the wool around the prepuce (ringing) is the most common method of control. Although surgical control through 'pizzle dropping' can be very effective (Marchant, 1986), the procedure does not appear to be warranted as a routine control method.

Strike secondary to other diseases

Sheep may become struck when suffering from other diseases such as acute conjunctivitis or foot rot. In these cases exudates render the sheep attractive to flies. Clean surgical wounds are rarely struck (Seddon, 1967). Post-partum staining of the udder and perianal region of lambing ewes renders this class of sheep highly susceptible to crutch strike, despite the fact that these ewes are often mulesed and crutched (Raadsma, unpublished). Chemicals are used for treatment.

Body strike

Strike involving any part of the body other than the breech, head and pizzle is termed 'body strike' Most commonly affected sites are the shoulder and back regions (Belschner, 1937; Raadsma *et al.*, 1989). Body strike outbreaks ('flywaves') develop rapidly after periods of prolonged and heavy rainfall during the warmer months of the year. Young sheep, regardless of sex, with 3–6 months' fleece growth are the most susceptible. The dependence of body strike on seasonal conditions means that the prevalence shows considerable variation, some years are free of body strike, whereas in exceptional years, up to 50% of young sheep may be affected. A prevalence of 20% is considered serious with significant production losses and

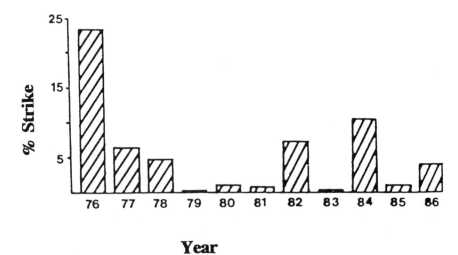

Year

Fig. 16.1. Yearly variation in the prevalence of body strike in young Merino hoggets at ARC Trangie. (Adapted from Raadsma, 1987.)

associated mortality. Outbreaks of body strike are unpredictable and chemical control and/or shearing when the risk of fly strike is high, are the most common control options. Neither of these control options is fully effective.

In summary, Raadsma and Paxton (1988, unpublished) observed from 1,183 strikes in 32,621 sheep observations over a 12-year period, that the majority of strikes (60%) were body strikes which were largely confined to sheep under 2 years of age (4% affected), compared to adult sheep (0.4% affected). Yearly variation in the prevalence of body strike in a fairly typical sheep-raising environment such as found at Trangie (500 mm average rainfall) suggest that a serious outbreak can be expected once every 3–4 years (Fig. 16.1).

The cost of flystrike has increased over the last 50 years from 2.6% to over 6% of gross wool industry earnings (Raadsma and Rogan, 1987). Treatment and prevention account for approximately 60–80% of this depending on seasonal conditions, while the remainder is due to production losses. Estimates by Beck *et al.* (1985) suggest that body strike is the most costly form of strike and without long-term control options. This chapter will focus on this form of strike and the potential of selective breeding.

Predisposing factors to body strike

A normal dry fleece is not a suitable environment for the development of flystrike. Susceptibility follows excessive wetting and bacterial proliferation. The two major predisposing conditions are bacterial dermatoses fleece rot and dermatophilosis (Belschner, 1937; Edwards *et al.*, 1985; Watts *et al.*, 1979).

Fleece rot

This condition is a mild superficial dermatitis induced by moisture and bacterial proliferation at skin level. Exudation, resulting in a matted band of fibres combined with extensive staining of the fleece are the major signs. A number of bacterial species occur in fleece rot but *Pseudomonas* spp, especially *Pseudomonas aeruginosa, Pseudomonas maltophilia,* are considered the causal agents (London and Griffith, 1984; MacDiarmid and Burrell, 1986; Merritt and Watts, 1978; Seddon, 1937). Involvement of other bacterial species such as *Proteus, Staphylococcus, Bacillus, Corynebacterium, Enterobacter* and *Flavobacterium* spp. as causal agents has not yet been fully resolved (Burrell, 1988; Jansen and Hayes, 1987; Mulcock, 1965).

Fleece rot should not be confused with the more severe *Pseudomonas* infection reported by Gumbrell (1988) in which *Ps. aeruginosa* occurs in a severe ulcerative dermatitis penetrating the skin, causing considerable skin damage and mortality in affected sheep. This condition is also associated with extensive wetting.

Dermatophilosis

This is a serious dermatitis resulting from skin wetting followed by infection by the bacterium *Dermatophilus congolensis.* A full description of the pathogenesis and epidemiology of this disease has been given by Roberts (1967).

Fleece rot and dermatophilosis are intricately linked to the development of body strike in that actively infected sheep attract gravid female blowflies and encourage oviposition. Exudative lesions provide the moisture essential for eggs to hatch and provide soluble protein for first instar larvae (Gheradi *et al.*, 1981, 1983; Merrit and Watts, 1978; Sutherland *et al.*, 1983; Watts and Merrit, 1981). Sandeman *et al.* (1987) demonstrated that first instar larvae were able to initiate flystrike on a wet skin without the need for obvious predisposing conditions. However, a wet fleece environment without bacterial proliferation is inconceivable. Furthermore, the high attractiveness of sheep which are predisposed, makes these sheep more likely to get struck than sheep which are merely wet. Raadsma *et al.* (1988) has demonstrated that both the severity of fleece rot and moisture status of the fleece were the two major variables which determined the likelihood of development of body strike.

Apart from the strong biological relationship between the predisposing conditions and body strike, it is important to have information on the genetic correlation between these conditions. For fleece rot this relationship is at least 0.75 and in some cases unity (Atkins and McGuirk, 1979; Raadsma *et al.*, 1989; Atkins and Raadsma, unpublished data). This relationship has not yet been established for dermatophilosis and body strike. This chapter will thus evaluate susceptibility to fleece rot and body strike concurrently from a breeding point of view, but cannot include susceptibility to dermatophilosis due to lack of genetic information for this disease in sheep.

Variation in resistance to fleece rot and body strike

Phenotypic variation

One particular attribute of disease traits is that their phenotypic expression follows a binomial distribution. Sheep are either affected with the condition or not. It is also clear that the prevalence of the condition following an outbreak is seldom 100%. Expression within the range from 0 to 100% depends on a large number of environmental and genetic factors. The hypothetical concept that the expression of fleece rot and body strike can be attributed to an underlying variable with a normal distribution in relation to a fixed reference point, termed the threshold, seems plausible. Dempster and Lerner (1950) and Falconer (1981) termed this variable liability and is the cumulative effect of all environmental and genetic factors contributing to the prevalence of the disease. Raadsma *et al.* (1988) provided experimental support for this theoretical concept in the case of fleece rot and body strike. McGuirk and Atkins (1984) showed that the heritability of liability to fleece rot remained constant, whereas the heritability on the observed scale was dependent on the prevalence of the condition, with the highest value at intermediate prevalence. This fits theoretical expectation of a trait with a binomial distribution.

Liability cannot be measured directly in individuals. Some understanding of the many environmental and genetic factors which contribute to liability are discussed below. The adoption of the liability scale has a number of advantages to allow comparison across data sets and make sensible predictions of genetic gain as the prevalence is likely to decrease.

Repeatability of resistance

Knowledge of the repeatability of susceptibility to fleece rot and body strike is useful in predicting scope for improvement in the current generation by culling those animals in which the disease is expressed early in life.

Theoretically it also sets the upper limit for the heritability of resistance to these diseases. The special problem of variable prevalence between years and a general increase in resistance with age, compounds the difficulty of obtaining reliable estimates from field and experimental records. Atkins and McGuirk (1979) estimated the repeatability for prevalence of fleece rot at 0.06 and 0.21 when mean prevalence was 12% and 36% respectively. Under high rainfall conditions with prevalence to fleece rot in the range 35–66%, Raadsma (1988, unpublished) estimated the repeatability of expression of fleece rot to be in the range $0.17 \pm 0.03 - 0.48 \pm 0.05$ for three separate data sets. Under experimental conditions, Raadsma *et al.* (1988) estimated the repeatability of liability to fleece rot at 0.43 ± 0.07. Given that fleece rot is extremely simple to measure, some gain in accuracy and effective heritability may be possible if multiple measurements are available early in an animal's lifetime (Falconer, 1981). Thus fleece rot could be measured as weaners (3–10 months) and the hogget stage (10–16 months) without increasing generation length.

Genetic variation within flocks

Heritability

Raadsma and Rogan (1987) reviewed in detail the sources and problems associated with estimating the heritability of resistance to fleece rot and body strike. From the many estimates, it was evident that the heritability for prevalence of fleece rot changed with the mean prevalence of the condition. As discussed previously this fits theoretical expectations of a trait with a binomial distribution. On the underlying liability scale it was concluded that the heritability of liability to fleece rot is likely to be in the order of 30–40%. Estimates for the heritability of liability to body strike are far more scarce, not only because a limited number of experimental flocks will tolerate a high level of flystrike, but often the flocks are unpedigreed and thus unsuitable for genetic analysis. Atkins and McGuirk (1979) presented two heritability estimates for the prevalence of body strike, which after transformation assuming an underlying scale, were 0.25 ± 0.15 and 0.53 ± 0.22. Additional information on the heritability of liability to body strike has been obtained by Raadsma *et al.* (1988), Raadsma (1991) who obtained estimates of 0.58 ± 0.42, 0.53 ± 0.25 and 0.10 ± 0.13 from three small data sets in which the prevalence of body strike was very high (22.7%, 19.8% and 18.6% respectively). This supports the view that the predisposition of sheep to both fleece rot and body strike is in part genetically determined, with potential scope for within-flock genetic improvement.

Genetic variation between breeds

There is no conclusive evidence of differences between breeds in their susceptibility to fleece rot and body strike. Lack of suitable studies which allow for comparisons across breeds in the same environment, and in which breed differences can be evaluated against genetic differences within breeds, are the major reason. Between breed comparisons are warranted especially in the comparison of Merino and breeds based on the Merino such as Corriedales, Polwarth, Comebacks and Zenith. It is unlikely that a Merino breeding enterprise would consider cross breeding with non-Merino breeds solely for the purpose of increasing resistance to fleece rot and flystrike, since this would drastically alter the nature of the enterprise where wool production is the sole objective.

Genetic variation between flocks within the Merino breed

There is now considerable evidence that differences between Merino flocks in their susceptibility to fleece rot and body strike are very large. Increased resistance from the strong-wool to medium-wool to the fine-wool strain is now well documented (Atkins and McGuirk, 1979; Dunlop and Hayman, 1958; Raadsma *et al.*, 1989). Resistance can also be attributed to specific bloodlines within a strain (Fig. 16.2). The implications of this large source of variation are of considerable importance to the Australian Merino

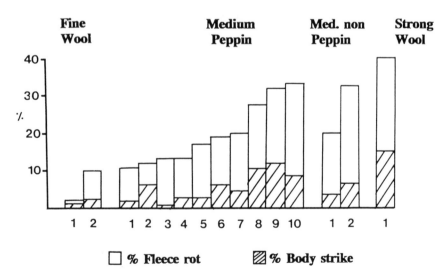

Fig. 16.2. Variation between Merino strains and bloodlines within strains in their susceptibility to fleece rot and body strike. (Adapted from Atkins and McGuirk, 1979.)

industry, since the majority of commercial breeders purchase their replacement rams from outside sources. The change of ram source may provide a
rapid means to improve resistance, provided no compromise is made in
other production traits of the new bloodline.

Genotype by environment interaction

Because rainfall has a major influence on the expression of fleece rot and
body strike, it is obvious that in dry environments the problem is small or
not existent. In wet years on the other hand some genotypes will perform
considerably better than others. On this basis, a large genotype by environment interaction has been observed involving strain and bloodline comparisons across a range of environments (Dunlop and Hayman, 1958).
Similarly, if differences across years are taken as different environments,
Atkins and McGuirk (1979) observed strong interactions between 15
different bloodlines in their susceptibility to fleece rot. A recent update of
this interaction is seen in Fig. 16.3, where the relative difference between
flocks increased as the mean prevalence increased to 50%. This interaction
is due to scale effects (Type I) and not due to a change in ranking (Type II).
Similarly when 40 different flocks were compared for four years under a
high rainfall environment, between flock rankings were high ($r = 0.96 \pm$
0.19) across years (Raadsma 1988, unpublished). A recent experimental
study by deKroes and Raadsma (1988, unpublished) demonstrated a
consistent ranking across two age groups of sheep from seven bloodlines
known to vary considerably in their susceptibility to fleece rot ($r = 0.82$). A
similar ranking was less consistent when the flocks were tested across two
different rainfall patterns ($r = 0.63$).

No formal estimates have been published of sire by environment
interactions, although it is likely that these exist for the same reason of flock
by environment interactions. This highlights the problem of making selec-
tions in a dry, low risk environment and breeding in a wet, high risk
environment.

Inclusion of resistance to fleece rot and body strike in breeding plans

Utilizing genetic variation between flocks

Complete replacement of a susceptible bloodline with a more resistant
bloodline, or introduction of a more resistant bloodline through continued
purchase of rams, are two options to make rapid increases in resistance.
The major limitations of exploiting this option are that resistant bloodlines
can not be readily identified unless they participate in formal and well-

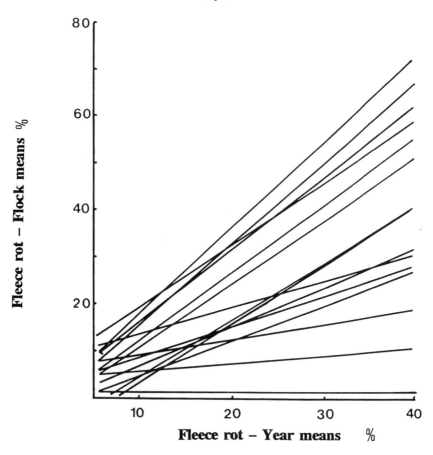

Fig. 16.3. Flock by year interaction in the susceptibility of 15 flocks to fleece rot. Flock means are regressed against the year mean prevalence. (Adapted from Atkins and McGuirk, 1979.)

organized flock comparisons, which need to be maintained in high risk environments (Roberts *et al.*, 1987). Second, the economic performance of a bloodline does not depend solely on resistance to fleece rot and body strike. Combined information is thus necessary for all important production traits at the flock level before sensible decisions can be made.

Utilizing genetic variation within flocks

Body strike as the sole objective

Projected progress   Raadsma and Rogan (1987) provided a brief overview on the scope and limitations of exploiting genetic variation within flocks in resistance to body strike. The options available to the breeder to improve resistance to body strike include direct selection, indirect selection using either fleece rot or other indicator traits, or combined direct plus indirect selection. More recently Raadsma (1991) demonstrated that under a high risk environment for body strike, with 20% of animals affected each year, progress under mass direct selection would only reduce the prevalence to 7% after 20 years (Fig. 16.4). If a more realistic selection environment is chosen, in which a flywave occurs once every 3–5 years, interspersed with a variable prevalence as seen in Fig. 16.1, progress would be considerably less. Approximately 17% of sheep would still be affected with body strike under flywave conditions after 20 years of direct selection (Fig. 16.4). Clearly, the influence of making selection decisions on the observed scale strongly limits practical progress as is demonstrated when hypothetical progress is calculated for direct selection on the underlying scale (Fig. 16.4).

Indirect selection   Indirect selection utilizing fleece rot is an attractive alternative solution since fleece rot is normally expressed at a higher prevalence than body strike. It is simple, cheap and easy to measure, and it has a high co-heritability (defined as $h_1 \, h_2 \, r_{g1.2}$) with liability to body strike (0.4). Figure 16.4 shows that under both conditions where either 5% of sheep are struck with the occurrence of a flywave every three years, or where a high prevalence of body strike is experienced (20%), utilizing fleece rot would lead to faster genetic progress. Fleece rot would be expressed at 30% and 80% respectively. Fleece rot is thus a moderately efficient indicator for selection against body strike.

Unfortunately, many sheep breeding studs which act as sources for genetic stock in Australia, go to considerable lengths to protect sale rams from wet conditions conducive to the development of fleece rot. Furthermore, the majority of studs are located in relatively low risk environments. It is possible to screen animals for resistance to fleece rot and body strike through inducing fleece rot through artificial wetting but this option is unattractive to ram breeders. Ram buyers (commercial breeders) have thus little information on the relative resistance of rams offered for sale. The search for alternative selection criteria, has been the focus of considerable research in this area (Raadsma and Rogan, 1987). Of the many fleece traits which have been suggested, it appears that at least four have a moderate co-heritability with liability to fleece rot. These traits include greasy wool colour (0.18), variability in fibre diameter (0.25), wax content (0.16), and staple length (0.16) (Atkins, 1987b; James *et al.*, 1987; Raadsma and Rogan, 1987; Raadsma and Wilkinson, 1990). If a similar relationship is observed with liability to body strike (Gilmour and Raadsma, 1986), progress using an indicator trait with a moderate co-heritability (0.18) would be as effective as fleece rot expressed at a high prevalence (Fig. 16.4).

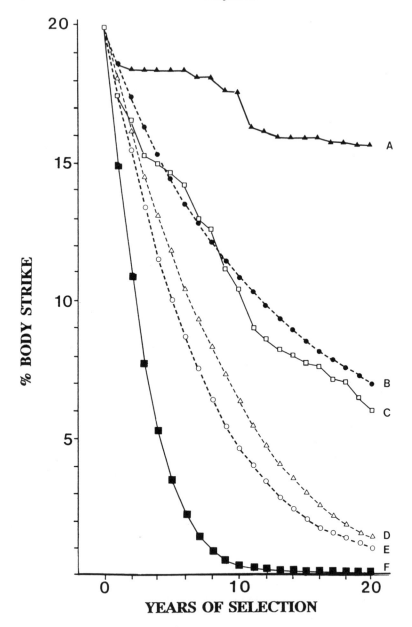

Fig. 16.4. Long-term changes in prevalence of body strike (%) under: (A) direct selection, low (5%) variable prevalence; (B) direct selection, high (20%) constant prevalence; (C) indirect selection, low (30%) variable prevalence of fleece rot; (D) hypothetical indirect selection, trait with moderate (0.18) co-heritability; (E) indirect selection, high (80%) constant expression of fleece rot; (F) hypothetical direct selection on the underlying scale. (Adapted from Raadsma, 1991.)

Observed responses The value of fleece rot, either through natural expression or under experimental induction (McGuirk *et al.*, 1978; Raadsma *et al.*, 1988, 1989) as an indirect selection criterion for resistance to body strike can be gauged from the following selection experiment. Two single character selection flocks were established by the NSW Department of Agriculture in 1974–1979 in which the sole selection criterion was either the presence or absence of fleece rot. The flocks are termed susceptible and resistant respectively. After 15 years of direct selection (approximately four generations), with selection predominantly on the ram side, the flocks have diverged at a rate of 0.13 σ of liability to fleece rot per year (Fig. 16.5a). The flocks also show a difference in susceptibility to body strike (Fig. 16.5b).

Projected advantages Under the hypothetical situation where resistance to body strike is the sole breeding objective, it is difficult to give exact estimates of economic returns for each unit of increase in resistance. This is because different management options are available to the breeder to deal with flystrike (Ponzoni, 1982a, 1985), and the prevalence varies from year to year. Evaluating the advantage of including body strike in breeding plans is further complicated when resistance to body strike is considered in light of other important breeding objectives (see below). In discussing the importance of body strike as a breeding objective, it is perhaps better to consider its long-term projected advantages. For instance, what are the economic advantages to management and overall profitability if the susceptibility of the flock is reduced to the extent that no serious control or preventive treatments are required after ten years of selective breeding? Clearly, increased labour costs would favour the development of such easy-care features in the Merino breed.

Piper and Barger (1988) highlighted similar problems when selecting for increased resistance to internal parasites and considered the concept of a biological and economic threshold below which parasitism had no adverse effects. In similar vein, resistance to body strike could be seen as an increase in the number of years which are free from strike and require no preventive treatment. It would also give sheep managers the freedom from anxiety associated with animal health and production problems attributed to body strike. It should be noted that once the level of resistance to body strike, has reached a defined economic threshold, no further selection would be warranted provided the pathogen does not change, or selection for other breeding objectives increases liability.

Figure 16.6 shows the change in prevalence of body strike over a 10-year cycle under various predicted selection strategies discussed previously. It highlights that body strike could be reduced to a minor problem if selection was feasible directly on the underlying scale, or through an efficient indirect selection strategy.

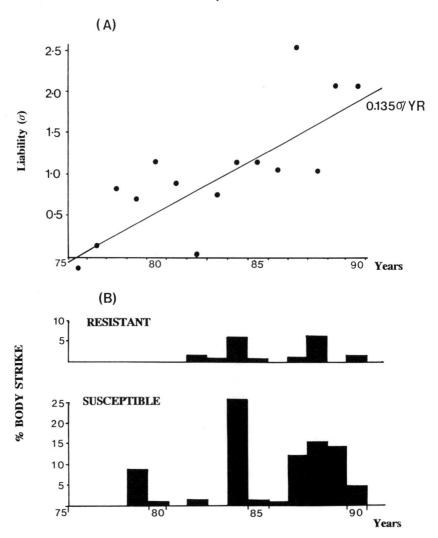

Fig. 16.5. Responses in flocks selected solely for (Susceptible) and against (Resistant) the expression of fleece rot. (A) divergence between flocks in their liability (σ) to fleece rot. (B) difference between flocks in their natural expression of body strike (%). (Raadsma, Nesa and Mortimer, unpublished data.)

Resistance to body strike in designed breeding plans

Over the last 20 years, and particularly in the last 10, there has been a massive upsurge in the design, development and to a lesser extent, adoption rate of breeding programmes based on measured performance for wool sheep (see Atkins, 1987a, 1988; Ponzoni, 1982b, 1987, 1988). At present,

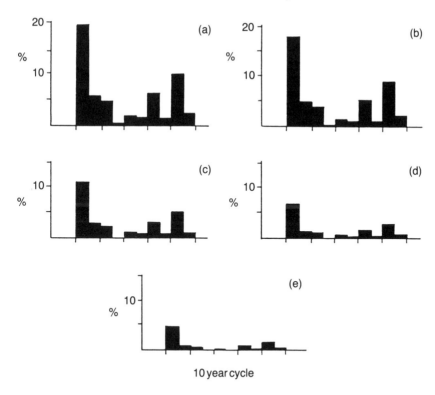

Fig. 16.6. Body strike patterns (for a 10-year cycle) following 10 years of selection based on: (a) no selection; (b) direct selection, low (5%) variable prevalence; (c) direct selection, high (20%) constant prevalence; (d) indirect selection, low (30%) variable prevalence of fleece rot; (e) hypothetical indirect selection, trait with moderate (0.18) co-heritability. (Adapted from Raadsma, 1991.)

resistance to fleece rot and body strike have not been formally included in such programmes. Lack of knowledge on the relative importance of fly strike in the breeding objective due to difficulties in estimating relative economic weights for this objective, combined with imprecise estimates of relevant genetic parameters are the major reasons (Atkins, 1987b; Ponzoni, 1982a, 1985).

Recently Atkins (1987b) considered resistance to fleece rot as a breeding objective for Merino sheep. A summarized version of his calculations has been given where the other two objectives were reducing fibre diameter and increasing clean fleece weight (Table 16.1). The predicted rates of progress in reducing the prevalence of fleece rot were calculated for four situations (a) no inclusion of fleece rot in the breeding objective, (b) direct selection utilizing fleece rot expressed at 15% prevalence, (c) indirect selection using

Table 16.1. Genetic changes per generation in the prevalence of fleece rot (ROT) due to (a) no selection against ROT, (b) direct selection, 15% prevalence, (c) indirect selection based on two indicator traits with moderate scope (IND 1, IND 2) and (d) combined indirect plus direct selection.

Selection criteria[a]	Reduction % fleece rot per generation
(a) CFW + FD	+0.3
(b) CFW + FD + ROT 15%	−0.9
(c) CFW + FD + IND 1	−0.2
CFW + FD + IND 2	0.0
CFW + FD + IND 1 + IND 2	−0.5
(d) CFW + FD + ROT 15% + IND 1	−1.1
CFW + FD + ROT 15% + IND 2	−1.1
CFW + FD + ROT 15% + IND 1 + IND 2	−1.4

[a]Index included clean fleece weight, fibre diameter and fleece rot as breeding objectives, production objectives.
Source: Adapted from Atkins (1987, unpublished data).

indicators with a moderate co-heritability with susceptibility to fleece rot, (d) combined direct and indirect selection. The selection index also included clean fleece weight and fibre diameter as selection criteria. It was evident that by not including fleece rot as an objective, selection for production traits would slowly increase the susceptibility of sheep to fleece rot. Direct selection, even at a low prevalence, was effective in reducing, albeit very slowly, the susceptibility of sheep to fleece rot. The use of indirect selection was not very effective in reducing susceptibility to fleece rot. In this case the importance of the relationship between the indicator traits and the other objectives in the index (fleece weight and fibre diameter) negated potential improvement in reducing susceptibility to fleece rot. Combined indirect plus direct selection resulted in marginally better progress than direct selection alone.

Atkins (1987b) pointed out that while the responses in fleece rot per generation were relatively small, if reduced susceptibility to fleece rot was the sole objective, responses would only be marginally higher, since the relatively low prevalence (15%) would limit progress. Hypothetical selection directly on the underlying scale should see a decrease of 9% after one generation.

It is thus unlikely that much progress would be made after 20 years of selection using the index described in (b) (from 15% to 10% at best). As pointed out by Ponzoni (1982a, 1985) the relative advantages of such a selection strategy could be minimal if alternative forms of control are still necessary. Should the problem of fleece rot and body strike be serious enough to warrant inclusion as a breeding objective it may be desirable to alter the design of breeding programmes to address the properties peculiar to disease resistance. First, little economic benefit is likely until body strike falls to below economic threshold. Second, production in a diseased flock may well be subject to adverse genotype by environment interactions. Third, once resistance has reached an economic threshold, importance of disease resistance is lessened provided it remains below the threshold. It may thus be desirable to improve resistance to disease as fast as is possible until an economic threshold has been reached. Genetic change in production objectives may have to be restricted during this phase. Once the desired disease resistance threshold has been reached, other production breeding objectives would adopt their relative importance and the importance of disease resistance would be restricted. One possible way that this could be achieved is to give resistance to body strike an inflated weighting. It would consider not only the appropriate economic weighting per unit increase in resistance, but also consider the cost of preventive management routine necessary over the period of selection until the desired threshold had been reached. For instance what advantage would resistance to body strike have if a prescribed threshold level was reached in 10 years instead of the projected 30 years using a traditional index. Similarly the cost or gain in production objectives could be calculated for the same 30-year cycle.

Before we can adopt resistance to fleece rot and body strike in breeding programmes, we need thus a better understanding of the economic and other advantages of doing so, and we need more efficient selection strategies for improving resistance. Estimates of genetic parameters concerning potential indirect selection criteria need to be accurate otherwise valuable selection effort may be misdirected (Sales and Hill, 1976). Projected advantages of using indirect selection criteria also need to consider the cost of additional measurement vs economic advantages of improving disease resistance. The identification of suitable indirect selection criteria remains thus a high priority.

Identification of new indirect selection criteria

Desirable properties of indirect selection criteria include low cost and simplicity of measurement, and a high efficiency for genetic improvement in the desired trait (McGuirk and Atkins, 1980). The latter not only includes a high genetic covariance with disease resistance but also, as has been highlighted previously, no antagonistic genetic relationships with other breeding objectives. Efficiency of indirect selection criteria also needs to include scope to optimize selection differentials in both males and females. The search for such traits is difficult at best. A full understanding of all factors contributing to liability of fleece rot and body strike combined with exploitation of suitable genetic resources is the most likely means to identify new, useful selection criteria.

Liability of sheep to fleece rot and body strike

To understand the nature of liability to fleece rot and body strike, it is essential to consider the interplay of all variables which comprise the liability variable. From current knowledge in the literature it is clear that the three major factors influencing liability to fleece rot and body strike are moisture, pathogenic challenge, and inherent susceptibility.

Moisture

This is possibly the single most important variable which influences the development of fleece rot and body strike, under both natural and experimental conditions. For a brief review see Raadsma (1990). Incidental information from experiments by deKroes and Raadsma (1988, unpublished data) could detect no difference in the prevalence of fleece rot when 749 Merino sheep were exposed to two wetting treatments (light drizzle rain 3.6 mm/day vs heavy rain 8.9 mm/day) for the same duration. It is thus possible that once a certain precipitation rate has been exceeded, the duration of wetting is more important in influencing the prevalence of fleece rot as was shown by McGuirk *et al.* (1978), and Raadsma *et al.* (1988). Variation between sheep in moisture status of the fleece during the development of fleece rot was a key variable which reflected differences between sheep in liability (Raadsma, 1990). Potential indicator traits should thus focus on factors responsible for wetting and drying characteristics of the fleece and skin.

Pathogenic variation

As discussed previously, there is little doubt about the role of *Ps. aeruginosa* in the development of fleece rot, and *L. cuprina* in the development of flystrike. Variation in the capacity of these pathogens to cause disease has not been considered in detail. Burrell and MacDiarmid (1984) reported the presence of at least 14 serogroups of *Ps. aeruginosa* in field isolates, but did not demonstrate variation between serogroups in their ability to cause fleece rot. This problem is further complicated by the lack of published knowledge on the key bacterial virulence factors in the development of fleece rot. London *et al.* (1984) and Burrell and MacDiarmid (1984) considered proteases to be important virulence factors and described considerable differences between isolates in their ability to produce these enzymes *in vitro*, but once again this was not correlated with development of disease. Chin and Watts (1988) considered phospholipase C (PLC) to be a prime virulence candidate but did not report on variation between isolates of *Ps. aeruginosa* to produce phospholipase C and corresponding disease. At this stage it is more than likely that a number of important enzymes of *Ps. aeruginosa* are potential virulence factors and that differences between strains could be responsible for differences between sheep or flocks in their liability to fleece rot. Raadsma and Burrell (1988, unpublished data) could not detect the presence or absence of specific *Ps. aeruginosa* or *Ps. maltophilia* serogroups when 350 sheep from 15 resistant and susceptible genotypes were compared under the same challenge. It may thus be that variation between flocks under field conditions is partly due to virulence of the microflora, but for sheep within a flock, differences in microbial challenge may be less variable. The importance of transmission of *Ps. aeruginosa* between wet sheep has not been demonstrated, but this would ensure that sheep within a flock are exposed to similar pathogens.

In similar vein, partial characterization of proteases and collagenases from *L. cuprina* larvae has been reported by Bowles *et al.* (1988) who suggested a role for these in the pathogenesis of body strike. Elliot *et al.* (1980) suggested larval antigen from wild strains was immunologically more reactive than antigen from laboratory strains, but variation between strains of *L. cuprina* in their capacity to produce body strike has not been reported. Also, when sheep are grazed as a single flock, it is highly likely that they are all exposed to the same fly pressure. The identification of the key virulence factors in both fleece rot and body strike remains thus a high priority.

Host susceptibility

The susceptibility of sheep should be seen in response to the three essential components: moisture, microbial activity and larval activity. All three components are necessary to cause disease, but the response to each of the challenge factors may well be limited by a different defence system within the host. McGuirk *et al.* (1978), Watts (1981) and Raadsma *et al.* (1988) speculated that the fleece, the skin surface and the skin itself could act as effective barriers to the disease challenge.

Fleece barrier   The possibility of the fleece as a mediator in the development of fleece rot was shown by Raadsma (1987) who exposed sheep with varying periods of wool growth to a fleece rot challenge and demonstrated that sheep within 4–6 months of wool growth were most susceptible. Variation in wetting and drying characteristics were the major variables which were influenced by the fleece barrier. The physical and chemical structural nature of the fleece, in particular its capacity to provide substrates for microbial activity, are possibly the major fleece contributions to the development of fleece rot. Despite the evaluation of a large number of fleece traits, no single trait can be held responsible for resistance within this barrier (James *et al.*, 1989; Raadsma and Rogan, 1987).

Fleece/skin interface   This is possibly the main site for operation of the threshold to both fleece rot and flystrike since the pathogens are non-invasive, and this barrier has to be broken to develop clinical signs of disease. James and Warren (1979) and James *et al.* (1984) reported on the possible importance of the wax layer on the skin surface. Chemical influence on the microenvironment, in particular pH, is another major variable which may influence the activity of a number of important enzymes in the pathogenesis of fleece rot and fly strike (Guerrini *et al.*, 1988). Further detailed studies in this area are warranted.

Skin/host immune barrier   In the case of fleece rot, few detailed studies on the scope for immunological control have been published. Hay *et al.* (1982), Hollis *et al.* (1982), Chapman *et al.* (1984) and Chin and Watts (1988) have reported on immunological changes at skin level in sheep exposed to moisture, bacterial activity and potential virulence components. Burrell *et al.* (1982) and Burrell (1985) successfully manipulated the host immune system to give protection against homologous fleece rot challenge. The exact role of a protective immune response in innately resistant animals has yet to be defined or published. The scope for productive research in this area must be substantial given major developments in immunology over the last decade.

   Reports by O'Donnell *et al.* (1980, 1981), Elliot *et al.* (1980), Sandeman *et al.* (1985, 1986) and Bowles *et al.* (1988) indicated the production of specific antibodies (largely IgG) to larval challenge, but could not demonstrate significant associations with resistance as indicated by the failure of flystrike to develop. Sandeman *et al.* (1985, 1986) and Bowles *et al.* (1987), attempted to induce acquired resistance through repeated challenge with larvae or larval products. Large variation between sheep, combined with low numbers precluded the demonstration of true acquired resistance (i.e. failure to become struck after challenge). Recent developments by Sandeman's group do suggest a role for reaginic antibodies and an immediate hypersensitivity reaction in sheep following challenge with larvae or larval products (Sandeman *et al.*, 1986; Bowles *et al.*, 1988). Manipulation of non-specific inflammatory responses may thus provide some scope to reduce or limit wound development during flystrike.

Developments in light of using immune response traits as indicators for genetic resistance are discussed later on.

Variation between sheep in their susceptibility to fleece rot and body strike is not only the net effect of all the barriers to provide protection, but should also include changes in relative importance of the barriers during actual development of the disease. For example, the value of the fleece to moisture challenge is positive during the initial wetting phase, but once the skin has been wet the presence of a wet fleece would actually promote the development of fleece rot rather than provide protection. Similar changes in the role of skin and immunological barriers may operate (Raadsma, 1990).

Should it be known which of the barriers makes the strongest contribution to resistance, it would be possible to focus research attention on components of this barrier to identify potential markers. In an attempt to evaluate the relative importance of the fleece vs non-fleece barriers, Raadsma (1988, unpublished), compared 800 sheep from 17 genotypes for their susceptibility to fleece rot with and without the presence of the barrier. It was concluded that inherent genetic resistance is likely to reside in the skin and or host- immune barrier. The presence of a fleece was found to exacerbate the severity of fleece rot and thus increase the susceptibility of sheep to body strike (Raadsma *et al.*, 1988).

Exploitation of suitable genetic resources for experimentation

The comparison of sheep which are known to be resistant or susceptible to fleece rot and body strike offers experimenters a powerful tool to 'home in' on either mechanisms of resistance and or genetic markers which could be used for selective breeding.

The differences between strains and bloodlines can be readily exploited in experimental studies to give an insight on possible mechanisms. The major drawback is that it may not be reliable to extrapolate from 'between flock' comparisons to factors which are responsible for differences between sheep 'within' flocks. Flock differences are likely to be a consequence of genetic drift and selection for traits other than specifically for resistance.

The evaluation of a large number of progeny groups from well-designed matings is an inefficient means to search for new markers. The establishment of single character selection flocks is a powerful means to amplify genetic differences between sheep within flocks. One of the major reasons for establishing such selection flocks is for the verification and identification of correlated responses. The responses need not only be production traits but could also be traits believed to be in part responsible for resistance. The author has compared the resistant and susceptible flocks for a wide range of fleece and skin characteristics (Raadsma, 1987). In general there was reasonable agreement between the direction and magnitude of predicted and observed differences. This shows that such comparisons could be used as a short cut in understanding the genetic basis for resistance. This short cut would be in the form of detailed measurements on a limited number of animals rather than the large number required in a random non-selected background. Hill (1985) and Henderson (1989) highlight the problems of interpreting and analysing differences between selection lines in correlated responses and making judgements on their potential suitability as indirect selection criteria. This is particularly relevant in the comparison of the 'resistant' and 'susceptible' fleece rot selection lines maintained at Trangie. The fact that the lines are not replicated, have no unselected control line and were based on an extremely small group of sires during their establishment, may invalidate some of the comparisons which are made. It is quite possible that some of the observed differences between the lines are solely due to random genetic drift. This could be particularly relevant for traits related to immunological competence and possibly the major histocompatibility complex (MHC) where a change in gene frequency between the lines due to genetic drift will present a misleading picture. The use of a second test of selection lines selected for essentially the same trait would give the experimenter a powerful and easy means to verify any potential association between resistance and any physiological or DNA marker.

Recent developments

Over the last three years there has been a renewed interest in the search for new immunologically based indirect selection criteria. Few details have been published as yet. O'Maera *et al.* (submitted) demonstrated a clear and consistent difference between sheep from the fleece rot resistant and susceptible selection lines in weal size response to larval secretory antigens (LESA) (Fig. 16.7). The results suggest an immediate type hypersensitivity reaction to what is clearly a strong mixture of inflammatory agents. The failure to show a difference between the lines in larval survival rates following implantation, supports the hypothesis that the inflammatory response is in relation to selection for fleece rot and that fleece rot is the essential predisposing agent before the onset of body strike. The author is not convinced that natural resistance to body strike is mediated through direct effect (rejection) of larvae, rather the degree of predisposition (fleece rot) holds the key to the development of body strike. Further detailed analysis awaits a better understanding of the possible immunological basis for resistance to bacterial and/or larval antigens at skin level.

The search for involvement of Class I or Class II genes from the major histocompatability complex in resistance to both fleece rot and body strike has been negligible. Sandeman *et al.* (1986) could not detect an association between ovine lymphocyte histoglobulins (OLA) and resistance to larvae

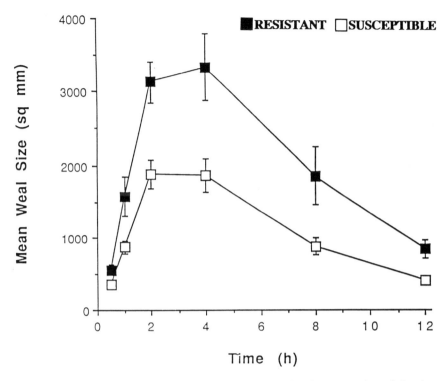

Fig. 16.7. Mean weal size (length × width) ± SEM versus time following intradermal injection of larval secretory antigen in lambs from fleece rot RESISTANT and SUSCEPTIBLE selection flocks. (Adapted from O'Meara *et al.* submitted).

under experimental conditions in a very small number of animals. The author, in conjunction with Litchfield, Nicholas, Egerton, Brown and Hulme, has subjected 74 sheep, to preliminary analysis of restriction fragment length polymorphism (RFLP) of genomic DNA within the Class II region of ovine MHC. The use of human Class II DR B and DQ A probes in conjunction with two restriction enzymes, showed considerable polymorphism. A total of 20% of the sheep were known to be affected with body strike. From a number of significant associations between the presence or absence of bands, one band (PvuII/DRB 1.4) showed a consistent association with both fleece rot and body strike ($P <$ 0.001). This association was maintained when sires were included in the model. Sheep without the band had 8% body strike (n=44) compared with 32% in sheep for which the band was present (n=30). These results are of pre-liminary nature only and considerably more animals need to be analysed before any firm recommendation can be made on the use of DNA RFLPs as indirect selection criteria.

Acknowledgements

The author is grateful to Drs K.D. Atkins and S. Mortimer and Miss Nesa to allow their unpublished results to be cited in this paper. This paper has had the benefit of comments made by Professor J.R. Egerton. Part of the work reported here was funded by the Wool Research and Development Fund on recommendation of the Australian Wool Corporation.

References

Atkins, K.D. (1987a) Potential responses to selection in Merino sheep given current industry structure and selection practices. In: McGuirk, B.J. (ed.), *Merino Improvement Programs In Australia.* Proceedings of the National Symposium, Leura, NSW, pp. 299–312.

Atkins, K.D. (1987b) Resistance to fleece rot and body strike: its role in a breeding objective for Merino Sheep. *Proceedings of the Sheep Blowfly and Flystrike Management Workshop,* Trangie, NSW, Department of Agriculture, pp. 3.1–7.

Atkins, K.D. (1988) Consequences of selection for wool production and implications for Merino improvement programmes. *Proceedings of the 3rd World Congress on Sheep and Beef Cattle Breeding.* vol. 2, pp. 409–28.

Atkins, K.D. and McGuirk, B.J. (1979) Selection of Merino sheep for resistance to fleece rot and body strike. *Wool Technology and Sheep Breeding,* 27, 15–19.

Beck, T., Moir, B. and Meppem, T. (1985) The cost of parasites to the Australian sheep industry. *Quarterly Review of Rural Economics* 7, 336–43.

Belschner, H.G. (1937) Studies on the sheep blowfly problem: II Observations on fleece rot and body strike in sheep, particularly in regard to their incidence, type of sheep susceptible and economic importance. *Department of Agriculture NSW Scientific Bulletin* no. 54, pp. 61–95.

Bowles, V.M., Carnegie, P.R. and Sandeman, R.M. (1987) Immunization of sheep against infections with larvae of the blowfly *Lucilia cuprina. International Journal of Parasitology* 17, 743–58.

Bowles, V.M., Carnegie, P.R. and Sandeman, R.M. (1988) Characterization of proteolytic and collagenolytic enzymes from the larvae of *Lucilia cuprina,* the sheep blowfly. *Australian Journal of Biological Research,* 41, 269–78.

Bull, L.B. (1931) Some observations of the folds in the breech of sheep, and its possible relationship to blowfly strike. *Australian Veterinary Journal,* 7, 143–8.

Burrell, D.H. (1985) Immunisation of sheep against experimental *Pseudomonas aeruginosa* dermatitis and fleece rot associated body strike. *Australian Veterinary Journal* 62, 55–7.

Burrell, D.H. (1988) Bacteriology and pathogenesis of fleece rot and flystrike. *Sheep Health and Refresher Course,* Post Graduate Committee in Veterinary Science, University of Sydney, vol. 110, pp. 231–47.

Burrell, D.H. and MacDiarmid, J.A. (1984) Characterisation of isolates of *Pseudomonas aeruginosa* from sheep. *Australian Veterinary Journal* 61, 277–9.

Burrell, D.H., Merritt, G.C., Watts, J.E. and Walker, K.H. (1982) The role of *Pseudomonas aeruginosa* in pathogenesis of fleece rot and the effect of immunization. *Australian Veterinary Journal* 58, 34–5.

Chapman, R.E., Hollis, D.E. and Hemsley, J.A. (1984) How quickly does wetting affect the skin of Merino sheep. *Proceedings of the Australian Society of Animal Production*, 15, 290–2.

Chin, J.C. and Watts, J.E. (1988) Biological properties of phospholipase C purified from a fleece rot isolate of *Pseudomonas aeruginosa*. *Journal of General Microbiology* 134, 2567–75.

Dempster, E.R. and Lerner, I.M. (1950) Heritability of threshold characters. *Genetics* 35, 212–36.

Dunlop, A.A. and Hayman, R.H. (1958) Differences among Merino strains in resistance to fleece rot. *Australian Journal of Agricultural Research* 9, 260–6.

Edwards, J.R., Gardiner, J.J., Morris, R.T., Love, R.A., Spicer, P., Bryant, R., Gwynn, R.V.R., Hawkins, C.D. and Swan, R.A. (1985) A survey of ovine dermatophilosis in Western Australia. *Australian Veterinary Journal* 62, 361–5.

Elliot, M., Pattie, W.A. and Dobson, C. (1980) The immune response of sheep to larvae of *L. cuprina*. *Proceedings of the Australian Society of Animal Production* 13, 500.

Falconer, D.S. (1981) *Introduction to Quantitive Genetics* 2nd edn. Longman, London.

Gherardi, S.G., Monzu, N., Sutherland, S.S., Johnson, K. and Robertson, G.M. (1981) The association between body strike and dermatophilosis of sheep under controlled conditions. *Australian Veterinary Journal* 57, 268–71.

Gherardi, S.G., Sutherland, S.S., Monzu, N. and Johnson, K.G. (1983) Field observations on body strike in sheep affected with dermatophilosis and fleece rot. *Australian Veterinary Journal* 60, 27–8.

Gilmour, A.R. and Raadsma, H.W. (1986) Estimating genetic variation and covariation for flystrike incidence in Australian Merino sheep on an underlying normal scale. *Proceedings of the 3rd World Congress of Genetics Applied to Livestock Production*, Vol. 12, 490–3.

Guerrini, R.H., Murphy, G.M. and Broadmeadow, M. (1988) The role of pH in the infestation of sheep by *Lucilia cuprina* larvae. *International Journal for Parasitology* 18, 407–9.

Gumbrell, R.C. (1988) More pelt defects in lambs. *New Zealand Veterinary Journal* 36, 99.

Hay, J.B., Mills, S.C. and Maddocks, J.G. (1982) The effect of exposure to rainfall on white blood cell counts of sheep. *Australian Veterinary Journal* 59, 60–1.

Hill, W.G. (1985) Detection and genetic assessment of physiological criteria of merit within breeds. In: Land, R.B. and Robinson, D.W. (eds), *Genetics of Reproduction in Sheep*, Butterworths, London, pp. 319–31.

Henderson, N.D. (1989) Interpreting studies that compare high – and low – selected lines on new characters. *Behavioural Genetics* 19, 473–502.

Hollis, D.E., Chapman, R.E. and Hemsley, J.A. (1982) Effects of experimentally induced fleece rot on the structure of the skin of Merino sheep. *Australian Journal of Biological Science* 35, 545–56.

James, P.J., Ponzoni, R.W., Walkley, J.R.W., Whiteley, J.E. and Stafford, J.E. (1987) Fleece rot in South Australian Merinos: heritability and correlations with fleece characters. In: McGuirk, B.J. (ed.), *Merino Improvement Programs in Australia. Proceedings of a National Symposium*, Leura, N.S.W, pp. 341–5.

James, P.J. and Warren, G.H. (1979) Effect of disruption of the sebabceous layer of the sheep's skin on the incidence of fleece rot. *Australian Veterinary Journal* 55, 333–5.

James, P.J., Warren, G.H. and Neville, A. (1984) The effect of some fleece characters on skin wax layer and fleece rot development in Merino sheep following wetting. *Australian Journal of Agricultural Research* 35, 413–22.

James, P.J., Warren, G.H., Ponzoni, R.W. and MacLachlan, H.G. (1989) Effect of early life selection using indirect characters on the subsequent incidence of fleece rot in a flock of South Australian Merino ewes. *Australian Journal of Experimental Agriculture* 29, 9–15.

Jansen, B.C. and Hayes, M. (1987) The relationship between the skin and some bacterial species occurring on it in the Merino. *Ondersterpoort Journal of Veterinary Research* 54, 107–11.

London, C.J. and Griffith, I.P. (1984) Characterization of Pseudomonads isolated from diseased fleece. *Applied and Environmental Microbiology* 47, 993–7.

London, C.J., Griffith, I.P. and Kortt, A. (1984) Proteinases produced by Pseudomonads isolated from sheep fleece. *Applied and Environmental Microbiology* 47, 75–9.

MacDiarmid, J.A. and Burrell, D.H. (1986) Characterization of *Pseudomonas maltophilia* isolates from fleece rot. *Applied and Environmental Microbiology* 51, 346–8.

Marchant, B. (1986) Improving wether production and reducing stains through pizzle dropping and testosterone implants. *Wool Technology and Sheep Breeding* 36, 67–71.

McGuirk, B.J. and Atkins, K.D. (1980) Indirect selection for increased resistance to fleece rot and body strike. *Proceedings of the Australian Society of Animal Production* 13, 92–5.

McGuirk, B.J. and Atkins, K.D. (1984) Fleece rot in Merino sheep. I. The heritability of fleece rot in unselected flocks of medium – wool Peppin Merinos. *Australian Journal of Agricultural Research* 35, 424–34.

McGuirk, B.J., Atkins, K.D., Kowal, E and Thornberry, K. (1978) Breeding for resistance to fleece rot and body strike. The Trangie programme. *Wool Technology and Sheep Breeding* 26, 17–24.

Merritt, G.C. and Watts, J.E. (1978) The changes in protein concentration and bacteria of fleece and skin during the development of fleece rot and body strike in sheep. *Australian Veterinary Journal* 54, 517–20.

Mulcock, A.P. (1965) The fleece as a habitat for micro-organisms. *New Zealand Veterinary Journal* 13, 87–93.

O'Donnell, I.J., Green, P.E., Connell, J.A. and Hopkins, P.S. (1980) Immunoglobulin G antibodies to the antigens of *L. cuprina* in the sera of fly struck sheep. *Australian Journal of Biological Science* 33, 27–34.

O'Donnell, I.J. Green, P.E., Connell, J.A. and Hopkins, P.S. (1981) Immunization of sheep with larvae antigens of *Lucilia cuprina*. *Australian Journal of Biological Science* 34, 411–17.

O'Maera, T.J., Nesa, M., Raadsma, H.W., Saville, D.G. and Sandeman, R.M. (19XX) Variation in skin inflammatory responses between sheep resistant and susceptible to fleece rot and blowfly strike. *International Journal for Parasitology* (submitted)

Piper, L.R. and Barger, J.A. (1988) Resistance to gastro-intestinal strongyles:

feasibility of a breeding programme. *Proceedings of the 3rd World Congress on Sheep and Beef Cattle Breeding* vol. 1, pp. 593–611.

Ponzoni, R.W. (1982a) Resistance to fleece rot and body strike in the selection objectives for Australian Merino sheep. *Proceedings of the Australian Association of Animal Breeding and Genetics* 3, 245–6.

Ponzoni, R.W. (1982b) Breeding objectives in sheep improvement programmes. *Proceeding of the 2nd World Congress of Genetics Applied to Livestock Production*, vol. 5, pp. 619–34.

Ponzoni, R.W. (1985) The importance of resistance to fleece rot and body strike in the breeding objective of Australian Merino sheep. *Wool Technology and Sheep Breeding* 32, 20–21, 33–40.

Ponzoni, R.W. (1987) WOOLPLAN – design and implications for the Merino industry. In: McGuirk, B.J. (ed.) *Merino Improvement Programs in Australia*. Proceedings of a National Symposium, Leura, NSW, pp. 25–40.

Ponzoni, R.W. (1988) On farm performance recording services for beef cattle and sheep in Australia – organisation and trends. *Proceedings of the 3rd World Congress on Sheep and Beef Cattle Breeding*, vol. 1, 239–58.

Raadsma, H.W. (1987) Flystrike control: an overview of management and breeding options. *Wool Technology and Sheep Breeding* 35, 174–85.

Raadsma, H.W. (1990) Fleece rot and body strike in Merino sheep. III. Significance of fleece moisture following experimental induction of fleece rot. *Australian Journal of Agricultural Research* 40, 207–20.

Raadsma, H.W. (1991) Fleece rot and body strike in Merino sheep. V. Heritability of liability to body strike in weaner sheep under flywave conditions. *Australian Journal of Agricultural Research* 42 (in press).

Raadsma, H.W., Gilmour, A.R. and Paxton, W.J. (1988) Fleece rot and body strike in Merino sheep. I. Evaluation of liability to fleece rot and body strike under experimental conditions. *Australian Journal of Agricultural Research* 39, 917–34.

Raadsma, H.W., Gilmour, A.R. and Paxton, W.J. (1989) Fleece rot and body strike in Merino sheep II. Phenotypic and genetic variation in liability to fleece rot following experimental induction *Australian Journal of Agricultural Research*, 40, 207–20.

Raadsma, H.W. and Rogan, J.M. (1987) Genetic variation in resistance to blowfly strike. In: McGuirk, B.J. (ed.), *Merino Improvement Programs in Australia*. Proceedings of a National Symposium, Leura, NSW, p. 32.

Raadsma, H.W. and Wilkinson, B.R. (1990) Fleece rot and body strike in Merino sheep. IV. Experimental evaluation of traits related to greasy wool colour for indirect selection against fleece rot. *Australian Journal of Agricultural Research*, 41, 139–53.

Roberts, D. (1967) Barriers to *Dermatophilis dermatonomus* infection on the skin of sheep. *Australian Journal of Agricultural Research* 14, 492–508.

Roberts, E.M., McCully, R., Morrison, A., Butt, J., Coy, J. and Lane, G. (1987) Exploiting between flock variation in Merinos by the use of the National Merino Reference Flock. In: McGuirk, B.J. (ed.), *Merino Improvement Programs in Australia*. Proceedings of a National Symposium, Leura, NSW, pp. 383–7.

Sales, J. and Hill, W.G. (1976) Effect of sampling errors on efficiency of selection indices. 2. Use of information on associated traits for improvement of a single trait. *Animal Production* 23, 1–14.

Sandeman, R.M., Bowles, V.M., Stacey, I.H. and Carnegie, P.R. (1986) Acquired resistance in sheep to infection with larvae of the blowfly *Lucilia cuprina*. *International Journal for Parasitology* 16, 69–75.

Sandeman, R.M., Collins, B.J. and Carnegie, P.R. (1987) A scanning electron microscope study of *L. cuprina* larvae and the development of blowfly strike in sheep. *International Journal for Parasitology* 17, 759–65.

Sandeman, R.M., Dowse, C.A. and Carnegie, P.R. (1985) Initial characterization of the sheep immune response to infections of *Lucilia cuprina*. *International Journal of Parasitology*, 15, 181–5.

Seddon, H.R. (1937) Studies on the sheep blowfly problem. III. Bacterial colouration of wool. *Department of Agriculture, NSW, Scientific Bulletin*, 54, 96–110.

Seddon, H.R. (1967) *Diseases of Domestic Animals*, Part 2. *Arthropod Infestations*, 2nd edn, Department of Health., Canberra, ACT pp. 34–66.

Seddon, H.R., Belschner, H.G. and Mulhearn, C.R. (1931) Studies on cutaneous myiasis of sheep (Sheep Blowfly Attack). *Science Bulletin, 37*, Department of Agriculture 42pp.

Sutherland, S.S., Gherardi, S.G. and Monzu, N. (1983) Body strike in sheep affected with dermatophilosis with or without fleece rot. *Australian Veterinary Journal* 60, 88–9.

Watts, J.E. (1981) 'Predisposing causes of fleece rot and body strike of sheep.' Unpublished Ph.D. Thesis, The University of Sydney.

Watts, J.E. and Merritt, G.C. (1981) Leakage of plasma proteins onto the skin surface of sheep during the development of fleece rot and body strike. *Australian Veterinary Journal* 57, 98–9.

Watts, J.E., Murray, M.D. and Graham, N.P.H. (1979) The blowfly strike problem of sheep in New South Wales. *Australian Veterinary Journal* 55, 325–34.

Section 5
Resistance to Viral Diseases

Several chronic virus diseases of farm animals are resistant to conventional means of prevention and treatment, e.g. scrapie and Maedi/Visna in sheep. The recent outbreak of bovine spongiform encephalopathy (BSE) has served as a singular reminder that such diseases may have repercussions well outside the species initially involved.

The susceptibility of sheep to the ovine lentivirus infections (e.g. Maedi/Visna) have long been known to be subject to distinct genetic variation. Several sheep breeds have a high degree of resistance, and within breed genetic variance has also been observed. The possibility of developing transgenic sheep with resistance genes incorporated is discussed and optimism of imminent progress in this field is evident. The difficulties of early diagnosis of diseases which have long incubation periods set substantial problems for research and preventive practice. The mechanism by which the pathogens infect the host is being clarified and this holds promise of application in developing blood tests for susceptibility and for the control and possible eradication of scrapie and BSE.

In relation to acute virus diseases, e.g. foot and mouth disease, the success of vaccines has tended to overshadow possibilities of genetic resistance. However, it is now established that genetic variation in susceptibility does exist in some species and the knowledge gained may be practically applied in the future.

Chapter 17

Strategies for the Genetic Control of Ovine Lentivirus Infections

J.C. DeMartini, R.A. Bowen, J.O. Carlson and
A. de la Concha-Bermejillo
Departments of Pathology, Physiology and Microbiology,
Colorado State University, Fort Collins, CO 80523, USA

Summary

Infection by ovine lentivirus (OvLV) is a worldwide cause of slowly progressive inflammatory disease of lungs, brain, joints, and mammary gland of adult sheep. Prevalence of OvLV in sheep flocks can be limited by using test and cull schemes and by prevention of vertical (milk) transmission of the virus, but these methods are labour intensive and expensive. There are two general ways to control OvLV in sheep through genetic means. One approach is to select and propagate sheep that are naturally resistant to the virus. This approach requires identification of genetic markers for traits linked to disease resistance and animal production characteristics in the targeted sheep breed. The second approach, development of OvLV-resistant transgenic sheep, is based on the introduction of viral and other genes into the sheep genome. Candidate genes for introduction include: viral structural protein genes such as *env*, mutated viral regulatory protein genes such as *tat* and *rev*, genes encoding viral antisense RNA, and genes encoding ribozymes capable of cleaving viral RNA within infected cells. Expression of introduced genes must be regulated in pertinent target cells of the host by using appropriate promoters, initiation and termination codons, and cell membrane anchoring sequences if necessary. After introducing the gene constructs into embryos by microinjection, infection with recombinant retroviruses, or transfection, transgenic sheep that stably express the desired gene can be selected and expanded and their ability to resist OvLV assessed by natural or experimental challenge. Currently available tools of biotechnology, reproductive physiology, and knowledge of OvLV molecular virology and pathogenesis provide a sense of optimism that progress in controlling retroviral diseases through these means is imminent.

Introduction

Retrovirus-induced diseases of horses and poultry were among the first animal diseases known to be caused by 'filterable' viruses. Early investigations concerning the transmission, pathogenesis, diagnosis and control of equine infectious anaemia and avian leukosis have served as a foundation for experimental studies of retrovirus infections of many species, including man. Similarly, studies of visna/maedi, ovine progressive pneumonia and other ovine lentivirus (OvLV)-induced diseases first led to the concept of 'slow virus infection'.

Because of their capacity to integrate into the genome of the host, to induce a persistent infection, and to vary the antigenicity of envelope glycoproteins, retrovirus infections have proved difficult to control in domestic animals. The limited success that has been achieved has been through prevention of virus spread by elimination of infected animals through various test and culling schemes. Immunoprophylactic approaches, with the possible exception of a feline leukaemia virus vaccine, have not proved useful for controlling either infection or disease caused by this group of viruses. Recently, much has been learned concerning the biology of lentiviruses and their interaction with the host. The purpose of this chapter is to integrate this knowledge with newer approaches in mammalian genetics, embryo manipulation, and molecular biology in order to develop innovative strategies for the control of lentivirus infections of sheep.

Characteristics of ovine lentiviruses

The viral genome and its products

Viruses of the subfamily Lentivirinae within the family Retroviridae cause slow, progressive diseases in sheep, goats, horses, cats, subhuman primates and man (reviews: Cheevers and McGuire, 1988; Narayan and Clements, 1989). The genome of these viruses consists of two identical positive-sense single-stranded polyadenylated RNA subunits of 9–10 kilobases. These RNA viruses replicate within cells through formation of a reverse transcriptase-dependent DNA intermediate which may become integrated into the genome of the host cell. The nucleotide sequences of two OvLV proviruses have been published and their genomic organization is similar to other lentiviruses (Querat *et al.*, 1990; Sonigo *et al.*, 1985) (Fig. 17.1). The DNA provirus contains long terminal repeats (LTR) at 5′ and 3′ ends and includes not only the typical retroviral genes *gag*, *pol*, and *env*, but also open reading frames (Q,S) between *pol* and *env* that encodes genes that regulate viral replication (*tat*, *rev* and others). The virion consists of double membrane-bound particles 120–140 nm in size that contain a cylindrical

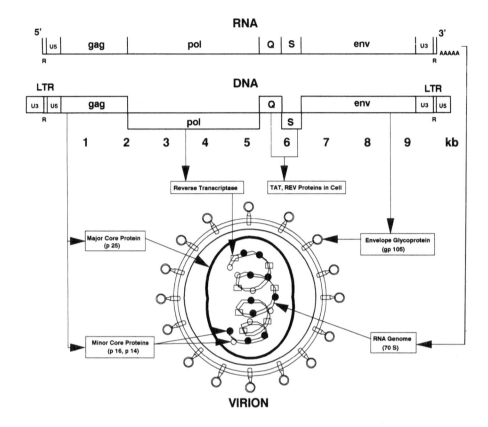

Fig. 17.1. Genomic organization and structure of OvLV. The genomic RNA is contained within the virion. Reverse transcriptase then copies the RNA into DNA in the infected cell. The rectangles in the DNA represent open reading frames (except for the 5′ and 3′ long terminal repeats [LTR]). The three structural genes are the *gag*, *pol*, and *env*. The proteins that they encode are indicated in the diagram of the virion. Two additional genes, Q and S, located between the *pol* and *env* genes are important in transcriptional and post-transcriptional regulation. The LTRs contain enhancer–promoter elements for the initiation of DNA transcription.

nucleoid and bud from the plasma membrane of the cell. The main viral proteins of immunological significance include a group antigen major core protein (p25–30, *gag*), minor core proteins (p14 and p16, *gag*) and an envelope glycoprotein (gp 105–135, *env*).

In vitro replication of OvLV

In vitro, OvLV infect and replicate in macrophages and in primary cells derived from the sheep choroid plexus, lung, trachea, cornea, and goat synovial membrane cells (Narayan and Clements, 1990). In contrast to restricted replication *in vivo*, the replication cycle *in vitro* is rapid, with formation of infectious virions within 72 hours. Cytopathic effects include formation of multinucleated syncytial cells followed by varying degrees of lysis depending on the viral strain. 'Rapid/high' (lytic) isolates, typified by visna virus, replicate rapidly to high titre ($1 \times 10^{6-7}$/ml), whereas slow/low (persistent) isolates replicate slowly with persistent syncytia formation and reach only low titres ($1 \times 10^{3-4}$)(Querat *et al.*, 1984). Viral *in vitro* phenotype has been shown to correlate with *in vivo* pathogenicity (Lairmore *et al.*, 1987). Variations in viral *in vitro* phenotype also reflect OvLV genomic diversity (Querat *et al.*, 1987).

Pathogenesis and immunology of ovine lentivirus infection

Clinicopathological syndromes associated with OvLV

Chronic respiratory disease (known as maedi, ovine progressive pneumonia, and other names in various countries) is the predominant clinical syndrome associated with OvLV infection (reviews: Cutlip *et al.*, 1988; Dawson, 1980; Narayan and Cork, 1985). Affected sheep have tachypnoea, dyspnoea, and occasionally cough, fever, and inappetence, especially if secondary bacterial pneumonia occurs. The primary pulmonary lesion is lymphoid interstitial pneumonia. OvLV infection often coexists with sheep pulmonary adenomatosis (jaagsiekte, ovine pulmonary carcinoma), which is thought to be caused by a type D retrovirus (DeMartini *et al.*, 1988). Lymphoid hyperplasia causes pulmonary lymph nodes to enlarge to 5–10 times normal size, a consistent feature of sheep with maedi. Lymphocytic indurative mastitis ('hardbag'), perhaps the most common lesion of OvLV infection, causes an enlarged, symmetrically firm udder and markedly reduced milk production. OvLV-associated non-suppurative, lymphoid synovitis and arthritis result in painful swelling of one or more joints and lameness. Demyelinating encephalomyelitis (visna), which occurred more commonly in Icelandic sheep than elsewhere in the world, is characterized by tremor, ataxia, and occasionally progressive paralysis involving the rear limbs. Other manifestations of OvLV infection include general loss of body condition, vasculitis and orchitis (Palfi *et al.*, 1989).

Because of a prolonged period of up to 3 or more years which may elapse between the time of infection and development of clinical disease in OvLV-infected sheep, OvLV-associated diseases are primarily recognized in adults. The onset of clinical disease is insidious and the course is slow and progressive, until death ensues. The incidence and organ-specific frequency of clinical disease in OvLV-infected sheep is poorly understood, and the question of variation in pathogenicity of OvLV strains is not completely resolved. In a sample of 30 cull ewes selected in two subsequent years from a flock of 10,000 sheep, all were OvLV seropositive, approximately 50% had lymphoid mastitis, and 75% had lesions of progressive pneumonia (DeMartini, unpublished). Experimental OvLV infection of 12 neonatal lambs by intratracheal injection resulted within 6 months in pulmonary lesions in 75%, mammary lesions in 33% of the females, and synovitis in 30% (Lairmore *et al.*, 1988). Differences in pathogenicity between lytic and persistent OvLV strains were found in this study.

Pathogenesis of OvLV infection

Once infected with OvLV, sheep apparently remain infected for life. Host cells infected *in vivo* primarily include monocytes and macrophages of the blood, lung, spleen and bone marrow. OvLV usually can be isolated from seropositive sheep, even years after the original infection. OvLV-induced persistent infection seems to be based on at least three factors. (1) Latent infection of host cells by OvLV DNA provirus *in vivo* with restriction of virus expression (Haase, 1986) and consequent failure of immune recognition of infected cells. Narayan *et al.* (1982) have described long-term non-productive infection of blood monocytes which express virus upon differentiation into macrophages. (2) Low induction and affinity of virus neutralizing antibody in infected animals (Narayan and Clements, 1989). (3) Virus mutation involving *env* genes with emergence of new antigenic variants which are not neutralized by pre-existing antibody (Clements *et al.*, 1988).

Immune responses to OvLV

Although not effective in eliminating viral infection, humoral and cell-mediated responses do occur in OvLV-infected sheep. Sequential development of serum antibodies to specific viral proteins were studied by immunoblotting in 16 lambs inoculated with lytic and persistent OvLV strains (Kajikawa *et al.*, 1990). Antibody to major core protein p25 appeared at about 3 weeks and was followed about 2 weeks later by anti-p16, -p14 and -gp105. Antibody titres to these proteins were higher in experimentally infected lambs than in naturally infected adult sheep with clinical disease. In sheep infected with Icelandic visna strains of OvLV, neutralizing antibodies develop to high titres, usually within 4–8 months of infection, but they are not effective in neutralizing new antigenic variants of virus which form *in vivo* (Clements *et al.*, 1988). On the other hand, North American OvLV isolates induce low or negligible titres of neutralizing antibody (usually 1:2–1:8) which are not protective. An agar gel immuno-diffusion test (Cutlip *et al.*, 1977) and an ELISA (Houwers *et al.*, 1982) based on antibodies to *env* glycoproteins or major core proteins, are probably the most widely used serological tests for OvLV infection. Cell-mediated immune responses in the form of viral antigen-induced lymphocyte blastogenesis have been described in sheep infected with OvLV (Larsen *et al.*, 1982) and in goats infected with CAEV lentivirus (DeMartini *et al.*, 1983). In the former study, there was no evidence of virus-induced immunosuppression as determined by mitogen-induced lymphyocyte responsiveness or by immune responses to BCG vaccination.

Epizootiology and methods of control of ovine lentivirus infection

Epizootiology

Ovine lentivirus infection has been reported in most sheep-raising countries of the world, with the notable exception of Australia and New Zealand (Dawson, 1980). Horizontal transmission of OvLV via aerosol or contact and vertical transmission via lactogenic infection are the most common means of OvLV field transmission. Prenatal infection is highly unusual (Cutlip *et al.*, 1981; DeBoer *et al.*, 1979). Evidence suggests that in flocks with mixed infected and uninfected ewes, approximately 37% of the off-spring of infected ewes and about 20% of the offspring of uninfected ewes become OvLV infected within one year (Houwers *et al.*, 1989). In one study, after introduction of two infected ewes into a flock of 22 seronegative ewes, 80% of the flock was infected within 5 years due to horizontal transmission (Houwers and van der Molen, 1987). Close confinement in winter housing, perhaps combined with genetic susceptibility of Icelandic sheep breeds, was suspected to have been a factor contributing to the explosive OvLV outbreak in Iceland (Palsson, 1976).

Breed susceptibility

If verified, differences in breed susceptibility to OvLV infection or OvLV-induced disease could be exploited as a means for limitation of OvLV-associated losses in areas where the virus infection is prevalent. This could be accomplished either by expansion of resistant breeds or by identification, cloning, and selecting for or introducing gene(s) responsible for OvLV resistance in breeding sheep.

Limited data are available for OvLV breed effects. Several studies of OvLV seroprevalence in mixed-breed flocks suggest that some breeds may be more resistant to OvLV infection than others. Reports from studies on different continents suggest that Finnish breeds have a greater tendency to become infected by OvLV than the Ile de France breed (Houwers *et al.*,

1989), Rambouillet or Columbia breeds (Gates *et al.*, 1978). Interpretation of such retrospective studies, however, is plagued by uncertain or inconsistent exposure histories and the commonly observed increase in rate of OvLV infection with age in field flocks.

Breed-related resistance to OvLV-induced disease was suggested by epidemiological studies in Iceland which indicated that crosses between Icelandic sheep and Border Leicester rams appeared to be particularly resistant, and that in these crosses the progression of pulmonary lesions was delayed (Palsson, 1976). In contrast to these findings, Border Leicester sheep were much more likely than Columbia sheep to develop multisystemic lesions in response to experimental or natural OvLV infection (Cutlip *et al.*, 1986). Interpretation of these results is complicated by variances in viral strains and host genetics within family lines. Thus, further rigorously controlled prospective research is required before a particular breed or group of sheep can be confirmed as resistant to OvLV infection or OvLV-induced disease.

Traditional methods of control

To date, the only effective means for controlling OvLV-induced disease has been through removal of infected sheep or prevention of spread of the virus. The most dramatic example of the former was the major effort undertaken in Iceland in the 1950s in which over 600,000 OvLV-diseased and exposed sheep were slaughtered and farms then repopulated with unexposed sheep (Palsson, 1976). After development of reliable serological tests, OvLV-infection has been eliminated from flocks by removing lambs from seropositive ewes before nursing and rearing them in isolation and by semi-annual or annual testing and culling of seropositive ewes and their progeny (Cutlip and Lehmkuhl, 1986; DeBoer *et al.*, 1979; Houwers *et al.*, 1983, 1984). Voluntary eradication programmes based on these methods have been established in the Netherlands, Great Britain and Belgium, and have achieved considerable success. However, because such methods are expensive and are not readily adopted by farmers, they may not be as successfully employed in countries with larger sheep populations where the seroprevalence of OvLV infection is high. This necessitates the development of alternative strategies for control of OvLV.

Prospects for immunization

Control of OvLV infection through immunization is a logical if difficult goal. The unique features of lentivirus replication and virus–host interaction, as cited above, are roadblocks to development of an efficacious vaccine. Attempts to prevent virus infection by OvLV or CAEV with formalin-inactivated virus in adjuvants have failed (Cutlip *et al.*, 1987;

Pearson *et al.*, 1989). Immunization attempts have even resulted in exacerbated signs of disease in immunized animals, suggesting an immuno-pathological basis for the lesions (McGuire *et al.*, 1986). The viral challenge in these experiments, however, may have overwhelmed the immune response. Recently, a formalin-inactivated whole simian immunodeficiency lentivirus vaccine has been shown to confer protection against a low dose virus challenge in macaques (Murphey-Corb *et al.*, 1989). Although intensive research may eventually lead to a suitable immunogen for lentivirus infections, this goal will probably not be realized soon.

Strategies for selection of sheep naturally resistant to OvLV

During the last few decades, animal breeders have been very successful in breeding animals that are superior in the production of milk, eggs, meat and wool. However, selection for disease resistance has been largely ignored (Gavora and Spencer, 1983). Historically, disease prevention has been through improvement of the environment, vaccination and drug use. These approaches allow retention of breeding animals with marginal natural resistance, thus raising the possibility of selecting indirectly for disease susceptibility (Warner *et al.*, 1987). For example, two simulation studies predict that mastitis prevalence in dairy cattle will increase 0.4%/year and 0.02 mastitis cases/year, respectively, as a result of current selection trends for milk yield. Both studies assumed a genetic correlation between yield and mastitis of 0.30. Although these rates of change in mastitis seem low, the increase is alarming when projected over the long term (reviewed in Shook, 1989). This is perhaps the most compelling reason for including selection for disease resistance in breeding programmes. Studies on laboratory rodents have shown that it is possible to improve resistance to specific diseases by selection (Stear, 1982).

Mechanisms of disease resistance

Development of disease is the result of an interaction between the genotype of the individual and the environment. A prerequisite for viral infection is the presence of specific cell surface receptors for the virus. Genetic control of the resistance to infection and disease is highly complex and involves an interaction among several body systems (Warner *et al.*, 1987), especially the immune system. It is well known that the immune response is under the control of several genetic systems, of which the most important is the major histocompatibility complex (MHC). The MHC is a complex of closely linked genes that control humoral and cellular immune responses (Nicholas, 1987; Stear, 1982).

In humans, the susceptibility to many diseases has been shown to be associated with the human MHC, the HLA (human leucocyte antigen) complex (Cooper and Clayton, 1988). Among domestic animals, the role of the MHC in disease resistance has been best studied in the chicken. The B locus of the chicken MHC has been shown to determine the humoral immune response to synthetic antigens, the level of serum complement and resistance to Marek's disease, Rous sarcoma, and lymphoid leukosis viruses. An impressive demonstration that resistance to disease can be increased through selective breeding comes from the work of Cole (1968) with Marek's disease. In only four generations of selection, the incidence of the disease was reduced from 51% to 7%. Further studies have shown that there is a strong association between the B^{21} allele and resistance to Marek's disease, and that B^{21} is almost completely dominant in its effect (Nicholas, 1987). In other species, there have been only a few reports of the relationship of the MHC to the immune response and disease resistance. For instance, in sheep, resistance to scrapie and response to vaccination against *Trichostrongylus colubriformis* have been shown to be associated with the sheep MHC (OLA) (Warner *et al.*, 1987).

Criteria for traits to be included in a breeding scheme

Genetic improvement requires genetic evaluation of all potential breeding stock. Accurate genetic evaluation uses performance records not only from the individual but also ancestors, descendants, and collateral relatives. Four major criteria should be considered when deciding which traits to include in the breeding programme: (1) trait must have a reasonably large genetic variability and heritability; (2) trait must have important economic value; (3) trait must be measurable at low cost, clearly defined and observations consistently recorded according to the definition, and (4) a marker trait may be used if it has either a lower recording cost, higher heritability, or can be measured earlier in life than the economically important trait it represents (Shook, 1989).

Marker traits for genetic improvement of disease resistance

Selection for genetic disease resistance requires the identification of specific genes or genetic markers linked to resistance to infection or disease. Selection for marker traits offers several advantages over selection on records of disease incidence. For example, laboratory tests are more objective and can be more easily standardized. Producers and veterinarians seem more willing to submit samples for laboratory testing than to record occurrence of disease. Marker traits may be inherited quantitatively or qualitatively. A quantitative marker trait must have a high genetic correlation with the disease trait. If the marker gene is at a single locus, the genotype at the locus must exhibit large differences for the disease trait. Beyond this requirement, the choice of marker traits depends on a combination of factors and how these compare with disease traits. A marker trait will be most useful if it can be measured in both males and females, has a higher heritability than the disease trait, indicates subclinical variations in the disease, and is measurable early in life and at low cost. Not all these advantages will exist for any one marker (Shook, 1989). Some of the more common markers that have been associated to resistance or susceptibility to disease include MHC and blood groups antigens, antibody responses to specific antigens, and restriction fragment length polymorphism markers. (Cooper and Clayton, 1988; Nicholas, 1987; Stear, 1982). The major disadvantages of the use of marker traits are the cost of measurement and their imperfect correlation with disease (Shook, 1989).

Selection approach for OvLV resistant animals

Establishing a selection programme for resistance to ovine lentivirus infection or disease is difficult because the OvLV viral receptor is unknown and the pathogenesis of the disease is poorly understood. The main problem is defining the measures or markers to be used. Because signs of OvLV-induced disease usually develop late in life and are often confused with other diseases, presence or absence of clinical disease would have to be corroborated with other laboratory analysis, making the selection system expensive and complicated.

The second problem would be to decide which animals should be considered resistant to OvLV. It is well known that a percentage of animals in a positive flock will remain seronegative for life. Although such animals may never have been exposed to OvLV or may have become infected and are tolerant to the virus, it is also possible that they possess innate resistance to OvLV infection, perhaps because of a lack of specific cell receptors for the virus. If this is true, the descendants of those animals that remain seronegative in an infected flock should be selected for breeding stock. An alternative strategy would be to select the descendants of those animals that become infected with OvLV (seroconvert) but do not develop disease. These animals could be considered 'high responders' in terms of acquired immunity against the virus and thus capable of controlling the infection. Since infected animals remain infected for life and OvLV is commonly transmitted to the offspring, by selecting the descendants of seropositive animals one could also perpetuate the infection in the breeding stock.

Once a group of OvLV-resistant sheep is identified, it would be necessary to define a genetic marker that correlates highly with the resistance trait and to assure that the animals exhibit good 'production' characteristics. Data on the association of production traits with disease resistance, immune response and the MHC are important for breeding programmes, but are still not adequately developed for domestic animal species. It is clear that many questions would have to be answered before a selection programme for OvLV resistant animals can be established. With recent advances in genetic engineering, the introduction of desired genes into the germplasm is an alternative.

Strategies for developing OvLV resistant transgenic sheep

Controlling an infectious disease by genetic modification of the host requires development of a 'resistance gene' and methods for introducing that DNA into the germ line of animals such that it will be expressed in appropriate target cells. Ultimately, hemizygous founder animals bearing such transgenes must be amplified to generate populations of resistant animals, most likely by production of homozygotes and use of embryo transfer and artificial insemination.

Methods for production of transgenic sheep

Since its introduction less than a decade ago, production of transgenic mice has become a widely used and powerful technique for studying normal and abnormal gene expression (reviewed by Camper, 1987; Palmiter and Brinster, 1985). Most commonly, a sequence of DNA containing both regulatory and coding regions is cut from a plasmid and purified, then microinjected into one pronucleus of a recently fertilized, one-cell embryo. In a fraction of the surviving embryos, the transgene is observed to have integrated into the genome and is carried in both somatic and germ cells. Assuming the transgene is integrated into a site that does not silence transcription (which happens with unfortunately high frequency), expression is dependent largely on the regulatory region of the transgene and thus, the transgene can be designed for relatively promiscuous or stringently regulated expression in specific cells and tissues. For example, the visna virus LTR has been shown to direct expression of a reporter gene in macrophages, lymphocytes and the central nervous system of transgenic mice (Small *et al.*, 1989).

In addition to microinjection, two additional approaches have been used for production of transgenic mice. Early embryos can be infected with recombinant retroviruses which efficiently integrate into genomic DNA as single proviruses (Soriano *et al.*, 1986; Stewart *et al.*, 1987). Alternatively, embryonal stem cells can be transfected with DNA, selected *in vitro* for a desired phenotype, then injected into the blastocoele of early embryos to generate germline chimaeric transgenics (reviewed by Capecchi, 1989).

Success in producing transgenic mice stimulated similar efforts in several species of domestic livestock, and indeed several research groups have reported birth of transgenic lambs (Clark *et al.*, 1989; Hammer *et al.*, 1985;

Pursel *et al.*, 1989; Simons *et al.*, 1988). Success toward production of transgenic sheep, as measured by the number of transgenic lambs obtained per injected ovum, has been much lower than that obtained with mice, generally less than 1%. This is not surprising, as embryos from sheep and other domestic species are much less well studied. For example, obtaining a group of synchronous sheep embryos at the pronuclear stage is more difficult, and such research is very much more costly than with mice. Nevertheless, efforts continue to optimize conditions for production of transgenic sheep (Walton *et al.*, 1987) and there is reason for optimism that success rates will improve.

Several additional constraints must be considered with regard to producing populations of any type of transgenic sheep. First, even with the use of modern reproductive technology, such projects will take considerable time. The generation interval of sheep dictates that it will require a minimum of approximately 2.5 years from the birth of a transgenic founder (hopefully a male) to the generation of a pubertal homozygous transgenic. Second, one must expect that a fraction of transgenic animals produced using currently available technology will not express the transgene at a level sufficient to alter its phenotype, and an additional number will have some type of defect, due to such problems as insertional mutagenesis. Finally, a disease-resistant transgenic sheep will be of marginal value if production characteristics are sacrificed or neglected during the years required to achieve homozygosity.

Strategies to interfere with virus infections in the host

Since virus-resistant transgenic sheep have not been constructed we will outline strategies that have been developed in other systems with varying degrees of efficacy.

Virus structural protein genes

One of the prominent strategies for developing virus resistance has been to develop transgenic organisms that express a critical structural protein of the virus. The rationale for this approach is that many viruses, including OvLV, show interference. That is, infection with one virus will prevent super-infection with related viruses. This has been most elegantly developed in several plant systems to develop virus-resistant plants. Plants expressing viral coat proteins are often resistant to infection by the virus. The strategy has worked with tobacco mosaic virus (TMV) (Powell *et al.*, 1986), tobacco rattle virus (TRV) (Angenent *et al.*, 1990), potato virus X (Hemenway *et al.*, 1988) and others. Of these, the protection of tobacco plants against tobacco mosaic virus is the most thoroughly studied. Plants expressing the coat protein of TMV are quite resistant to infection by complete virions.

However, when the plants were inoculated with TMV genomic RNA or with virions partially disrupted by incubation at pH 8, they were not protected (Register and Beachy, 1988). This suggests that the expression of TMV coat protein interferes with an event early in the infection such as the initial uncoating of the viral RNA. This is probably the case for most of the plant viral systems. An exception is that the expression of potato virus X coat protein does protect against inoculation with RNA and thus the mechanism of inhibition may be somewhat different in this system (Hemenway *et al.*, 1988).

Expression of viral structural proteins has also been used to protect chickens against avian leukosis retroviruses. Robinson *et al.* (1981) showed that some strains of chickens that carried defective endogenous proviruses with a type E envelope were resistant to infection with other type E retroviruses. This resistance correlated with expression of the envelope glycoprotein of the endogenous virus. Recently transgenic chickens have been developed that express the type A envelope from a defective provirus (Crittenden *et al.*, 1989; Salter and Crittenden, 1989). These chickens are resistant to infection by subgroup A but not subgroup B avian leukosis virus. The resistance correlates with expression of the envelope glycoprotein and not with expression of gag protein. The molecular basis of this is thought to be due to interaction of the envelope protein with receptor proteins either on the surface of the cell or in the endoplasmic reticulum (Delwart and Panganiban, 1989; Federspiel *et al.*, 1989). This interaction blocks the binding of the virus to receptor protein and therefore interferes with entry of the virus into the cell.

Thus, in both plant and animal systems, expression of the appropriate viral structural protein in transgenic organisms has been shown to interfere with virus infection. By analogy with the avian leukosis virus system it might be possible to produce transgenic sheep carrying the gene for OvLV envelope glycoprotein or a portion of it. Expression of this gene might then interfere with OvLV infection. Replication interference between OvLV subtypes has recently been described (Jolly and Narayan, 1989). OvLV strains related by phenotype and genotype cross-interfered whereas different subtypes did not, thus permitting superinfection of cultured cells or animals by different lentivirus subtypes; the interference seemed to occur at the level of binding of the virus to the cell. One potential problem with this approach is that the lentivirus glycoproteins tend to cause cell fusion and syncytium formation which could be detrimental to an animal. However, it might be possible by careful genetic engineering to eliminate the fusion function and retain the receptor binding function of the glycoprotein.

Enthusiasm for this approach must be tempered by recent studies with transgenic mice that express viral structural proteins. In one of these studies, mice expressing hepatitis B virus surface antigens were shown to be predisposed to the development of hepatocellular carcinoma (Chisari *et al.*,

1989). This was shown to be correlated with the level of expression of the protein. In another study, transgenic mice expressing the vesicular stomatitis virus G glycoprotein were shown to develop autoantibodies against this protein when infected with wild type VSV (Zinkernagel *et al.*, 1990). Thus, one must be aware that complications such as tumours or autoimmune disease might arise from expression of viral structural protein genes in transgenic animals.

Virus regulatory protein genes

A second strategy might be to express a viral non-structural protein in the transgenic animal. Many viruses, lentiviruses included, have regulatory proteins that modulate viral gene expression during an infection. Generally these are transacting factors that control the temporal expression of the viral genes at either a transcriptional or post-transcriptional level. Cell culture studies suggest that expression of certain types of mutant regulatory proteins can interfere with the normal course of a viral infection. These mutants must be transdominant and thus able to disrupt the function of normal regulatory proteins in an infection.

Three examples of expression of these transdominant regulatory proteins are presented below. Their effects have been demonstrated only in cultured cells but could be adapted to transgenic animals. The first example involves herpes simplex virus 1. A viral protein, VP16, is required for transactivation of the immediate early class of genes. The protein has a binding domain and an activating domain. The binding domain presumably interacts with other cellular proteins that interact with the promoter. The activating domain then activates transcription from the promoters. A truncated VP16 was constructed which retains the binding domain but which has lost the activating domain (Friedman *et al.*, 1988). Cells that express this truncated protein are significantly more resistant to infection by HSV-1 than cells that do not. Thus the transdominant mutant protein prevents the normal transcriptional programme.

The second and third examples involve the regulatory proteins of the *tat* and *rev* genes of the lentivirus human immunodeficiency virus (HIV-I). The tat protein is a transacting transcriptional regulatory protein required for efficient transcription from the LTR promoter. *Rev* protein seems to be necessary for transport of singly spliced and unspliced HIV RNAs from the nucleus to the cytoplasm where they can act as mRNAs. During the course of an HIV infection the rev protein must accumulate in order for the synthesis of viral structural proteins to proceed. It seems to function by binding to a site in these RNAs to allow transport into the cytoplasm. Certain *tat* and *rev* mutants are deficient in their functions and are transdominant and thus prevent either synthesis or transport of RNA from the nucleus (Green *et al.*, 1989; Malim *et al.*, 1989). Thus cells that express these mutant proteins would not allow the normal growth of the virus because of aberrant regulation at either the transcriptional or post-transcriptional level.

Visna virus is known to have both *tat* and *rev* proteins and thus the possibility exists that transgenic sheep expressing transdominant mutations might be resistant to ovine lentivirus infection. Again, however, enthusiasm must be tempered because of an experiment in which the HIV *tat* protein is expressed in transgenic mice. Mice expressing this protein exhibit skin lesions that resemble Kaposi's sarcoma, a cancer often associated with AIDS (Vogel *et al.*, 1988). Whether transdominant *tat* mutants would cause these lesions is unknown.

Viral antisense RNA

Another strategy for interfering with viral infections is to express antisense RNA representing portions of viral RNA. Antisense RNA presumably forms a double-stranded RNA with the target RNA which can disrupt normal function of the target in the cell. Several mechanisms have been described. (1) It can prevent the normal folding of an RNA molecule. The frequency of initiation of plasmid Col E1 DNA replication is controlled by an antisense mechanism (Polisky, 1988). In this case the RNA primer for replication must be folded into a certain conformation to act as a primer. An antisense RNA to a portion of the primer prevents this folding and thus regulates the frequency of initiation. (2) It can interfere with translation of an mRNA. The frequency of translation of the mRNA for the Tn10 transposase is controlled by an antisense RNA that binds to the ribosome binding site of the mRNA and prevents initiation of translation (Simons and Kleckner, 1983). (3) It can target an RNA molecule for modification and perhaps degradation. In *Xenopus* a portion of the gene for basic fibroblast growth factor is transcribed in both directions. This allows association of the two transcripts in this region which is then a substrate for a modifying enzyme that converts adenine residues in double-stranded RNA into inosines. This disrupts the double strands and results in inactivation of the mRNA (Kimelman and Kirschner, 1989).

The use of the antisense mechanisms described above for altering a virus infection has not been extensively explored. Transgenic plants expressing antisense portions of the TMV and potato virus X genomes are protected against low levels of infecting virus (Hemenway *et al.*, 1988; Powell *et al.*, 1989). However, the level of protection is much lower than that provided by expression of the respective coat proteins as described above. Antisense oligonucleotides have been shown to enter cells and inhibit HIV replication in tissue culture (Agrawal *et al.*, 1989; Buck *et al.*, 1990) suggesting that it may be possible to interfere with ovine lentiviral infection by expression of antisense RNA in the cells.

Ribozymes

A variation on the antisense strategy would be to design and express ribozymes that cleave viral RNA within the infected cell. Certain RNA molecules have been shown to be capable of performing enzymic reactions. Among these are several plant satellite virus and virusoid and viroid RNAs that cleave their own replicative form RNAs during replication to generate progeny genomes. Based on studies to locate and define the active sites of these self-cleaving RNAs (Forster and Symons, 1987; Hampel and Tritz, 1989), it has been possible to design ribozymes that bind RNA targets by base pairing and carry the active site for cleaving the RNA at the binding site (Hampel *et al.*, 1990; Haseloff and Gerlach, 1988; Uhlenbeck, 1987). Since these ribozymes are RNA molecules themselves, they could be expressed in cells and would cleave any substrate RNA which they encounter. One problem has been to design ribozymes that are active under physiological conditions since the optimal reaction conditions for many of them are far from those found in cells. Nevertheless cleavage of HIV-1 genomes by ribozymes has been reported in HIV-infected cell cultures (Sarver *et al.*, 1990), and there is no theoretical reason to rule out ribozymes against ovine lentiviruses as well.

Other strategies

An additional strategy to develop resistant transgenic animals would be to design a toxin gene which would be expressed only in virus infected cells (for example: under control of virus specific transactivating factors). Such cells would commit suicide by expressing the toxin before allowing the virus to multiply. This would prevent the spread of the virus to other cells. Another possibility would be to express a soluble form of the virus receptor which would circulate in the bloodstream and bind virus before it could infect cells and thus protect the animal. This receptor could also be coupled to antibody Fc regions to assist in destruction of virus or infected cells by the immune system (Byrn *et al.*, 1990). These types of experiments are highly speculative and little has been done to test these concepts on the organism level.

Conclusions

Ovine lentivirus causes a persistent infection in sheep and induces multi-systemic disease in a proportion of infected animals. Because of complexities in the host–lentivirus interaction and cell-associated transmission of the virus to the offspring of infected ewes, control of infection by means other than culling or prevention of spread is problematic. Therefore, exploration of alternative control strategies employing selection and expansion of animals genetically resistant to OvLV or transgenic for certain viral genes merits consideration.

Sheep naturally resistant to OvLV could be selected on the basis of resistance to infection or resistance to OvLV-induced disease. Selection for resistance to infection would be more practical because of the difficulties in detecting animals resistant to disease, and would have the additional advantage of eliminating the virus from sheep flocks. However, resistance to OvLV infection has not yet been experimentally demonstrated to occur and, to become useful on a large scale, would require identification of an associated phenotypic marker that could be easily assessed. Elucidation of the molecular basis for natural resistance mechanisms would assist in development of rational strategies for design of resistant transgenic sheep.

Strategies for developing transgenic sheep which resist OvLV infection include introduction of viral structural or non-structural regulatory genes, viral RNA antisense-encoding oligonucleotides, and ribozymes that cleave viral RNA within infected cells. Success of any of these approaches will depend on optimizing expression of the desired gene in appropriate target cells without adversely affecting cell or host physiological functions. Despite the technical and political problems associated with introducing a transgene into populations of sheep, there is considerable reason for optimism that this approach will succeed. It is very likely that many aspects of this infant technology will be improved substantially in the years to come, and the promise of controlling major diseases by genetic means will surely stimulate additional research. In the final analysis, the difference between cost and return probably will determine which method will be used. Efforts expended in attempts to control lentivirus-associated infection or disease transgenetically will undoubtedly add to our understanding of disease pathogenesis and ultimately to additional methods of disease control.

Acknowledgments

This work was supported by Public Health Service grant R01 AI 25770 from the National Institutes of Health and by grant AID/DSAN/XII-6-0049 from the US AID Title XII Small Ruminant Collaborative Support Program.

References

Agrawal, S., Ikeuchi, T., Sun, D., Sarin, P.S., Konopka, A., Maizel, J. and Zamecnik, P. (1989) Inhibition of human immunodeficiency virus in early infected and chronically infected cells by antisense oligodeoxynucleotides and their phospho-thioate analogues. *Proceedings of the National Academy of Science USA* 86, 7790–4.

Angenent, G.C., Vanden Ouweland, J.M.W. and Bol, J.F. (1990) Susceptibility to virus infection of transgenic tobacco plants expressing structural and non-structural genes of tobacco rattle virus. *Virology* 175, 191–8.

Buck, H.M., Koole, L.H., van Genderen, H.P., Smit, L., Gellen, J.L.M.C., Juvriaans, S. and Goudsmit, J. (1990) Phosphate-methylated DNA aimed at HIV-1 RNA loops and integrated DNA inhibits viral infectivity. *Science* 248, 208–12.

Byrn, R.A., Mordenti, J., Lucas, C., Smith, D., Maesters, S.A., Johnson, J.S., Cossum, P., Chemow, S.M., Wurm, F.M., Gregory, T. Groopman, J.E. and Capon, D.J. (1990) Biological properties of a CD4 immunoadhesion. *Nature* 344, 667–70.

Camper, S.A. (1987) Research applications of transgenic mice. *BioTechniques* 7, 638–50.

Capecchi, M.R. (1989) Altering the genome by homologous recombination. *Science* 244, 1288–92.

Cheevers, W.P. and McGuire, T.C. (1988) The lentiviruses: Maedi/visna, caprine arthritis–encephalitis, and equine infectious anemia. *Advances in Virus Research* 34, 189–215.

Chisari, F.V., Klopchin, K., Moriyami, T., Pasquinelli, C., Dunsford, H.A., Sell, S., Pinkert, C.A., Brinster, R.L. and Palmiter, R.D. (1989) Molecular pathogenesis of hepatocellular carcinoma in hepatitis B virus transgenic sheep. *Cell* 59, 1145–56.

Clark, A.J., Bessos, H., Bishop, J.O., Brown, P., Harris, S., Lathe, R., McClenaghan, M., Prowse, C., Simons, J.P., Whitelaw, C.B.A. and Wilmut, I. (1989) Expression of human anti-hemophiliac factor IX in the milk of transgenic sheep. *Bio/Technology* 7, 487–92.

Clements, J.E., Gdovin, S.L., Montelaro, R.C. and Narayan, O. (1988) Antigenic variation in lentiviral diseases. *Annual Review of Immunology* 6, 139–59.

Cole, R.K. (1968) Studies on the genetic resistance to Marek's disease. *Avian Diseases* 12, 9.

Cooper,. D.N. and Clayton, J.F. (1988) DNA polymorphism and the study of disease association. *Human Genetics* 78, 299–312.

Crittenden, L.B., Salter, D.W. and Federspiel, M.J. (1989) Segregation, viral phenotype, and proviral structure of 23 avian leukosis virus inserts in the germ line of chickens. *Theoretical and Applied Genetics* 77, 505–15.

Cutlip, R.C., Jackson, T.A. and Laird, G.A. (1977) Immunodiffusion test for ovine progressive pneumonia. *American Journal of Veterinary Research* 38, 1081–4.

Cutlip, R.C. and Lehmkuhl, H.D. (1986) Eradication of ovine progressive pneumonia from sheep flocks. *Journal of the American Veterinary Medical Association* 188, 1026–7.

Cutlip, R.C., Lehmkuhl, H.D., Brogden, K.A. and Sacks, J.M. (1986) Breed susceptibility to ovine progressive pneumonia (maedi-visna) virus. *Veterinary Microbiology* 12, 283–8.

Cutlip, R.C., Lehmkuhl, H.D., Brogden, K.A. and Schmerr, M.J.F. (1987) Failure of experimental vaccines to protect against infection with ovine progressive pneumonia (maedi/visna) virus. *Veterinary Microbiology* 13, 201–4.

Cutlip, R.C., Lehmkuhl, H.D. and Jackson, T.A. (1981) Intrauterine transmission of ovine progressive pneumonia virus. *American Journal of Veterinary Research* 42, 1795–7.

Cutlip, R.C., Lehmkuhl, H.D., Schmerr, M.J.F. and Brogden, K.A. (1988) Ovine progressive pneumonia (maedi-visna) in sheep. *Veterinary Microbiology* 17, 237–50.

Dawson, M. (1980) Maedi/visna: A review. *Veterinary Record*, 212–16.

DeBoer, G.F., Terpstra, C., Houwers, D.J. and Hendriks, J. (1979) Studies in epidemiology of maedi/visna in sheep. *Research in Veterinary Science* 26, 202–8.

Delwart, E.L. and Panganiban, A.T. (1989) Role of reticuloendotheliosis virus envelope glycoprotein in superinfection interference. *Journal of Virology* 63, 273–80.

DeMartini, J.C., Banks, K.L., Greenlee, A., Adams, D.S. and McGuire, T.C. (1983) Augmented T lymphyocyte responses and abnormal B lymphocyte numbers in goats chronically infected with the retrovirus causing caprine arthritis-encephalitis. *American Journal of Veterinary Research* 44, 2064–69.

DeMartini, J.C., Rosadio, R.H. and Lairmore, M.D. (1988) The etiology and pathogenesis of ovine pulmonary carcinoma (Sheep pulmonary adenomatosis). *Veterinary Microbiology* 17, 219–36.

Federspiel, M.J., Crittenden, L.B. and Hughes, S.H. (1989) Expression of avian reticuloendotheliosis virus envelope confers host resistance. *Virology* 173, 167–77.

Forster, A.C. and Symons, R.H. (1987) Self cleavage of plus and minus RNAs of a virusoid and a structural model for the active sites. *Cell* 49, 211–20.

Friedman, A.D., Trizenberg, A.J. and McKnight, S.L. (1988) Expression of a truncated viral *trans*-activator selectively impedes lytic infection by its cognate virus. *Nature* 335, 452–4.

Gates, N.L., Winward, L.D., Gorham, J.R. and Shen, D.T. (1978) Serologic survey of prevalence of ovine progressive pneumonia in Idaho range sheep. *Journal of the American Veterinary Medical Association* 173, 1575–7.

Gavora, J.S. and Spencer, J.L. (1983) Breeding for immune responsiveness and disease resistance. *Animal Blood Groups Biochemistry Genetics* 14, 159.

Green, M., Ishina, M. and Lowenstein, P.M. (1989) Mutational analysis of HIV-1 tat minimal domain peptides: Identification of *trans*-dominant mutants that suppress HIV-LTR-driven gene expression. *Cell* 58, 215–23.

Haase, A.T. (1986) Pathogenesis of lentivirus infections. *Nature* 322, 130–6.

Hammer, R.E., Pursel, V.G., Rexroad, C.E., Wall, R.J., Bolt, D.J., Ebert, K.M., Palmiter, R.D. and Brinster, R.L. (1985) Production of transgenic rabbits, sheep and pigs by microinjection. *Nature* 315, 680–3.

Hampel, A. and Tritz, R. (1989) RNA catalytic properties of the minimim (-)sTRSV sequence. *Biochemistry* 28, 4929–33.

Hampel, A., Tritz, R., Hicks, M. and Cruz, P. (1990) 'Hairpin' catalytic RNA model: evidence for helices and sequence requirement for substrate RNA. *Nucleic Acids Research* 18, 299–304.

Haseloff, J. and Gerlach, W.L. (1988) Simple RNA enzymes with new and highly specific endoribonuclease activities. *Nature* 334, 585–91.

Hemenway, C., Fang, R.X., Kaniewski, W.K., Chua, N.H. and Tauer, N.E. (1988)

Analysis of the mechanism of protection in transgenic plants expressing potato virus X coat protein or its antisense RNA. *EMBO Journal* 7, 1273–80.

Houwers, D.J., Gielkens, A.L. and Schaake, J.Jr (1982) An indirect enzyme-linked immunosorbent assay (ELISA) for the detection of antibodies to maedi-visna virus. *Veterinary Microbiology* 7, 209–19.

Houwers, D.J., Konig, C.K., DeBoer, G.F. and Schaake, J.Jr (1983) Maedi-visna control in sheep. I. Artificial rearing of colostrum-deprived lambs. *Veterinary Microbiology* 8, 179–85.

Houwers, D.J., Schaake, J.Jr and DeBoer, G.F. (1984) Maedi-visna control in sheep II. Half-yearly serological testing with culling of positive ewes and progeny. *Veterinary Microbiology* 9, 445–51.

Houwers, D.J. and van der Molen, E.J. (1987) A five-year serological study of natural transmission of maedi-visna virus in a flock of sheep, completed with post mortem investigation. *Journal of Veterinary Medicine* 34, 412–31.

Houwers, D.J., Visscher, A.H. and Defise, P.R. (1989) Importance of ewe lamb relationship and breed in the epidemiology of maedi-visna virus infections. *Research in Veterinary Science* 46, 5–8.

Jolly, P.E. and Narayan, O. (1989) Evidence for interference, coinfections, and intertypic virus enhancement of infection by ovine-caprine lentiviruses. *Journal of Virology* 63, 4682–8.

Kajikawa, O., Lairmore, M.D. and DeMartini, J.C. (1990) Analysis of antibody responses to phenotypically distinct lentiviruses. *Journal of Clinical Microbiology* 28, 764–70.

Kimelman, D. and Kirschner, M.W. (1989) An antisense in RNA directs the covalent modification of the transcript encoding fibroblast growth factor in Xenopas oocytes. *Cell* 59, 687–96.

Lairmore, M.D., Akita, G.Y., Russell, H.R. and DeMartini, J.C. (1987) Replication and cytopathic effects of ovine lentivirus strains in alveolar macrophages correlate with in vivo pathogenicity. *Journal of Virology* 61, 4038–42.

Lairmore, M.D., Poulson, J.M., Adducci, T.A. and DeMartini, J.C. (1988) Lentivirus-induced lymphoproliferative disease. *American Journal of Pathology* 130, 80–9.

Larsen, H.J., Hyllseth, B. and Krogsrud, J. (1982) Experimental maedi virus infection in sheep: cellular and humoral immune response during three years following intranasal inoculation. *American Journal of Veterinary Research* 43, 384–9.

Malim, M.H., Bohnlein, S., Hauber, J. and Cullen, B.R. (1989) Functional dissection of the HIV-1 rev *trans*-activator – Derivation of a *trans*-dominant repressor of rev function. *Cell* 58, 205–14.

McGuire, T.C., Adams, D.S., Johnson, G.C., Klevjer-Anderson, P., Barbee, D.D. and Gorham, J.R. (1986) Acute arthritis in caprine arthritis-encephalitis virus challenge exposure of vaccinated or persistently infected goats. *American Journal of Veterinary Research* 47, 537–40.

Murphey-Corb, M., Martin, L.N., Davison-Fairburn, B., Montelaro, R.C., Miller, M., West, M., Ohkawa, S., Baskin, G.B., Zhang, J.Y., Putney, S.D., Allison, A.C. and Epstein, D.A. (1989) A formalin-inactivated whole SIV vaccine confers protection in Macaques. *Science* 246, 1293–7.

Narayan, P. and Clements, J.E. (1989) Biology and pathogenesis of lentiviruses.

Journal of General Virology 70, 1617–39.

Narayan, P. and Clements, J.E. (1990) Biology and pathogenesis of lentiviruses of ruminant animals. In: Gallo, R.C. and Wong-Staal, F. (eds), *Retrovirus Biology and Human Disease.* Marcel Dekker, New York, pp. 117–46.

Narayan, O. and Cork, L.C. (1985) Lentiviral diseases of sheep and goats: chronic pneumonia leukoencephalomyelitis and arthritis. *Review of Infectious Diseases* 7, 89–98.

Narayan, O., Wolinsky, J.S., Clements, J.E., Strandberg, J.D., Griffin, D.E. and Cork, L.C. (1982) Slow virus replication: the role of macrophages in the persistence and expression of visna viruses in sheep and goats. *Journal of General Virology* 59, 345–356.

Nicholas, F.W. (1987) *Veterinary Genetics,* Clarendon Press, Oxford. Chapter 8, pp 232–72.

Palfi, V., Glavits, R. and Hajtos, I. (1989) Testicular lesions in rams infected by maedi/visna virus. *Acta Veterinaria Hungarica* 37, 97–102.

Palmiter, R.D. and Brinster, R.L. (1985) Transgenic mice. *Cell* 41, 343–5.

Palsson, P.A. (1976) Maedi and visna in sheep. In: Kimberlin, R.H. (ed.) *Slow Virus Diseases of Animals and Man.* North-Holland, Amsterdam, pp. 17–43.

Pearson, L.D., Poss, M.L. and DeMartini, J.C. (1989) Animal lentivirus vaccines: problems and prospects. *Veterinary Immunology and Immunopathology* 20, 183–212.

Polisky, B. (1988) Col E1 replication control circuitry: sense from antisense. *Cell* 55, 929–32.

Powell, P.A., Nelson, R.S., De, B., Hoffman, N., Rogers, S.G., Fraley, R.T. and Beachy, R.N. (1986) Delay of disease development in transgenic plants that express the tobacco mosaic virus coat protein gene. *Science* 232, 738–43.

Powell, P.A., Stark, D.M., Sanders, P.R. and Beachy, R.N. (1989) Protection against tobacco mosaic virus in transgenic plants that express tobacco mosaic virus antisense RNA. *Proceedings of the National Academy of Science USA* 86, 6949–52.

Pursel, V.G., Pinkert, C.A., Miller, K.F., Bolt, D.J., Campbell, R.G., Palmiter, R.D., Brinster, R.L. and Hammer, R.E. (1989) Genetic engineering of livestock. *Science* 244, 1281–8.

Querat, G., Audoly, G., Sonigo, P. and Vigne, R. (1990) Nucleotide sequence analysis of SA-OMVV, a visna-related ovine lentivirus: phylogenetic history of lentiviruses. *Virology* 175, 434–47.

Querat, G., Barban, V., Sauze, N. Filippi, P., Vigne, R., Russo, P. and Vitu, C. (1984) Highly lytic and persistent lentiviruses naturally present in sheep with progressive pneumonia are genetically distinct. *Journal of Virology* 52, 672–9.

Querat, G., Barban, V., Sauze, N., Vigne, R., Payne, A., York, D., de Villiers, E.M. and Verwoerd, D.W. (1987) Characteristics of a novel lentivirus derived from South African sheep with pulmonary adenocarcinoma (jaagsiekte). *Virology* 158, 158–67.

Register, J.C. III and Beachy, R.N. (1988) Resistance to TMV in transgenic plants results from interference with an early event in infection. *Virology* 166, 524–32.

Robinson, H.L., Astrin, S.M., Senior, A.M. and Salazar, F.H. (1981) Host susceptibility to endogenous viruses: defective glycoprotein-expressing proviruses interfere with infections. *Journal of Virology* 40, 745–51.

Salter, D.W. and Crittenden, L.B. (1989) Artificial insertion of a dominant gene for resistance to avian leukosis virus into the germ line of the chicken. *Theoretical and Applied Genetics* 77, 457–61.

Sarver, N., Cantin, E.M., Chang, P.S., Zola, J.A., Ladne, P.A., Stephens, D.A. and Rossi, J.J. (1990) Ribozymes as potential anti-HIV-1 therapeutic agents. *Science* 247, 1222–5.

Shook, G.E. (1989) Selection for disease resistance. *Journal of Dairy Science* 72, 1349–62.

Simons, J.P., Wilmut, I., Clark, A.J., Archibald, A.L., Bishop, J.O., Lathe, R. (1988) Gene transfer into sheep. *Bio/Technology* 6, 179–83.

Simons, R.W. and Kleckner, N. (1983) Translational control of ISIO transposition. *Cell* 34, 683–91.

Small, J.A., Bieberich, C., Ghotbi, Z., Hess, J., Scangos, G.A. and Clements, J.E. (1989) The visna virus long terminal repeat directs expression of a reporter gene in activated macrophages, lymphyocytes, and the central nervous system of transgenic mice. *Journal of Virology* 63, 1891–6.

Sonigo, P., Alizon, M., Staskus, K., Klatzmann, D., Cole, S., Danos, O., Retzel, E., Tiollsais, P., Haase, A. and Wain-Hobson, S. (1985) Nucleotide sequence of the visna lentivirus: relationship to the AIDS virus. *Cell* 42, 369–82.

Soriano, P., Cone, R.D., Mulligan, R.C. and Jaenisch, R. (1986) Tissue-specific and ectopic expression of genes introduced into transgenic mice by retroviruses. *Science* 234, 1409–13.

Stear, M.J. (1982) The future role of immunogenetics in animal breeding. In: Barker, J.S.F., Hammond, K. and McClintock, A.E. (eds), *Future Developments in the Genetic Improvement of Animals*. Academic Press, New York.

Stewart, C.L., Schuetze, S., Vanek, M. and Wagner, E.F. (1987) Expression of retroviral vectors in transgenic mice obtained by embryo infection. *The EMBO Journal* 6, 383–8.

Uhlenbeck, O.C. (1987) A small catalytic oligoribonucleotide. *Nature* 328, 596–600.

Vogel, J., Hinricks, S.H., Reynolds, R.K., Luciw, P.A. and Joy, G. (1988) The HIV *tat* gene induces dermal lesions resembling Kaposi's sarcoma in transgenic mice. *Nature* 335, 606–11.

Walton, J.R., Murray, J.D., Marshall, J.T. and Nancarrow, D.C. (1987) Zygote viability in gene transfer experiments. *Biology of Reproduction* 37, 957–967.

Warner, C.M., Meeker, D.L. and Rothschild, M.F. (1987) Genetic control of immune responsiveness: a review of its use as a tool for selection for disease resistance. *Journal of Animal Science* 64, 394–406.

Zinkernagel, R.M., Cooper, S., Chambers, J., Lazzarini, R.A., Hengartner, H. and Arnheiter, H. (1990) Virus-induced autoantibody response to a transgenic viral antigen. *Nature* 345, 68–71.

Chapter 18
Genetic Control of Acute Viral Diseases

R.P. Kitching

AFRC Institute for Animal Health, Ash Road, Pirbright,
Woking, Surrey GU24 0NF, UK

Summary

There has been little incentive to investigate genetic determinants of resistance to acute viral disease. The success of vaccines against these diseases
and the cost of breeding programmes to develop both highly productive
and disease-resistant breeds has discouraged serious research. However,
techniques are now available which, theoretically, could identify resistance
genes in farm animals and mediate their transfer to a new host. The
prospect emerges of animals engineered to withstand acute viral disease,
together with high production potential. There is ample evidence for a
genetic component in disease resistance and some of these resistance genes
have already been isolated from the mouse genome. The move from inbred strains of mice to the genetically highly divergent breeds of cattle, pigs
and sheep will be a major challenge which could make the picture of resistant breeds a mirage. The probability that disease resistance is associated
with groups of genes, and not single genes, and that resistance to different
viruses requires very different sets of resistance genes will further
complicate the investigation. However, even without the certainty of
success, a study of resistance to acute viral disease could result in valuable
insights into the mechanism of natural resistance and the variability
between individual response to infection. The problems both technical and
ethical may prove insurmountable but, at worst, there is the assurance of a
better understanding of response to infection and vaccination and the
possibility that vaccines could be engineered to suit the animals, even if the
animals cannot be engineered to resist the diseases.

Introduction

It has long been recognized that certain breeds within an animal species are more susceptible to acute viral infections than others. The fine wool and mutton breeds of sheep are more susceptible to bluetongue than are the haired and fat-tailed breeds (Taylor, 1986), and Soay sheep are so suscept-ible to virulent strains of capripoxvirus that they die before showing the characteristic papules of the disease and fail to transmit infection (Kitching and Taylor, 1985a). Previous contact with a virus over generations would seem to select breeds which are partially or totally resistant to clinical disease, indicating that viruses are themselves potent evolutionary forces. That such breeds tend to be of poor productive capacity reflects more the rapid development of highly productive breeds by selective breeding in countries free of these acute virus diseases, than the implication that only unimproved animals have the capacity to resist disease. When, at the end of the last century, rinderpest was taken into Africa, 95% of the cattle and susceptible wildlife was destroyed, evidence that without prior contact, native animals are as susceptible as highly bred imported animals.

The question is, can those genes which have evolved over generations in animals in constant contact with viruses causing acute infections in more susceptible stock be identified, isolated and ultimately transferred? The theoretical answer is yes, although in practice, little has been achieved except by empirical cross-breeding experiments between local and imported animals. These experiments have been directed towards producing high yielding animals capable of surviving in a hostile environ-ment. There have been no specific attempts to breed for resistance to acute viral disease, attention being predominantly on the ability to resist drought, poor nutrition and difficult climatic conditions. Protection against viral disease has relied more on preventing infection by isolation or vaccination. What experiments have been undertaken have progressed little further than those carried out using mice.

Pox virus infection

Studies on the virulence of ectromelia in mice showed that certain labora-tory strains of mice had a significantly higher case-fatality rate (Briody *et al.*, 1956). Following inoculation into the footpad, C57BL mice developed both a humoral and cell-mediated immune response one to two days earlier than did WEHI mice and this was related to their high survival rate. The infective dose for both strains of mice was similar (Schell, 1960). The results of cross-breeding experiments between susceptible strains such as BALB/c and resistant strains such as AKR and C57BL indicated the presence of an autosomal dominant gene, the Rmp-1 or resistance to mousepox gene (Wallace *et al.*, 1985). This gene did not reduce susceptibility to infection, but appeared to influence the ability of the mouse to limit the spread of the virus following footpad inoculation, possibly by more rapidly mobilizing cytotoxic T cells (O'Neill and Brenan, 1987). Some of the experimental results indicated that other genes were also involved in resistance, such as H-2 genes (O'Neill *et al.*, 1983).

If populations exposed to acute viral disease do eventually develop a degree of genetic resistance, it would be reasonable to assume that the human population would have become less susceptible to the effects of smallpox. Unfortunately for the argument there is very little evidence for this. Attempts to relate resistance to the dominant blood groups in countries at risk were unsuccessful, and many of the studies were confounded by the difficulty of separating the effects due to different strains of smallpox from differences in host response. In retrospect, if tissue typing technology had been available when smallpox was prevalent, some of the observations recorded could have been the basis of rewarding study. However, capripoxvirus infection in sheep and goats is still a major problem in most of Africa, the Middle East and Asia, and although capripoxviruses are a different genus from that of smallpox, the epidemiology and pathology of capripox have a remarkable number of similarities to smallpox (Kitching and Taylor, 1985b).

Capripox is a malignant pox infection of sheep and goats characterized by fever and generalized pocks. It is endemic in most of Africa, the Middle East and India and is a major constraint to sheep and goat production, particularly as the pressure to sustain more animals on marginal land increases, and improved breeds are introduced. Capripox has contributed significantly to the failure of a number of feedlot enterprises in the Middle East. It is not difficult to find animals with clinical disease in the villages and markets and recent surveys have identified different epidemiological situations from the endemic where animals contact the disease as lambs or kids, to the epidemic where the disease is reintroduced into isolated villages to the pandemic where capripox is introduced into a totally susceptible country (Kitching *et al.*, 1986, 1987a). However, although severe clinical disease is not uncommon in the endemic areas, mortality due to uncomplicated capripox was observed to be lower than expected by comparison with that in British sheep and goats experimentally infected with the same isolates of virus. A virulent isolate of capripox collected in North Yemen was used in a trial of an experimental vaccine as a challenge virus to test the efficacy of the vaccine. In British goats this challenge virus always caused severe generalized skin and internal lesions and death on day 10–14 after inoculation. When the same challenge dose was used in North Yemen using fully susceptible indigenous black goats, in each case a severe local reaction developed at the inoculation site, but the disease failed to generalize, and the animals fully recovered. Similar results were obtained in a trial in

Oman, using a highly virulent isolate (in British goats) collected in Oman and inoculated into susceptible Dhofari goats (Kitching *et al.*, 1987b). In further studies, post mortems carried out on indigenous animals, killed while affected with generalized capripox, showed the multiple skin lesions, but considerably fewer of the lung and abomasal lesions seen in British sheep and goats. The inference from these observations is that the indigenous animals, while clearly susceptible to the disease, are better able to control the pathological effects.

Capripoxvirus also affects cattle, causing lumpy skin disease (Neethling). This disease causes a very variable clinical response even in cattle within the same herd, and of apparently similar history (Kitching *et al.*, 1989). In a 100 cow dairy herd, 80 may seroconvert without showing any clinical signs, 10 may develop only a few or even single lesions, and the remaining 10 may develop generalized lumpy skin disease of which five may die. It is possible that this variable response relates to hormonal status or the presence of other pathogens, but that explanation is not consistent with descriptions of the disease in the field. The severity of the response more closely resembles a hypersensitivity reaction which may be associated with previous contact with antigenically related pox viruses, or could as likely reflect a genetic susceptibility. The response to capripoxviruses in general could provide a valuable insight into host susceptibility.

Foot-and-mouth disease virus infection

Subak-Sharpe *et al.* (1963) compared the effect of intramuscular and intraperitoneal inoculation of foot-and-mouth disease (FMD) virus into two strains of mice, AX and P. The AX strain was considerably more susceptible in terms of the development of clinical disease and death. This difference was described as being due to differences in viral replication at the inoculation site and differences in the rate at which the virus spread from the inoculation site in the different mouse strains. In an earlier paper (Subak-Sharpe, 1961a) comparing the different susceptibility to FMD virus in 14 different strains of mice, it was suggested that several genes were responsible for the observed differences.

More recently (Brown, 1989), evidence has been presented that the response of mice to vaccination with FMD virus peptide is related to MHC restriction. Mice with the H-2k haplotype responded better to uncoupled peptide by producing neutralizing antibody than did mice with the H-2d haplotype. By coupling the peptide to carriers which contained T helper cell epitopes, the response in the two mouse strains became similar. This result further indicated that antibody production by the specific B cells required T helper cell activation.

The variable response in cattle to conventional FMD vaccine has long been acknowledged (Pay, 1989), as has the inability to control FMD by vaccination alone. Commonly within a herd there is a small percentage of animals which fail to respond with protective levels of antibody, while the remainder of the herd are well protected. Undoubtedly part of this is due to some animals being missed due to administrative or vaccinator errors. Failure to respond may be related to interference by high levels of maternal antibody which subsequently wane leaving the animal fully susceptible to infection. However, the results from mice studies could indicate a genetic component in the animal's ability to respond to vaccination. Certainly, careful monitoring of FMD virus antibody levels in well vaccinated dairy cattle in herds in the Middle East suggest that not all the variation can be blamed on poor vaccine, or bad management (Kitching, unpublished).

Indigenous Zebu and Zebu-grade cross cattle in Kenya are more resistant to FMD infection than imported European cattle. Problems were encountered when potency testing locally produced vaccine because the unvaccinated control animals frequently failed to show generalized lesions of FMD when given the conventional challenge dose of 10,000 bovine ID_{50} of virus by tongue inoculation. It was necessary to increase the challenge dose by a factor of ten (Anderson *et al.*, 1981).

Bluetongue virus infection

Bluetongue virus is restricted in distribution to between the latitudes 40°N and 35°S by the distribution of its insect vector. Within this area the virus cycles between its vertebrate host, which in farmed land is predominantly cattle and sheep, and the culicoides vector species, which varies from continent to continent. However, bluetongue disease is rarely seen, and is usually only associated with the importation of European or Australian breeds of sheep or cattle. There are at least 24 serologically distinct types of bluetongue virus, and in countries such as Sudan, up to 18 serotypes may be circulating, with no clinical disease in the indigenous animals being seen. In 1956, bluetongue infected insects entered southern Spain and Portugal and reportedly 180,000 sheep died of bluetongue. There have been numerous other reports of the disease being taken over the borders of its normal distribution, almost certainly by unusual wind currents taking infected insects and causing high mortality in a previously unchallenged sheep population.

The picture is complicated by the different virulence of different strains of virus within the serotypes, but generally the indigenous sheep, although fully susceptible to infection, appear better able to resist the pathological effects. Cattle are susceptible to bluetongue virus, and because many of the vectors preferentially feed on cattle, it has been suggested that they are the primary host. Only in North America, South Africa, Portugal and Israel has disease in cattle been reported, and then in only a very small percentage of the total number of animals infected. In 1950/51 bluetongue was reported in Israel affecting imported Austrian, American and French cattle, but since that time there have been very few reports of disease in cattle, in spite of serological evidence of up to 95% seroconversion in some herds (Shimshony, 1987). The disease in cattle appears to be due to a hypersensitivity reaction, and as discussed with lumpy skin disease of cattle could be related to a host genetic predisposition.

Experiments with British sheep, using the same strain of virus, indicated that Soay sheep were the most susceptible, dying before showing any clinical signs. Dorset Horn and Merino sheep developed severe clinical signs before death, while Scottish Halfbreds survived the longest. It was not possible to correlate these differences with differences in levels of viraemia or development of antibody (Gard, 1987). Jeggo *et al.* (1986) were unable to show any clear differences between the response in British sheep inoculated with a virulent strain of bluetongue virus. A more fruitful investigation would have been to compare the clinical and histopathological response between British and Middle Eastern breeds of sheep.

Rinderpest virus infection

Most of the comparative work on rinderpest has been on the different virulence of strains of virus rather than on differences in host resistance. Sonoda (1983), however, did review the safety of different rinderpest vaccines in different hosts. A virulent strain of rinderpest virus which had been attenuated by passage in goats was safe and effective in protecting cattle of Southeast Asian origin, Zebu cattle and Egyptian buffalo, but proved fatal in European and Asian cattle and Asian buffalo. Rinderpest virus attenuated by passage in rabbits was safe to use in these latter breeds, but even this proved fatal for Japanese and Korean cattle and some local African breeds such as Ankole and N'Dama. For these particularly susceptible species rinderpest virus passaged in rabbits and chicken eggs was required. Similarly with pigs, Yorkshire pigs were the most resistant to rinderpest, whereas the native pigs of Indonesia and Cambodia (Hai-Nam line) were a little more susceptible, and the native Thailand pigs were the most susceptible.

The susceptibility of different breeds of cattle to attenuated strains of rinderpest virus is also seen in their susceptibility to field strains. Scott and Brown (1961) contrasted the clinical disease seen in susceptible breeds such as Japanese Black and Guernsey with the subclinical or inapparent infection seen in cattle with low susceptibility such as East African Shorthorn.

Although of little importance to the epidemiology of rinderpest, but of relevance to any future laboratory study on the genetics of resistance to rinderpest, is the observation reported by Scott (1964) that researchers attempting to develop lapinized rinderpest vaccine strains found a small percentage of rabbits within their experimental groups which appeared to be naturally resistant.

Other acute viral diseases

There is considerable anecdotal evidence for natural resistance of certain breeds of animals to other acute viral diseases, unfortunately most with very little indication of any attempt to take advantage of the attribute. Davies (1981), for instance reported that in Kenya the indigenous cattle and sheep were relatively resistant to Rift Valley fever compared with the exotic animals imported to upgrade the local animals. Of interest was his comment that the main effects of the epizootic of Rift Valley fever in Kenya were seen in these imported animals and their crosses with the indigenous animals, suggesting that the natural resistance of the native animals was not evident in their improved offspring.

Sheep and goats reared in areas where Nairobi sheep disease is endemic are also reported to be more resistant to the clinical disease than imported animals, frequently having only a mild or even symptomless infection.

The list is extensive, and with few exceptions such as possibly African swine fever, animals reared in a disease endemic area appear to have developed a degree of natural, or genetic resistance to that particular disease.

Definition of disease resistance

Acute viral disease is not difficult to recognize in susceptible animals. However, when it is necessary to define resistance, a number of variable host determinants must be considered in addition to the natural or genetic resistance under consideration in this chapter. The sex, age and nutritional status of the host can influence the response to infection. Young mice are considerably more susceptible to FMD virus than adult mice (Subak-Sharpe, 1961b), and to a number of other viral diseases. Pregnancy and lactation can increase the susceptibility of mice, as defined by mortality, to FMD infection (Campbell, 1960), and there is considerable field evidence that pregnant heifers in vaccinated herds appear to be more susceptible to clinical FMD than other age groups (Kitching, unpublished). Weinberg (1987) listed 13 viral diseases which had increased virulence in pregnant women. This was not due to an increased susceptibility to infection, but to a decreased ability of the cell-mediated immune system to respond. The hormonal status of the pregnant animal is clearly implicated, but so may be the nutritional status, particularly towards the end of pregnancy when the developing fetus is most demanding.

Disease may result not only because the immune system is unable to respond sufficiently rapidly, but because it may over-respond. C57 BL mice are more resistant to the disease caused by lymphocytic choriomeningitis virus because of a reduced cell-mediated response. The presence of other pathogens can also increase host susceptibility to disease. Rinderpest virus can cause immunosuppression, and transform chronic subclinical ana-plasmosis in cattle into a fatal infection.

To define disease resistance in terms of a genetic component which could be of use in a programme to increase resistance in a strain of animal will be extremely difficult. It will be necessary to identify the mechanism by which the viral agent infects the host, and causes disease. In the majority of acute animal viral diseases this information is lacking. The observations described above of natural resistance could be due to a number of genes, and the experiments described in genetically defined strains of mice will be impossibly expensive to replicate in farm animals. In addition, genes selected for resistance to one disease could increase susceptibility to another. An alternative approach with viral diseases, against many of which good vaccines already exist, may be to examine how animals respond to vaccination, and select genes which enhance this response.

Viral pathogenesis

The understanding of how viruses cause disease is still in its infancy, particularly for those diseases of major importance in domesticated animals. Pathogenicity was equated by Mims and White (1984) to virulence and the ability to cause disease. Determinants important in virulence include dose, route of infection, speed of replication, temperature sensi-tivity of the virus, and genetic components which can confer on the virus resistance to certain non-specific host defences. The M2 gene in reovirus type 1 enables it to resist intestinal proteases. When this gene is transferred to a strain of reovirus type 3, which is neurovirulent but protease sensitive, it enables the virus to infect the host by the oral route.

Viruses use a variety of mechanisms to gain entry to the host cell. The majority appear to take advantage of normal cell membrane receptors and components to attach, and then exploit normal membrane transport mechanisms such as those used to take in nutrients, communicate, maintain ionic balance or secrete to gain entry (Marsh and Helenius, 1989). Viral attachment to specific cells is defined by the specific receptors, proteins, lipids or oligosaccharides on the cell surface. According to the distribution of these attachment sites, some viruses have a very broad host cell range, whereas others are very restricted. Many viruses such as Semliki Forest virus (SFV) and influenza virus enter host cells in a clathrin-coated vesicle. The vesicle fuses with the endosomal wall through which the virus must still penetrate before entering the cytoplasm. SFV, influenza virus, West Nile virus, vesicular stomatitis virus, rabies virus and African swine fever virus all require acid conditions for endocytic host cell entry, whereas para-myxovirus and herpes virus entry is pH independent. This vesicle-mediated transport system is probably the main mechanism by which macromolecules are transported between cells.

The major rhinovirus receptor is the cell adhesion molecule, ICAM-1, which is the receptor for lymphyocyte function associated molecule 1 (LFA-1) that mediates adhesion to leucocytes in certain immune reactions. Rhinovirus infection induces an inflammatory response which stimulates further expression of the ICAM-1 site, which thereby increases the number of rhinovirus attachment sites (Staunton *et al.*, 1989).

Many natural adhesive proteins contain the tripeptide recognition site arginine–glycine–aspartic acid (RGD). This site is recognized by certain receptors, also of the integrin family which includes ICAM-1, present on the cell surface. Fox *et al.* (1989) were able to block the attachment of FMD virus to BHK cells using RGD-containing peptides; attachment was also blocked by antibody against the RGD region of the FMD surface protein, VP1.

These and other examples demonstrate the versatility of viruses and their ability to utilize natural host cell functions to initiate infection. Any attempt to prevent infection by blocking these processes could have serious consequences for the host. How, therefore, and at what stage of infection, can genetic resistance in certain individuals within a susceptible species manifest itself? There is increasing evidence that some of these differences reside in the efficiency of the immune response to initial infection.

Mechanism of genetic resistance

Immunity to infection can for convenience be described under a number of different headings which can obscure the totally integrated functioning of these separate components. The route of entry into the host of many of the acute viral infections is across the mucous membranes of the intestinal or respiratory tracts. Usually, following an initial replicative cycle at the site of entry and/or the local draining lymph node, free virus or cell-associated virus is taken in the bloodstream to its main target organs. At different stages of its life cycles, the virus may become susceptible to the different mechanisms of the host immune system (reviewed by Roitt *et al.*, 1985). Certain of these mechanisms have been identified as being of particular relevance to the understanding of genetic resistance.

The Mx gene is a single autosomal dominant gene assigned to mouse chromosome 16, which confers resistance to pneumotropic, neurotropic and hepatotropic strains of influenza virus. The gene is present in certain inbred laboratory strains of mice, in particular A2G in which the resistance trait was first noticed, but absent from the majority of laboratory strains. It is thought that the Mx gene is a wild mouse gene, partially deleted in strains such as BALB/cj, A/J, 129/J, C57BL/6J and CBA/J, but reintroduced into other strains by unintentional mixing with wild mice (Bang, 1978). Mx$^+$ cells in tissue culture respond to treatment with alpha or beta interferons by producing Mx protein, which gives them specific resistance to orthomyxovirus infection. That the resistance was due to the Mx protein, and not to the other interferon-induced changes in the cell, was elegantly shown by Staeheli *et al.* (1986). Interferons are produced by many cell types following viral infection. Knowledge about their function is increasing, and although the list of effects of interferon is long, it is possible that they are all manifestations of only a few changes in biochemical pathways within the cell.

Unlike the innate immune responses, such as interferon production, the adaptive immune response depends on recognizing antigenic determinants on the virus or infected cell. The theory of clonal selection of B lymphocytes postulates that each antibody-producing lymphocyte is capable of responding to only a single antigen, the receptor for which is present on the cell surface. When stimulated by the presence of this antigen, the lymphocyte proliferates as a clone, producing more antigen-specific antibody. Included in the theory is that no clones of lymphocytes will persist that recognize self antigens. If, therefore, the host is born with a repertoire of preprogrammed responses to specific antigens, and if there are differences in emphasis within this repertoire between individuals, as there undoubtedly are, those individuals that are born with clones of lymphocyte whose receptors recognize a more complete spectrum of the antigenic determinants of a prevalent viral pathogen, would be at an evolutionary advantage.

The same could be said to be true of the major histocompatibility complex (MHC). The antigen-presenting cells present processed viral antigen in combination with the MHC class II molecule to subsets of T lymphocytes. Considerable polymorphism has been shown in the MHC molecules of different individuals by the variation in the specific antigen selected for presentation, and the subsequent subset of T lymphocytes stimulated to respond. According to the subset involved, the outcome could be a successful elimination of the infection, a hypersensitivity reaction, an autoimmune response, a suppressed response or any other of the range of responses typically seen in an infected group. Therefore, as with antibody production, those individuals whose MHC presents the viral antigen most significant to a successful immune response will have an advantage. In addition, those individuals whose self antigens have characteristics in common with certain of the viral antigens would be at a distinct disadvantage, as their ability to respond to these antigens may have been eliminated.

By identifying and exploiting the genetic basis for this polymorphism of the adaptive immune response, it would perhaps be possible to transfer disease resistance.

Manipulation of genetic resistance to disease

A transgenic animal is one that has had new genes introduced by genetic engineering techniques. New genes may be inserted by microinjection of DNA into the pronucleus, injection of chromosome fragments into the pronucleus, or by using retroviral vectors with or without cultured stem cells. Transgenic technology has obvious potential for transferring genes for improved production, growth and reproductive efficiency. However, the potential for enhanced disease resistance may be ethically the most attractive, although probably the most difficult to achieve.

Some recent success in transferred resistance to a virus disease has been achieved with the transfer of the Mx gene into Mx⁻ mice embryos (Arnheiter *et al.*, 1989). Differences were seen in the amount of Mx protein expression in individual mice challenged with influenza virus, and none performed as well as the control A2G mice which naturally express the Mx gene. A gene that codes for an enzyme 2-5 oligo A synthetase, and confers resistance to picornavirus infection (FMD virus is a picornavirus) has recently been transferred to hamster ovary cells in tissue culture (Chebath *et al.*, 1987).

Transgenic technology also opens up the possibility of introducing genes from one species of animal to another, an impossibility in conventional breeding programmes. The human beta 1 interferon gene has been inoculated with suitable promoters into a mouse germ line. The survivors showed an increased resistance to challenge with pseudorabies (Aujeszky's) virus when compared with non-transgenic control mice (Chen *et al.*, 1988).

There are few other examples of genetically transferred resistance to acute viral disease in animals other than poultry, although considerable money and effort is being directed towards this objective (Bloomfield, 1990). The force behind the effort is not surprisingly the possibility of major profits, and an application to human health with even greater profits. Already this new technology has become a legal minefield with patents on techniques such as antisense RNA, 'Gene Shears' and transgenic animals. However, with these incentives progress has not been particularly rapid, and the deficiencies in our understanding of basic immunological processes, and the pathogenesis of common viral diseases are becoming apparent, and an important constraint. Nevertheless, Jenner was able to take advantage of an empirical observation that cowpox protected against smallpox, a phenomenon which was not explained until almost 200 years later. If the initial enthusiasm does not produce the results that the venture capitalists would like, one would hope that they may continue to finance the investigation as to why the transfer experiments were unsuccessful and thereby create a base for continued, and ultimately rewarding research leading to practical applications.

References

Anderson, E.C., Doughty, W.J. and Crees, H.J.S. (1981) A review of the epidemiology and control of foot-and-mouth disease in Kenya since 1968. *Report of the Ministry of Livestock Development*, Kenya.

Arnheiter, H., Skuntz, S., Noteborn, M., Chang, S., Meier, E. and Weissmann, C. (1989) Transgenic mice expressing Mx protein are resistant to influenza virus. *Journal of Interferon Research* 9, S100.

Bang, F.B. (1978) Genetics of resistance of animals to viruses. 1. Introduction and studies on mice. *Advances in Virus Research* 23, 269–348.

Bloomfield, G. (1990) *Trends in Veterinary Research and Development* 2. *Transgenic Animals*. PJB Publications Ltd., Richmond, Surrey, UK.

Briody, B.A., Hauschka, T.S. and Mirand, E.A. (1956) The role of genotype in resistance to an epizootic of mouse pox (ectromelia). *American Journal of Hygiene* 63, 59–70.

Brown, F. (1989) The development of chemically synthesized vaccines. In: Bittle, J.L. and Murphy, F.A. (eds), *Advances in Veterinary Science and Comparative Medicine*, vol. 33, *Vaccine Biotechnology*, Academic Press, London, pp. 173–93.

Campbell, C.H. (1960) The susceptibility of mother mice and pregnant mice to the virus of foot-and-mouth disease. *Journal of Immunology* 84, 469–74.

Chebath, J., Benech, P., Revel, M. and Vigneron, M. (1987) Constitutive expression of (2-5) oligo A synthetase confers resistance to picornavirus infection. *Nature* 330, 587–8.

Chen, X-Z., Yun, J.S. and Wagner, T.E. (1988) Enhanced viral resistance in transgenic mice expressing the human beta 1 interferon. *Journal of Virology* 62, 3883–7.

Davies, F.G. (1981) *Rift Valley fever in Kenya*. Office International des Epizooties, Technical Series No. 1., Paris, France.

Fox, G., Parry, N.R., Barnett, P.V., McGinn, B., Rowlands, D.J. and Brown, F. (1989) The cell attachment site on foot-and-mouth disease virus includes the amino acid sequence RGD (Arginine-Glycine-Aspartic Acid). *Journal of General Virology* 70, 625–37.

Gard, G.P. (1987) *Studies of bluetongue virulence and pathogenesis in sheep*. Technical Bulletin No. 103, Dept. of Industries and Development, Darwin, Northern Territories, Australia.

Jeggo, M.J., Corteyn, A.H., Taylor, W.P., Davidson, W.L. and Gorman, B.M.

(1986) Virulence of bluetongue virus for British sheep. *Research in Veterinary Science* 42, 24–8.

Kitching, R.P., Bhat, P.P. and Black, D.N. (1989) The characterization of African strains of capripoxvirus. *Epidemiology and Infection* 102, 335–43.

Kitching, R.P., Hammond, J.M. and Taylor, W.P. (1987b) A single vaccine for the control of capripox infection in sheep and goats. *Research in Veterinary Science* 42, 53–60.

Kitching, R.P., McGrane, J.J., Hammond, J.M., Miah, A.H., Mustafa, A.H.M. and Majumder, J.R. (1987a) Capripox in Bangladesh. *Tropical Animal Health and Production* 19, 203–8.

Kitching, R.P., McGrane, J.J. and Taylor, W.P. (1986) Capripox in the Yemen Arab Republic and the Sultanate of Oman. *Tropical Animal Health and Production* 18, 115–22.

Kitching, R.P. and Taylor, W.P. (1985a) Clinical and antigenic relationship between isolates of sheep and goat pox viruses. *Tropical Animal Health and Production* 17, 64–74.

Kitching, R.P. and Taylor, W.P. (1985b) Transmission of capripoxvirus. *Research in Veterinary Science* 39, 196–9.

Marsh, M. and Helenius, A. (1989) Virus entry into animal cells. *Advances in Virus Research* 36, 107–51.

Mims, C.A. and White, D.O. (1984) *Viral Pathogenesis and Immunology,* Blackwell Scientific Publications, Oxford.

O'Neill, H.C., Blanden, R.V. and O'Neill, T.J. (1983) H-2-linked control of resistance to ectromelia virus infection in B10 congenic mice. *Immunogenetics* 18, 255–65.

O'Neill, H.C. and Brenan, M. (1987) A role for early cytotoxic T cells in resistance to ectromelia virus infection in mice. *Journal of General Virology* 68, 2669–73.

Pay, T.W.F. (1989) Control of foot-and-mouth disease in cattle by vaccination. *Foot-and-Mouth Disease Bulletin* 27, 1–9. Coopers Animal Health, Pirbright, Surrey, UK.

Roitt, I.M., Brostoff, J. and Male, D.K. (1985) *Immunology,* Churchill Livingstone and Gower Medical Publishing, London.

Schell, K. (1960) Studies on the innate resistance of mice to infection with mousepox. II Route of inoculation and resistance; and some observations on the inheritance of resistance. *Australian Journal of Experimental Biological and Medical Research* 38, 289–307.

Scott, G.R. (1964) Rinderpest. *Advances in Veterinary Research* 9, 113–224.

Scott, G.R. and Brown, R.D. (1961) Rinderpest diagnosis with special reference to the agar gel double diffusion test. *Bulletin of Epizootic Diseases of Africa* 9, 83–125.

Shimshony, A. (1987) Bluetongue activity in Israel, 1950–1985: the disease, virus prevalence, control methods. In: Taylor, W.P. (ed.), *Bluetongue in the Mediterranean Region,* for the Commission of the European Communities, EUR 10237.

Sonoda, A. (1983) Production of rinderpest tissue culture live vaccine. *Japanese Archives of Research Quarterly* 17, 191–8.

Staeheli, P., Haller, O., Boll, W., Lindenmann, J. and Weissmann, C. (1986) Mx protein: constitutive expression in 3T3 cells transformed with cloned Mx cDNA confers selective resistance to influenza virus. *Cell* 44, 147–58.

Staunton, D.E., Merluzzi, V.J., Rothlein, R., Barton, R., Marlin, S.D. and Springer, T.A. (1989) A cell adhesion molecule, ICAM-1, is the major surface receptor for rhinoviruses. *Cell* 56, 849–53.

Subak-Sharpe, H. (1961a) The quantitative study of foot-and-mouth disease virus in unweaned mice. II Studies with additional mouse strains and comparison of some methods of titration. *Archiv Fur Die Gesamte Virusforschung* 11, 39–63.

Subak-Sharpe, H. (1961b) The effect of passage history, route of inoculation, virus strain and host strain on the susceptibility of adult mice to the virus of foot-and-mouth disease. *Archiv Fur Die Gesamte Virusforschung* 11, 373–99.

Subak-Sharpe, H., Pringle, C.R. and Hollom, S.E. (1963) Factors influencing the dynamics of the multiplication of foot-and-mouth disease virus in adult mice. *Archiv Fur Die Gesamte Virusforschung* 12, 600–19.

Taylor, W.P. (1986) The epidemiology of bluetongue. *Revue Scientifique et Technique de Oficina Internacional de Epizootias* 5, 351–6.

Wallace, G.D., Buller, R.M.L. and Morse, H.C. (1985) Genetic determinants of resistance to ectromelia (mousepox) virus-induced mortality. *Journal of Virology* 55, 890–1.

Weinberg, E.D. (1987) Pregnancy-associated immune suppression: risks and mechanisms. *Microbial Pathogenesis* 3, 393–7.

Chapter 19

The Genetics of Scrapie Susceptibility in Sheep

(and its implications for BSE)

Nora Hunter and James Hope
*Institute for Animal Health, AFRC/MRC Neuropathogenesis
Unit, Ogston Building, West Mains Road, Edinburgh
EH9 3JF, UK*

Summary

Scrapie is a transmissible disease which affects the central nervous system of sheep and goats. Natural scrapie has been endemic in Europe for over 250 years and can be transmitted from affected to healthy sheep (and other species) by inoculation or feeding of diseased tissues. Scrapie and related diseases are invariably fatal and are characterized by long, largely asymptomatic, incubation periods which may last for months or years.

Scrapie is caused by an unusual pathogen. This pathogen has some of the properties of a conventional virus but can survive normal virucidal procedures such as prolonged exposure to formalin, dry heat and some regimes of autoclaving. The molecular structure of the pathogen is unknown. It can only be detected by transmission to other animals, hence it is very difficult to monitor and eradicate. Biochemical studies of scrapie infectivity have indicated that a protein may be the sole or, at least, an integral part of the infectious particle and a candidate host protein (PrP or prion protein) has been identified. Prion, virino and slow virus are among the many names used for the causative agent of scrapie and related diseases. Scrapie-associated fibrils (SAF), and other modified forms of PrP, accumulate in the brain and some peripheral tissues (spleen, lymph nodes) during the development of these diseases. To date, these aggregated forms of a normal host protein are the only candidate antigens which either specifically (as part of a virino or prion structure) or non-specifically (occluding a normal virus) co-purify with the infectious particle of scrapie.

Diagnosis of the disease is based on clinical signs and post-mortem brain pathology. Some of the neurological lesions of scrapie are found in the other members of this group of transmissible spongiform encephalopathies:

common features include the widespread formation of membrane-bound vesicles or vacuoles in cell bodies of nerve cells, vacuolation of the extracellular space and a proliferation of astroglial cells in the absence of demyelination or other overt inflammatory response. These features develop in the later stages of the incubation period, and it is very difficult to detect those animals with scrapie which are not visibly affected by the disease. This has greatly hindered the effectiveness of approaches for the control and management of scrapie.

The overall rate of replication of the transmissible agent, and hence the incubation period of the disease, depends primarily on the interaction between the strain(s) of the pathogen and the genotype of the affected animal. Over the past 20 years, the problem of scrapie has been tackled by culling and selection of sheep which are relatively resistant or susceptible to natural or experimental scrapie infection. This has been expensive and time-consuming. This chapter reviews our understanding of the biology and molecular genetics of scrapie susceptibility and how the application of a simple blood test may facilitate animal health programmes aimed at the control and eradication of scrapie and BSE.

Introduction

Scrapie is a transmissible disease which affects the central nervous system of sheep and goats. Natural scrapie has been endemic in Europe for over 250 years – indeed it was considered such a major problem for sheep farmers in Britain in the middle of the eighteenth century that a petition was sent to the House of Commons asking for restrictions to be placed on breeders dealing in sheep in the hope of controlling the spread of the disease. Scrapie can be transmitted from affected to healthy sheep (and other species) both vertically and horizontally, by inoculation or by feeding of diseased tissues. The scrapie-like diseases are characterized by long, largely asymptomatic, incubation periods which may last for months or years (Dickinson, 1976; Parry, 1984) and there is no known cure.

Scrapie has been reported in most parts of the world although a few countries are regarded as free from the disease, notably Australia and New Zealand. Reports of scrapie still appear in countries previously thought to be free of the disease; for instance in Sweden (Elvander *et al.*, 1988) and Cyprus (Toumazos, 1988). Many countries prohibit the import of sheep from areas with scrapie, and this has a major effect on the UK export trade. Import of sheep to Australia requires their quarantine in Tasmania for at least 5 years and New Zealand has adopted similar long-term quarantine procedures. Unfortunately, even this does not guarantee exclusion of the disease. In part, this is because of the long incubation time of scrapie and hence the long survival of infected sheep. The lack of a preclinical diagnostic test or any means of detecting the transmissible agent adds to the problem; it is not possible to be absolutely certain that an apparently healthy animal is scrapie-free. There are tens of millions of sheep in the UK and in some flocks up to a tenth of the sheep succumb to the disease in any one year. The economic loss to the UK of scrapie in sheep was estimated to be £1.7 million in 1980; the loss is probably much more and is well appreciated by the sheep industry. Concern over damage to the livestock and meat trade has heightened in recent years following the apparent spread of scrapie from sheep to cattle (to produce bovine spongiform encephalopathy or BSE) (Wells *et al.*, 1987; Wilesmith *et al.*, 1988). The cost of BSE to European trade has been predicted to be £265 million in 1990.

The scrapie pathogen (known variously as a virino, prion or slow virus) is unusual. It has some of the properties of a conventional virus but can survive normal virucidal procedures such as prolonged exposure to formalin, dry heat and some regimes of autoclaving. The molecular structure of the pathogen is unknown. It can only be detected by bioassay (transmission to other animals), hence it is very difficult to monitor and eradicate. Biochemical and physical studies of scrapie infectivity have indicated that protein is an essential part of the infectious particle – indeed some have suggested on this basis that it contains only protein. A candidate host protein (PrP or prion) protein has been found to be closely associated with infectivity and it has been postulated that PrP (or an abnormal form of it) is, in one way or another, the cause of scrapie-like diseases (McKinley *et al.*, 1983; Prusiner, 1982). Scrapie-associated fibrils (SAF) are disease-specific structures and are made up of modified forms of PrP. These and other PrP aggregates accumulate in the brain and some peripheral tissues (spleen, lymph nodes) during development of the disease. To date, these aggregated forms of a normal host protein are the only candidate antigens which either specifically or non-specifically co-purify with the infectious particle of scrapie (Diringer *et al.*, 1983; Doi *et al.*, 1988; Hope *et al.*, 1986, 1988a; Kitamota *et al.*, 1989; Merz *et al.*, 1981) and BSE (Hope *et al.*, 1988b).

Diagnosis of the disease is based on clinical signs and post-mortem brain pathology (Fraser, 1976). Some of the neurological lesions of scrapie are found in the other members of this group of transmissible spongiform encephalopathies: common features include the widespread formation of membrane-bound vesicles or vacuoles in cell bodies of nerve cells, vacuolation of the extracellular space and a proliferation of astroglial cells in the absence of demyelination or other overt inflammatory response. These features develop in the later stages of the incubation period, and it is very difficult to detect (by histopathology) those animals with scrapie which are not visibly affected by the disease. This has greatly hindered the effectiveness of approaches for the control and management of scrapie and indeed of BSE.

The advice to farmers facing an outbreak of scrapie in their flocks has remained unchanged for a very long time. From the time of the first recording of scrapie in Britain in the eighteenth century it has been noticed that the disease 'runs in families'. The recommendation is therefore to cull all animals in an affected line whether or not they appear healthy. This drastic and often expensive measure will not always succeed in eradicating scrapie – contagious spread undoubtedly exists (Dickinson *et al.*, 1974) and not every susceptible animal will necessarily succumb in any one year (Gordon, 1966). The development of sheep lines with some resistance to disease has met with some success (Dickinson, 1976; Hoare *et al.*, 1977; Nussbaum *et al.*, 1975) but has been expensive and time-consuming. In the main part of this chapter, we review the biology and molecular genetics of scrapie susceptibility in sheep and outline how the application of a simple blood test may facilitate animal health programmes aimed at the control and eradication of scrapie. Our latest information on the transmission and molecular genetics of BSE is presented in an appendix.

Development of sheep lines with predictable response to scrapie

It has been known for over 200 years that the development of natural scrapie in sheep depended on genetic factors (Anon, 1755). Early experimental studies of transmission of scrapie using sheep often gave confusing and contradictory results. One of the studies most often quoted is the 'twenty four breed' experiment described by W.S. Gordon in 1966. Twenty four different breeds of sheep (approximately 40 animals per breed) were injected intracerebrally (i.c.) with the source of scrapie known as SSBP/1 (Dickinson, 1976). The animals were observed for two years and the number developing scrapie varied from an incidence of 78% clinical cases in the Herdwick breed to no cases at all in the Dorset Downs. These results were thought to demonstrate breed differences in susceptibility but were not reproducible – on testing in a different group of Herdwick sheep, Pattison (1966) found only 30% were susceptible. There were also problems with the 'resistant' Dorset Downs some of which did eventually develop SSBP/1 scrapie at much later dates (Dickinson, 1976). It became clear that there was as much variation in the 'take' of scrapie inoculation between different flocks of the same breed as there was between different breeds and there was a perceived need to develop lines of sheep with more predictable response to induced scrapie if any progress in studying the disease was to be made. There are three such flocks in Britain – Cheviots and Herdwicks (both at Neuropathogenesis Unit (NPU), Edinburgh) and Swaledales (at Redesdale, Northumberland).

NPU Cheviots

The Cheviot flock selection was started in 1961 and the sheep were split into two lines (positive and negative) on the basis of their response to subcutaneous (s.c.) injection with SSBP/1 (sheep scrapie brain pool 1) which is thought to be a mixture of scrapie strains. It was derived from a pool of three brains taken from pure or cross-bred Cheviot sheep at the terminal stage of scrapie (Dickinson, 1976). The survival time of the sheep is mainly determined by a single gene, *Sip*, with two alleles sA and pA; sA is dominant. When injected subcutaneously, the positive line (*Sip* sAsA or *Sip* sApA) develop scrapie in around 300 days whereas the negative line (*Sip* pApA) survive a natural lifespan. The negative line are not thought to be truly resistant as they can contract scrapie after about 1,000 days following intracerebral (i.c.) injection. Positive-line sheep die within 200 days of intracerebral injection of SSBP/1 (Dickinson *et al.*, 1968). Isolates (and perhaps natural outbreaks) of scrapie may be classified according to their relative effects on sheep of the different *Sip* genotypes. Most isolates (A group) produce the disease in carriers of the sA allele faster than in pApA sheep, but with at least one isolate (CH1641) (C group) the ranking of *Sip* type (in respect of survival time) is not clear cut and may even be reversed (Foster and Dickinson, 1988a) (Table 19.1).

NPU Herdwicks

Also in 1961, W.S. Gordon began a similar selection process using SSBP/1 and Herdwick sheep (Nussbaum *et al.*, 1975). Positive and negative lines were produced in the same way and susceptibility differences were again shown to be under the control of a single gene with two alleles, the dominant allele being that conferring high susceptibility to SSBP/1. There

Table 19.1. Incubation period of SSBP/1 and CH1641 in Cheviot sheep of differing *Sip* genotype.

Source of scrapie	Route of injection	Incubation period (days ± SEM)*	
		Positive line (Sip^{sAsA} or Sip^{sApA})	Negative line (Sip^{pApA})
SSBP/1	Intracerebral	197 ± 7	917 ± 90
	Subcutaneous	313 ± 9	—
CH1641	Intracerebral	595 ± 122	360 ± 15

*Data adapted from Dickinson *et al.* (1968) and Foster and Dickinson (1988a).

is, as yet, no formal proof that the *Sip* gene is acting in the Herdwicks as well as the Cheviots, but the results of crossing experiments suggest that it does (Foster, unpublished). Unfortunately only a few of these selected Herdwick sheep now remain.

Swaledales

A low susceptibility Swaledale flock was started at Compton in 1973 (Davies and Kimberlin, 1985; Hoare *et al.*, 1977) by the relatively simple means of inoculating the animals subcutaneously with scrapie and breeding only from the survivors. The sources of scrapie brain were different – SW73 and SW75 (both pools of Swaledale scrapie brain) – however, susceptibility was again found to be controlled by a single gene with two alleles, the dominant one resulting in a shorter incubation period but this time the dominance appeared to be only partial. Homozygous and hetero-zygous 'susceptible' animals seemed to have shorter and longer incubation periods respectively. The progeny of the survivors had a greatly reduced susceptibility to experimental challenge with scrapie and were used to form a nucleus flock of scrapie 'resistant' (or negative line) Swaledales now at the MAFF Experimental Husbandry Farm, Redesdale, Northumberland. We believe, but have not formally proved, that these sheep are also of the genotype *Sip* pApA in which replication or the deleterious effects of the scrapie pathogen may be sufficiently inhibited to allow survival.

Mode of action of *Sip*

It is not known where *Sip* acts in the replication cycle of the scrapie pathogen to determine the fate of the infected animal. Extensive studies by Hadlow, Ecklund and colleagues from the late 1950s and more recently by Kimberlin and Walker have helped define the spread of infection within an animal following natural or experimental infection with scrapie (Hadlow *et al.*, 1982; Kimberlin and Walker, 1988a). Infection via the peripheral routes appeared to mimic more closely the natural disease, where the earliest rises in titre were in parts of the alimentary tract, the spleen and various lymph nodes. Only much later was infection detected in the central nervous system. These data were consistent with the idea of maternal transmission of the disease from ewe to lamb via an oral route (Hadlow *et al.*, 1982). Kimberlin and Walker (1988b) have suggested that *Sinc*, the murine homologue of *Sip*, acts by controlling the pathogen's route from sites of peripheral replication into the central nervous system. If true, and *Sip* acts similarly, then the 'low susceptibility' flocks cannot be regarded as scrapie-free and may act as carriers of infection. Unfortunately, data on the 'carrier status' of *Sip* pApA animals is lacking as there is neither a convenient assay for infection nor has there been (until now) an assay for *Sip* genotype.

Biochemical markers for susceptibility

The expense and time involved in selecting sheep by their response to exposure to scrapie has hindered its practical application. Over the years this has stimulated a search for *in vitro* methods for predicting the suscept- ibility or resistance of sheep to natural or experimental scrapie. Towards this end, various biochemical markers have been tested in the Herdwick selected lines. No correlation was found with susceptibility and phenotypes of albumin, prealbumin, esterase, haemoglobin, transferrin, reduced glutathione and α-mannosidase or with OLA haplotypes (Collis and Millson, 1975; Collis *et al.*, 1977; Cullen *et al.*, 1984). A linkage with OLA haplotypes was described in Ile de France sheep with natural scrapie (Millot *et al.*, 1988) but this has been disputed (Cullen, 1989). This linkage should now be reassessed using unrelated animals.

Originally a search was made for *in vitro* markers of scrapie resistance/ susceptibility using the two lines of NPU Cheviot sheep which were select- ively bred for increased (positive line) or decreased (negative line) incidence of SSBP/1-induced scrapie. Paradoxically, a candidate product of the *Sip* locus was discovered in studies on the molecular structure of the rodent scrapie pathogen. This was the PrP protein. Mutations in or around the PrP gene were found to be linked to the alleles of *Sinc* (Carlson *et al.*, 1986; Hunter *et al.*, 1987) and we have reported *Eco*RI and *Hind*III restriction fragment length polymorphisms (RFLPs) of the sheep PrP gene which, in a small sample of animals from our Cheviot flock, correlated with the alleles of *Sip* (Hunter *et al.*, 1989). The *Sip* sA allele was linked to 6.8kb *Eco*RI and 3.4kb *Hind*III fragments of the PrP gene, and the *Sip* pA allele was linked to 4.4kb *Eco*RI and 5.0kb *Hind*III fragments. This correlation has been further validated in the NPU Cheviot flock and extended to the Swaledale 'negative line' sheep.

Linkage of PrP gene RFLPs in the NPU Cheviot flock

Using one RFLP (*Eco*RI) the original study has been extended to 77 positive line and 31 negative line sheep from the NPU Cheviot flock. (The number of animals in each line is variable but usually around 100.) All 31 animals have a PrP-*Eco*RI fragment of 4.4kb. Of the positive line, 25% (24 animals) have one 6.8kb fragment and 65% (50 animals) have both fragments (Table 19.2a). This strengthens the preliminary results suggesting that the 6.8kb fragment is a marker for *Sip* sA and the 4.4kb fragment a marker for *Sip* pA.

Table 19.2.   *Eco*RI polymorphism.

	*Eco*RI fragment size		
	6.8kb	Both	4.4kb
(a) NPU Cheviot flock			
Positive line (77)	27 (35%)	50 (65%)	0
Negative line (31)	0	0	31 (100%)
(b) Swaledale flock			
Redesdale (79)	2 (3%)	18 (23%)	58 (74%)
'Controls' (31)	12 (39%)	10 (32%)	9 (30%)

Linkage of PrP gene RFLPs in the Redesdale Swaledale flock

We have tested 79 Swaledale sheep DNA samples so far for the *Eco*RI RFLP. The 4.4kb fragment is homozygous in 74% (58 animals); the 6.8kb fragment is homozygous in 3% (2 animals) and 23% (18 animals are heterozygous, having both PrP *Eco*RI fragments) (Table 19.2b). There is, therefore, in this low susceptibility line, a large number (97%) carrying the putative Cheviot pA marker. Less than one-third of these animals are heterozygotes but, if the data obtained by Hoare *et al.* (1977) and Davies and Kimberlin (1985) are correct, and the linkage is correct, heterozygotes have a lower susceptibility to scrapie than homozygote susceptible sheep. There are very few animals homozygous for the putative *Sip* sA marker.

These results would fit very well with the idea that the 4.4kb fragment is linked to low susceptibility to experimental scrapie. To assess the significance of this finding, 31 animals from a separate, unselected Swaledale flock were also tested. These animals were of a similar age range to the Redesdale flock but came from a flock which has had a number of natural scrapie cases. Of these sheep 39% (12 animals) had the 6.8kb fragment, 30% (nine animals) had the 4.4kb fragment and 32% (10 animals) had both the 6.8kb and the 4.4kb fragments (Table 19.2b). Despite the low numbers, there is clearly a different frequency from that found in the Redesdale flock.

This is encouraging, although we need to know much more about the relationship between the PrP and *Sip* genes, and their mechanism of action before regarding this RFLP linkage analysis as a fool-proof predictor of a sheep's response to challenge with scrapie.

The natural disease

Results obtained from selected lines of sheep are open to doubts because of the potential problems of inbreeding. Great care has been taken to reduce this to a minimum with the NPU Cheviot flock but nevertheless it is necessary to test the RFLPs in the 'real world' of the farm with unselected and unrelated sheep. This work is now underway and will continue for some time. However, the relevance of the selected sheep lines has been underlined by the fact that outbreaks of natural scrapie have occurred in the positive lines of both Cheviots and Herdwicks but not in the negative lines (Dickinson, 1974; Pattison, 1974) nor in the Swaledale nucleus flock (A.J. Chalmers, personal communication). The *Sip* gene is also thought to control the host response to natural scrapie although the sA allele is thought to be recessive in this case. Crossbreeding studies (Foster and Dickinson, 1988b) followed by observation of the progeny for the development of natural or induced scrapie have given good evidence that the *Sip* genotype is just as important following natural challenge as it is after experimental challenge. The apparent differences in dominance of the alleles of *Sip* in the Cheviot, Herdwick and Swaledale experimental flocks (full or partial dominance of sA) and natural scrapie (sA believed to be recessive) have yet to be explained. However, it is known from studies of murine scrapie that low doses of inoculum and also different routes of infection can greatly change the incubation period recorded (Chandler, 1963; Kimberlin and Walker, 1979). In the field, route of infection and dose of scrapie may differ considerably from that used for inducing the disease in experimental sheep, so care is needed in interpretation and comparison of the natural and experimental diseases.

Sip and the structure of ovine PrP protein

Greater confidence in the use of the type of RFLP test described above will come from understanding the involvement of the ovine PrP gene and its protein in the development of scrapie. The genetic linkage of the PrP gene and incubation time control genes has been demonstrated now in mice, sheep and humans and has given rise to the idea that PrP protein might be the *Sip* (or *Sinc*) gene product. To investigate the link between PrP and host genes controlling the incubation period of natural scrapie, we have isolated and sequenced clones of genomic DNA carrying most of the ovine PrP gene and searched for amino acid substitutions in ovine PrP which might give clues to the molecular basis of *Sip* allelism. Two clones, encoding predicted PrP proteins of 256 amino acids, were found to differ by three nucleotides in the open reading frame of the gene. This results in a single amino acid variant: an arginine/glutamine polymorphism at codon

171 (Goldmann *et al.*, 1990a). These clones also differed in the *Eco*RI restriction enzyme site outside the PrP coding region which we had previously shown to be linked to the alleles of *Sip* (Hunter *et al.*, 1989). The correlation of amino acid variations in the PrP protein itself with the alleles of *Sip* (Goldmann *et al.*, 1990a) and *Sinc* (Westaway *et al.*, 1987), and with the incidence of related diseases in man, familial Crutzfeldt–Jakob disease (F-CJD) (Owen *et al.*, 1989) and the Gerstmann–Straussler syndrome (GSS) (Hsiao *et al.*, 1989) lends support to the notion that PrP is a product of the host 'survival time' locus (Table 19.3). Indeed the transfer of the hamster PrP gene into mice renders the transgenic mouse much more susceptible to hamster scrapie (Scott *et al.*, 1989). However, there are one or two reports of mice in which there may have been recombination between the *Sinc* gene and PrP gene RFLPs (Carlson *et al.*, 1989; Race *et al.*, 1990). Further investigations are needed to explain exactly how such apparently minor changes in the molecular structure of a single protein might make the difference between life and death following exposure to scrapie-like agents.

Potential and pitfalls

Genetic selection of sheep for low or high susceptibility to scrapie may be greatly facilitated by the availability of a rapid, laboratory test for *Sip* genotype. The RFLP test for *Sip* alleles requires only a small blood sample and the results are available within days rather than the years required for bioassay. This linkage test has yet to be validated in other flocks and breeds of sheep but it is potentially useful to pedigree livestock breeders. Selection for scrapie 'resistance' by this means or any other might not ensure a scrapie free flock. Selection strategies need to be tested in parallel with more fundamental studies on the mechanism of *Sip* action and on the dynamics of spread of infection in *Sip* pApA, *Sip* sApA and *Sip* sAsA sheep. However, maintenance of mixed populations of *Sip* alleles in a selection line may be useful in preventing high incidence outbreaks of scrapie. Similarly, mutations in or around the bovine PrP gene may also indicate high or low susceptibility to BSE and that would have implications for the control and management of scrapie throughout the world.

Transmission of BSE to rodents and ruminants

To date (November, 1990) there is no evidence for the natural transmission of BSE but the disease has been transmitted to mice (Fraser *et al.*, 1988), cattle (Dawson *et al.*, 1990a), a pig (Dawson *et al.*, 1990b), sheep and goats (Foster and Hope, unpublished) by intracranial injection, and to mice by

Table 19.3. Linkage of the alleles of *Sip* and homologous genes to polymorphisms in or around the PrP gene.

Species	Disease	Gene	Restriction fragment length polymorphisms	Site(s) of PrP polymorphism	References
Sheep	Natural and Experimental Scrapie	*Sip*	*Hind*III, *Eco*RI	171 Gln/Arg	Dickinson and Outram (1988) Hunter *et al.* (1989) Goldmann *et al.* (1990)
Mouse	Experimental Scrapie	*Sinc*	*Bst*E11	108 Leu/Phe 189 Val/Thr	Dickinson and Outram (1988) Carlson *et al.* (1986) Hunter *et al.* (1987) Westaway *et al.* (1987)
Man	F-CJD	?	*Msp*I	Insert in repeat region (68–99)	Owen *et al.* (1989)
	GSS	?	*Dde* 1	102 Leu/Pro	Hsiao *et al.* (1989)

oral dosing (Barlow and Middleton, 1990).

Investigation of the effect of mouse genes on the survival time of mice challenged with BSE-infected cattle tissues is in progress. To date, four experiments have been completed, each using the brains from one of four cases of BSE from different parts of England. Four inbred mouse strains were used: two homozygous for the s7 allele of *Sinc* (C57BL and RIII/ FaDk), and two homozygous for the p7 allele (VM/Dk and IM/DK) (Fraser *et al.*, 1988; Fraser, unpublished data). Between 24 and 36 mice of each strain were each injected at two sites (intracerebrally and intraperitoneally) with a 10-fold dilution of brain homogenate in saline. All groups of mice became affected but with differing mean incubation times (Table 19.4). While the ranking of mouse genotypes with respect to incubation period roughly accords to their *Sinc* genotype, comparison of the survival time of different strains of *Sinc* s7 (RIII versus C57BL) or *Sinc* p7 (VM versus IM) mice indicates other genes affect the incubation period of interspecies transmission of BSE. The importance of these genes diminishes rapidly on further (intraspecies) passage of BSE infectivity, when the *Sinc* genotype becomes the single, major host determinant of incubation period (Fraser, unpublished data).

For technical reasons, these data are not comparable to results of sheep-to-mouse transmissions carried out in the 1970s and differ significantly from the mean survival times of mice similarly injected with brain homogenate from a scrapie-affected Greyface sheep in a contemporary experiment (Fraser, unpublished data). The relative contribution of difference in titre, strain of BSE pathogen, route of injection and donor species to differences in survival time are under investigation.

Knowledge of the factors affecting the transmission of BSE to mice has become of vital importance, as transmission of mice provides a way of measuring the amount of infectivity in cattle tissues and their by-products. This information is necessary to assess the relative risks of handling/eating these materials, to gauge the efficiency of disinfection protocols, and to allow monitoring of food and pharmaceutical processes for the distribution

Table 19.4 Incubation period of mice infected with BSE (days ± SE).

| | Mouse strain | | | |
	RIII	C57BL	VM	IM
Case 1	328 ± 3	438 ± 7	427 ± 8	538 ± 7
Case 2	327 ± 4	407 ± 5	497 ± 9	545 ± 11
Case 3	321 ± 4	434 ± 5	510 ± 5	559 ± 8
Case 4	316 ± 3	423 ± 6	523 ± 8	563 ± 8

of infectivity in product and by-product. More speculatively, the properties of BSE in mice may allow strain typing and, following comparison with sheep-derived scrapie strains, identify possible sources of origin of BSE.

Identification of different forms of the bovine PrP gene

The discovery of fibrils similar to scrapie-associated fibrils (SAF) in BSE-affected cow brain supported the clinical and pathological diagnosis of its scrapie-like nature (Wells *et al.*, 1987), and we have shown the major protein of BSE fibrils is the bovine PrP homologue (Hope *et al.*, 1988b). Different forms of the mouse, hamster and sheep PrP genes are linked to the relative survival time of these species following inoculation of isolates of scrapie, BSE or Creutzfeldt–Jakob disease (see above), and this led to a search for mutations of the bovine PrP gene which might predispose cattle to BSE. We have found and sequenced different forms of the bovine PrP gene which contain either five or six copies of a 24-nucleotide element encoding the octapeptide, Pro-His/Gln-Gly-(Gly)-Gly-Gly-Trp-Gly-Gln (Goldmann *et al.*, 1991). Apart from rare human alleles linked to familial CJD which contain six or nine extra copies (Owen *et al.*, 1989), PrP genes from all other species contain five copies of this highly-conserved octa-peptide motif in the *N*-terminal region of the protein. Out of twelve cattle, eight animals were found to be homozygous for genes with six copies of the repeat peptide (6:6), while four were heterozygous (6:5). Two confirmed cases of BSE occurred in (6:6) homozygous animals (Goldmann *et al.*, 1991). The amino acid sequences of the five and six copy variants were otherwise identical and this highly-conserved protein showed only six or seven differences from ovine PrP. The frequency of these PrP alleles in the UK national herd and their possible linkage to the incidence of BSE is under investigation.

Acknowledgements

We thank Dr Hugh Fraser and Dr Wilfred Goldmann, AFRC and MRC Neuropathogenesis Unit, Edinburgh for data on the transmission and molecular genetics of BSE, James Foster for his diligence on the sheep susceptibility project, and Mike Dawson and Trevor Martin at the MAFF Central Veterinary Laboratory, Weybridge for bovine DNA.

References

Anon (1755) *Proceedings of the House of Commons*, London.
Barlow, R.M. and Middleton, D.J. (1990) Dietary transmission of bovine spongi-

form encephalopathy to mice. *Veterinary Record* 126, 111–12.

Carlson, G.A., Kingsbury, D.T., Goodman, P.A., Coleman, S., Marshall, S.T., DeArmond, S., Westaway, D. and Prusiner, S.B. (1986) Linkage of prion protein and scrapie incubation time genes. *Cell* 46, 503–11.

Carlson, G.A., Goodman, P.A., Lovett, M., Taylor, B.A., Marshall, S.T., Petersa-Forchia, M., Westaway, D. and Prusiner, S.B. (1989) Genetics and polymorphism of the mouse prion gene complex: control of scrapie incubation time. *Molecular Cellular Biology* 8, 5528–40.

Chandler, R.L. (1963) Experimental scrapie in the mouse. *Research in Veterinary Science* 4, 276–85.

Collis, S.C. and Millson, G.C. (1975) Transferrin polymorphism in Herdwick sheep. *Animal Blood Groups Biochemistry and Genetics* 6, 117–20.

Collis, S.C., Millson, G.C. and Kimberlin, R.H. (1977) Genetic markers in Herdwick sheep: no correlation with susceptibility or resistance to experimental scrapie. *Animal Blood Groups Biochemistry and Genetics* 8, 79–83.

Cullen, P.R. (1989) Scrapie and the sheep MHC: claims of linkage refuted. *Immunogenetics* 29, 414–16.

Cullen, P.R., Brownlie, J. and Kimberlin, R.H. (1984) Sheep lymphocyte antigens and scrapie. *Journal of Comparative Pathology* 94, 405–15.

Davies, D.C. and Kimberlin, R.H. (1985) Selection of Swaledale sheep of reduced susceptibility to experimental scrapie. *Veterinary Record* 116, 211–14.

Dawson, M., Wells, G.A.H. and Parker, B.N.J. (1990a) Preliminary evidence of the experimental transmissibility of bovine spongiform encephalopathy to cattle. *Veterinary Record* 126, 112–13.

Dawson, M., Wells, G.A.H., Parker, B.N.J. and Scott, A.C. (1990b) Primary parenteral transmission of bovine spongiform encephalopathy to the pig. *Veterinary Record* 127, 338.

Dickinson, A.G. (1974) Natural infection, 'spontaneous generation' and scrapie. *Nature* 252, 179–80.

Dickinson, A.G. (1976) Scrapie in sheep and goats. In: Kimberlin, R.H. (ed.), *Slow Virus Diseases of Animals and Man.* North-Holland Publishing Company, Amsterdam, Oxford, pp. 209–41.

Dickinson, A.G. and Outram, G.W. (1988) Genetic aspects of unconventional virus infections. In: *Novel Infectious Agents and the Central Nervous System.* Wiley, Chichester, pp. 63–83.

Dickinson, A.G., Stamp, J.T., Renwick, C.C. and Rennie, J.C. (1968) Some factors controlling the incidence of scrapie in Cheviot sheep injected with a Cheviot-passaged scrapie agent. *Journal of Comparative Pathology* 78, 313–21.

Dickinson, A.G., Stamp, J.T. and Renwick, C.C. (1974) Maternal and lateral transmission of scrapie in sheep. *Journal of Comparative Pathology* 84, 19–25.

Diringer, H., Gelderblom, H., Kascsak, R. and Wisniewski, H.M. (1983) Scrapie infectivity, fibrils and low molecular weight protein. *Nature* 306, 476–8.

Doi, S., Ito, M., Shinagawa, M., Sato, G., Isomura, H., and Goto, H. (1988) Western blot detection of scrapie-associated fibril protein in tissues outside the central nervous system from preclinical scrapie infected mice. *Journal of General Virology* 69, 955–60.

Elvander, M., Engvali, A. and Klingeborn, B. (1988) Scrapie sheep in Sweden. *Acta Veterinaria Scandinavica* 29, 509–10.

Foster, J.D. and Dickinson, A.G. (1988a) The unusual properties of CH1641, a sheep passaged isolate of scrapie. *Veterinary Record* 123, 5–8.

Foster, J.D. and Dickinson, A.G. (1988b) Genetic control of scrapie in Cheviot and Suffolk sheep. *Veterinary Record* 123, 159.

Fraser, H. (1976) The pathology of natural and experimental scrapie. In: Kimberlin, R.H. (ed.) *Slow Virus Diseases of Animals and Man*, North-Holland Publishing Co., Amsterdam, 44, pp. 267–305.

Fraser, H., McConnell, I., Wells, G.A.H. and Dawson, M. (1988) Transmission of bovine spongiform encephalopathy to mice. *Veterinary Record* 123, 472.

Goldmann, W., Hunter, N., Foster, J.D., Salbaum, J.M., Beyreuther, K. and Hope, J. (1990) Two alleles of a neural protein gene linked to scrapie in sheep. *Proceedings of the National Academy of Sciences* 87, 2476–80.

Goldmann, W., Hunter, N., Martin, T., Dawson, M. and Hope, J. (1991) Different forms of bovine PrP gene have five or six copies of a short G-C-rich element within the protein coding exon. *Journal of General Virology* 72, 201–4.

Gordon, W.S. (1966) Variation in susceptibility of sheep to scrapie and genetic implications. In: *Report of Scrapie Seminar*, 1964 ARS 91-53 US Department of Agriculture, pp. 53–67.

Hadlow, W.T., Kennedy, R.C. and Race, R.E. (1982) Natural infection of Suffolk sheep with scrapie virus. *Journal of Infectious Diseases* 146, 657–64.

Hoare, M., Davies, D.C. and Pattison, I.H. (1977) Experimental production of scrapie-resistant Swaledale sheep. *Veterinary Record* 101, 482–4.

Hope, J., Morton, L.J.D., Farquhar, C.F., Multhaup, G., Beyreuther, K. and Kimberlin, R.H. (1986) The major polypeptide of scrapie-associated fibrils (SAF) has the same size, charge distribution and N-terminal protein sequence as predicted for the normal brain protein (PrP) *EMBO Journal* 5, 2591–7.

Hope, J., Multhaup, G., Reekie, L.J.D., Kimberlin, R.H. and Beyreither, K. (1988) Molecular pathology of scrapie associated fibril protein (PrP) in mouse brain affected by the ME7 strain of scrapie. *European Journal of Biochemistry* 172, 271–7.

Hope, J., Reekie, L.J.D., Hunter, N., Multhaup, G., Beyreuther, K., White, H., Scott, A.C., Stack, M.J., Dawson, M. and Wells, G.A.H. (1988) Fibrils from brains of cows with new cattle disease contain scrapie-associated protein. *Nature* 336, 390–2.

Hsiao, K., Baker, H.F., Crow, T.J., Poulter, M., Owen, F., Terwilliger, J.D., Westaway, D., Ott, J. and Prusiner, S.B. (1989) Linkage of a prion protein missense variant to Gerstmann–Straussler syndrome. *Nature* 338, 342–5.

Hunter, N., Hope, J., McConnell, I. and Dickinson, A.G. (1987) Linkage of the scrapie-associated fibril protein (PrP) gene and *Sinc* using congenic mice and restriction fragment polymorphism analysis. *Journal of General Virology* 68, 2711–16.

Hunter, N., Foster, J.D., Dickinson, A.G. and Hope, J. (1989) Linkage of the gene for the scrapie-associated fibril protein (PrP) to the *Sip* gene in Cheviot sheep. *Veterinary Record* 124, 364–6.

Kimberlin, R.H. and Walker, C.A. (1979) Pathogenesis of mouse scrapie: dynamics of agent replication in spleen, spinal cord and brain after infection by different routes. *Journal of Comparative Pathology* 89, 551–62.

Kimberlin, R.H. and Walker, C.A. (1988a) Pathogenesis of experimental scrapie. In:

Novel Infectious Agents and the Central Nervous System. Wiley, Chichester, pp. 37–62.

Kimberlin, R.H. and Walker, C.A. (1988b) Incubation periods in six models of intraperitoneally injected scrapie depend mainly on the dynamics of agent replications within the nervous system and not the lymphoreticular system. *Journal of General Virology* 69, 2953–60.

Kitamota, T., Mohri, S. and Tateishi, J. (1989) Organ distribution of proteinase-resistant prion protein in humans and mice with CJD. *Journal of General Virology* 70, 3371–9.

McKinley, M.P., Bolton, D.C. and Prusiner, S.B. (1983) A protease-resistant protein is a structural component of the scrapie prion. *Cell* 35, 57–62.

Merz, P.A., Somerville, R.A., Wisniewski, H.M. and Iqbal, K. (1981) Abnormal fibrils from scrapie-infected brain. *Acta Neuropathology* 54, 63–74.

Millot, P., Chatelain, J., Dautheville, C., Salmon, D. and Cathala, F. (1988) Sheep major histocompatbility (OLA) complex: linkage between a scrapie susceptibility/resistance locus and the OLA complex in Ile-de-France sheep progenies. *Immunogenetics* 27, 1–11.

Nussbaum, R.E., Henderson, W.M., Pattison, I.H., Elcock, N.V. and Davies, D.C. (1975) The establishment of sheep flocks of predictable susceptibility to experimental scrapie. *Research in Veterinary Science* 18, 49–58.

Owen, F., Poulter, M., Lofthouse, R., Collinge, J., Crow, T.J., Risby, D., Baker, H.F., Ridley, R.M., Hsiao, K. and Prusiner, S.B. (1989) Insertion in prion protein gene in familial Creutzfeldt–Jakob disease. *Lancet* i, 51–2.

Parry, H.B. (1984) *Scrapie.* Academic Press, London.

Pattison, I.H. (1966) The relative susceptibility of sheep, goats and mice to two types of the goat scrapie agent. *Research in Veterinary Science* 7, 207–12.

Pattison, I.H. (1974) Scrapie in sheep selectively bred for high susceptibility. *Nature* 248, 594–5.

Prusiner, S.B. (1982) Novel proteinaceous infectious particles cause scrapie. *Science* 216, 136–44.

Race, R.E., Graham, K., Ernst, D., Caughey, B. and Chesebro, B. (1990) Analysis of linkage between scrapie incubation period and the prion protein gene in mice. *Journal of General Virology* 71, 493–7.

Scott, M., Foster, D., Mirenda, C., Serban, D., Confal, F., Walchli, M., Torchia, M., Groth, D., Carlson, G., DeArmond, S.J., Westaway, D. and Prusiner, S.B. (1989) Transgenic mice expressing hamster prion protein produce species-specific scrapie infectivity and amyloid plaques. *Cell* 59, 847–57.

Toumazos, P. (1988) First report of ovine scrapie in Cyprus. *British Veterinary Journal* 1144, 98–9.

Wells, G.A.H., Scott, A.C., Johnson, C.T., Gunning, R.F., Hancock, R.D., Jeffrey, M., Dawson, M. and Bradley, R. (1987) A novel progressive spongiform encephalopathy in cattle. *Veterinary Record* 121, 419–20.

Westaway, D., Goodman, P.A., Mirenda, C.A., McKinley, M.P., Carlson, G.A. and Prusiner, S.B. (1987) Distinct prion proteins in short and long scrapie incubation period mice. *Cell* 51, 651–62.

Wilesmith, J.W., Wells, G.A.H., Cranwell, M.P. and Ryan, J.B.M. (1988) Bovine spongiform encephalopathy: epidemiological studies. *Veterinary Record* 123, 638–44.

Section 6
Breeding for Resistance to Bacterial/Production Diseases

Lameness and mastitis are complex disorders of farm animals, which often involve severe bacterial infections. The economic cost to the agricultural industry, and the management problems stemming from these disorders is enormous. In spite of the decades of research and of the investment of millions of pounds into treatment and drug therapy, the problems are intransigent and increasing in incidence. This section provides an excellent review of the nature of the problems, their complex causation and of the, as yet, relatively modest attempts to find an alternative way to combating the problems through breeding for resistance.

Resistance to 'footrot' in sheep varies both between and within breeds but it is not yet clear to what extent resistance can be incorporated into sheep improvement schemes either in terms of minimizing its incidence or increasing responsiveness to vaccination. The review reported here will help progress in this field.

Lameness in cows is possibly even more complex than in sheep and is clearly a production disease with multifactorial causes. The detailed description of the causes of lameness and of the heritability of the component traits suggest that the time is ripe for a re-evaluation of selection methods in cattle.

Mastitis seems just as intractable as lameness, especially in sheep. However, new understanding of the disease in sheep is discussed with a view to breeding for resistance. In cattle the Swedish initiative in incorporating mastitis resistance in their national breeding scheme is described and provides a model which others may emulate in the future.

One clear triumph of recent research is in the difficult war against *Escherichia coli* diarrhoea in young pigs. The discovery of the *E. coli* K88 receptor gene has enabled the goal of breeding for resistance to be achieved. The fruits of this exciting discovery are still to be reaped.

Chapter 20

Breeding Sheep for Resistance to Footrot

J.R. Egerton and H.W. Raadsma

Department of Animal Health, University of Sydney, Private Mailbag 3, Camden NSW 2570 Australia

Summary

Footrot is a contagious disease of sheep. Its occurrence results from an interaction between a specific transmitting agent, *Bacteroides nodosus*, and other microorganisms, the innate and acquired resistance of the host and the environment in which sheep are kept.

The economic impact of footrot is determined by the prevalence of severe infections and the duration of those infections. The cost of the disease is increased by treatment and control methods which are not completely effective and, under Australian conditions, the decreased market value of all sheep from affected farms. It has been estimated that in New South Wales the annual cost of control of footrot on affected farms is between Aus$5,000 and Aus$8,000 depending on the suitability of different parts of the State for the spread and persistence of footrot. Costs of this magnitude justify alternative approaches to the management of footrot in the Australian sheep industry. We have, therefore, addressed the possibility of incorporating resistance to footrot into breeding plans. Our objectives are either to render sheep sufficiently resistant naturally to footrot so that its economic and social impact is minimal or to breed sheep which are much more responsive to vaccination.

We have reviewed all the information known to us on resistance to footrot and concluded that insufficient data were available to allow inclusion of resistance to footrot as a breeding objective in designed breeding programmes. Information is needed on relative economic values of resistance, scope for genetic improvement between and within breeds and genetic covariance between resistance and other breeding objectives for sheep. We have described a study which addresses these points for Merino sheep in Australia.

Introduction

Breeding sheep for resistance to footrot needs to be considered in the context of how and why the disease develops and its economic significance in flocks. This chapter considers some aspects of the bacterial, host and environmental factors which contribute to its occurrence. Estimates for losses attributable to different forms of the disease are presented and the current knowledge on the scope for genetic improvement is reviewed.

Factors determining the expression of footrot in sheep

Bacterial factors

Footrot of sheep is an infectious and contagious disease resulting from invasion of epidermal tissue of the hooves by a mixed group of bacteria (Egerton *et al.*, 1969; Roberts and Egerton, 1969). An essential component of this mixture is *Bacteroides nodosus*, a Gram-negative anaerobic bacterium which, so far as is known, occurs naturally only in the feet of ruminants affected by footrot. Its capacity for survival in the environment is limited to a few days (Beveridge, 1941).

Because *B. nodosus* is a strict parasite, the initiation of outbreaks requires either the continuing presence of affected animals or introduction of one or more infected animals to a susceptible flock.

Among isolates of *B. nodosus* there is a wide spectrum of virulence. Two virulence determinants for *B. nodosus* have been identified – extracellular fimbriae (pili) and capacity for production of proteases (Skerman, 1989). The fimbriae of *B. nodosus* are antigenically diverse and nine major serogroups have been identified (Claxton *et al.*, 1983). Fimbriae are a major protective antigen but immunity is serogroup specific. The most virulent isolates are highly pilate and also elaborate proteases which are qualitatively distinct from those produced by benign isolates. Virulent isolates cause severe infections in a high proportion of sheep at risk. Many of these severe infections persist for a long period even when the environment is unfavourable for the development of new cases. The prevalence of infection induced by benign isolates can be high, but neither the severity of the disease nor its persistence under dry conditions is as pronounced as in virulent outbreaks (Egerton and Parsonson, 1969). In recent years in Australia other outbreaks of footrot have been observed which are clearly neither benign nor virulent (Stewart *et al.*, 1986). These have been described as intermediate footrot.

Environmental conditions

Water maceration

Before new cases of footrot develop following contact between affected and non-affected sheep, the interdigital skin (IDS) of the latter must be suitably predisposed (Egerton *et al.*, 1969). Under grazing management conditions, predisposition usually follows extensive exposure to wet pasture. The capacity of pasture to provide adequate moisture to induce maceration of the IDS is governed by rainfall, pasture type and probably evaporation rates.

Ambient temperature

There is some evidence that consistently low ambient temperatures inhibit transmission, perhaps through induction of vascular shunts which result in marked decrease in temperature of the extremities of sheep (Graham and Egerton, 1968). By contrast, consistently high ambient temperatures occurring in the absence of adequate precipitation result in dehydration of both pasture and interdigital skin. The latter prevents predisposition and, thus, infection.

Sheep factors

Observations in naturally infected flocks suggest that sheep vary in their resistance to footrot. The fact that the prevalence is seldom 100% suggests that some sheep are less likely to contract footrot. Egerton *et al.* (1983) also found that in outbreaks under field conditions four categories of sheep could be recognized: sheep which never became infected, sheep which were infected but showed signs of spontaneous healing, sheep which responded to therapeutic vaccination, and sheep which remained chronically infected.

Variation in resistance to infectious disease is the consequence of a combination of innate and acquired resistance. In the case of footrot, innate resistance may be responsible for preventing invasion of the epidermis by the bacteria responsible for the disease. The capacity of bacteria to invade the interdigital skin may depend on the response of epithelial cells to chronic exposure to water and environmental bacteria. It is obvious that, in sheep exposed to the same wet environmental conditions, there is considerable difference in the degree of inflammatory response observed (Egerton, unpublished). Whether or not this response is genetically determined is not known. The capacity of *B. nodosus* to invade and multiply in the epidermis of interdigital skin is dependent on skin temperature being above 35°C. Skin temperature in turn is governed by the maintenance of arterial blood flow through the digital arteries. The vascular shunts which restrict blood flow to the extremities of mammals may also be genetically determined. If so, susceptibility to footrot would depend in part on the physiological response of the host to cold environments.

Acquired resistance is a sequel either to naturally acquired infection, colostral transfer of immunity or immunization. There is no evidence that infection with footrot results in immunity to the disease (Egerton and Roberts, 1971; Raadsma and Egerton, unpublished). Resistance to infection follows active immunization (Egerton, 1970). This resistance is usually serogroup specific although Stewart (1989) demonstrated the existence of some cross protection between serogroups. Passively acquired resistance has resulted from injection of immunoglobulins (Egerton and Merritt, 1973) and from the transfer of colostrum of immunized ewes (Claxton, 1981).

The evidence for genetic variation in response to either passive or active immunization against footrot is unclear. All available published information on different aspects of genetic variation to footrot will be reviewed later in this chapter.

Factors determining the expression of footrot in flocks

The expression of footrot in a flock of sheep is the consequence of the interaction of all those factors (see above) which affect its occurrence, its duration and its severity.

Prevalence of footrot

If one ignores any variation in resistance between flocks, the likely prevalence of sheep affected with footrot in different environments exposed to agents of varying virulence will be as shown in Table 20.1. The prevalence estimates in each circumstance are those likely to be observed where there is no managerial interference with expression of the disease.

Table 20.1. Prevalence of footrot in sheep exposed to *B. nodosus* of different virulence in different environments.

Virulence of footrot	Environmental risk		
	Low	Moderate	High
Benign	5–15	15–25	25–60
Intermediate	10–20	25–45	40–60
Virulent	25–40	45–75	75–90

Prevalence of severe infections

Prevalence of *B. nodosus* infection by itself is not necessarily an indication of the severity and hence the economic impact of the disease. The prevalence of severe cases is more important. A system of scoring severity, based on the progression of infection within and beyond the interdigital skin, has been developed for measuring semi-objectively the effect of footrot in sheep (Egerton and Roberts, 1971). In this system, a score of four is assigned to any digit where infection has progressed to the sensitive laminae underlying hard keratin components of the hoof.

The prevalence of (a) affected sheep and (b) severely affected sheep resulting from infections with different isolates in different environments is suggested in Fig. 20.1.

When large numbers of sheep are involved, field diagnosis must be based on inspection of sufficient sheep to enable confident judgements to be made about the overall severity of the outbreak. Egerton (1989) has suggested that benign, intermediate and virulent footrot be defined and distinguished according to the prevalence of severe infections observed in affected flocks rather than the characteristics of *B. nodosus* isolated from one or more affected sheep.

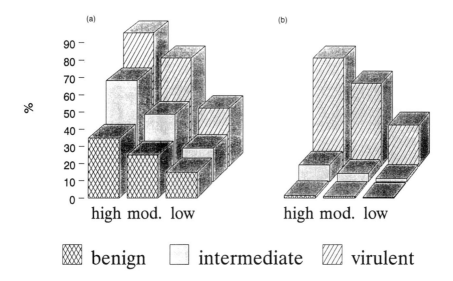

Fig. 20.1. Prevalence of (a) affected sheep and (b) severely affected sheep in low, moderate and high risk environments after outbreaks of different types of footrot.

Economic impact of footrot

Losses from footrot

In Merino flocks, there is evidence that severely affected sheep produce less wool of lower quality. Similarly, body weight is reduced by about 10% (Marshall and Walker, unpublished). Affected ewes are considered less likely to rear lambs. Severely affected feet, apart from causing lameness and/or recumbency, are likely to be infested with *Lucilia cuprina*, the primary sheep blowfly. Sheep with footrot have thus been known to have a higher prevalence of bodystrike which in turn results in more wool loss and increased treatment costs.

Because it is recognized as a serious problem, most owners of affected farms attempt either control of eradication. Traditional systems for footrot management are either laborious or expensive or both and are not completely effective. Before resistance to footrot could be included in breeding plans it is important to evaluate both the genetic variation in resistance and the likely economic and other benefits which would be derived from such an approach. The next section provides our analysis of the economic impact of footrot in New South Wales.

Regional differences in New South Wales

There are approximately 57,000,000 sheep in New South Wales but different regions of the State vary in their suitability for the occurrence and persistence of footrot. It is possible, after considering annual rainfall and ambient temperatures, to divide New South Wales into environmental regions which are low, moderate and high risk for the occurrence of footrot (Fig. 20.2).

Estimates have been made on the flock prevalences of footrot in these different regions (Anon, 1990). These estimates are for properties with either virulent or intermediate footrot. No data are available on the ratio of virulent to intermediate infections among affected flocks but our experience is that it is not less than 4 to 1. Estimates of losses attributable to footrot are calculated on this basis. Using these data and the model for prevalence of infection and severe infections in different environments (Fig. 20.1) it is possible to estimate costs of footrot for affected farms in the different zones of New South Wales (Appendix II, p. 370).

We have recognized that, in the absence of regulatory control, the occurrence of footrot on a farm does not always justify treatment of some form. We have made estimates of the losses likely to arise where owners take one of three options: (a) not to take any action, (b) implement control, or (c) proceed through control to eradication. We have based our estimates of losses on the likely proportion of severe cases arising from different

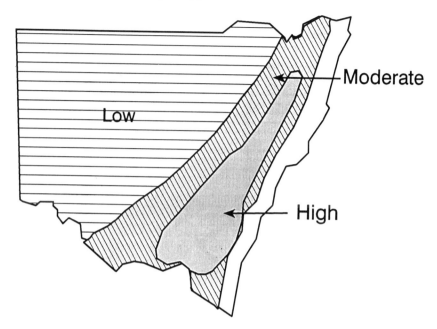

Fig. 20.2. Distribution of low, moderate and high risk environments for the outbreak of footrot in New South Wales, Australia.

forms of the disease in different regions (Fig. 20.1). We accept that fluctuations in climatic conditions and in the value of wool will alter these estimates from season to season. They are, however, the only attempted analysis for the economic justification of breeding for resistance to footrot (Table 20.2). Appendix I (p. 369) presents the assumptions made to estimate the costs presented in Table 20.2.

There are 413, 1,102 and 4,323 affected farms in the low, moderate and high risk zones in New South Wales with a mean number of sheep per property of 1,130, 1,300 and 1,093 respectively. The combined cost of production loss and control annually is thus about Aus$43,000,000 (Appendix II). The cost of footrot to the industry in circumstances where eradication is undertaken is difficult to estimate because there are insufficient data on the likelihood of achieving eradication. The results do demonstrate that in the case of benign footrot the cost of control would exceed that which could be directly attributed to the disease if left untreated, irrespective of the environment in which it is expressed (Table 20.2). It is also clear that the cost of virulent and intermediate footrot could be halved through conventional control techniques but this cost still greatly exceeds the losses associated with benign footrot. Control of virulent footrot is dependent on costly recurrent management techniques which are not fully effective. The scope for improving the resistance of the host to minimize the impact of virulent

Table 20.2. Footrot costs: Aus$ per sheep.

Management options	Environmental risk		
	Low	Moderate	High
(a) No control			
Virulent	6.75	9.60	14.35
Intermediate	2.40	2.95	3.90
Benign	0.05	0.10	0.20
(b) Control			
Virulent	5.45	6.40	8.30
Intermediate	1.20	1.40	2.15
Benign	(1.20)	(1.20)	(1.20)
(c) Eradication			
Virulent	8.45	8.55	9.60
Intermediate	5.51	5.71	5.92
Benign[a]	–	–	–

Figures in parentheses indicate minimum costs for control of footrot with no direct losses attributable to the disease.

[a] Benign footrot is not subject to regulatory control programmes.

footrot through selective breeding has not been seriously considered until recently.

Genetic improvement of resistance to footrot

Breeding strategies should aim at reducing the impact of virulent or intermediate footrot to that of benign footrot so that no specialized control strategies are warranted and the disease has minimum impact on production (Fig. 20.1, Table 20.2). An alternative strategy could be to improve the responsiveness of sheep to vaccination so that footrot could be managed similarly to the clostridial diseases with annual booster vaccinations offering long-term effective protection.

After reviewing all the information known to us on resistance to footrot in sheep it is abundantly clear that the few studies which have been conducted were incomplete, often small and sometimes confounded by other experimental objectives. At present, it is not possible to include resistance to footrot in well-designed breeding programmes due to insufficient information.

Definitions of traits indicating resistance to footrot

On the basis of observations by Egerton *et al.* (1983), Raadsma *et al.* (1990) defined the key indicators of resistance to footrot as:

1. Failure to develop clinical footrot following challenge.
2. Ability to show spontaneous healing.
3. Accelerated healing after therapeutic vaccination.
4. Failure to develop footrot during a challenge period following vaccination.

The appropriate measurements are semi-objective scores of footrot lesions and K-agglutination antibody levels following vaccination (Egerton, 1970; Egerton and Roberts, 1971).

Repeatability of resistance indicators

The accuracy by which we can measure resistance in individuals is determined by two criteria: consistency of the measurement or assay itself, and consistency of the performance of the individual. Information is required in both these areas since the basic measurements of resistance are semi-quantitative, and a complex disease such as footrot is likely to be subjected to considerable environmental influences and decrease the accuracy by which desirable/undesirable phenotypes and, hence, genotypes can be identified.

Accuracy of measurement

The scoring of sheep for clinical footrot has been the fundamental measurement used in many comparative studies. Yet the accuracy by which we can assess sheep is often ignored. On small subsamples from larger flocks we have shown our footrot grading system for individual feet to be highly repeatable for the same observer and between observers. Good positioning and lighting are the key variables which influence the accuracy of scoring.

Schwartzkoff and Handley (1985) discussed major sources of variation in the assessment of K-agglutinin assays. The major sources of variation in the assay, cell concentrations, quality and type were discussed, but their effect on accuracy of measurement was not quantified. Of some concern was the large inconsistency between different laboratories performing assays on identical sera (Schwartzkoff and Handley, 1985). Unfortunately, no estimates of reproducibility between assays within the laboratory were reported. We have been able to standardize the technique sufficiently well to obtain high correlations for repeat observations between and within observers ($r > 0.90$) and between repeat assays on the same samples ($r > 0.90$). It is, therefore, possible to rank accurately the titre of individual sheep with the same flock for the same serogroup. It is not known, however, if protective antibody levels are the same for different serogroups of *B. nodosus* or for individual sheep. Furthermore, it may be difficult directly to equate K-agglutination Ab levels across different serogroups.

Repeatability of animal performance

Within seasons of uncontrolled outbreaks of footrot, a range of events occurs which gives rise to a dynamic change of footrot scores within a flock (Egerton *et al.*, 1983). As a consequence, assessment of footrot at one particular time may be poorly correlated with later footrot scores or even previous footrot scores. Indeed, Norman and Hohenboken (1979) reported the repeatability of successive foot soundness scores over a 2-year period to be extremely low (0.0 − 0.02). Unfortunately, their assessment of foot soundness included a large number of variables/traits other than footrot. More recently, Skerman *et al.* (1988), in a large footrot study in New Zealand Romney sheep, reported a repeatability of footrot status of approximately 0.10 ± 0.02 for successive inspections spaced 6–12 months apart. Within a semicontrolled outbreak of footrot, in which footrot was experimentally induced at the same time, the repeatability was higher for successive readings taken at shorter 3-week time intervals ($r = 0.35$) (Raadsma and Egerton, unpublished data). It is thus necessary to take repeat readings when monitoring the dynamic footrot status of a flock and/or individual sheep.

Repeatability of foot scores between feet within an observation period could also provide some estimate of how consistent an animal's 'resistance' or 'susceptibility' to the disease might be reflected by individual foot scores. Parker *et al.* (1983) reported a correlation of approximately 0.40 for two feet subjected to the same challenge. Data obtained by Marshall (personal communication) and by us, suggests that the intraclass correlation for footrot under both field conditions and experimental challenge is slightly lower at 0.2 − 0.34. This suggests that feet under footrot challenge do not behave independently and measures such as '% feet infected' in field or clinical trials are not a valid means of expressing amount of footrot infection or making comparisons between treatments. The low correlation on the other hand suggests that all four feet should be used in assessing the status of individual animals. The reason for the apparent low repeatability is not known, but it suggests that factors at the local level either environmental (microflora, microenvironmental) or host mediated (i.e. skin inflammation, local cellular responses etc.) have a strong influence on the expression of footrot.

Of more interest in the repeatability of resistance to footrot is how consistently the performance of an animal can be assessed upon repeated, independent challenge with footrot. This measure would indicate the true repeatability of the trait (t) and would set the upper limit of the heritability of resistance to footrot. There have been no specific studies on this criterion published. Our preliminary analyses suggest that the repeatability of performance is 0.15 ± 0.05. Similarly, for vaccine responsiveness no estimates of the repeatability of vaccine response have been published. The use of repeatability experiments is unfortunately difficult since the second challenge/response is always confounded by the first and, particularly in the case of vaccine response, considerable carryover (memory) effects are likely.

Closely related to the repeatability of an individual's performance is the question of how well resistance involving one serogroup of *B. nodosus* indicates resistance to the other eight serogroups. Once again, no information has been published to describe intraclass correlations at this level.

Potential genetic differences in resistance to footrot

Within sheep populations it is possible to group potential sources of genetic variation in resistance to disease

1. between breeds,
2. between strains within breeds,
3. between bloodlines (stud lines) within strains,
4. between families (sire lines) within bloodlines (flocks),
5. between individuals within sirelines or flocks.

Although for most major production traits the variation between these sources is reasonably well known, for footrot there is a paucity of relevant information.

Differences between breeds

When making inferences about differences in the susceptibility of breeds to footrot it is important to make them on well-designed comparisons. Given the nature of the disease, it is obvious that nothing can be concluded from isolated studies involving a single breed. The few remaining published studies involving multiple breeds also show major deficiencies because the number of sires used to generate the 'breed sample' was either too small or not specified. It is thus highly likely that suggested breed differences in susceptibility to footrot and response to vaccination may have been due to genetic differences between sires used to generate the experimental sample. This is likely to be the case in studies such as those reported by Skerman *et al.* (1982), Emery *et al.* (1984), Stewart *et al.* (1985), Lewis *et al.* (1988), where the number of animals/breed ranged from 4 to 45.

In an abstract published by Baker *et al.* (1986) it was suggested that both

Chapter 20

Romney and Perendale purebred sheep were more resistant than their respective crosses involving the Merino breed. This would support observations by Egerton *et al.* (1972) that British breeds crossed with Merinos were in turn more resistant than purebred Merino sheep.

In an adequately designed breed comparison involving two dam breeds and four sire breeds Norman and Hohenboken (1979) could not report a consistent breed difference for foot soundness scores under repeated field observations. Unfortunately, their assessment of foot soundness included other foot problems which should have been distinguished from footrot. The poor definition of resistance to footrot also precludes conclusive evidence of the breed comparison reported by Cumlivski (1988).

Differences between strains and bloodlines

Within the Australian Merino breed a number of distinct strains exist. These consist of the fine-wooled strain, developed in high rainfall areas of Australia, a medium-wooled (non-Peppin) strain, using the same source of foundation stock as the fine wools and developed in slightly drier areas of Western New South Wales. A second major medium-wooled strain, also developed in New South Wales, known as the Peppin strain, was derived separately from the non-Peppin strain. Finally, South Australian strong-wool strain was developed in that state from Merino-based crossings and selective breeding. As the basis of classification suggests, these strains have retained their major Merino breed characteristics, but vary considerably in the average fibre diameter of their fleece wool. As a consequence of the strain developments, the strains also vary considerably in other major production characteristics, including fleece weights, fleece length, body weight and resistance to disease such as fleece rot and susceptibility to flystrike (Atkins and McGuirk, 1979; Mortimer and Atkins, 1989; Raadsma and Rogan, 1987; Raadsma, 1991).

There is no information on how these strains differ in their resistance to footrot, although the fine-wool and medium-Peppin strains were developed in areas of high and moderately high footrot risk, whereas the strong-wool strain, in particular, was developed in dry, low risk footrot areas in South Australia. Emery *et al.* (1984) and Stewart *et al.* (1985) report on a small comparison involving a fine-wool flock and a medium Peppin flock but could not detect a consistent response to either footrot or vaccination, which could be separated from the sire effects 'within' the strains. Further studies on the comparison of strain differences are needed since it has proved to be a major source of variation within the Merino breed for important production characteristics.

Within strains of Merinos a number of distinct bloodlines, or studlines, can be identified. The main characteristic of these bloodlines is their base foundation within one of the major strains. However, many bloodlines have been kept genetically distinct and evolved under specific selection programmes. Combined with the effect of genetic drift, large differences have been reported between bloodlines, managed under the same environmental conditions, in almost all major production characteristics (Atkins and McGuirk, 1979; Mortimer and Atkins, 1989) and resistance to some diseases (Atkins and McGuirk, 1979; Raadsma and Rogan, 1987). For the Merino breed no estimates have been published on possible differences between bloodlines in their susceptibility to footrot.

Skerman (1985), Skerman and Moorhouse (1987) report on the development of bloodlines in the Romney Marsh and Corriedale breeds with increased resistance to footrot for each of the two breeds, respectively. Both bloodlines evolved through direct selection under natural outbreaks of footrot and extensive use of sires whose progeny showed increased resistance over their contemporaries. Both studs also claim now that footrot is an insignificant problem in their flocks (Skerman and Moorhouse, 1987; Warren *et al.*, 1990). Although reports of this nature highlight the potential for genetic control of this disease, formal genetic comparisons such as those described by Skerman and Moorhouse (1987) are needed, but are usually lacking, in on-property experiments. The follow-on benefit of increased usage of breeding stock from more resistant bloodlines in the Merino industry still awaits evaluation, but has already prompted certain stud breeders to place heavy selection pressure on the search for, and development of, bloodlines with increased resistance (Patterson and Patterson, 1989).

Differences between sheep within flocks

Phenotypic

Differences between sheep have been reported for all traits indicating resistance to footrot in a wide range of breeds and a wide range of *B. nodosus* serogroups. These results arise from field observations, where it was often reported that a sample of animals failed to become affected following natural outbreaks or as control sheep in experiments evaluating vaccine efficacy. Similarly, considerable variation has been reported between agglutinin titres of sheep following vaccination.

Genetic variation

The estimation of additive genetic variance (heritability) of resistance to footrot has received very little scientific attention. It was not until recently that genetic differences between sire lines were reported in studies by Parker *et al.* (1983, 1985), Skerman and Moorhouse (1987) and Bulgin *et al.*

(1988). Baker *et al.* (1986), Skerman (1985) and Skerman *et al.* (1988) reported the first heritability estimates for resistance to footrot (Table 20.3).

At this point some of the problems associated with traits describing resistance to footrot become clear. First, to diagnose 'footrot' only when underrunning of soft horn occurs and 'footscald' when only interdigital skin inflammation has been recorded is erroneous, unless individual intradigital lesions were specifically confirmed to be free from *B. nodosus* infection. Failure to recognize this mars previous studies. Second, it is usual to present heritability estimates for prevalence of the condition. It is thus likely that differences in heritability estimates are a function of the mean prevalence rather than a real difference in magnitude of genetic variation. Where appropriate, it is possible to obtain estimates of the heritability of liability to footrot, independent of prevalence of the condition. It is normal to convert the h^2 of prevalence to h^2 of liability by $z^2/(p.q)$ where z is the ordinate of a normal distribution above which lies a proportion of 'p' affected animals (i.e. prevalence) (Falconer, 1981). It is also possible to obtain h^2 estimates of liability directly on the underlying scale by procedures described by Gilmour *et al.* (1985).

Table 20.3 describes the known heritability estimates of two different

Table 20.3.   Heritability ($\pm$ SE) estimates for resistance to footrot reported in three studies published to date.

	Study 1	Study 2	Study 3
Breed	–	–	Romney
Affected			
Prevalence	–	0.40	–
h^2 prevalence	0.53	0.34 ($\pm$ 0.16)	–
h^2 liability	–	0.55 ($\pm$ 0.26)	0.28
Severely affected			
Prevalence	–	0.23	–
h^2 prevalence	0.23	0.03 ($\pm$ 0.12)	–
h^2 liability	–	0.07 ($\pm$ 0.29)	0.17
Footrot severity	–	–	
h^2 score	–	–	0.14 $\pm$ 0.03

Source: Study 1, Alwan (1983).
            Study 2, Baker *et al.* (1986).
            Study 3, Skerman *et al.* (1988).

levels of severity of footrot expressed as either lesions with or without interdigital skin inflammation and lesions with or without underrunning of the soft and hard horn of the foot.

Current investigations for Merino sheep

For Merino sheep, Raadsma *et al.* (1990) outlined a study which has specifically been designed to obtain estimates of the heritability of resistance to footrot for a range of conditions, including experimental challenge, natural development on pasture and following vaccination. In brief, the experiment utilizes the half-sib design where approximately 1,200 progeny from 120 sires will be screened for resistance to footrot (both clinical scores and antibody levels), for 6 months following experimental challenge with a single isolate of *B. nodosus* of moderate virulence. After 2 months of natural development of the disease (three inspections) all sheep are vaccinated with an homologous recombinant DNA pilus vaccine. Healing and persistence of antibody titres are monitored for a further 4 months after vaccination. Vaccine responses to decavalent whole cell vaccines are monitored in non-challenged half-sibs ($n = 1,200$).

Preliminary results from our first challenge in 1989 included 410 progeny from 38 sires. Prevalence of footrot is shown in Fig. 20.3.

It was evident that not all sheep succumbed to footrot after initial challenge. Resistant sheep remained free from footrot for the duration of the experiment, susceptible sheep on the other hand showed clinical signs of severe footrot at all inspections. Sheep with intermediate resistance contracted footrot but were able to overcome the disease through self cure or after vaccination.

The heritability of resistance was estimated at 0.31 ± 0.15 directly after challenge. Subsequent observations of the sheep on pasture resulted in a lower heritability in the range of 0.04 to 0.11 ± 0.12, reflecting increased variability, possibly due to non-genetic (environmental) influences. Over the entire observation period the heritability of footrot was 0.18 ± 0.10. We intend to repeat this for a further 2 years to screen another 800 progeny from a further 80 sires to obtain accurate estimates of the relevant genetic parameters.

From our first large-scale challenge in 1989 it was evident that homologous vaccine was highly effective in promoting healing in infected sheep. The prevalence of footrot decreased from over 60% to less than 5% after vaccination in 1989 (Fig. 20.3). Despite the effectiveness of vaccination, some sheep failed to respond and remained chronically infected. This residual group of sheep is believed to be highly susceptible and unable to reject the disease even after immunostimulation. The heritability of the healing response to footrot following vaccination was $0.19 \pm$

Chapter 20

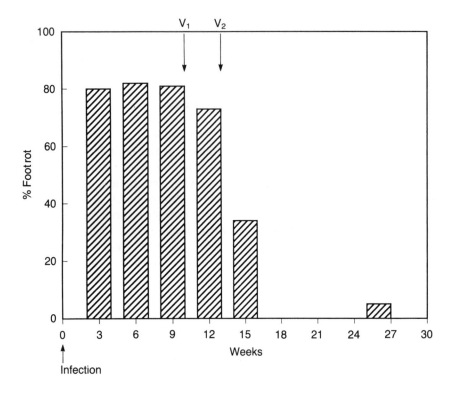

Fig. 20.3. Change in footrot prevalence (%) in Merino hoggets screened for resistance following experimental challenge with footrot (weeks) and two vaccinations (V_1, V_2) with an homologous rDNA pilus vaccine. (Adapted from Raadsma *et al.*, 1990.)

0.14 for the first group of 410 progeny tested by us (Raadsma *et al.*, 1990).

Variation in antibody titres during the experiment and between sheep is shown in Figs 20.4 and 20.5(a–c) respectively. During the first 9 weeks after challenge, the small rise in antibody titre was related to the duration and severity of infection. It is thought to be another indicator of susceptibility to footrot with higher titres indicating more susceptible sheep (Raadsma *et al.*, 1990). Elevated antibody levels after vaccination, on the other hand, greatly exceed those normally seen during infection. Peak titre responses and persistence of antibody titres are important indicators of vaccine efficacy.

At this stage, it can be stated that estimates of the heritability of resistance to footrot are scarce and those which exist are derived from studies of questionable validity. In due course we hope to provide more reliable estimates for Merino sheep.

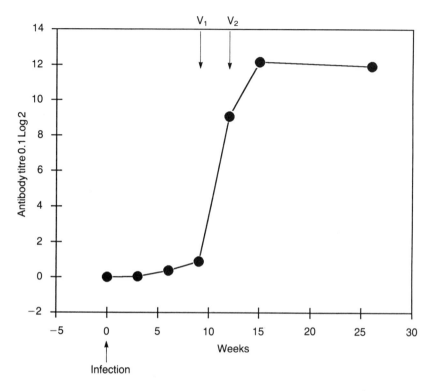

Fig. 20.4. Change in K-agglutination antibody titre (0.1 log2) in Merino hoggets following experimental challenge with footrot (week 0) and two vaccinations (V_1, V_2) with an homologous rDNA pilus vaccine. (Adapted from Raadsma *et al.*, 1990.)

Genetic markers for resistance

The option of exploiting genetic differences in footrot through direct selection is clearly not practical as it would require challenging all animals before selection. We are therefore examining the possibility of using indirect selection strategies. To minimize the effect of challenge on vaccine responsiveness, we are monitoring serological responses in the non-challenged progeny group following a full course of vaccination with a decavalent vaccine. Routine serological procedures based on K-agglutination and ELISA techniques form the basis of this screen. The other options for indirect selection strategies include the use of genetic markers linked to resistance.

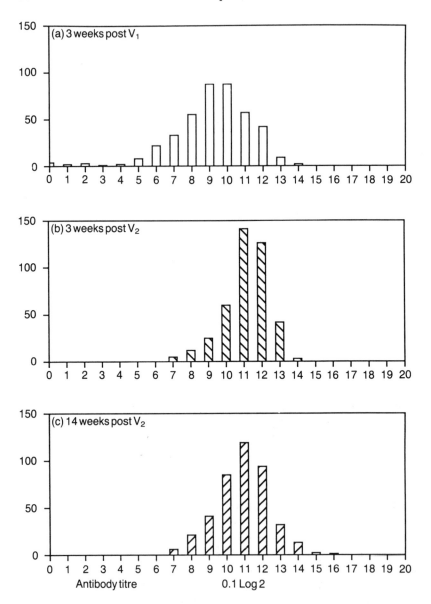

Fig. 20.5. Distribution of sheep vaccine responses (N) in (a) primary – 3 weeks post V_1, (b) secondary – 3 weeks post V_2, and (c) persistence – 12 weeks post V_2, of K-agglutination antibody titres (0.1 log2) following vaccination with an experimental monovalent rDNA pilus vaccine (Adapted from Raadsma *et al.*, 1990.)

OLA typing

The role of the major histocompatibility complex (MHC) in modulating immune responses and subsequently disease resistance is well documented for a number of species. Ovine lymphocyte histoglobulins (OLA) are glycoproteins which are present on the surface of most cells and are divisible into two types, class I and class II histoglobulins and are controlled by genes of the MHC. Class I MHC histoglobulins can now be serologically typed, utilizing a panel of 17 typing sera as outlined by Outteridge *et al.* (1985). Two particular histoglobulins are of interest, namely SY1b and SY6, since they have been implicated in both resistance to footrot and responses following vaccination in previous studies using different flocks (Outteridge *et al.*, 1989). However, a comprehensive analysis awaits the evaluation of OLA typing before any recommendations can be made on the use of OLA markers and resistance to footrot.

The use of OLA typing has some drawbacks in that it is a difficult test which has proved difficult to standardize. The fact that cells have to be fresh and that only a limited number of sheep can be typed at one time might limit final application in the field. Although class II MHC genes are believed to have a key role in the regulation of immune responses, it has proved even more difficult to type animals for class II histoglobulins with a serological assay. The use of cDNA probes for class II genes from human, mouse and other species has opened the possibility of screening the sheep genome for corresponding class II genes.

RFLP analysis

Analysis of restriction fragment length polymorphism (RFLP) of genomic DNA offers a routine, straightforward and accurate method for detecting genetic variation within the MHC which is much favoured over serological analysis. The basic technique for characterizing genomic DNA from sheep is now well established.

A member of our group (A. Litchfield), has established the basic procedures of this technique in 84 progeny from a pilot experiment. A significant association was detected between increased susceptibility to footrot and the presence of a 14.5 kb band following restriction with TaqI enzyme and probing with a human class II MHC probe (DR beta). Full details of these results will be presented elsewhere (Litchfield *et al.*, in preparation).

It may thus be possible in due course to screen important genes directly at the DNA level, not only for genetic variation within the MHC, or for resistance to footrot, but for other major disease and, possibly, production traits. There simply is no other potential indirect selection technique which can offer this power of screening sheep for a large number of disease resistance or production traits.

Resistance and other breeding objectives

Should it prove to be feasible to improve resistance to footrot through selective breeding, it is unlikely to be the sole breeding objective in well-designed programmes. It is important to have accurate information on resistance to footrot and all the major production traits which are recommended as breeding objectives and/or selection criteria for sheep and wool production. Programmes of this nature are particularly relevant in high rainfall areas where resistance to other important diseases such as flystrike and internal parasites may also need to be considered.

No estimates of the relevant phenotypic or genetic parameters are available for any of these diseases or their relationship to production traits. Our own project and those of Gray (1991) and Windon (1991) are designed to provide answers to these important questions.

References

Alwan, M. (1983) 'Studies of the flock mating performance of Booroola Merino crossbred ram lambs and the foot conditions in the Booroola Merino crossbreds and Perendale sheep grazed on hill country.' M.Agric.Sci.Thesis, Massey University.

Anon (1990) Working Paper, Footrot Strategic Plan Steering Committee. Department of Agriculture and Fisheries, Sydney.

Atkins, K.D. and McGuirk, B.J. (1979) Selection of Merino sheep for resistance to fleece rot and bodystrike. *Wool Technology Sheep Breeding* 27, 15–19.

Baker, R.L., Clark, J.N., Harvey, T.G. and Meyer, H.H. (1986) Inheritance of foot and jaw abnormalities in sheep. *Proceedings of the New Zealand Society of Animal Production* 46, 93–100.

Beveridge, W.I.B. (1941) Council for Scientific and Industrial Research, Melbourne, Bulletin 140.

Bulgin, M.S., Lincoln, S.D., Parker, C.T., South, P.J., Dahmen, J.J. and Lane, V.M. (1988) Genetic-associated resistance to footrot in selected Targhee sheep. *Journal of American Veterinary Medicine Association* 192, 512–15.

Claxton, P.D. (1981) 'Studies on *Bacteroides nodosus* vaccines.' PhD thesis, University of Sydney.

Claxton, P.D., Ribeiro, L.A. and Egerton, J.R. (1983) Classification of *Bacteroides nodosus* by agglutination tests. *Australian Veterinary Journal* 60, 331–4.

Cumlivski, B. (1988) Resistance of sheep kept together of the breeds Kent (Romney Marsh), Lincoln, Cotswold and Leicester. In: *Proceedings, 3rd World Congress on Sheep and Beef Cattle Breeding*, vol. 1, Paris, France.

Egerton, J.R. (1970) Successful vaccination of sheep against footrot. *Australian Veterinary Journal* 46, 114–15.

Egerton, J.R. (1989) Control and eradication of footrot at the farm level – the role of veterinarians. *Proceedings of the Second International Congress for sheep veterinarians*, Palmerston North, pp. 215–22.

Egerton, J.R. and Merritt, G.C. (1973) Serology of footrot: antibodies against *Fusiformis nodosus* in normal affected vaccinated and passively immunised sheep. *Australian Veterinary Journal* 49, 139–45.

Egerton, J.R. and Parsonson, I.M. (1969) Benign footrot – a specific interdigital dermatitis of sheep associated with infection by strains of *Fusiformis nodosus*. *Australian Veterinary Journal* 45, 345–9.

Egerton, J.R., Morgan, I.R. and Burrell, D.H. (1972) Footrot in vaccinated and unvaccinated sheep. Incidence, severity and duration of infection. *Veterinary Record* 91, 447–53.

Egerton, J.R., Ribeiro, L.A., Kieran, D.J. and Thorley, C.M. (1983) Onset and remission of ovine footrot. *Australian Veterinary Journal* 60, 334–6.

Egerton, J.R. and Roberts, D.S. (1971) Vaccination against ovine footrot. *Journal of Comparative Pathology* 81, 179–85.

Egerton, J.R., Roberts, D.S. and Parsonson, I.M. (1969) The aetiology and pathogenesis of ovine footrot. I. A histological study of the bacterial invasion. *Journal of Comparative Pathology* 79, 207–16.

Emery, D.L., Stewart, D.J. and Clark, B.L. (1984) The comparative susceptibility of five breeds of sheep to footrot. *Australian Veterinary Journal* 61, 85–8.

Falconer, D.S. (1981) *Introduction to Quantitative Genetics* 2nd edn. Longman, London, pp. 340.

Gilmour, A.R., Anderson, R.D. and Rae, A.L. (1985) The analysis of binomial data by a generalised linear mixed model. *Biometrika* 72, 593–9.

Graham, N.P.H. and Egerton, J.R. (1968) Pathogenesis of ovine footrot: the role of some environmental factors. *Australian Veterinary Journal* 44, 235–40.

Gray, G.D. (1991) Breeding for resistance to trichostrongyle nematodes in sheep. In: Owen, J.B. and Axford, R.F.E. (eds), *Breeding for Disease Resistance in Farm Animals*. CAB International, Wallingford, pp. 139–61.

Lewis, R.D., Meyer, H.H., Gradin, J.L. and Smith, A.W. (1988) Effectiveness of vaccination in controlling ovine footrot. *Journal of Animal Science* 67, 1160–6.

Mortimer, S.I. and Atkins, K.D. (1989) Genetic evaluation of production traits between and within flocks of Merino sheep. I. Hogget fleece weight, body weight and wool quality. *Australian Journal of Agricultural Research* 40, 433–43.

Norman, L.M. and Hohenboken, W. (1979) Genetic and environmental effects on internal parasites, foot soundness and attrition in crossbred ewes. *Journal of Animal Science* 48, 1329–37.

Outteridge, P.M., Stewart, D.J., Skerman, T.M., Duffy, J.H., Egerton, J.R., Ferrier, G. and Marshall, D.J. (1989) A positive association between resistance to ovine footrot and particular lymphocyte antigen types. *Australian Veterinary Journal* 66, 175–9.

Outteridge, P.M., Windon, R.G. and Dineen, J.K. (1985) An association between a lymphocyte antigen in sheep and the response to vaccination against the parasite *Trichostrongylus colubriformis*. *International Journal of Parasitology* 15, 121–7.

Parker, C.T., Cross, R.T. and Hamilton, K.L. (1983) Selection for footrot resistance in Targhee sheep. *Journal of Animal Science* 62, 164.

Parker, C.T., Cross, R.T. and Hamilton, K.L. (1985) Genetic resistance to footrot in sheep. *Proceedings of the Sheep Veterinary Association*, vol. 9, pp. 16–19.

Patterson, R.G. and Patterson, H.M. (1989) A practical approach to breeding footrot resistant Merinos. Review, *Journal of the New Zealand Mountain Lands Institute* 46, 64–75.

Raadsma, H.W. (1991) Genetic variation in resistance to fleece rot and flystrike in sheep. In: Owen, J.B. and Axford, R.F.E. (eds), *Breeding for Disease Resistance in Farm Animals.* CAB International, Wallingford, pp. 263–90.

Raadsma, H.W., Egerton, J.R., Outteridge, P.M., Nicholas, T.W., Brown, S.C. and Litchfield, A.M. (1990) An investigation into genetic aspects of resistance to footrot in Merino sheep. *Wool Technology and Sheep Breeding* 38, 7–12.

Raadsma, H.W. and Rogan, J.M. (1987) Genetic variation in resistance to blowfly strike. In: McGuirk, B.J. (ed.), *Merino Improvement Programmes in Australia Proceedings, National Symposium,* Leura, pp. 321–40.

Roberts, D.S. and Egerton, J.R. (1969) The aetiology and pathogenesis of ovine footrot. II. The pathogenic association of *Fusiformis nodosus* and *F. necrophorus. Journal of Comparative Pathology* 79, 217–26.

Schwartzkoff, C.L. and Handley, B.G. (1985) Analytical criteria for footrot vaccines. In: Stewart, D.J., Petersen, J.E., McKern, N.M. and Emery, D.L. (eds), *Footrot in Ruminants, Proceedings of Workshop,* Melbourne, pp. 181–5.

Skerman, T.M. (1985) In: Stewart, D.J., Petersen, J.E., McKern, N.M. and Emery, D.L. (eds), *Footrot in Ruminants, Proceedings of Workshop,* Melbourne, pp. 77–8.

Skerman, T.M. (1989) Isolation and identification of *Bacteroides nodosus.* In: Egerton, J.R., Yong, W.K. and Riffkin, G.G. (eds), *Footrot and Foot Abscess of Ruminants,* CRC Press, Boca Raton, Florida, pp. 85–105.

Skerman, T.M., Erasmuson, S.K. and Morrison, L.M. (1982) Duration of resistance to experimental footrot infection in Romney and Merino sheep vaccinated with *B. nodosus* oil adjuvant vaccine. *New Zealand Veterinary Journal* 30, 27–31.

Skerman, T.M., Johnson, D.L., Kane, D.W. and Clarke, J.N. (1988) Clinical footscald and footrot in a New Zealand Romney flock; phenotypic and genetic parameters. *Australian Journal of Agricultural Research* 39, 907–16.

Skerman, T.M. and Moorhouse, S.R. (1987) Broomfield Corriedales: a strain of sheep selectively bred for resistance to footrot. *New Zealand Veterinary Journal* 35, 101–6.

Stewart, D.J. (1989) Vaccination against footrot and foot abscess. In: Egerton, J.R., Yong, W.K. and Riffkin, G.G. (eds), *Footrot and Foot Abscess of Ruminants,* CRC Press, Boca Raton, Florida.

Stewart, D.J., Emery, D.L., Clark, B.L., Peterson, J.E., Iyer, H. and Jarrett, R.G. (1985) Differences between breeds of sheep in their response to *Bacteroides nodosus* vaccines. *Australian Veterinary Journal* 62, 116–20.

Stewart, D.J., Petersen, J.E., Vaughan, J.A., Clark, B.L., Emery, D.L., Caldwell, J.B. and Kortt, A.A. (1986) The pathogenicity and cultural characteristics of virulent, intermediate and benign strains of *Bacteroides nodosus* causing ovine footrot. *Australian Veterinary Journal* 63, 317–26.

Warren, H., Daniel, W. and Parker, T. (1990) Where are we at in sheep breeding? *Wool Technology and Sheep Breeding* 38, 27–8.

Windon, R.G. (1991) Genetic control of host responses involved in resistance to gastrointestinal nematodes of sheep. In: Owen, J.B. and Axford, R.F.E. (eds), *Breeding for Disease Resistance in Farm Animals,* CAB International, Wallingford, pp. 162–86.

APPENDIX I

Details of costs and assumptions associated with footrot in sheep

(a) Cost of footrot per affected sheep. Prevalence will vary according to conditions shown in Fig. 20.1.

10% reduction in wool production	$ 2.00
15% flystrike in affected sheep	
@ Aus$5.00/struck sheep	$ 0.75
10% reduction in weaning rate	$ 2.00
Increased feeding cost for affected sheep	$ 0.25
Decrease sale value of diseased sheep	$ 14.00
	$ 19.00

(b) Cost of footrot for all sheep on properties affected with footrot, irrespective of prevalence.

Reduced market values of sheep for sale	$ 2.00
Vaccination if in control programme (2 ×)	$ 2.00
Labour (2 × mustering for vaccination)	$ 0.50
Labour entailed in eradication (4 × mustering)	$ 1.00
Reduced stocking rate for sheep in eradication	
programmes (virulent only) (0.4 DSE/ha)	$ 2.00

(c) Variable cost of control affected sheep.

Antibiotic treatment if prevalence > 10%	$ 1.00
Footbathing if prevalence < 10% (3 ×)	$ 1.20

(d) Effectiveness of control is at best 70%, leaving a proportion of severely affected sheep despite treatment and/or prevention.

(e) Thus, for virulent footrot in a high risk environment the cost of footrot would be

(i) No control – prevalence (65% severely infected)	
cost/affected sheep = 0.65 × Aus$19.00	$ 12.35
cost/sheep (lost market opportunity)	$ 2.00
Cost per sheep	$ 14.35

(ii) Control – 2 vaccinations, reduction in prevalence from
65% → 20%

cost/affected sheep = 0.20 × Aus$19.00	$ 3.80
cost/sheep (lost market opportunity)	$ 2.00
cost of control – all sheep vaccinated	$ 2.50
Cost per sheep	$ 8.30

(iii) Eradication – 2 vaccinations, + 1 × antibiotics + 3 inspections,
reduction in prevalence (65% → 10% – culls)

cost/affected sheep = 0.10 × Aus$21.00	$ 2.10
vaccination	$ 2.50
eradication (labour for inspection)	$ 1.00
reduced market opportunities	$ 2.00
reduced stocking rate (eradication)	$ 2.00
Cost per sheep	$ 9.60

APPENDIX II

Estimated cost (Aus$) of controlling virulent and intermediate footrot on affected farms in New South Wales, Australia

	Suitability of environment for footrot		
	Low	Moderate	High
Affected properties[a]	413	1102	4323
Sheep population on affected properties[a]	567,000	1,431,250	4,729,000
Average sheep population[a]	1373	1300	1094
Cost/sheep[b]	$ 3.80	$ 5.40	$ 7.07
Cost/affected farm	$ 5,217	$ 7,020	$ 7,734
Annual loss in NSW	$ 2,154,600	$ 7,728,750	$ 33,434,030

[a] Figures from Anon (1990).
[b] Cost of controlling footrot on affected farms assuming a ratio of 4:1 of virulent to intermediate infections with cost figures from Table 20.2 weighted accordingly.

Chapter 21

A Review of Factors Predisposing to Lameness in Cattle

Paul R. Greenough

Department of Veterinary Anaesthesiology, Radiology and
Surgery, The Western College of Veterinary Medicine,
University of Saskatchewan, Saskatoon, Saskatchewan
S7N 0WO, Canada

Summary

Lameness is the clinical sign of many diseases for some of which there may be a heritable predisposition. Some of the data used in the literature are based on lesions that have been inaccurately determined.

Poor characteristics of limb conformation, claw shape, size and quality may predispose to some diseases or lesions that cause lameness. The evaluation of limb and claw conformation, expressed as 'feet and legs' in the literature, is often based on descriptive traits. These traits may be defined in too general terms which may have led to some studies producing misleading findings.

This chapter reviews the literature and in so doing suggests that because developments in understanding bovine lameness have been so rapid during recent years a re-evaluation of past findings would be appropriate.

Introduction

Lameness is a clinical sign or symptom of a disorder that causes a disturbance in locomotion. This disturbance is observed as an aberration of gait that may be caused by metabolic or systemic disturbance, injury to the musculoskeletal system, or infection.

In Europe the economic importance of lameness is now recognized to have reached a serious level. One worker estimated that the average annual financial loss due to hoof problems in the United Kingdom amounts to approximately $2,000 (Canadian) for every 100 cows (Whitaker *et al.*, 1983). Other workers in the UK have estimated that lameness costs

£15,000,000 annually (Baggott and Russell, 1981; Pinsent, 1981). Digital disease in dairy cattle in Quebec is estimated to cost $10,000,000 per annum (Choquette-Levy *et al.*, 1985).

A historical perspective on the evolution of studies on bovine locomotory disorders

Progress in understanding the causes of lameness in cattle has been slow and is still, in 1990, very incomplete.

A simple review of current knowledge about lameness was published in English by Greenough in 1962 and was followed by Nilsson's elegant monograph on laminitis in 1963. The 1960s became a period during which veterinarians grappled with new concepts concerning the pathogenesis of claw disease in individual animals (Maclean, 1965, 1966; Morrow, 1966; Smedegaard, 1964a, b). The first English language text on lameness in cattle was by Greenough *et al.* (1972). Also during the 1970s several other reports (Maclean, 1971b; Toussaint-Raven, 1971), were published. However, the most significant event of the decade was the First Symposium on Disorders of the Ruminant Digit which was held in Utrecht in 1976. Symposia have been held approximately every two years since, the sixth occurring in Liverpool in 1990 and they have been responsible for the introduction of accurate and appropriate terminology for digital diseases (Espinasse *et al.*, 1984), meaningful terminology for digital anatomy (Greenough, 1978) and an examination for the descriptive terminology for hoof deformities (Greenough, 1980).

In the early 1980s, Toussaint-Raven introduced the important hypothesis that many diseases, previously believed to be specific conditions that occurred randomly, were more frequently encountered in animals affected with chronic laminitis. Peterse (1979, 1982, 1986a, 1987) and Peterse *et al.* (1984, 1986) working at Utrecht have helped to substantiate the nutritional aetiology of the disease which has become known as the 'subclinical laminitis syndrome' (Greenough, 1985). In brief, it is believed that intensive management, particularly involving nutrition, interferes with the blood supply in the digit (Boosman and Mutsaers, 1988). Bone and tissue abnormality results (Maclean, 1970, 1971a). The histopathological changes compromise the quality of the claw horn produced and renders it more susceptible to 'specific' diseases.

Once the concept of subclinical laminitis became accepted, not only was the earlier literature reviewed but new histopathological studies were conducted (Anderssen and Bergman, 1980; Boosman *et al.*, 1989; Mortensen *et al.*, 1986). The advent of the subclinical laminitis hypothesis resulted in a significant number of publications dealing with the topic (Bergsten, 1988; Bergsten *et al.*, 1986; Bradley *et al.*, 1989; Greenough and Gacek, 1986). Of particular note were those suggesting that a high incidence of lameness can be correlated with the incidence of production diseases (Mortensen and Hesselholt, 1986; Moser and Divers, 1987; Russell *et al.*, 1986) or reproductive inefficiency (Collick *et al.*, 1989; Lucey *et al.*, 1986). The explanation for these correlations appears to be that the pathophysiological conditions that cause laminitis affect other important body organs thereby causing animals to fail to reach their expected productivity. Perhaps the most valuable thought is that subclinical laminitis is a multifactoral disease involving management and nutrition, thereby justifying an epidemiological approach to the control of lameness (Mortensen and Hesselholt, 1982). This approach would include considerations of genetic predisposition (Berger, 1979; Nilsson, 1963; Smedegaard, 1964a).

Lameness caused by infectious agents

The most common cause of lameness is referred to as interdigital phlegmon (footrot, foul in the foot) which is an acute cellulitis of the interdigital region caused by *Fusobacteriun necrophorum* (Clark *et al.*, 1985) and *Bacteroides melaninogenicus* (Berg and Loan, 1975). The spread of the disease is predominantly related to the contamination and quality of the environment.

Interdigital dermatitis is an infectious disease caused by *Bacteroides nodosus* (Clark *et al.*, 1986; Egerton and Laing, 1978/1979; Laing and Egerton, 1978; Morgan, 1969; Thorley *et al.*, 1977). This organism appears to be capable, under ideal conditions, of invading the horn of the heel bulb. When this occurs the resulting lesion is referred to as erosion of the heel. An alternative aetiology for erosion of the heel may involve subclinical laminitis. All three conditions are loosely referred to in the literature as 'footrot'. Needless to say, misunderstanding, confusion and inaccurate observations result.

It is probable that some breeds have greater resistance to *B. nodosus* infection than others (Andersson and Lundstrum, 1981). Frisch (1976) reported that *Bos indicus*–cross cattle are less susceptible to 'footrot' than animals of *Bos taurus* origin whereas Hollon and Branton (1975) reported a contrary finding. McDowell and McDaniel (1968) found no evidence that crossbreds were superior to purebreds during first lactation in resistance to 'footrot' (not defined). In contrast, Hollon and Branton (1975) reported higher incidences of 'footrot' (not specified) in first lactation among Holsteins than in crossbreds. Toussaint-Raven and Cornelisse (1971) suggest an individual, possibly hereditary, sensitivity to interdigital dermatitis.

A heritability for 'footrot' (not defined) in Denmark was reported to be

0.13 in first to fifth parity cows (Nielsen and Smedegaard, 1984). A second Danish study (Petersen *et al.*, 1982) found that the heritability of 'footrot' (inadequately defined) was 0.27 in heifers. In the Netherlands, much lower figures of 0.08 have been reported for interdigital dermatitis in heifers (Peterse, 1986a, Peterse and Antonisse, 1981).

From the foregoing review it would seem to be unwise to draw definitive conclusions about the genetic resistance to interdigital infections based on work that has been reported in which the term 'footrot' has not been clearly defined.

Interdigital hyperplasia (interdigital fibroma)

An interdigital fibroma (corns) is a condition resulting from weakness of the distal interphalangeal ligament in Herefords (Wegner, 1970) and other breeds (Cirlan, 1982). There is no evidence that this condition is heritable in Holsteins, however, it is believed among clinicians, to be more commonly observed when the incidence of interdigital dermatitis is high.

Lameness producing lesions of the proximal limb

Lameness caused by lesions in the proximal limb have mostly resulted from trauma. (Vitamin E deficiency is a notable excepton.)

Lesions indistinguishable from osteochondritis dissecans occur more frequently as the animal increases in age (Neher and Tietz, 1959) and is associated with a 'post-legged' conformation (Bailey, 1985). In one study nearly 20% of AI bulls with steep hocks became lame with advanced age (Feher *et al.*, 1968). The joint affected most commonly is the femorotibial articulation, next commonly the coxofemoral joint and least commonly the tibiotarsal joint (Nesbitt *et al.*, 1975). Degenerative joint disease of the hip may be a heritable trait (Howlett, 1972; Kendrick and Sittmann, 1964).

A second condition that has been observed (subjectively) by the author in father/son/brother relationships is serous tarsitis (bog spavin) which is an irreversible distension of the hock joint caused by fluid. This condition is considered to be heritable in Danish red cattle (Sittmann and Kendrick, 1964).

The accepted criteria for judging the straightness of a limb (post-legged) is to evaluate the angle of the hock which, according to Hansen (1964) averaged 155° ± 3 in Simmental bulls. Habel (1948) quoted that as early as 1896 an animal was defined as having a straight limb if the hock angle exceeded 170°. Clearly the subject has been a concern for a very long time and it is all the more remarkable that there is still no standard method for measuring this angle. An objective system used by the author would suggest that the angle of the hock of four breeds of modern bull calves aged between 8 and 12 months is between 162° and 165°.

The literature records that 'feet and legs' is a trait that has very low heritability (Diers and Swalve, 1987; Doormaal and Burnside, 1987; Hay *et al.*, 1983; Meyer *et al.*, 1985; Placke *et al.*, 1983; Smith *et al.*, 1985a, 1985b). The trait is descriptive, that is to say the data on limb angulation are subjective. The only objective study of limb angulation was conducted by Habel (1948). He measured the metatarsal inclination in dairy cattle and found this characteristic to be highly heritable (0.66). Furthermore, Cassel *et al.* (1973) while confirming the low heritability of 'feet and legs' added the reservation that 'further research should determine if some alternative scale or procedure for measurement would result in higher heritabilities'.

Unfortunately, show judges and some pedigree beef breed societies appear to have been promoting selective breeding to increase hip height. One strategy for increasing hip height is to select for straight hock. There may be a mistaken belief that improving hip height will improve meat yield from the limbs. Berg and Butterfield (1976) established that the meat to bone ratio remains constant irrespective of the angulation of the limb and consider the influence of the show ring to be counterproductive.

A relationship does exist between height and weight (Hand *et al.*, 1986) but unqualified selection for size in dangerous. The writer has observed one animal that achieved remarkable size but showed evidence of abnormality.

Show judges use descriptive traits to evaluate animals. The skill to evaluate morphological characteristics with consistent accuracy is shared by a limited number of individuals (Fisher *et al.*, 1981). Bonsma (1973) criticizes the influence of the show ring because it fails to recognize the importance of 'functional efficiency'.

Limbs (legs) are often evaluated separately from feet but even the term limbs is insufficiently specific. In the hind (pelvic) limb for example, the angle of the hock is the most important to evaluate. The angle of the stifle is also important but because of the anatomical arrangement of muscles the stifle is said to 'reciprocate' the angle of the hock. The slope of the pastern is another limb angle that may be significant.

Over the past 30 years, selective breeding has probably increased the average angle of the hock in beef breeds by 10 degrees. Stature is a trait of which hind limb angulation may be a component. The heritability of stature is relatively high, therefore, it can be argued that even if leg traits have a low heritability, the heritability of joint angles should be re-evaluated.

The concept of functional efficiency discussed by Bonsma (1973) must be remembered. If changes in conformation are planned, care must be taken to ensure that they do not interfere with the functional efficiency of the animal. Breeding for straight legs for example could be leading towards selection for a sublethal trait.

Posture

This is a phenomenon that is poorly understood by those who evaluate cattle. An abnormal posture is frequently confused with a weakness in conformation. Abnormal posture is acquired. The phenomenon of 'cow hock' (turning in of the hocks) is caused by an overburdening of the heel of the lateral claw by a buildup of horn. Skilled hoof trimming can correct this defect. 'Camping under' is a posture that results from pain and is easily misinterpreted as 'sickle hock'. 'Camping back' is even more common, being due to pain in the heels, usually sole ulcer. In some instances, bow leg or 'standing narrow' is a postural accommodation for pain in the medial claws which occurs quite frequently with laminitis.

The claw

About 86% of all claw disorders occur in the hind claws and of these 75–85% are located in the lateral claw (Andersson and Lundstrum, 1981; Arkins, 1981; Prentice, 1970; Russell *et al.*, 1982). Most claw lesions are associated with or predispose to a breach in the horn that allows infection to enter. Pododermatitis circumscripta (sole ulcer) is one of the more common conditions that affects the claw. It is caused by abnormal pressure being exerted on the horn in the central part of the sole. Another example is avulsion to the zona alba (white line disease) which is splitting of the wall away from the sole in the region of the heel. Erosion of the heel has already been mentioned. This condition places greater strain on other regions of the sole.

These diseases are more common when the claw horn is soft. Soft horn wears more rapidly than hard horn, the sole becomes flat and very prone to direct injury.

It follows that the quality of the horn is directly related to claw disease. In the most practical sense high claw quality is defined as a low susceptibility to claw disorders and a low need for claw care (Politiek *et al.*, 1986). The quality of a claw is assessed to some extent by its shape and size. The resistance to wear or hardness is another important factor that is determined by the chemical and physical makeup of the horn, to which there is a genetic component (Distl *et al.*, 1982).

Functional anatomy of the claw

The claw has two major functions: to absorb part of the shock of locomotion and to protect the terminal structures of the digit from disease. Diseases affecting the claw horn diminish the ability of the claw to function normally.

The wall of the normal claw acts as a horse-shoe shaped spring. An elastic pad (digital cushion/heel bulb) is located between the branches of the spring. When the elastic pad is compressed, it expands sideways transferring pressure to the spring. Thus if the quality of the horn in the wall is reduced by disease and/or the heel bulb is eroded, the shock absorber will no longer function correctly with devastating results for other structures in the region.

Claw quality

Horn quality becomes more important in animal production as confinement increases. Abrasive surfaces such as concrete increase rate of wear and new horn of high quality must replace it or animal performance will suffer (Hahn *et al.*, 1986; McDaniel *et al.*, 1982).

Improving claw quality in the short term can be achieved by management procedures that reduce claw problems in confinement (Vermunt, 1990). In the long term, genetic improvement may make an important contribution. Although analyses of claw scores have failed to identify genetics as a major factor, belief persists among dairy producers and advisors that heredity plays an important role in claw abnormality. Selection of bulls based on lower susceptibility to claw disorders in their daughters may reduce the incidence of lameness (Russell *et al.*, 1986).

The wall of the claw is a living diary of the events in the life of an animal. The wall of the claw at the dorsal surface grows at a rate of approximately 0.5 cm per month. Events such as weaning, calving and febrile disease cause a temporary setback in horn growth. This results in the formation of a 'hardship groove'. As new horn is produced, the groove moves distally until it is worn away. Laminitis produces pronounced ripples or rough bands in the horn of the wall. Rough ridges can be produced by very sudden changes in feed. In groups of animals kept under identical conditions some animals are affected more severely than others (Greenough *et al.*, 1990).

Claw conformation

This is evaluated by the industry as a score for claw traits which is used in the process of genetic selection. However, the relation between visual scores and actual measurements is low (Huber *et al.*, 1983; Politiek *et al.*, 1986). Visual scores do not provide useful data for improving claw quality and their genetic relationship with claw disorders is low (Hahn *et al.*, 1984a, b; Huber *et al.*, 1983; Peterse, 1986b; Politiek *et al.*, 1986; Smit *et al.*, 1986a, b).

Objective claw measurements

In a study of Angus cattle, coefficients of variation for subjective claw scores were high (Morris *et al.*, 1985). Lower variability was obtained when claws from slaughtered animals were measured (Andersson and Bergman, 1980). Lower coefficients of variation are found with objectively measured claw traits in mature cows (Hahn *et al.*, 1977a).

Many attempts have been made to define the ideal parameters of claw measurements. (Distl *et al.*, 1984; Fessl, 1968, 1982a, b; Schleiter and Gunther, 1967). Vermunt (1990) obtained useful claw measurements using a computerized photogrammetric system. The following are the most commonly used claw measurements:

1. The angle between dorsal wall and the sole (angle of the toe);
2. The toe to heel ratio, i.e. vertical height from ground to the skin horn junction of the dorsal border to the vertical height from the ground to the skin horn junction at the heel;
3. The surface area of the sole;
4. The length of the dorsal border.

Heritability of claw measurements

Moderate to high heritabilities for the angle at the toe, the length of the dorsal border and height at the heel have been recorded. Claw traits which can be measured accurately with high repeatabilities are length and, to a lesser degree, angle at the toe but height at the heel has a lower repeatability (Andersson and Lundstrum, 1981; Distl *et al.*, 1982, 1984; Hahn *et al.*, 1977b, 1978b, 1984a, b; Huber *et al.*, 1983; McDaniel *et al.*, 1984b; Smit *et al.*, 1986b).

Drodz (1980) reported significant differences in morphological claw traits among ten strains of Friesian cattle.

Claw characteristics and production

Claw measurements of young Holstein cows in North Carolina were associated with their future economic value (McDaniel *et al.*, 1984a). Cows with shorter, steeper-angled claws have better reproductive efficiency, higher milk yield in later life, and higher longevity than cows with longer, smaller-angled claws. An increased milk production is associated with higher heels, as well as with a greater difference in heel depth between lateral and medial claws (Smit *et al.*, 1986a).

There are dangers in relating conformational traits to production. It is not wise to make these correlations in animals older than twenty months of age because the combined effects of age and management can change shape and size (Vermunt, 1990). Furthermore, extreme care has to be taken to avoid confusing cause with effect which might result in selecting for susceptibility to claw disease.

Changes in claw conformation with age

Age affects all claw measurements in heifers, indicating changes in claw shape and dimensions during growth (Hahn *et al.*, 1978b, 1984b). It has also been shown that claw characteristics of young cows were significantly associated with longevity, among other parameters (McDaniel *et al.*, 1984a).

Claw angles decrease and lengths of the dorsal border increase with age. Claw size increases from first to second lactation. There is a difference of opinion regarding the extent to which size increases in later lactations (Andersson and Lundstrum, 1981; Fessl, 1968; Hahn *et al.*, 1984b). Cows surviving three to four lactations have shorter claws with steeper angles than non-survivors, based on measurements of length and angle of the dorsal border and heel height.

Claw conformation and the incidence of lameness

Eddy and Scott (1980) and Russell *et al.* (1982) assessed the relationship between claw conformation and the occurrence of lameness. A 21% and 42% incidence of claw lesions were reported in abnormally shaped claws, respectively. Overgrowth was the most commonly recorded abnormal shape and was most frequently associated with sole ulcers and white zone lesions.

The length and angle of the dorsal of the hind lateral claw in Holstein–Friesian and Dutch Friesian heifers in the Netherlands are correlated to the severity of claw disorders (Peterse, 1986b).

1. Steep claws have less numerous and less serious sole lesions (Smit *et al.*, 1986b).
2. The length of the dorsal border has less influence on the incidence of sole lesions than the angle of the claw.
3. High heels are associated with sole ulcers.
4. Long claws are associated with more sole ulcers and higher total lesion scores.
5. Short toe lengths and steep angles are needed for good locomotion (Manson and Leaver, 1989).
6. Cows affected by digital disease have longer toes and deeper heels (Andersson and Lundstrum, 1981).
7. Abnormally shaped claws predispose to claw disorders.

Surface area of the sole of claw

Under natural conditions the cow bears weight in the abaxial wall and heel. The zone for weight bearing extends no more than 2 cm from the outer surface of the wall. As the wall is worn down by abrasive surfaces the zone of weight bearing extends further into the sole until the sole becomes flat and the whole surface is in contact with the ground. Under conditions of intensive management the sole is usually worn flat. Therefore, it may be misleading to attempt to measure the surface area of the sole too accurately. However, an indication of claw size is needed. For the time being it may be wise to calculate the surface area of the claw by using a formula involving the width and length (Fessl, 1968).

The size of the bearing surface of the claw, the rate at which it wears, weight of the animal and the angulation of the limb are interrelated.

Hardness of claw horn

Hardness of the horny capsule depends on several factors. It is not known which, if any, of the physical characteristics including chemical composition are heritable.

The moisture content

This is probably the most important and variable factor that determines the hardness of the horn. Martig *et al.* (1980) found a negative correlation between hardness and water content of horn. A low water content makes horn less prone to abrasion (wear) and to lesions (Dietz and Prietz, 1981).

Microarchitecture

Horn quality is determined to a large degree by the number of tubules per unit area. Good quality horn has a large number of horn tubules per mm^2; i.e. averages of 80 and 16 tubules/mm^2 for the dorsal border and sole area, respectively. The tubules have a thick cortex in relation to the medulla and are clearly demarcated from the intertubular substance (Dietz and Prietz, 1981; Politiek *et al.*, 1986). Fewer tubules per mm^2 means that more moisture can be taken up by the intertubular space.

Pigmentation of claws

Pigmentation of horn improves the quality (Dietz and Prietz, 1981). Pigmented horn being up to 30% harder (resistance to grinding) than non-pigmented horn. Petersen *et al.* (1982) demonstrated a negative correlation between the colour of claw horn and the occurrence and severity of sole contusion. Less pigmentation resulted in more severe contusions (Chesterton *et al.*, 1989).

Deformities of the bovine claw

Deformities of the claws have been described (Dietz and Prietz, 1981; Greenough, 1980; Greenough *et al.*, 1981; Mahin and Addi, 1982). These abnormalities fall into three major categories.

Corkscrew claw is an abnormality that affects the lateral hind claws of Holsteins (Bouckaert *et al.*, 1958). It is reported to affect 3% of Holsteins and to be caused by strain on the collateral ligament of the distal interphalangeal joint. Bony periarticular exostoses form and stimulate excessive production of abaxial wall which displaces the sole and forms a corkscrew (Greenough, 1962). It is the anecdotal opinion of knowledgeable veterinarians that this condition is heritable.

The condition occurs in cows that are older than five years of age. Bulls rarely show the abnormality in its most extreme form. It is the writer's firm opinion that no bull should be selected for breeding purposes if the axial surface of a lateral hind claw is more concave than that of the medial claw.

Slipper foot was named in the German literature because the claw resembles a Persian slipper. The dorsal surface is heavily ridged, flat, wider than normal and square at the toe. One or more claws may be affected on any or all limbs. It is the anecdotal opinion of many veterinarians that this condition is a sequel to chronic laminitis.

The third group of claw abnormalities has been referred to as normal overgrowth. The claws are longer than normal. Many of these abnormalities result from reduced wear either because the claw is drier and harder than normal (climatic influence?) or the animal has been confined on soft pack with little opportunity for the horn to wear at a normal rate. Excessive claw horn production is also seen in animals that are fed heavily and have high daily weight gain (Greenough *et al.*, 1990). Hind claws are more often overgrown than front claws.

Glicken and Kendrick (1977) suggest that there is a genetic component in the expression of claw deformity. The illustration in their paper is one of slipper foot and the description could include corkscrew claw. Normal overgrowth is not defined. This paper is an example of the need to define the data accurately before subjecting them to elaborate statistical analysis. Eddy and Scott (1980), consider that some claw deformities may be hereditary. Russell *et al.* (1982) suggested that the high rate of lesions in abnormally shaped claws is probably caused by the recent emphasis on conformation of legs rather than claws in breeding stock. Animals affected with corkscrew claw or slipper foot are less able to compete for feed and consequently lose body condition.

Laminitis

Laminitis has been known to occur in horses for many years in which species the aetiology has been extensively investigated (Garner, 1975; Garner *et al.*, 1975; Hood and Stephens, 1981; Obel, 1948; Robinson *et al.*, 1975). In the horse, the disease varies in severity from a very painful life-threatening condition (acute) to one that produces mild lameness (subacute). Chronic laminitis also occurs in the horse.

A classic monograph on the topic of laminitis in cattle was published by Nilsson in 1963.

The traditional explanation of the aetiology of the disease involves an excess carbohydrate intake, followed by ruminal acidosis. As a result of an environment that is hostile to the normal rumen flora, toxic substances, that act on small blood vessels, are produced. When absorbed into the bloodstream these toxins have a devastating effect on the network of tiny vessels that nourish the horn-producing tissues of the claw. At first, blood stagnates causing increased pressure and pain in the claw. This process finally causes irreversible damage to some of the structures that produce horn.

The presence of receptors for epidermal growth factor have been demonstrated in the horn-producing tissues of the claw (Ekfalk *et al.*, 1988) and this forms the basis of an alternative hypothesis regarding the aetiology of laminitis. Current concepts suggest that a combination of factors influence the occurrence and severity of laminitis (Bazeley and Pinsent, 1984; Bergsten, 1988; Brochart, 1987; Colam-Ainsworth *et al.*, 1989; Greenough and Vermunt, 1990; Livesey and Fleming, 1984; Mortensen and Hesselholt, 1982, 1986; Peterse, 1980, 1982, 1985, 1986a; Smart, 1986; Weaver, 1979).

An inherited form of laminitis in Jersey cattle has been reported in South Africa (De Boom *et al.*, 1968), in the USA (Merrit and Riser, 1968), and in the UK (Edwards, 1972). A familial predisposition to laminitis has been observed by Nilsson (1963) and Maclean (1965). Bazeley and Pinsent (1984) mentioned a possible inherited predisposition to laminitis.

Swedish Friesians are more-often affected by sole haemorrhages and sole ulcer than Swedish Red and White cattle (Andersson and Lundstrum, 1981). Brochart (1987) noticed a greater susceptibility to laminitis in Friesian cattle than in Holstein Friesian-cross cows.

The lesion most commonly associated with subclinical laminitis is podermatitis circumscripta usually referred to as 'sole ulcer'. Peterse and Antonisse (1981) and Peterse (1986b) estimated the heritability for sole ulcer in Friesian heifers at 0.14. Nielsen and Smedegaard (1984) gave a similar estimate. Heritability for sole lesions has been estimated at 0.1 and 0.2 by Petersen *et al.* (1982) and Smit and Verbeek (1984), respectively.

Subclinical laminitis

As the name would imply, this is unrelated to an obvious clinical sign such as lameness.

The concept of subclinical laminitis is relatively new. Although several groups of scientists are investigating the condition in Europe, there is very little interest in the disease among clinical research scientists in North America (Colam-Ainsworth *et al.*, 1989; Greenough and Vermunt, 1990; Greenough *et al.*, 1990; Livesey and Fleming, 1984; Maclean, 1971b; Peterse, 1982, 1986a; Toussaint-Raven, 1973). Veterinary practitioners, who work mainly with dairy cows that are managed intensively, are fully aware of the magnitude of the problem. In practical situations the disease should be suspected if more than 10% of a herd were to become lame annually for reasons other than deformity or interdigital phlegmon. Further confirmation would come from an analysis of the lesions causing the lameness. Any occurrences of ulceration of the sole (pododermatitis circumscripta) or white zone lesions should be investigated with subclinical laminitis in mind. The most consistent sign that subclinical laminitis exists is staining of the sole of the claw with blood (Peterse, 1982, 1986a, 1987).

It is not the purpose of this chapter to discuss the clinical or therapeutic aspects of subclinical laminitis. There seems to be no doubt that the disease is related to a high energy intake, the frequency and quantity of consumption and the period of time over which the increased intake occurs. The quantity and quality of roughage intake has also been suggested as major factors influencing the occurrence of the disease.

The hypothesis of a multifactoral aetiology extends beyond nutritional factors. It has been suggested above that there is a genetic susceptibility to the disease. Anecdotally, there is a consensus that trauma is a precipitating factor. In a previous section of this chapter, it was implied that mechanical forces applied to the sole of the claw results from a permutation of factors, claw size, body weight, architecture of limb angles, claw hardness, pigmentation of the claw and the quality of the surface over which the animal walks.

The recognition of the condition in heifers has created an interest in the role of growth.

Exercise is essential for normal blood flow through the claw, therefore, it has been suggested that management practice that interferes with exercise may be a factor that influences the occurrence of lameness (Weaver, 1979; Zeeb, 1987).

Finally, in recent years behaviour has been suggested as a predisposing factor in lameness (David, 1986; Irps, 1987; Kempens and Boxberger, 1987). Furthermore, the effects of confrontation (between submissive heifers and dominant cows in the dry herd) could be considered a 'personality trait' that may be heritable. Mixing heifers and dry cows causes some younger animals to slip more frequently and to rest less. Animals walking roadways can also have problems associated with behaviour (Chesterton *et al.*, 1989).

Discussion

It is difficult to provide irrefutable evidence, *per se*, that there is a genetic resistance to 'lameness'. Lameness is a clinical sign that is common to a number of diseases affecting the locomotory system.

Nevertheless, there is evidence that degenerative joint disease has a heritable component and furthermore the condition is exacerbated in animals with straight hind limbs. Selection for this particular characteristic is favoured by some breed societies and is regarded as desirable in the show ring.

The heritability of 'feet and legs' is regarded as low. This evaluation is based almost entirely on descriptive traits. Feet (claws) have at least five separate characteristics, toe/heel ratio, heel height, pigmentation, hardness, area of the bearing surface. Each can be measured objectively and each can influence the incidence of disease. One or more of the factors might have a higher heritability than others, therefore, it may be wise to review the validity of past reports.

The limb is a propulsive/supporting device, the mechanical parameters of which can be measured and the transmission of mechanical stress can be calculated. Functional efficiency is a permutation of body weight, mechanical characteristics of the limb and the ability of the claw to absorb mechanical stress.

Subclinical laminitis is possibly the least understood and most important (economically) disease that threatens the progress of high intensity dairy farming at this time. The requirement for high producing cows for a high intake of carbohydrate will not change. Intensive management systems are unlikely to be abandoned. The genetic selection of cattle for characteristics of high production will not stop. Consequently, it is no longer realistic to select physical characteristics on the basis of tradition, aesthetics or bias. Descriptive traits may have a place if they have relevance, if they are repeatable and if objective measurement is impracticable.

In reviewing the scientific literature, it is apparent that data collected about diseases of the locomotory system have, in some instances, been based on a misinterpretation of the lesions observed. Most of the data collected on body and limb conformation are based on non-scientific descriptions of the point being measured. The accuracy of some of the work that has been reported needs to be reassessed.

The decade of the 1990s could, should and can be the decade of object-ivity in the study of the angularity of the limbs and body size in cattle. This objectivity should extend to the study of those diseases of the locomotory system that may be affected by the angularity and size of the limbs and body.

There is a pressing need for veterinary and animal scientists to work more closely together than they have in the past. Some of the most thoughtful publications have been generated by joint consultation between the animal scientists and veterinarian. How can veterinary and animal scientists face the twenty first century if they continue passively to accept the 'mystique' that is associated with the evaluation of conformation. Admittedly, in the world of animal health research 'he who pays the piper has been calling the tune'. Maybe the time has come to suggest that a new melody could be developed, the finale of which would be based on 'the crackle of money'.

Acknowledgements

My appreciation of the contemporary complexities of bovine lameness has been greatly enhanced by my close working association with Jos Vermunt over the past two years. To him must go much of the credit for the content of this chapter. I would also like to acknowledge the cheerful patience of Ms Janna Boymook in preparing the numerous drafts of this paper.

References

Andersson, L. and Bergman, A. (1980) Pathology of bovine laminitis especially as regards vascular lesions. *Acta Veterinaria Scandinavia* 21, 559–66.
Andersson, L. and Lundstrum, J. (1981) The influence of breed, age bodyweight and season on digital diseases and hoof size in dairy cows. *Zentralblatt fur Veterinarmedizin* 28, 141–51.
Arkins, S. (1981) Lameness in dairy cows. Parts I and II. *Irish Veterinary Journal* 35, 135–40, 163–70.
Baggott, D.G. and Russell, A.M. (1981) Lameness in cattle. *British Veterinary Journal* 137, 113–32.
Bailey, J.V. (1985) Bovine arthritides; classification, diagnosis, prognosis and treatment. In: *The Veterinary Clinics of North America: Food Animal Practice* 1, 39–51.
Bazeley, K. and Pinsent, P.J.N. (1984) Preliminary observations on a series of outbreaks of acute laminitis in dairy cattle. *Veterinary Record* 115, 619–22.
Berg, J.N. and Butterfield, R.M. (1976) *New Concepts of Cattle Growth.* Halsted Press, John Wiley, New York.
Berg, R.T. and Loan, R.W. (1975) *Fusobacterium necrophorum* and *Bacteroides melaninogenicus* as etiologic agents of foot rot in cattle. *American Journal Veterinary Research* 36, 1115–22.

Berger, G. (1979) Klauenkrankungsrate bei Khen verschiedener Genotypen unter den Bedingungen der einstreuiosen Laufstallhaltung. *Monatshefte für Veterinär-medizin* 34, 161–4.

Bergsten, C. (1988) Sole bruising as an indicator of laminitis in cattle. A field study. *Proceedings of the XVth World Buiatrics Congress.* Palma de Mallorca, pp. 1072–6.

Bergsten, C., Andersson, L. and Wiktorson, L. (1986) Effect of feeding intensity at calving on the prevalence of subclinical laminitis. *Proceedings of the Vth Symposium on Disorders of the Ruminant Digit,* pp. 33–7.

Bonsma, J.C. (1973) In: Cunha, T.J., Warwick, A.C. and Kroger, A.C. (eds), *Factors Affecting Calf Crop,* University of Florida Press, pp. 197–231.

Boosman, R., Koeman, J. and Nap, R. (1989a) Histopathology of the bovine pododerma in relation to age and chronic laminitis. *Zentralblatt Veterinamedizin A* 36, 438–46.

Boosman, R. and Mutsaers, C.W.A.A.M. (1988) Arteriography of the bovine claw in relation to chronic laminitis. *Proceedings of the XVth World Buiatrics Congress.* Palma de Mallorca, pp. 1077–82.

Boosman, R., Nemeth, F,. Gruys, E. and Klarenbeek, A. (1989b) Arteriographical and pathological changes in chronic laminitis in dairy cattle. *Veterinary Quarterly* 11, 144–55.

Bouckaert, J., Oyaert, W. and Deloddere, E. (1958) The corkscrew claw. *Vlaams Dierg* 27, 149.

Bradley, H.K., Shannon, D. and Neilson, D.R. (1989) Subclinical laminitis in dairy heifers. *Veterinary Record* 125, 177–9.

Brochart, M. (1987) Foot lameness of the cow, a multifactorial disease. In: Wierenga, H.K. and Peterse, D.J. (eds) *Cattle Housing Systems, Lameness and Behaviour.* Martinus Nijhoff, Boston, pp. 159–65.

Cassel, B.G., White, J.M., Vinson, W.E. and Kliewer, R.H. (1973) Genetic and phenotypic relationships among type traits in Holstein–Friesian cattle. *Journal of Dairy Science* 58, 1171–7.

Chesterton, R.N., Pfeiffer, D.U., Morris, R.S. and Tanner, C.M. (1989) Environment and behavioural factors affecting the prevalence of foot lameness. *New Zealand Veterinary Journal* 37, 135–42.

Choquette-Levy, L., Baril, J., Levy, M. and St Pierre, H. (1985) A study of foot disease of dairy cattle in Quebec. *Canadian Veterinary Journal* 26, 278–81.

Cirlan, M. (1982) Hyperplasia interdigitalis in bulls. A genetic and epizootologic 6-year study. *Proceedings of the IVth International Symposium Disorders of the Ruminant Digit,* Paris, France.

Clark, B.L., Stewart, D.J. and Emery, D.L. (1985) The role of *Fusobacterium necrophorum* and *Bacteroides melaninogenicus* in the etiology of interdigital necrobacillosis in cattle. *Australian Veterinary Journal* 62, 47.

Clark, B.L., Stewart, D.J., Emery, D.L. and Dufty, J.H. (1986) Immunization of cattle against interdigital dermatitis (foot rot) with an autogenous *Bacteroides nodosus* vaccine. *Australian Veterinary Journal* 63, 61–2.

Colam-Ainsworth, P., Lunn, G.A., Thomas, R.C. and Eddy, R.G. (1989) Behaviour of cows in cubicles and its possible relationship with laminitis in replacement dairy heifers. *Veterinary Record* 125, 573–5.

Collick, D.W., Ward, W.R. and Dobson, H. (1989) Associations between types of lameness and fertility. *Veterinary Record* 125, 103–6.

David, G.P. (1986) Cattle behaviour and lameness. *Proceedings of the Vth International Symposium of Disorders of the Ruminant Digit*, Dublin, Ireland, pp. 79–86.

De Boom, H.P.A., Adelaar, T.F. and Terblanche, M. (1968) Hereditary laminitis of Jersey cattle, a preliminary report. *Proceedings of the IIIrd Congress South African Genetic Society.*

Diers, H. and Swalve, H. (1987) Estimation of genetic parameters and breeding values for linear scored type traits. *Proceedings of the 38th Annual Meeting of the European Association for Animal Production*, Lisbon.

Dietz, O. and Prietz, G. (1980) *Proceedings IIIrd International Symposium Disorders Ruminant Digit*, Vienna, Austria, pp. 78–86.

Dietz, O. and Prietz, G. (1981) Klauenhornqualitt – Klauenhornstatus. *Monatshefte für Veterinärmedizin* 36, 419–22.

Distl, O., Graf, F. and Krausslich, H. (1982) Genetische Variation von morphologischen, histologischen und elektrophoretischen Parametern bei Rinderklauen und deren phnotypischen und genetischen Beziehungen. *Zchtungskunde* 54, 106–23.

Distl, O., Huber, M., Graf, F. and Krusslich, H. (1984) Claw measurements of young bulls at performance testing stations in Bavaria. *Livestock Production Science* 11, 587–98.

Doormaal, B.J. Van and Burnside, E.B. (1987) Impact of selection on components of variance and hereditabilities of Canadian Holstein conformation traits. *Journal of Dairy Science* 70, 1452–7.

Drozdz, A.A. (1980) Hoof box traits of 10 groups of crossbred Friesian cattle. *31st Annual Meeting Study Comm EAPP*. Munich, West Germany.

Eddy, R.G. and Scott, C.P. (1980) Some observations on the incidence of lameness in dairy cattle in Somerset. *Veterinary Record* 113, 140–4.

Edwards, G.N. (1972) Hereditary laminitis in Jersey cattle. *Proceedings of the VIIth World Buiatrics Congress*, London, UK. pp. 663–8.

Egerton, J.R. and Laing, E.A. (1978/1979) Characteristics of *Bacteroides nodosus* isolated from cattle. *Veterinary Microbiology* 3, 269–79.

Ekfalck, A., Funkquist, B., Jones, B. and Obel, N. (1988) Presence of receptors for epidermal growth factor (EGF) in the matrix of the bovine hoof – a possible new approach to the laminitis problem. *Zentralblatt fur Veterinamedizin* 35, 321–30.

Espinasse, J., Savey, M., Thorley, C.M., Toussaint-Raven, E. and Weaver, A.D. (1984) Colour atlas on disorders of cattle and sheep digit – international terminology. Editions du point veterinaire, 25 mede Bourgelat, 94704 Maisons-Alfort, France.

Feher, G., Haraaszti, J. and Meszaros, I. (1968) Steep posture of the hind legs of bulls and its harmful consequences. *Magyar Allatorosok Lapja* 23, 277–84.

Fessl, L. (1968) Biometric studies on ground surface of bovine claws and the distribution of weight on the extremities. *Zentral blatt fur Veterinarmedizin* 15, 844–60.

Fessl, L. (1982a) Claw size of cattle in relationship to dunggrid and slotfloor housing. *Proceedings of the IVth International Symposium Disorders Ruminant Digit*, Paris, France.

Fessl, L. (1982b) On standardization of claw measurements in cattle. *Proceedings of the IVth International Symposium Disorders Ruminant Digit*, Paris, France.

Fisher, A.V., Harries, J.M. and Robinson, J.M. (1981) Sensory assessment of fatness and conformation of beef steers. *Meat Science* 283–95.

Frisch, J.E. (1976) The comparative incidence of foot rot in *Bos taurus* and *Bos indicus* cattle. *Australian Veterinary Journal* 52, 228–9.

Garner, H.E. (1975) Pathophysiology of equine laminitis. *Proceedings of the American Association of Equine Practitioners* 21, 384–7.

Garner, H.E., Coffman, J.R., Hahn, A.W., Hutcheson, D.P. and Tumbleson, M.E. (1975) Equine laminitis of alimentary origin: an experimental model. *American Journal of Veterinary Research* 36, 441–4.

Glicken, A. and Kendrick, J.W. (1977) Hoof overgrowth in Holstein–Friesian dairy cattle. *Journal of Heredity* 68, 386–90.

Greenough, P.R. (1962) Observations on some of the diseases of the bovine foot. Part I and Part II. *Veterinary Record* 74, 1–9, 53–63.

Greenough, P.R. (1978) The nomenclature of anatomical features of the bovine digits. *Proceedings of the IInd Symposium on Disorders of the Ruminant Digit.* Skara.

Greenough, P.R. (1980) Claw deformities. *Proceedings of the IIIrd International Symposium Disorders Ruminant Digit*, Vienna, Austria. pp. 56–61.

Greenough, P.R. (1985) The subclinical laminitis syndrome. *Bovine Practitioner* 20, 14–149.

Greenough, P.R. and Gacek, Z. (1986) A preliminary report on a laminitis like condition occurring in bulls under feeding trials. *Vth International Symposium on Disorders of the Ruminant Digit.* Dublin, pp. 63–8.

Greenough, P.R. and Vermunt, J.J. (1990) Evaluation of subclinical laminitis and associated lesions in dairy cattle. *Proceedings of the VIth International Symposium Disorders Ruminant Digit.* Liverpool, UK.

Greenough, P.R., Vermunt, J.J., McKinnon, J.J., Fathy, F.A., Berg, P.A. and Cohen, R.H. (1990) Laminitis-like changes in the claws of feedlot cattle. *Canadian Veterinary Journal* 31, 202–8.

Greenough, P.R., Weaver, A.D. and McCallum, F.J. (1972) *Lameness in Cattle.* 1st ed., Oliver and Boyd, Edinburgh.

Greenough, P.R., Weaver, A.D. and McCallum, F.J. (1981) *Lameness in Cattle*, 2nd ed. John Wright, Bristol, p. 107.

Habel, R.E. (1948) On the inheritance of metatarsal inclination in Ayrshire cattle. *American Journal of Veterinary Research* 9, 131–9.

Hahn, M.V., McDaniel, B.T. and Wilk, J.C. (1977a) Variation in and relationships among various hooves in two breeds of dairy cattle. *Journal of Dairy Science* 60 [Suppl 1], 146–7 (Abstr.).

Hahn, M.V., McDaniel, B.T. and Wilk, J.C. (1977b) Repeatability of measurements of variation in feet of dairy cattle. *Journal of Diary Science* 60 (Suppl 1), 146 (Abstr.).

Hahn, M.V., McDaniel, B.T. and Wilk, J.C. (1978) Heritabilities of objectively measured hoof traits of Holsteins. *Journal of Dairy Science* 61 (Suppl 1), 83–4 (Abstr.).

Hahn, M.V., McDaniel, B.T. and Wilk, J.C. (1984a) Description and evaluation of objective hoof measurements of dairy cattle. *Journal of Dairy Science* 67, 229–36.

Hahn, M.V., McDaniel, B.T. and Wilk, J.C. (1984b) Genetic and environmental variation of hoof characteristics of Holstein cattle. *Journal of Dairy Science* 67, 2986–98.

Hahn, M.V., McDaniel, B.T. and Wilk, J.C. (1986) Rates of hoof growth and wear in Holstein cattle. *Journal of Dairy Science* 69, 2148–56.

Hand, R.K., Gould, S.R., Basarab, J.A., and Engstrom, D.F. (1986) Condition score, body weight and hip height as predictors of gain in various breed crosses of yearling steers on pasture. *Canadian Journal of Animal Science* 63, 447–52.

Hansen, K.M. (1964) Arthritis of the hocks in the progeny of an A.I. bull: arthritis serosa tarsi. *Asberetn Inst Sterilitesforsk (Copenhagen)* 199.

Hay, G.M., White, J.M., Vinson, W.E. and Kliewer, R.H. (1983) Components of genetic variation for descriptive type traits of Holsteins. *Journal of Dairy Science* 66, 1962–6.

Hollon, B.F. and Branton, C. (1975) Performance of Holstein and crossbred dairy cattle in Louisiana. III. Health and Viability. *Journal of Dairy Science* 58, 93–101.

Hood, D.M. and Stephens, K.A. (1981) Physiopathology of equine laminitis. *Compendium Continuing Education Veterinary Practice* 3, S454–9.

Howlett, C.R. (1972) Inherited degenerative arthropathy of the hip in young beef bulls. *Australian Veterinary Journal* 48, 562.

Huber, M., Distl, O. and Graf, F. (1983) Claw measurements of young bulls at performance testing stations in Bavaria. *34th Annual Meeting Study Comm EAAP*, Madrid, Spain.

Irps, H. (1987) The influence of the floor on the behaviour and lameness of beef bulls. In: Wierenga, H.K., Peterse, D.J. (eds), *Cattle Housing Systems, Lameness and Behaviour*. Martinus Nijhoff, Boston, pp. 73–86.

Kempens, K. and Boxberger, J. (1987) Locomotion of cattle in loose housing systems. In: Wierenga, H.K. and Peterse, D.K. (eds), *Cattle Housing Systems, Lameness and Behaviour*. Martinus Nijhoff, Boston, pp. 107–18.

Kendrick, J.W. and Sittmann, K. (1964) Inherited osteoarthritis of dairy cattle. *Journal of the American Veterinary Medicine Association* 148, 535.

Laing, E.A. and Egerton, J.R. (1978) The occurrence, prevalence and transmission of *Bacteroides nodosus* infection in cattle. *Research in Veterinary Science* 24, 300–4.

Livesey, C.T. (1984) Importance of laminitis in dairy cows. *Veterinary Record* 114, 22.

Livesey, C.T. and Fleming, F.L. (1984) Nutritional influences on laminitis, sole ulcer and bruised sole in Friesian cows. *Veterinary Record* 114, 510–12.

Lucey, S., Rowlands, G.J. and Russell, M. (1986) The association between lameness and fertility in dairy cows. *Veterinary Record* 118, 628–31.

Maclean, C.W. (1965) Observations on acute laminitis of cattle in south Hampshire. *Veterinary Record* 77, 662–72.

Maclean, C.W. (1966) Observations on laminitis in intensive beef units. *Veterinary Record* 78, 223–31.

Maclean, C.W. (1970) A post-mortem X-ray study of laminitis in barley beef animals. *Veterinary Record* 86, 457–62.

Maclean, C.W. (1971a) The pathology of laminitis in dairy cows. *Journal of Comparative Pathology* 81, 563–9.

Maclean, C.W. (1971b) The long-term effects of laminitis in dairy cows. *Veterinary Record* 89, 34–7.

Mahin, L. and Addi, A. (1982) Les maladies digites des bovins. *Annales de Médicine Vétérinaire* 126, 597–620.

Manson, F.J. and Leaver, J.D. (1989) The effect of concentrate, silage ratio and hoof trimming on lameness in dairy cattle. *Animal Production* 49, 15–22.

Martig, J., Leuenberger, W.P. and Dozzi, M. (1980) Quality and alterations of bovine claws as a function of different variables. *Proceedings of the IIIrd International Symposium Disorders of the Ruminant Digit*, Vienna, Austria. pp. 40–55.

McDaniel, B.T., Hahn, M.V. and Wilk, J.C. (1982) Floor surface and effects upon feet and leg soundness. *Proceedings of the Symposium on Management of Food Producing Animals*, Purdue University, W. Lafayette, Indiana, USA. Vol. II, p. 816.

McDaniel, B.T., Verbeek, B., Wilk, J.C., Everett, R.W. and Keown, J.R. (1984a) Relationships between hoof measures, stayabilities, reproduction and changes in milk yield from first to later lactations. *79th Annual Meeting ADSA*, Texas. A & M University (also: *Journal of Dairy Science*) [Suppl 67], 198–9 (Abstr.).

McDaniel, B.T., Verbeek, B., Hahn, M.V., Wilk, J.C. and Keown, J.F. (1984b) Genetics of hoof measurements: repeatabilities, heritabilities and correlations with yields by lactation. *79th Annual Meeting ADSA*, Texas A & M University. (also: *Journal Dairy Science*, 1984b; [Suppl 67], 199 (Abstr.).

McDowell, R.E. and McDaniel, B.T. (1968) Interbreed matings in dairy cattle. II. Herd health and fertility. *Journal of Dairy Science* 51, 1275–81.

Merrit, A.M. and Riser, W.H. (1968) Laminitis of possible hereditary origin in Jersey cattle. *Journal of American Veterinary Medical Association* 153, 1074–5.

Meyer, K., Burnside, E.B., Hammond, K. and McClintock, A.E. (1985) Evaluating dairy sires for conformation of their daughters: use of first classification records. *Australian Journal of Agricultural Research* 36, 509–25.

Morgan, I.R. (1969) A survey of cattle feet in Victoria for *Fusiformis nodosus*. *Australian Veterinary Journal* 45, 264–6.

Morris, C.A., Cullen, N.G. and Packard, P.M. (1985) Foot scores of cattle. Variation among subjective scores of Angus cattle from different herds and sire groups. *New Zealand Journal of Experimental Agriculture* 13, 235–40.

Morrow, D.A. (1966) Laminitis in cattle. *Veterinary Medicine* 61, 138–40.

Mortensen, K. and Hesselholt, M. (1982) Laminitis in Danish dairy cattle – an epidemiological approach. *IVth International Symposium on Disorders of the Ruminant Digit*. Paris.

Mortensen, K. and Hesselholt, M. (1986) The effects of high concentrate diet on the digital health of dairy cows. *Proceedings of the International Production Congress*. Belfast.

Mortensen, K., Hesselholt, M. and Basse, A. (1986) Pathogenesis of bovine laminitis (diffuse aspectic pododermatitis). Experimental models. *Proceedings XIVth World Congress on Diseases of Cattle*. Dublin. pp. 1025–30.

Moser, E.A. and Divers, T.J. (1987) Laminitis and decreased milk production in first-lactation cows improperly fed a dairy ration. *Journal of American Veterinary*

Medicine Association 190, 1575–6.

Neher, G.M. and Tietz, W.J. (1959) Observations on the clinical signs and gross pathology of degenerative joint disease in aged bulls. *Laboratory Investigation* 8, 1218.

Nesbitt, G.H., Amstutz, H.E. and Lewis, R.E. (1975) Lameness in cattle: survey of 102 cases including history, radiographic findings, etc. *Bovine Practitioner* 10, 39.

Nielsen, E. and Smedegaard, H.H. (1984) Disease in legs and hooves with Black and White dairy cattle in Denmark. *568 Report National Institute Animal Science.* Copenhagen, Denmark.

Nilsson, S.A. (1963) Clinical, morphological and experimental studies of laminitis in cattle. *Acta Veterinaria Scandinavia* [Suppl] 4, 9–304.

Obel, N. (1948) Studies on the histopathology of acute laminitis. Diss, Stockholm. (Uppsala: Almquist and Wiksells Boktryckeri).

Peterse, D.J. (1979) Nutrition as a possible factor in the pathogenesis of ulcers of the sole in cattle. *Tijdschr Diergeneesk* 104, 966–70.

Peterse, D.J. (1980) Judgment of bovine claws by the occurrence of sole lesions. Thesis Rijksuniversiteit Utrecht.

Peterse, D.J. (1982) Prevention of laminitis in Dutch dairy herds. *Proceedings of the IVth Symposium on Disorders of the Ruminant Digit.* Paris.

Peterse, D.J. (1985) Laminitis and interdigital dermatitis and heel horn erosion: a European perspective. In: *Veterinary Clinics of North America: Food Animal Practice.* 1, 83–91.

Peterse, D.J. (1986a) Lameness in cattle. *Proceedings of the XIVth World Congress Diseases in Cattle*, Dublin, Ireland. pp. 1015–23.

Peterse, D.J. (1986b) Claw measurements as parameters for claw quality in dairy cattle. *Proceedings Vth International Symposium Disorders Ruminant Digit.* Dublin, Ireland. pp. 87–91.

Peterse, D.J. (1987) Aetiology of claw disorders in dairy cattle. In: Wierenga HK, Peterse, D.J. (eds), *Cattle Housing Systems, Lameness and Behaviour.* Boston: Martinus Nijhoff Publishers. pp. 3–7.

Peterse, D.J. and Antonisse, W. (1981) Genetic aspects of feet soundness in cattle. *Livestock Production Science* 8, 253–61.

Peterse, D.J., Korver, S., Oldenbroek, J.K. and Talmon, F.P. (1984) Relationship between levels of concentrate feeding and incidence of sole ulcers in dairy cattle. *Veterinary Record* 115, 629–30.

Peterse, D.J., Van Vuuren, A.M. and Ossent, P. (1986) The effects of daily concentrate increase on the incidence of sole lesions in cattle. *Proceedings of the Vth Symposium of Disorders of the Ruminant Digit.* Dublin. pp. 39–46.

Petersen, P.H., Nielsen, A.S., Buchwald, E. and Thysen, I. (1982) Genetic studies on hoof characters in dairy cows. *Z Tierzchtg Zchtgsbiol* 99, 286–91.

Pinsent, P.J.N. (1981) The management and husbandry aspects of foot lameness in dairy cattle. *Bovine Practitioner* 16, 61–4.

Placke, K.H., Claus, J. and Kalm, E. (1983) Type classification of German Black Pied cattle. II Factors affecting type traits. *Zuchtungskunde* 55, 258–64.

Politiek, R.D., Distl, O., Fjeldaas, T., Heeres, J., McDaniel, B.T., Nielsen, E., Peterse, D.J., Reurink, A. and Standberg, P. (1986) Importance of claw quality: review and recommendations to achieve genetic improvement. Report of the EAAP working group of 'Claw Quality in Cattle'. *Livestock Production Science*

15, 133–52.

Prentice, D.E. (1970) 'Some investigations into foot lameness in cattle'. MVSc Thesis, University of Liverpool.

Robinson, N.E., Dabney, J.M., Weidner, W.J., Jones, G.A. and Scott, J.B. (1975) Vascular responses in the equine digit. *American Journal of Veterinary Research* 36, 1249–53.

Russell, A.M., Bloor, A.P. and Davies, D.C. (1986) The influence of sire on lameness in cows. *Proceedings of the Vth International Symposium Disorders Ruminant Digit.* Dublin, Ireland. pp. 92–9.

Russell, A.M., Rowlands, G.J., Shaw, S.R. and Weaver, A.D. (1982) Survey of lameness in British dairy cattle. *Veterinary Record* 111, 155–60.

Schleiter, H. and Gunther, M. (1967) Ein Beitrag zur Definition einiger Klauenformen des Rindes. *Maghréb Vétérinaire* 22, 886–92.

Sittmann, K. and Kendrick, J.W. (1964) Hereditary osteoarthritis in dairy cattle. *Genetics* 35, 132.

Smart, M.E. (1986) Relationship of subclinical laminitis and nutrition in dairy cattle: a Canadian experience. *Vth International Symposium on Disorders of the Ruminant Digit.* Dublin. pp. 51–62.

Smedegaard, H.H. (1964a) Contusion of the sole in cattle. *Veterinarian* 2, 119–32.

Smedegaard, H.H. (1964b) Foot rot and chronic foot rot in cattle. *Veterinarian* 2, 299–307.

Smit, H. and Verbeek, B. (1984) Genetic aspects of claw disorders: measurements and scores in Friesian dairy cattle. *Proceedings EAAP Cong,* The Hague, Netherlands.

Smit, H., Verbeek, B., Peterse, D.J. Jansen, J., McDaniel, B.T. and Politiek, R.D. (1986a) The effect of herd characteristics on claw disorders and claw measurements in Friesians. *Livestock Production Science* 15, 1–9.

Smit, H., Verbeek, B., Peterse, D.J., Jansen, J., McDaniel, B.T. and Politiek, R.D. (1986b) Genetic aspects of claw disorders, claw measurements and 'type' scores for feet in Friesian cattle. *Livestock Production Science* 15, 205–17.

Smith, S.P., Allaire, F.R., Taylor, W.R., Kaeser, H.E. and Conley, J. (1985a) Genetic parameters and environmental factors associated with type traits scored on an ordered scale during first lactation. *Journal of Dairy Science* 68, 2058–71.

Smith, S.P., Allaire, F.R., Taylor, W.R., Kaeser, H.E. and Conley, J. (1985b) Genetic parameters associated with type traits scored on an ordered scale during second and fourth lactations. *Journal of Dairy Science* 68, 2655–63.

Thorley, C.M., Clader, H.A.M. and Harrison, W.J. (1977) Recognition in Great Britain of *Bacteroides nodosus* in foot erosions of cattle. *Veterinary Record* 100, 137.

Toissaint-Raven, E. (1971) Lameness in cattle and foot care. *Tijdschr Diergeneeskd* 96, 1244–64.

Toussaint-Raven, E. and Cornelisse, J.L. (1971) The specific, contagious inflammation of the interdigital skin in cattle. *Veterinary Medicine Review* 2/3, 223–47.

Toussaint-Raven, E. (1973) Lameness in cattle and footcare. *Netherlands Journal of Veterinary Science* 5, 105–11.

Vermunt, J. (1990) Lesions and physical and morphological characteristics of the claws of dairy heifers raised under two management systems. Master of Science Thesis. University of Saskatchewan, Canada.

Weaver, A.D. (1979) The prevention of laminitis in dairy cattle. *Bovine Practitioner* 14, 70–2.

Wegner, V.W. (1970) Neue Erkenntnisse Zur Atiologie Fur Lima. *Deutcher Tierarztlich Wokenschrift* 77, 229–32.

Whitaker, D.A., Kelly, J.M. and Smith, E.J. (1983) Incidence of lameness in dairy cows. *Veterinary Record* 113, 60–2.

Zeeb, K. (1987) The influence of the housing system on locomotory activities. In: Wierenga, H.K. and Peterse, D.J. (eds) *Cattle Housing Systems, Lameness and Behaviour*. Martinus Nijhoff, Boston, pp. 101–6.

Chapter 22

Mastitis in Cattle

Jan-Åke Eriksson
Swedish Association for Livestock Breeding and Production,
S-631 84 Eskilstuna, Sweden

Summary

A breeding programme for resistance to mastitis in Sweden is described and the practical experience, based on a five-year period, is evaluated.

Mastitis is defined as veterinary treated mastitis or mastitis as a reason for culling in the period 10 days before to 150 days after first calving. This trait shows a heritability of 0.02. A correlation between the bulls' breeding values for protein yield and resistance to mastitis of −0.07 and −0.21, for Swedish Red and White Breed (SRB) and Swedish Friesian Breed (SLB), respectively, indicates an unfavourable genetic correlation between the two traits.

Breeding values of bulls are based on progeny tests of about 150 daughters and are predicted by a BLUP-procedure. In order to increase the accuracy, an index combining breeding values for mastitis and somatic cell counts was introduced at the beginning of 1990. This improved breeding value should theoretically increase the accuracy by 15% which in practice were shown by the increased standard deviations of breeding value by an average of 13%.

The correlation of repeated breeding values between proven and unproven bulls were 0.68 and 0.67, for 28 SRB and 25 SLB bulls respectively, which is exactly as theoretically expected. The breeding values of the unproven bulls were based on 145 and 150 effective daughters and the breeding value as proven bulls on 2,039 and 1,604 effective daughters for SRB and SLB bulls respectively.

The regression of the sons' breeding value on the sires' breeding value were at the expected level, as well as the correlation of breeding values.

The economic impact of mastitis is discussed with relevance to the inclusion of resistance to mastitis in the bull-index. In Sweden the economic weight used in the bull-index seems to be an overestimation of the economic losses due to clinical mastitis compared to some recently published studies from other countries. The weight might, however, be justified taking into account the effect of subclinical mastitis, animal welfare and milk quality, that are hard to evaluate economically.

Finally, the breeding programme for resistance against mastitis seems to work to theoretical expectation. The effect on the genetic change in resistance against mastitis is not known, as the input of the dam pathway is unknown as well as the impact of importation of semen from foreign populations.

Introduction

Breeding for increased resistance to mastitis in dairy cows is one possible way to decrease the incidence of mastitis, a disease that is very frequent. Statistics from the Nordic countries show that the percentage of cows receiving veterinary treatment for mastitis in 1987 in Finland was 19.9 (Anon., 1987), in Norway 21.1 (Solbu, 1988) and in Sweden 20.6 (SHS, 1989a). Surveys from California, Michigan and Ohio report incidences of mastitis of 30–37 cases per 100 cows per year (Gardner *et al.*, 1990; Kaneene and Hurd, 1990a; Miller and Dorn, 1990) but these estimates include both veterinary treated and owner-reported mastitis. The figures from the Nordic countries and USA are not comparable because of different definitions but still show that mastitis is a great problem in dairy cattle generally.

Mastitis is a very costly disease as, apart from veterinary treatment costs, milk has to be discarded, milk production in the remaining lactation decreases, the risk of culling increases and extra labour is involved. Estimates of costs of a mastitic cow vary depending on assumptions. Östergaard (1989) estimated the Danish costs per case of mastitis to be equal to 240 kg 4% FCM (fat corrected milk). The incidence rate is not reported but Jensen (1988) reported an incidence of 0.55 in Danish cows. Costs reported by US farmers per case of mastitis vary from 103 to 122 US$ (Kaneene and Hurd, 1990b; Miller and Dorn, 1990; Sischo *et al.*, 1990), which is equivalent to the value of 400–470 kg of milk, assuming a milk price of $11.7 per 45 kg of milk (Sischo *et al.*, 1990). Recent Swedish studies (Clason and Eweritt, 1986) estimated the cost as the value of 2,167 kg of milk per veterinary treated cow with subclinical mastitis. This last estimate seems to be quite high and might be due to overestimation of milk loss and other costs, but it also includes the hidden costs of subclinical mastitis that were not included in the Danish and US figures. The cost figures will depend on the definition of mastitis. A recalculation of the above figures to average cost per cow-year gives an approximate value of 130 kg of milk per cow in Denmark and USA and 430 kg in Sweden.

The genetics of mastitis have been studied in several projects and have shown that the heritability of clinical mastitis recorded in the field in first lactation cows is in the range of 0.1–5% (Emanuelsson *et al.*, 1988; Lindstrom and Syvajarvi, 1978; Philipsson *et al.*, 1980; Solbu, 1984; Syvajarvi *et al.*, 1986). This level of heritability is often considered to be too low for breeding purposes, but model calculations done in some of the above-mentioned investigations show that progeny testing should give an acceptable repeatability of breeding values. Also heritability is only one ingredient in a successful breeding scheme. A second very important factor is the genetic standard deviation and this is relatively high for clinical mastitis.

Somatic cell count is often used as a phenotypic indicator of mastitis (Brolund, 1985). It has also been shown that somatic cell count is genetically correlated to clinical mastitis in the range of 0.5–0.8 (Emanuelsson *et al.*, 1988; Madsen *et al.*, 1987). Somatic cell count shows relatively high heritability in Swedish data, 0.04–0.11 (Emanuelsson *et al.*, 1988), levels that are supported by other studies on large field data (for example, Coffey *et al.*, 1985; Madsen *et al.*, 1987; Monardes and Hayes, 1985). These findings could be used to breed for resistance to mastitis, although model calculations show that using somatic cell count alone will not be as effective as using records of clinical mastitis (E. Wretler, SHS, unpublished). Another problem might be the genetic correlation with milk production, reported to be 0.1–0.4 by Emanuelsson *et al.* (1988) and by Monardes and Hayes (1985). This could cause problems because a breeding value based on somatic cell count might have a higher unfavourable correlation with milk production than a breeding value based on clinical mastitis.

Clinical mastitis also seems to have an unfavourable genetic correlation with milk production, in the range 0.2–0.3. (Emanuelsson *et al.*, 1988; Eriksson *et al.*, 1987; Madsen *et al.*, 1987; Syvajarvi *et al.*, 1986; Wilton *et al.*, 1972). Model calculations by Christensen (1989) showed an expected increase of 4.3 cases of mastitis per 100 cow years as a correlated response to a genetic increase in milk production of 500 kg, achieved by single trait selection on milk production.

The high frequency, high costs and unfavourable genetic correlation of mastitis with milk production are the main reasons for including mastitis in a breeding scheme. Practical experience of breeding for resistance to mastitis in Sweden is reviewed in this chapter.

Materials and methods

Recording system

A recording system for veterinary treatment of diseases in farm animals, compulsory for all veterinary surgeons, was introduced in the whole country on 1 January 1984. The system had operated in the province of Skaraborg since 1971 (Lindhe *et al.*, 1978). The veterinarians record diagnoses of disease and the identification number of treated animals on a special form. This information is matched with information about pedigree, milk production and fertility. The disease records are tested to check their validity according to, for example, species and diagnoses, and to ensure that all compulsory information is recorded. About 4% of all records do not pass this test (SHS, 1990). A second test was started in 1988 (SHS, 1990), where the disease record is matched with the central database and in this test some 10% of the disease records are lost because of the lack of an identification number.

These figures can be considered reasonable since disease records are not only from milk recorded herds, but also from herds involved only in AI-recording. The latter herds, comprising about 20% of all recorded animals (SHS, 1989b), tend to have a less accurate individual identification of their animals.

The quality of the disease records has probably improved since the start of the recording system, but there are no figures available confirming this. Prior to 1988 there was no possibility of checking the loss of records because of errors in animal identification numbers.

Information from the disease-recording system since 1988 has been sent monthly to the farmers with the milk-recording report. This feedback will probably even further improve the quality of the disease records as the farmers now have the possibility of checking and correcting the information.

Definition of mastitis

The trait mastitis includes all mastitis recorded as treated by a veterinarian and any mastitis reported as a reason for culling in the milk recording scheme. Mastitis is considered as an all-or-none trait. Mastitis is scored as 0 if the cow has no record of mastitis or as 1 if she has at least one record of mastitis.

The time period considered is 10 days before to 150 days after first calving, a period that includes 75% of all mastitis reported during the first lactation. This short time-period minimizes the risk of culling for reasons other than mastitis, as culling for other reasons takes place mainly in the later part of the lactation. Variation in culling may cause problems, as a culled cow has a lesser chance of getting mastitis than a surviving cow. A second reason for this short time period is to speed the final evaluation of the bulls as the official breeding value (BV) for resistance to mastitis is based on at least 100 daughters. In comparison an official BV for milk production is based on only 15 daughters with complete 305-day first lactation records.

Somatic cell counts taken monthly or bimonthly on all cows in the milk-recording scheme have been used since 1990 to improve the prediction of breeding value for mastitis. The trait is defined as the mean of the somatic cell count during the same time period as for mastitis and log-transformed to the base 10, as suggested by Ali and Shook (1980).

Data

The disease-recording system contains data for cows inseminated as heifers for the first time in 1981 or later and is continuously up-dated with new records three times per year. In January 1990 the total number of completed five month first lactations records, after editing, were 435,027 and 253,602 for Swedish Red and White Breed (SRB) and Swedish Friesian Breed (SLB) respectively. The editing excludes records of cows less than 22 or more than 36 months of age at calving, cows not bred by AI-sires, cows in herds not using AI, cross-bred cows, except SRB crosses with SLB, heifers culled before first calving and cows not old enough to have had the possibility of completing a five-month lactation. Daughter groups of less than 10 daughters and records in herd-years (HY) with only a single record are also excluded in order to reduce computer costs.

The first evaluation of bulls was made in September 1985 and it has been run three times a year since then. Breeding values are available on disc for every evaluation since January 1986.

Models

Separate BVs of mastitic and log-transformed cell count are calculated using a linear single trait mixed model procedure. Theoretically the categorical trait mastitis should have been evaluated by a threshold model, but several studies (Meijering, 1985; Meijering and Gianola, 1985; Weller *et al.*, 1988) of traits with similar heritabilities, incidence and progeny group size, suggest that it gives similar results to a linear model. A linear model was therefore chosen, because of simpler and cheaper computations. The following mixed model is used for the routine evaluation of bulls within breed of bulls.

$$Y_{ijklmn} = HY_i + CM_j + CA_k + DB_l + BULL_m + e_{ijklmn}$$

where:
Y_{ijklmn} = mastitis or cell count record on cow ijklmn,
HY_i = fixed effect of herd-year i,
CM_j = fixed effect of calving month j,
CA_k = fixed effect of calving age k,
DB_l = fixed effect of dam-breed 1,
$BULL_m$ = random effect of the sire m of cow, and
e_{ijklmn} = random residual.

Year refers to the 12 month period July to June when calving occurred. Twelve classes were defined for calving month. Calving age consists of the 15 classes in the interval 22–36 months of age. Dam of breed included three classes, namely SRB, SLB and crosses between SRB and SLB. Accepting crossbred daughters increases the daughter group size by 5.5 and 24.6% for SRB and SLB bulls respectively (Wretler, 1985).

Heritabilities assumed for mastitis and somatic cell count are 0.02 and 0.08 respectively for both sire breeds according to the analyses by Wretler (1985) and Emanuelsson *et al.* (1988). Relationships between bulls is taken into account by the relationship matrix considering sire and maternal grandsire of the bull (Henderson, 1975).

Improved index for resistance against mastitis

An improved breeding value for resistance against mastitis is predicted by combining BVs for mastitis and somatic cell count in an index. The selection index weights are calculated using the computing algorithm described by Eriksson *et al.* (1978) and is individually calculated for each bull. The aggregate genotype consists only of mastitis, which means that the BV of somatic cell count is only used as a correlated trait. A genetic correlation of 0.70 is assumed between the two traits, as found by Emanuelsson *et al.* (1988) and Wretler (1987). The genetic variance assumed for mastitis is 0.00141 and 0.00188 for SRB and SLB respectively; for somatic cell count a value of 0.0121 is taken for both breeds. Repeatability of the BV for mastitis and somatic cell count is calculated using selection index theory as suggested by Willmink and Dommerholt (1985), taking in to account a bull's own, his sire's and his half-brother's effective number of daughters. Bull sires' repeatability is based on his own and sons' effective number of daughters.

The BVs for the separate traits and the improved BV for mastitis are all published as relative values with a mean of the base group equal to 100, and a genetic standard deviation equal to 7 relative units. A moving base is used which includes three year batches of proven bulls. The definition of bulls included in the base group is based on birth year and actual year of running the bull evaluation. Number of effective daughters has to be at least 70 in order to be officially published or included in the base group.

Results and discussion

Practical experience of breeding

Mean and standard deviation of BVs of all bulls evaluated in January 1990 are shown in Table 22.1. A comparison of the mean of unproven and proven bulls indicates that the daughters of the two categories are treated equally by the dairyman as there are no differences in the mean of the BVs. Lower BVs of proven bulls would be expected if the dairyman called the veterinarian to treat less severe mastitis in daughters of proven bulls than in daughters of unproven bulls. BVs were available for the first time in September 1985 and it takes at least 3.5 years after selection of a bull for his second batch of daughters to get records for mastitis, and 5.5 years after selection of a bull-sire for his sons to be evaluated for mastitis. All bulls, both proven and unproven, are therefore an unselected sample of bulls as far as direct selection for mastitis is concerned.

The standard deviation of BVs is increased by increasing the number of effective daughters, as shown by comparing unproven and proven bulls, indicating an increased repeatability of the BVs.

Table 22.1.   Statistics of official breeding value for mastitis traits in January 1990.

Breed	Bull category	Number	Breeding value	Mean	Standard deviation	Average number of effective daughters
SRB	Unproven[a]	874	Improved mastitis	99.8	5.2	
			Mastitis	100.0	4.4	126
			Somatic cell count	99.5	6.4	109
	Proven[b]	144	Improved mastitis	99.7	6.7	
			Mastitis	99.8	6.7	1,470
			Somatic cell count	100.0	6.9	1,294
SLB	Unproven	520	Improved mastitis	100.9	6.4	
			Mastitis	100.9	6.0	131
			Somatic cell count	100.4	6.3	110
	Proven	100	Improved mastitis	100.0	7.6	
			Mastitis	100.1	7.7	1,175
			Somatic cell count	99.1	7.2	1,001

[a]Bulls with 70 ⩽ effective daughters ⩾ 220.
[b]Bulls with more than 220 effective daughters.

The standard deviation of BVs is a function of heritability, number of effective daughters and genetic standard deviation. Using the actual figures assumed in this evaluation, the standard deviations for mastitis in the SRB-breed at 4.4 are at, or just below, the expected level of 4.8, if the extra information gained from the relationship matrix (about 50 effective daughters) is taken into account. The SLB-breed shows the opposite, the standard deviations are higher than expectation even if the genetic standard deviation initially is assumed to be 15% higher than for SRB. This could be explained by a higher heritability and/or genetic standard deviation in the population than those assumed in the calculations. This change might be explained by an increased influence of US-Holstein in the SLB breed, although they have not shown any dramatic differences in BVs compared to the Swedish bulls.

Another probable explanation is that repeatability should be calculated using the heritability based on the units of the underlying scale. A simulation study by Danell (1980) has shown that the efficiency of progeny testing for all-or-none traits is nearly as effective as progeny testing on the underlying scale, if the progeny group size × incidence level × (1 − incidence level) exceeds 10. This means that in these data, heritability ought to be transformed to the underlying scale before calculation of repeatability. Transformation of heritabilities to the underlying scale, assuming a frequency of 0.080 and 0.116 for SRB and SLB, respectively, gives 0.066 for SRB and 0.056 for SLB.

Using this heritability and 180 effective daughters the expected standard deviation for BV of mastitis for SLB is 5.9 units or quite close to the level found in the unproven bulls. On the other hand this result indicates that the real heritability or genetic standard deviation for mastitis is lower in the SRB-breed than had been assumed.

Increasing the relative BV for mastitis by one unit decreases the number of cows with mastitis in the first five months of first lactation by 0.53 and 0.62 cows per hundred cows in SRB and SLB respectively. Expressed in standard deviations of BVs for unproven bulls, the figures are 2.8 and 4.0 cows per hundred cows for the SRB and SLB breed respectively. These figures indicate the large genetic difference that exists among unproven bulls.

Regression of son's index on sire's index

A comparison of BVs for mastitis of young bulls and their sires is possible, as several of the bull sires have their second batch of daughters in the material. A problem is to make a comparison of independent BVs as the relationship matrix enforces an autocorrelation between the BVs of sire and son.

A selection of son groups, in which the oldest son had his first official

BV in September 1986 or later and the sires that had their first official BV published at least eight months before the first sons' official BV, was used to study the regression of sons and sires. This selection of data should at least give BVs of sires that are unaffected by the performance of their sons' daughters although the sons are affected by their paternal half-sisters' performance.

The expected regression coefficient is close to 0.50 because the relationship matrix imposes an autocorrelation between sire's and son's BV. The exact expectation is not known as maternal grandsires in the relationship matrix also affect the regression. Without this autocorrelation the expected regression is about 0.19 assuming a heritability of 0.02 and the effective number of daughters of sons shown in Table 22.2. The regression coefficient in this case is a function only of the repeatability of the son's BV (Danell and Eriksson, 1982). This autocorrelation also increases the expected correlation between BV to around 0.66, a level that is much higher than the maximum correlation expected from an independent BV of sires and sons. This correlation will decrease to 0.5, the theoretical additive relationship coefficient, by increasing repeatability for the two BVs. The regression and correlations fit quite well with the expectations, as shown in Table 22.2, from which we may conclude that the evaluation procedure so far works according to the theory.

A comparison of repeated estimates of the same bull

Breeding values for mastitis of bulls with at least 800 effective daughters in January 1990 were selected and compared with their earlier BVs based on at least 70 and not more than 220 effective daughters. The upper limit of 220 effective daughters is chosen because some young bulls have daughter groups of that size. The selected bulls are a mixture of bulls with their first and second batch of daughters and bulls with only their second batch of daughters, in order to get as many bulls as possible. This means that the BVs used as the first available BV is sometimes based on the second batch

Table 22.2. Regression of sons' breeding value on sires' breeding value for mastitis.

Breed	Number of		Regression	Correlation	Average number of effective daughters	
	Sires	Sons			Sires	Sons
SRB	35	368	0.43±0.036	0.54	1,045	126
SLB	25	197	0.52±0.037	0.71	876	124

of daughters but in most cases is based on the first batch of daughters. A strict comparison of BVs as a young bull and as a proven bull would of course have been the most correct comparison. Unfortunately, the data do not allow that yet, as the time period covered by bull evaluation for disease is too short.

The result (Table 22.3) shows that the mean of the BVs has not changed between the two evaluations. The correlation between the BVs as an unproven and a proven bull is the same in the two breeds and is consistent with an expected correlation of 0.67 and 0.69 for SRB and SLB respectively, assuming a heritability of 0.02 (calculated according to Clay *et al.* (1979), referred to by Danell and Eriksson, 1982). Expected correlations using heritabilities on the underlying scale are 0.85 and 0.83 for SRB and SLB bulls respectively. Compared to these the estimated correlations are too small. It can, however, be concluded that the repeatability of BVs of the same bull is fairly accurate and promising for the future.

Correction for environmental effects

The corrections for environmental effects are done simultaneously with the predictions of bulls' transmitting abilities. Herd-year effects are highly significant and reduce error variance for mastitis by 2.5 and 4.1% for SRB and SLB respectively. Other important effects are month of calving (Fig. 22.1) and age of calving (Fig. 22.2). Mastitis increases with calving age in a nearly linear fashion. The increase is a little higher for SLB than SRB. This might be an effect of the higher average frequency in SLB, 0.1248 versus 0.0836 in SRB, which means that SLB has a 50% higher frequency of mastitis compared to SRB. This difference is not a comparison between breeds, as breeds are evaluated separately, but includes breed effects and environmental differences between herds having different breeds. The two breeds are not randomly distributed across Sweden. The SLB cows are mainly distributed in the southern, western and northern part of Sweden and the SRB cows in the middle and eastern part.

Table 22.3. Comparison between estimated breeding values for mastitis for the same bull as unproven and proven bull.

Breed	No of bulls	As unproven bull		As proven bull		Correlation
		No of effective daughters	Breeding value	No of effective daughters	Breeding value	
SRB	28	145	100.8	2,039	100.4	0.68
SLB	25	150	100	1,604	100.3	0.67

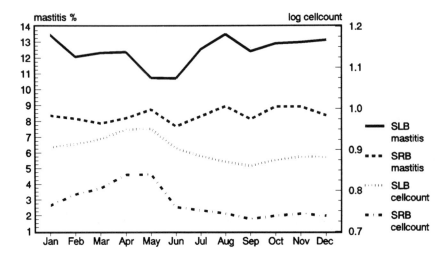

Fig. 22.1. Effect of month of calving on percentage mastitis and somatic cell count.

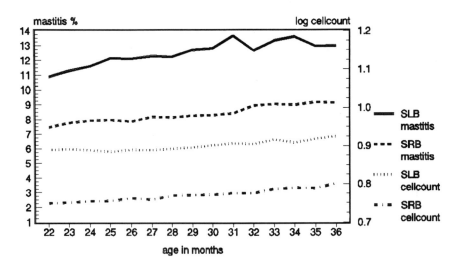

Fig. 22.2. Effect of age of calving on percentage mastitis and somatic cell count.

Month of calving effects (Fig. 22.1) show similar distinct seasonal effects in both breeds, although the magnitude is higher for SLB. Mastitis decreases when the cows are turned out to pasture in May–June. The increase during the summer months might be explained by summer mastitis affecting heifers on pasture, usually causing most problems in August.

Effect of breed of dam

Crossbred heifers are included in the sire evaluation in order to utilize the available information as effectively as possible. The breed of dam effects are assumed to correct for additive breed effects, heterosis effects and recombination effects. They are not included in order to estimate these effects but to correct the data for them. The estimates show that SRB dams have lower frequencies of mastitis in their daughters irrespective of whether it is estimated from the SRB or SLB data (Table 22.4). The difference, around one percentage point, supports the direction of the differences found in overall frequency between the two breeds but does not give an unbiased estimate of the breed differences.

Breeding values, correlations

Correlations between BVs (Table 22.5) might be used as an estimate of genetic correlation between traits. It has, however, to be adjusted for repeatability. The average number of effective daughters for SRB bulls is 126 and 109 for mastitis and somatic cell counts respectively and corresponding figures for SLB bulls 131 and 110.

Table 22.4. Estimates of breed of dam effects for mastitis and somatic cell count.

| Breed of dam | Breed of sire | | | |
| | SRB | | SLB | |
	Mastitis frequency	Somatic cell count	Mastitis frequency	Somatic cell count
SRB	−0.0055	−0.020	−0.0071	−0.017
SLB	0.0041	0.013	0.0050	0.013
SRB × SLB, SLB × SRB	0.0000	0.000	0.0000	0.000

Table 22.5. Correlations between breeding values of unproven bulls, SRB above the diagonal and SLB below the diagonal.

	Improved mastitis	Mastitis	Somatic cell count
Improved mastitis		0.87	0.81
Mastitis	0.92		0.41
Somatic cell count	0.76	0.45	

Adding 50 and 20 extra effective daughters to the average number of effective daughters for mastitis and somatic cell count, gives an estimate of 0.70 and 0.77 for the correlation between mastitis and somatic cell count for SRB and SLB respectively (because of the extra information through the relationship matrix). Theoretically including the somatic cell count will improve the standard deviation of BV for resistance to mastitis by 15% as compared to a BV based only on mastitis. This is the same as increasing the effective number of daughters from 150 to 265 per bull. The utilization of somatic cell count in an improved BV for resistance to mastitis therefore seems justified.

So far practical experience does not contradict this hypothesis, but it will take some years before it is possible to compare repeated BVs and sire-son regressions based on the improved BV.

Combining resistance to mastitis with other selection objectives

The BV for resistance to mastitis is included in a bull index combining several other traits in Table 22.6. The separate traits are weighted according to their economic weights. These are not always easy to assess, because knowledge about the impact of different traits is sometimes quite small. It is also difficult to decide on the procedure for the estimation of the economic values which is often based on a mixture of marginal incomes and cost savings. Examples are milk price – marginal feed cost for the last kg of milk produced and savings by decreased number of inseminations per pregnancy. Another complication is the genetic correlation between traits that is sometimes unfavourable. Some indications are shown in Table 22.6. The negative correlation with protein yield is a problem. Adjusting for repeatability, assuming 150 effective daughters in the progeny groups, h^2_{mast} = 0.02 and $h^2_{protein}$ = 0.25, gives estimates of genetic correlation between protein production and resistance to mastitis of 0.11 and 0.34 for SRB and SLB, respectively. The positive correlations between udder traits supports a selection for better udder conformation to counteract, at least partly, the negative influence of selection for increased milk production, especially in breeding programmes without evaluation for resistance against mastitis. It also supports a selection on udder traits in cows as mastitis records have very little value in individual selection.

In order to achieve effects in practical breeding, it is important to use weights that are not only accurate but acceptable to all involved. Otherwise the farmer will change breeds or import breeding material from other populations. The latter is a problem as bulls evaluated for several traits are often handicapped compared to bulls evaluated only for milk production and conformation traits.

The weight used in the bull index in Sweden is, per case of mastitis as defined in this paper, equivalent to the value of 1,000 kg of milk, a value

Table 22.6. Correlations[a] between improved breeding values of resistance against mastitis[b] and breeding value of other traits included in the bull index.

Breeding value[c] of	Improved breeding values of resistance against mastitis			
	SRB	(n)	SLB	(n)
Protein, kg	−0.07*	(873)	−0.21***	(520)
Daily gain	−0.06	(853)	−0.19***	(495)
Fertility of daughters	0.06	(873)	−0.09*	(520)
Stillbirth, sire of calf	0.00	(873)	0.14**	(520)
Stillbirth, maternal grandsire of calf	0.04	(873)	0.08	(520)
Resistance against other diseases	0.09**	(873)	0.11*	(520)
Udder total	0.27***	(193)	0.45***	(114)
Udder–floor distance	0.25***	(193)	0.43***	(114)
Teats length	0.02	(193)	0.17	(114)
Teats placement	0.14*	(193)	0.22***	(114)
Legs total	−0.08	(193)	0.20*	(114)
Extra teats	−0.03	(193)	0.16	(114)
Temperament	−0.03	(193)	−0.09	(114)
Bull index	0.31***	(193)	0.26**	(114)

* $P < 0.05$; ** $P < 0.01$; *** $P < 0.001$ under the hypothesis $r = 0$.
[a] Positive values are favourable.
[b] Repeatability = 0.6.
[c] Repeatabilities vary from 0.4–0.9 for different traits.

that is higher than estimates found elsewhere in the literature. A higher value may be justified to account for the effects of subclinical mastitis, milk quality and animal welfare that are hard to evaluate economically.

The correlation between the bull index and resistance to mastitis is positive in both breeds and guarantees on average a positive selection for resistance to mastitis in bulls. The total effect in the population also depends on the selection of cows and their contribution is not quite clear. They are mainly selected on milk yields and conformation but a certain amount of natural selection operates since mastitis is a major criterion of early culling.

Discussion

Evaluation of a practical breeding programme is not easy, as it usually has to be done by indirect methods. A common approach is to estimate genetic

changes from field data, a method that takes into account both the repeatability of BVs and selection of breeding animals. This method cannot be
applied in this study, because the breeding programme for resistance to
mastitis has not been running for more than 4 years. The approach used
was to test whether the theoretical expectations are fulfilled in the practical
evaluation of bulls.

Standard deviations of BVs were a little higher than expected for the
SLB-breed, which could be explained by higher heritability and/or genetic
standard deviation than found in earlier material. This explanation is the
most probable and is supported by the relatively high correlation found
between sire and son's BV. The increase corresponds to a 30% increase in
the genetic standard deviation for SLB from 0.042 to 0.055 frequency of
mastitis, a value about 50% above SRB, which might partly be explained
by the higher phenotypic standard deviation. This high genetic standard
deviation would imply a better possibility of genetic improvement in the
SLB breed.

The comparison of repeated estimates of the same bull showed clearly a
repeatability of BVs at the expected level. This is quite promising as it is
unusual to find values of this magnitude in Swedish field data. A similar
study of milk production traits in Sweden by Danell and Eriksson (1982),
showed lower repeatability than expected.

The repeatability of BVs for bulls and the regression of the BV of bulls
on the BV of their sires, shows that it is possible to change resistance to
mastitis by practical breeding. All parameters studied in the field indicate a
heritability of 0.02. This does not contradict the result by Danell (1980)
that heritability on the underlying scale should be used in connection with
progeny testing, but rather supports it, as realized heritabilities in the field
are usually smaller than expected.

In practice there are several problems in incorporating breeding for
resistance to mastitis in a breeding scheme, especially taking the international aspect of the bull semen market into account. In Sweden, for
example, all bull sire semen of the SLB breed last year was imported from
the US. This could place a barrier to the genetic improvement of resistance
to mastitis in the Swedish population of SLB cows. Under these conditions
it is actually a handicap for a bull to have BV on traits that are unique to a
specific country, especially when some of the functional traits, for example
female fertility and resistance to mastitis, seems to be unfavourably
correlated with milk production.

On the other hand putting too much weight on mastitis in a breeding
scheme could also be a problem. Farmers and AI-studs sometimes refuse to
use bulls with poor values for mastitis resistance although the overall merit
may be very high. In the long run this could have the effect that breeds with
good resistance to mastitis might disappear because of a lower overall
efficiency, thus increasing the frequency of mastitis in the national herd. It

is, therefore, important to assess accurately the economic value of this trait including aspects such as impact on milk quality, milk production, labour, consumer attitude and animal welfare. If done properly breeding for increased resistance to mastitis has great possibilities.

References

Ali, A.K.A. and Shook, G. (1980) An optimum transformation for somatic cell concentration in milk. *Journal of Dairy Science* 63, 487–90.

Anon. (1987) Husdjurens Hälsokontroll Finlands Husdjursavelsförening, Vanda, Finland, 44 pp.

Brolund, L. (1985) Cell counts in bovine milk. Causes of variation and applicability for diagnosis of subclinical mastitis. *Acta Veterinaria Scandinavica*, Suppl. 80, 1–123.

Christensen, L.G. (1989) Health at different genotype. Report 660. National Institute of Animal Science, Tjele, Denmark, Chapter 4. 20 pp.

Clay, J.S., Vinson, W.E. and White, J.M. (1979) Repeatability as an indicator of stability in contemporary comparison sire evaluations. *Journal of Dairy Science* 62, 1132–9.

Clason, A. and Eweritt, B. (1986) Ekonomiska Förluster vid Mastit. I: Juverinflammationer och deras Bekämpande. Swedish Association for Livestock Breeding and Production, Eskilstuna, Sweden.

Coffey, E.M., Vinson, W.E. and Pearson, R.E. (1985) Heritability for lactation average of somatic cell counts in first, second and third or later parities. *Journal of Dairy Science* 68, 3360.

Danell, B. and Eriksson, J.A. (1982) The direct sire comparison method for ranking of sires for milk production in the Swedish dairy cattle population. *Acta Agriculturae Scandinavica* 32, 47–64.

Danell, Ö.E. (1980) A note on the efficiency of progeny testing for all-or-none traits. Report 42. Department of Animal Breeding and Genetics, Swedish University of Agricultural Sciences, S-750 07 Uppsala, Sweden.

Emanuelsson, U., Danell, B. and Philipsson, J. (1988) Genetic parameters for clinical mastitis, somatic cellcounts, and milk production estimated by multiple-trait restricted maximum likelihood. *Journal of Dairy Science* 71, 467–76.

Eriksson, J.A., Wilton, J.W. and Henningsson, T. (1978) Estimating breeding values for rate of gain of beef bulls in Sweden. Report 23. Department of Animal Breeding and Genetics, Swedish University of Agricultural Sciences, Uppsala, Sweden.

Eriksson, J.A. and Wretler, E. (1987) Sire evaluation for diseases in Sweden. *38th Annual Meeting of European Association of Animal Production*, Lisbon.

Gardner, I.A., Hird, D.W., Christiansen, K.H., Sischo, W.M., Utterback, W.W., Danaye-Elmi, H. and Horn, B.R. (1990) Mortality, morbidity, case-fatality, and culling rates for California dairy cattle as evaluated by the National Animal Health Monitoring System, 1986–87. *Preventive Veterinary Medicine* 8, 2–3, 157–70.

Henderson, C.R. (1975) Inverse of a matrix of relationship due to sires and maternal grandsires. *Journal of Dairy Science* 58, 1917–21.

Jensen, N.E. (1988) Muligheder og begraensninger vedr mastitis – Definition og diagnostik. Rapport fra Nordisk Seminar om avlsvaerdivurdering af sygdomme den 29–30 September, Skejby, pp. 16–27.

Kaneene, J.B. and Hurd, H.S. (1990a) The National Animal Health Monitoring System in Michigan. I. Design, data and frequencies of selected dairy cattle diseases. *Preventive Veterinary Medicine* 8, 103–14.

Kaneene, J.B. and Hurd, H.S. (1990b) The National Animal Health Monitoring System in Michigan. III. Cost estimates of selected dairy cattle diseases. *Preventive Veterinary Medicine* 8, 127–40.

Lindhe, B., Hedebro, I. and Brolund, L. (1978) Efforts to include disease resistance in breeding programmes for dairy cattle. *29th Annual Meeting of European Association of Animal Breeding and Production*, Stockholm, 1978.

Lindström, U.B. and Syväjärvi, J. (1978) Use of field records in breeding for mastitis resistance in dairy cattle. *Livestock Production Science* 5, 29–44.

Madsen, P., Nielsen, S.M., Rasmussen, M.D., Klastrup, O., Jensen, N.E., Jensen, P.T., Madsen, P.S., Larsen, B. and Hyldgaard-Jensen, J. (1987) Undersögelser over genetisk betinget resistens mod mastitis. Report 621. National Institute of Animal Science, Copenhagen, Denmark, 227 pp.

Meijering, A. (1985) Sire evaluation for calving traits by best linear unbiased prediction and nonlinear methodology. *Zeitschrift für Tierzüchtung und Züchtunsbiologie* 102, 95–105.

Meijering, A. and Gianola, D. (1985) Observations on sire evaluation with categorical data using heteroscedastic linear models. *Journal of Dairy Science* 68, 1226–32.

Miller, G.Y. and Dorn, C.R. (1990) Costs of dairy cattle diseases to producers in Ohio. *Preventive Veterinary Medicine* 8, 171–82.

Monardes, H.G. and Hayes, J.F. (1985) Genetic and phenotypic relationships between lactation cellcount and milk yield and composition of Holstein cows. *Journal of Dairy Science* 68, 1250–6.

Östergaard, V. (1989) The economics of fertility and health traits. *40th Annual Meeting of the European Association of Animal Production*, Dublin.

Philipsson, J., Thavfelin, G. and Hedebro-Velander, I. (1980) Genetic studies on disease recordings in first lactation cows of Swedish dairy breeds. *Acta Agriculturae Scandinavica* 30, 327–35.

Sischo, W.M., Hird, D.W., Gardner, I.A., Christiansen, K.H., Carpenter, T.E., Utterback, W.W., Danaye-Elmi, C. and Heron, B.R. (1990) Economics of disease occurrence and prevention on California dairy farms: a report and evaluation of data collected for the National Animal Health Monitoring System 1986–87. *Preventive Veterinary Medicine* 8, 141–56.

Solbu, H. (1988) Helsekortordningen 1987. *Buskap og Avdrått* 2, 96–7.

Solbu, H. (1984) Disease recording in Norwegian dairy cattle. II. Heritability estimates and progeny testing for mastitis, ketosis and 'all Diseases'. *Zeitschrift für Tierzüchtung und Züchtungsbiologie* 101, 51.

SHS (1989a) Djurhälsovården 87/88. SHS meddelande 158, 50 pp.

SHS (1989b) Årsstatistik från SHS 1987/88. SHS meddelande 159, 66 pp.

SHS (1990) Djurhälsovård 88/89. SHS meddelande 161, 4 pp.

Syväjärvi, J., Saloniemi, H. and Gröhn, Y. (1986) An epidemiological and genetic study on registered diseases in Finnish Ayrshire cattle. IV. Clinical mastitis. *Acta Veterinaria Scandinavica* 27, 223–34.

Weller, J.W., Misztal, U. and Gianola, D. (1988) Genetic analysis of dystocia and calf mortality in Israeli-Holsteins by threshold and linear models. *Journal of Dairy Science* 71, 2491–501.

Wilmink, I.B.M. and Dommerholt, J. (1985) Approximate reliability of best linear unbiased prediction in models with and without relationships. *Journal of Dairy Science* 68, 946–52.

Wilton, J.W., van Vleck, L.D., Everett, R.W., Guthrie, R.S. and Roberts, S.J. (1972) Genetic and environmental aspects of udder infections. *Journal of Dairy Science* 55, 183.

Wretler, E. (1985) Mimeo. Analys av Sjukdatamaterial. Swedish Association for Livestock Breeding and Production, Eskilstuna, Sweden.

Wretler, E. (1987) Genetisk Analys av Mastit. Mimeo. Swedish Association for Livestock Breeding and Production, Eskilstuna, Sweden.

Chapter 23

Mastitis in Sheep

J.E.T. Jones

Department of Animal Health, Royal Veterinary College,
Bolton's Park, Potters Bar, Herts EN6 1NB, UK

Summary

Ovine mastitis occurs in all countries of the world in which sheep are kept but the disease has been studied mostly in countries in which there are milking flocks. In the UK there have been no systematic studies on mastitis since 1928 and for this reason the present field investigation was carried out on a population of 30,000 ewes in lowland flocks, between 1985 and 1987. During this period the annual incidence of acute mastitis in individual flocks ranged between 0 and 24% with a mean of about 5%. Most cases occurred in the first week and between the third and fourth weeks of lactation. *Staphylococcus aureus* and *Pasteurella haemolytica* occurring alone, together, or with other bacteria were the predominant pathogens isolated from the milk of almost 90% of 730 affected ewes which were sampled.

Subclinical mastitis occurs in ewes and results in reduced milk yield and decreased growth rate in lambs. Over 50% of the cases were caused by coagulase-negative staphylococci. Studies on cell counts indicate that a value of 10^6/ml is satisfactory as a provisional upper limit of normality on which to base further work. The Whiteside and California tests are effective screening tests for subclinical mastitis.

Possible methods of preventing mastitis include vaccination, removal of predisposing causes, when these are understood, and breeding for resistance. There are several deficiencies that need to be remedied before genetic control of mastitis can be contemplated. There is a need for more data on incidence so that the degree of variation can be assessed. We need an effective recording system in the national flock. A considerable effort is required to obtain information on possible genetic influences on the occurrence of mastitis; for example, on the effects of breed, of conformation of the udder and teats and of genetically controlled mammary gland defence mechanisms.

Introduction

Mastitis in ewes occurs in all countries of the world in which sheep are kept but most attention has been paid to the disease in those countries in which there are milking flocks, e.g. France, Italy, Spain and the Balkans. Investigations in meat flocks have been reported occasionally from a number of countries, e.g. New Zealand (Quinlavan, 1968) and the USA (Shoop and Myers, 1984).

In the UK there have been sporadic reports on particular aspects of mastitis, e.g. as a cause of culling (Herrtage *et al.*, 1974; Madel, 1981), dry ewe therapy (Buswell and Yeoman, 1976; Gibson and Hendy, 1976) and the estimation of the number of cells in milk for the detection of subclinical mastitis (Green, 1984; Mackie and Rodgers, 1986). There has been no systematic study of the clinical disease since that reported by Leyshon (1929) who investigated mastitis among flocks in East Anglia. Interestingly, the principal bacteria isolated by Leyshon 60 years ago are still regarded as the most common infectious agents causing ovine mastitis in the UK.

The economic loss in sheep flocks in the UK resulting from the various forms of mastitis has not been quantified but is likely to be considerable. The possible adverse effects of mastitis include:

1. Acute disease followed by death of the ewe.
2. Acute disease followed by systemic recovery and the necessity to cull the ewe because of a permanently damaged udder.
3. Replacement costs.
4. Veterinary costs.
5. Reduced milk production which may result in suboptimal growth of lambs.
6. Lamb mortality.

Despite the importance of mastitis, worldwide, there have been few comprehensive studies of the disease and, consequently, little information is available on genetic resistance to the introduction and establishment of infection in the ovine mammary gland. Consideration of possible genetic and other means of reducing the incidence or minimizing the effects of mastitis must be based on an understanding of the aetiology, pathogenesis and epidemiology of the disease. The account which follows is an attempt to summarize our own investigations and to review briefly existing information relevant to breeding for resistance to ovine mastitis.

A three-year field survey of mastitis in lowland flocks in England and

Wales began in 1985 and concentrated mainly on the clinical disease but also investigated subclinical mastitis.

The main objectives were:

1. To provide data on the incidence of mastitis.
2. To establish the relative importance of the various bacteria causing the disease.
3. To establish the period during lactation at which mastitis most commonly occurs.
4. To establish whether subclinical mastitis is important, especially in relation to the growth of lambs.

An outline of the preliminary results was reported at the end of the first year (Jones, 1985) and the account which follows is based largely on the subsequent work.

Survey procedures

In conjunction with the Veterinary Investigation Service a population of ewes varying between 20,000 and 30,000 in up to 70 flocks was kept under surveillance during the period 1985–1987. All cases of mastitis detected by the flock owner were recorded and, wherever possible, samples of milk from infected ewes were examined bacteriologically.

At the time of weaning, ewes in many flocks were examined physically so that chronic forms of mastitis could be detected.

In some flocks, samples of milk from a proportion of healthy ewes were examined bacteriologically and by counting the number of somatic cells present in the milk. Simple screening tests – the Whiteside Test (WST) and the California Mastitis Test (CMT) – for evidence of subclinical mastitis were also applied.

Results

Clinical disease

Acute mastitis

The nature and appearance of the milk samples submitted for examination together with the case history provided, confirmed that all samples originated from the acute form of mastitis. In the vast majority of ewes only one 'half' of the udder was affected.

Incidence   The incidence ranged between 0 and 24%; there was much variation between flocks and between years. The mean incidence was 5%; the majority of flocks had an incidence of less than 6%. In 1987, 13% of flocks had an incidence exceeding 10%.

Submission rate of milk samples for laboratory examination was highest during the first week and between the third and fourth week of lactation. It is a reasonable assumption that this reflected the periods of peak incidence of acute mastitis. Of 595 ewes for which the stage of lactation at which they were affected was known, samples from 319 (53.6%) were submitted during the first 5 weeks of lactation and almost 85% of all samples were submitted during the first 8 weeks. Some cases were not observed during lactation but were detected when the ewes were examined after weaning.

Bacteriological findings   Of 730 milk samples from ewes affected with acute mastitis, 647 (89%) were bacteriologically positive (Table 23.1).

The results of the examination of the bacteriologically 'positive' samples are summarized in Table 23.2 from which it can be seen that *Staphylococcus aureus* and *Pasteurella haemolytica* occurring alone, together, or in

Table 23.1.   The results of the bacteriological examination of milk samples from 730 ewes affected with acute mastitis (1985–1987).

	1985	1986	1987	Total
Bacteriologically positive samples	197 (88%)	286 (91%)	164 (87%)	647 (89%)
Bacteriologically negative samples	26 (12%)	30 (9%)	27 (13%)	83 (11%)
Total examined	223 (100%)	316 (100%)	191 (100%)	730 (100%)

Table 23.2.   The results of the examination of 647 bacteriologically 'positive' milk samples.

Bacterial isolates	*n*	%
Staphylococcus aureus	236	36.5
Pasteurella haemolytica	212	32.8
Staph. aureus + other bacteria	32	4.9
P. haemolytica + other bacteria	29	4.5
Staph. aureus + *P. haemolytica*	20	3.1
Other bacteria	118	18.2
	647	100.0

combination with other bacteria were isolated from over 80% of bacterio-logically positive samples. These results clearly demonstrate that *P. haemoloytica* is a major mastitis pathogen, a fact not previously appre-ciated because the traditional view has been that most cases of ovine mastitis are caused by *Staph. aureus.*

Chronic mastitis

In recent years it has become common practice to refer to chronic mastitis which may be observed at the premating examination of ewes as 'post-weaning' mastitis. The term implies, erroneously, that the mastitis is a consequence of weaning. Several thousand ewes were examined in this study at the time of weaning and at intervals of many weeks thereafter and very few cases of mastitis were observed to develop after weaning. The majority of cases of so-called 'post-weaning' mastitis are examples of pre-weaning mastitis detected post-weaning. They represent the chronic end point of acute mastitis that developed during lactation. For example, in one investigation 3,829 ewes, distributed among seven flocks were examined and 320 (8.4%) were found to be affected with chronic mastitis (usually manifested by multiple abscesses). Most of these chronic cases represented the sequelae of acute disease during lactation that had not been detected; a few would have been detected and been retained in the flock because one-half of the udder was functioning satisfactorily.

Subclinical mastitis

It has been generally assumed that the main adverse economic effects of ovine mastitis are those associated with clinical disease. The possible economic impact of subclinical mastitis has been largely ignored. Its importance as a limiting factor in milk production in cows is well docu-mented but there is little information on its effect on milk yield in the ewe.

In one of the present studies, bacteriological and cellular examination of milk samples obtained from ewes in several flocks showed that the pre-valence rate of subclinical mastitis ranged from 10% to 31%. Most sub-clinical infection was associated with coagulase-negative staphylococci (accounting for about 50% of all cases), streptococci, *Staph. aureus,* *P. haemolytica,* and *Actinomyces pyogenes.*

Somatic cell counts of over 2,000 milk samples from clinically healthy ewes indicated that the adoption of an upper limit of 1.0×10^6/ml for apparently normal milk, would provide, tentatively, a satisfactory working standard (El-Masannat, 1987).

Examination of the correlation between electronic milk cell counts and bacteriological findings during the period 2–18 weeks of lactation revealed that when a value of 1.0×10^6 cells/ml was taken as the upper limit, the sensitivity of this correlation was 83.6% and the specificity 87.5%. Counts above 1.0×10^6 cells/ml were correlated with a high proportion of infections.

Study of the relationship between the cell counts and the scores of the WST and the CMT showed that when a value of 1.0×10^6 cells/ml was taken as the upper limit for normal milk samples, 91.6% and 93.0% of samples with scores of + or more in the WST and CMT respectively were above this limit. (WST and CMT results were scored on a range of + to ++++.)

The WST and CMT are simple techniques for the detection of sub-clinical mastitis; they have proved to be effective screening tests in the ewe and can be employed under field conditions. In both tests, milk samples yielding scores of + and above should be examined bacteriologically.

The effects of experimentally induced staphylococcal subclinical mastitis on milk production in ewes and the growth of their lambs

Coagulase-negative staphylococci were isolated in pure culture in a small proportion of cases of acute mastitis and in a substantial proportion (almost 50%) of cases of subclinical mastitis. These organisms are usually regarded as non-pathogenic or of low pathogenicity and therefore warranted further study.

In each of two experiments (Fthenakis and Jones, 1990), subclinical mastitis was induced in ewes by inoculating both mammary glands with approximately 10^7 cfu of *Staph. simulans*, a coagulase-negative staphylo-coccus. In the first experiment, 20 primiparous Welsh Mountain ewes were divided into four groups: ewes in group W6 ($n=6$) were inoculated on day 6, those in group W16 ($n=6$) on day 16 after lambing; ewes of group WC6 ($n=4$) and in WC16 ($n=4$) were uninoculated controls. Lambs were not allowed solid food. In the second experiment, ten secundiparous Dorset Horn ewes were divided into two groups: ewes in group D6 ($n=5$) were inoculated on day 6 after lambing and those in group DC6 ($n=5$) were uninoculated controls. Lambs were given creep feed *ad libitum* and the amount consumed was measured.

Staphylococci were isolated consistently from the inoculated glands and the somatic cell counts on milk from these glands increased significantly ($P<0.001$). Milk yield decreased.

The mean volume of milk collected from ewes of group WC6 was 3,329 ml and that from ewes of group W6 was 2,421 ml, i.e. 27.3% less; that from ewes of group WC16 was 3,708 ml and that from ewes of W16 was 2,861 ml, i.e. 22.8% less. The volume from ewes of group DC6 was 8,564 ml and that from ewes of group D6 was 5,365 ml, i.e. 37.3% less. These differences were significant ($P<0.01$). No significant differ-ences in milk composition were recorded.

Table 23.3.   The effect of subclinical mastitis on the growth of lambs.

Group	Lamb weight (kg)		Daily weight gain (g)
	at 2nd day	at 52nd day	
W6	3.1 (SE 0.04)	8.1 (SE 0.33)	97 (SE 6.7)
WC6	3.1 (SE 0.18)	11.3 (SE 0.67)	163 (SE 9.7)
W16	3.0 (SE 0.26)	9.8 (SE 0.54)	139 (SE 8.8)
WC16	3.2 (SE 0.33)	10.8 (SE 0.31)	149 (SE 5.7)
D6	3.4 (SE 0.05)	16.2 (SE 0.10)	256 (SE 1.1)
DC6	3.3 (SE 0.07)	17.7 (SE 0.17)	290 (SE 2.6)

Source: Fthenakis and Jones (1990).

The weight of the lambs and their weight gain during the experiments are summarized in Table 23.3. There were differences in growth rate between the inoculated and control groups in both experiments ($P < 0.001$).

In these experiments additional information was obtained on somatic cell counts. Of 1,408 samples of apparently normal milk, 98.2% had a somatic cell count of less than 10^6/ml whereas 85.5% of 254 bacteriologically positive samples had a count greater than 10^6/ml.

The results showed the potential adverse economic effects of subclinical mastitis. The practicalities of management and husbandry make it unlikely that subclinical mastitis will ever be investigated in commercial meat flocks. However, the potential importance of subclinical mastitis needs to be considered in at least two circumstances. First, any nutritional research on milk production in ewes or on growth rate of lambs should take into account the possible presence of subclinical mastitis, because these experiments have shown how milk yield and lamb growth may be significantly diminished when ewes are affected. Second, subclinical mastitis in ewes could undoubtedly affect the productivity of dairy flocks and, for this reason, it requires investigation. The occurrence of clinical and subclinical mastitis is currently being monitored in 3,000 dairy ewes.

The clinical forms of mastitis may cause serious economic loss in individual flocks and probably cause substantial loss nationally. The extent to which subclinical disease may have adverse effects on milk production in meat flocks would be difficult to quantify and many would consider that the required effort would not be worthwhile. However, it should be possible to calculate the loss of production in dairy flocks when more information on prevalence becomes available.

The control of mastitis

Treatment

The treatment of acute mastitis in ewes is generally ineffective in restoring normal function of the affected gland. Often, the most that can be expected is that prompt treatment with appropriate antibiotics will ensure the survival of the ewe. In many instances, partially recovered ewes constitute a source of infection and should be culled.

Prevention

There are three possible approaches: (1) vaccination, (2) the modification of techniques of management and husbandry when factors predisposing to mastitis are defined and (3) breeding for resistance.

Vaccination   There are no commercially available vaccines for the prevention of mastitis in any of our domesticated animals. Despite the difficulties inherent in the production of effective vaccines, outlined by Anderson (1978), Watson (1988) in Austrialia is obtaining promising results, experimentally.

Defining predisposing causes   The principal bacteria causing mastitis are present in the tissues of healthy sheep. *Staph. aureus* may inhabit the skin, the oropharynx, the nasal passages and the vagina; it may be present in the vicinity of the teat orifice and in the teat canal without necessarily causing disease. *P. haemolytica* is present in the oropharynx and tonsils of lambs; the mouth of the lamb is the likely source of infection for the mammary gland. How do these pathogens gain access to the teat canal and thence to mammary tissue?

Factors considered to facilitate infection of the gland include conformation of the udder and the teats, anatomical defects of the teat; lesions on the skin of the teats or the gland, e.g. contagious ecthyma (orf), staphylococcal dermatitis, abrasions caused by lamb's teeth during vigorous sucking; faecal contamination of the teats, and transmission of pathogens by vectors. Establishment of infection in mammary tissue and development of the disease are influenced by the effectiveness of local defence mechanisms within the gland and the immunological status of the ewe which may, in turn, be affected by external factors tending to increase susceptibility, e.g. climate (wind chill and rain), population density. Breed and age may influence susceptibility (Watson *et al.*, 1990). More evidence is needed about the possible effects of these putative factors and on the mechanisms by which they act.

Breeding for resistance   Bacterial diseases of domesticated animals have not been significantly reduced in incidence or eliminated by genetic means.

Nevertheless, there are examples of genetic influences on the occurrence of bacterial diseases of sheep that could be exploited. Sheep of the Border Leicester, Romney and Dorset Horn breeds seem more resistant to foot rot (*Bacteroides nodusus*) than are Merinos; even vaccinated Merinos are less resistant than are vaccinated sheep of other breeds. In this example different susceptibility may be related to difference in the properties of the epidermis of the interdigital skin (Lantier and Vu Tien Khang, 1988). There is experimental evidence from France that lambs of the Ile-de-France breed are more resistant to caseous lymphadenitis, a disease caused by *Corynebacterium pseudotuberculosis*, than those of the Romanov and Prealpes du Sud breeds (Pepin *et al.*, 1988).

No evidence is available of possible genetic influences on the occurrence of ovine mastitis in the UK but some progress has been made in acquiring a better understanding of the disease. More is now known about incidence and considerably more about the stage of lactation at which mastitis is most prevalent; most of all there is now better information about which pathogens cause the disease. However, there remain several deficiencies which need to be remedied before any genetic control of mastitis can be contemplated.

First, more data are needed on incidence because of the need to know whether the disease displays sufficient genetic variation to determine whether a breeding programme, designed to enhance resistance, would be possible or, indeed, desirable. Second, an effective national recording system is needed. The maintenance of accurate, well-devised records of production and disease in individual animals is the cornerstone of any scheme for improvement in animal husbandry and health. The difficulty in obtaining records necessary to evaluate epidemiological characteristics of mastitis was well exemplified in our survey work. It was impossible to obtain the records required to allow general statements to be made about breed, families, parity, number of lambs suckled, milk yield in relation to the occurrence of mastitis. In many flocks individual ewes cannot be identified so that linking together, for example, breed, parity, number of lambs born and number reared, to provide information relating the individual to the whole population is not possible. This would be a major constraint on breeding programmes. Third, more information is needed about interactions and overlap between environmental and genetic factors influencing susceptibility.

In dairy sheep flocks, the potential for accurate recording is much greater than in meat flocks. Ewes are seen and handled daily during lactation; records of individuals can be similar to those maintained for dairy cows. The milk yield can be recorded and subclinical infection of the mammary gland can be easily detected by cell counts and bacteriological examination.

As bovine mastitis is so widespread and of such substantial economic importance, there is a voluminous literature on factors influencing susceptibility and resistance; the literature on these factors in sheep is scanty. Does information obtained about genetic resistance to bovine mastitis provide guidelines which, with appropriate modification, could be used for sheep?

There are few estimates of genetic variability in resistance to ovine mastitis. In contrast, many exist for bovine mastitis (Emanuelson, 1988) based on evaluation of clinical and subclinical disease. Any selection against clinical mastitis should be on an all-or-none basis; experience shows that the acute and chronic forms of mastitis are readily detected and the frequency of diagnostic errors would be low. In dairy sheep, information on subclinical mastitis could easily be augmented if there is a will to do so. It may be that reducing the occurrence of subclinical mastitis by genetic means is a realistic proposition and should be the subject of enquiry and discussion. Torres-Hernandez and Hohenboken (1979) reported a difference in incidence in subclinical mastitis between various breeds, Finnsheep and Romney crossbreds having a lower incidence than Cheviot and Dorset crossbreds. Charon and Skolasinski (1988) have reported differences in the prevalence of subclinical mastitis in Polish sheep, the Merino having a lower rate than the Polish Mountain. Recently, Watson *et al.* (1980) showed that there can be a significant difference in prevalence of infection between breeds. They found the prevalence in Border Leicester × Merino to be 10% whereas in Border Leicester × Booroola Merino and in the Merino, it was only 4%. In dairy cows there is a positive correlation between rate of infection and milk yield. It is known that in the ewe the genotype significantly influences milk yield (Owen, 1976). This is a salutary reminder that desirable production traits and resistance traits may not be favourably related; indeed they may be antagonistic as reported by Dario and Bufano (1991) who found that in Italy the incidence of clinical mastitis in Sarda ewes was much higher (17.7%) than in Altamurana ewes (1.43%); the milk yield of the Sarda was consistently greater than that of the Altamuranas.

The possible relationship between morphological features of the udder and teats of cows and resistance to mastitis has been reviewed by Seykora and McDaniel (1985) who concluded that firmly attached, higher udders are less susceptible to mastitis than pendulous udders. Teat end shape was also found to correlate with resistance to mastitis; in cows that have inverted or disk-shaped teat ends the incidence of mastitis is significantly higher than in cows with round teat ends.

In sheep, Mavrogenis *et al.* (1988) have shown that udder depth and udder circumference, traits that indicate the type and volume of the udder, have moderate to high estimates of heritability; these features were correlated positively with milk production. Teat length was highly correlated with teat diameter. Some of these characteristics could be related to resistance to mastitis but there are no reports providing evidence of such a relationship.

In cows, Madsen (1989) has shown that a high cell count in milk is genetically linked to a higher incidence of clinical mastitis. Information on cell counts in ewe's milk is accumulating steadily and it should soon be possible to demonstrate whether a similar link applies in sheep.

The role of local defence mechanisms in the mammary gland of sheep needs further investigation but it is likely that the genetic variation in the activity of lysozyme, lactoferrin, immunoglobulins and phagocytes, shown to operate in cows, also operates in ewes.

Ultimately, when the pathogenesis of mastitis is better understood, genetic analysis of the mechanisms of host resistance (or susceptibility) to the sequential process of infection, the establishment of pathogens in tissues and of the mode of action of their toxic products may be possible. Such knowledge would aid selection by indirect means, but it must be borne in mind that pathogens have the capability of adaptation and of developing mechanisms for subverting host defences. However, we are far from understanding the activities of mastitis pathogens at a cellular and molecular level so that our immediate aims should be to assess the desirability and feasibility of increasing resistance to mastitis in sheep by direct methods of selection. To do this there is a need to investigate the incidence and causes of mastitis in pedigree flocks or in well-defined cross-bred populations in which individuals are identifiable and in which flock records are sufficiently comprehensive and reliable for worthwhile conclusions to be drawn about possible breed and family susceptibility.

References

Anderson, J.C. (1978) The problem of immunization against staphylococcal mastitis. *British Veterinary Journal* 134, 412–20.

Buswell, J.F. and Yeoman, G.H. (1976) Mastitis in dry ewes. *Veterinary Record* 99, 221–2.

Charon, K.M. and Skolasinski, W. (1988) The genetic variability of the resistance of sheep to mastitis and the possibility of its application in breeding. *Proceedings of the 3rd World Congress on Sheep and Cattle Breeding*, Paris, vol. 1, pp. 640–2.

Dario, C. and Bufano, G. (1991) Investigation of mastitis occurrence in purebred and crossbred ewes. In: Owen, J.B. and Axford, R.F.E. (eds), *Breeding for Disease Resistance in Farm Animals*, CAB International, Wallingford, p. 479 (Abstr.).

El-Masannat, E.T.S. (1987) Ovine mastitis with special reference to mastitis caused by *Pastuerella haemolytica*. PhD thesis, University of London.

Emanuelson, U. (1988) Recording of production diseases in cattle and possibilities for genetic improvements: A review. *Livestock Production Science* 20, 89–106.

Fthenakis, G.C. and Jones, J.E.T. (1990) The effect of experimentally induced subclinical mastitis on milk yield of ewes and on the growth of lambs. *British Veterinary Journal* 146, 43–9.

Gibson, I.R. and Hendy, P.G. (1976) Mastitis in dry ewes. *Veterinary Record* 99, 511–12.

Green, T.J. (1984) Use of somatic cell counts for detection of subclinical mastitis in ewes. *Veterinary Record* 114, 43.

Herrtage, M.E., Saunders, R.W. and Terlecki, S. (1974) Physical examination of cull ewes at point of slaughter. *Veterinary Record* 95, 257–60.

Jones, J.E.T. (1985) An investigation of mastitis in sheep: preliminary phase. *Proceedings of the Sheep Veterinary Society* 10, 48–51.

Lantier, F. and Vu Tien Khang, J. (1988) Genetic variability of resistance to infectious disease. *Proceedings of the 3rd World Congress on Sheep and Cattle Breeding*, Paris. Vol. 1, 531–52.

Leyshon, W.J. (1929) An examination of a number of cases of ovine mastitis. *Veterinary Journal* 85, 286–300; 331–44.

Mackie, D.P. and Rodgers, S.P. (1986) Mastitis and cell content in milk from Scottish Blackface ewes. *Veterinary Record* 118, 20–1.

Madel, A.J. (1981) Observations on the mammary glands of culled ewes at the time of slaughter. *Veterinary Record* 109, 362–3.

Madsen, P. (1989) Genetic resistance to bovine mastitis. In: Van der Zijpp, A.J. and Sybesma, W. (eds), *Improving Genetic Disease Resistance in Farm Animals*, Kluwer Academic Publishers, Dordrecht.

Mavrogenis, A.P., Papachristoforou, C., Lysandrides, P. and Roushias, A. (1988) Environmental and genetic factors affecting udder characteristics and milk production in Chios sheep. *Genetique Selection Evolution* 20, 477–88.

Owen, J.B. (1976) *Sheep Production.* Baillière Tindall, London, p. 141.

Pepin, M., Lantier, F., Pardon, P. and Marly, J. (1988) Breed susceptibility to experimental *Corynebacterium pseudotuberculosis* infection in lambs. *Proceedings of the 3rd World Congress on Sheep and Cattle Breeding*, Paris, Vol. 1, 646–8.

Quinlivan, T.D. (1968) Survey observations on ovine mastitis in New Zealand stud Romney flocks. *New Zealand Veterinary Journal* 16, 149–60.

Seykora, A.J. and McDaniel, B.T. (1985) Udder and teat morphology related to mastitis resistance: a review. *Journal of Dairy Science* 68, 2087–93.

Shoop, D.S. and Myers, L.L. (1984) Serologic analysis of isolates of *Pasteurella haemolytica* and *Staphylococcus aureus* from mastitic ewes. *American Journal of Veterinary Research* 45, 1944–6.

Torres-Hernandez, G. and Hohenboken, W. (1979) Genetic and environmental effects on milk production, milk composition and mastitis incidence in crossbred ewes. *Journal of Animal Science* 49, 410–17.

Watson, D.L. (1988) Vaccination against experimental mastitis in ewes. *Research in Veterinary Science* 45, 16–21.

Watson, D.L., Franklin, N.A., Davies, H.I., Kettlewell, P. and Frost, A.J. (1990) Survey of intramammary infections in ewes on the New England Tableland of New South Wales. *Australian Veterinary Journal* 67, 6–8.

Chapter 24
Escherichia coli Resistance in Pigs

Inger Edfors-Lilja

Department of Animal Breeding and Genetics, Swedish University of Agricultural Sciences, Box 7023, S-750 07 Uppsala, Sweden

Summary

Diarrhoea in the young pig is an important problem in pig production throughout the world. The main aetiological agents in neonatal diarrhoea are *E. coli* strains possessing the K88 antigen. These strains have the ability to adhere to the intestinal mucosa which greatly facilitates the colonization of the small intestine, thereby causing diarrhoea. There exist various porcine phenotypes, one of which lacks the intestinal receptor for K88 pili and is thus more resistant to infection with *E. coli* K88. The receptor phenotype has been found in several pig populations, but the frequency varies among populations. The presence of the receptor that promotes adherence of K88*ac* pili is believed to be dominantly inherited, whereas the inheritance of the receptors for K88*ab* and K88*ad* is less clear. The use of receptor-free boars might be beneficial in herds with diarrhoea problems. Any receptor phenotype piglet born would then be the offspring of a receptor phenotype sow that secretes colostrum and milk containing antibodies to K88 protecting the young piglet against *E. coli* infection. Even if it is possible to breed for receptor free animals, knowledge is still lacking of the general function and significance of the receptor.

Genetic differences in resistance to *E. coli* infection after weaning have also been found, that would make it possible to improve resistance by breeding.

Introduction

Diarrhoea is a common problem in pig production. Enteropathogenic *Escherichia coli* is regarded as the main aetiological agent in neonatal diarrhoea (Sojka, 1965), although other infectious agents such as transmissible gastroenteritis virus, rotavirus, *Clostridium perfringens* and *Isosopora suis* are known to cause diarrhoea in newborn piglets (Glock and Whipp, 1986). Post-weaning diarrhoea and the oedema disease are commonly caused by enteropathogenic *E. coli*, although the primary cause might be the abrupt change in diet of the young pig (Svendsen, 1979).

This chapter focuses mainly on the resistance to *E. coli* infection in the newborn pig.

Neonatal diarrhoea

E. coli adhesion

The ability of porcine enterotoxigenic *E. coli* to adhere to the intestinal mucosa greatly facilitates the colonization of the small intestine, thereby causing diarrhoea. The adhesion is promoted by pili which are filamentous surface antigens, also termed adhesins, such as the K88 antigen (Jones and Rutter, 1972; Smith and Linggood, 1971; Stirm *et al.*, 1967). Three different antigenic variants of K88 antigen have been described. They all contain a common *a* type antigen and *b*, *c* or *d* type antigen (Guinee and Jansen, 1979). Other pili types of porcine ETEC are the 987P antigen (Isaacson *et al.*, 1977; Nagy *et al.*, 1977), the K99 antigen (Moon *et al.*, 1977), and the F41 antigen (Morris *et al.*, 1982).

Prevalence of K88 positive E. coli

Strains possessing the K88 pili are commonly isolated from cases of neonatal colibacillosis in swine. In Britain, 46% of the isolated strains were K88 positive (Brinton *et al.*, 1983). In two studies from Sweden, 24% and 53%, respectively, of isolated strains from one to six day old piglets with diarrhoea had the K88 antigen (Söderlind and Möllby, 1978; Söderlind *et al.*, 1982). Approximately 30% of isolated strains from older non-weaned pigs (one to six weeks old) were of the K88 type. In a more recent study in Sweden, the frequency of K88-positive strains isolated from pigs less than one-week old had declined to 19%, whereas the frequency of K99-positive strains had increased to the same level (Söderlind *et al.*, 1988). K88 positive strains still dominated in pigs one to six weeks old.

Nature of the K88 receptor

K88 *E. coli* adhere to specific receptors on pig intestinal epithelial cell brush borders depending on the antigenic variants of K88 (K88*ab*, K88*ac*, or K88*ad*). The specific receptors on the target cells are composed of different sugars, such as D-galactoside (Kearns and Gibbons, 1979; Sellwood, 1980). Other sugars that might be important in the K88 receptor are N-acetylglucosamine, N-acetylgalactosamine, and D-galactosamine (Sellwood, 1984).

Detection of receptor phenotype pigs

Receptor phenotype pigs can be identified by examining intestinal epithelial cell brush borders. Sellwood *et al.* (1975) have described an assay in which brush borders from intestinal specimens collected after slaughter are studied. In this assay, the brush borders are incubated together with *E. coli* in a phosphate buffer. The results of the test, adhesion or non-adhesion, are examined by interference contrast microscopy. The assay has also been applied on brush borders prepared from intestinal biopsies (Snodgrass *et al.*, 1981). A variant of the assay in which whole enterocytes are used, instead of brush borders, has also been described (Rapacz and Hasler-Rapacz, 1986).

Screening for receptor phenotype pigs by the *in vitro* microscopic adhesion test is cumbersome in that pigs must be euthanatized or intestinal biopsies must be done. Efforts to develop an assay which uses enterocytes obtained from faecal samples have been performed, but so far, no assay that can be easily used has been reported. A different approach, used by Atroshi *et al.* (1983) showed that cell membranes obtained from the sow milk expressed the K88 receptor and could be used for typing.

The procedure of brush border isolation, as well as the microscopic examination, is time consuming. However, an enzyme immunoassay and an ELISA that gave results that agreed well with the microscopic adhesion assay have recently been described (Chandler *et al.*, 1986; Valpotic *et al.*, 1989).

Genetic polymorpism and frequency of the K88 receptor

Genetic influence on resistance to enteropathogenic *E. coli* has been described by Sweeney (1968). The existence of two porcine phenotypes, one of which lacked the intestinal receptor for the K88*ac* pili of ETEC was first described by Sellwood *et al.* (1975). It was found that the presence of the receptor that promoted adherence of K88*ac*, and thereby susceptibility to K88 diarrhoea, was dominantly inherited (Gibbons *et al.*, 1977). However, the results from some studies make the recessive inheritance of the non-receptor phenotype questionable (Bijlsma and Bouw, 1987; Hu, 1988; Rapacz and Hasler-Rapacz, 1986; Welin-Berger, 1989). The inheritance of the receptors for K88*ab* and K88*ad* is less studied, although it has been suggested that the genes for K88*ab* and K88*ac* receptors are linked (Bijlsma and Bouw, 1987; Duval-Iflah *et al.*, 1987).

The existence of the two phenotypes, receptor–no receptor, has been demonstrated in several pig populations. The frequency of the receptor phenotype varied from 10 to 91% in the studied populations (Table 24.1). In a small sample of Chinese pigs (Chappuis *et al.*, 1984), no pigs of the receptor phenotype were found. Some pigs are susceptible to adhesion by all three types of K88 ETEC (K88*ab*, K88*ac*, and K88*ad*), whereas others are susceptible to only one or two of the types (Table 24.2).

Performance of receptor phenotype pigs

Few studies have been reported comparing production traits of pigs with or without the receptor. In one study comprising 57 Large White boars, no difference in performance between the receptor (K88*ac*) phenotypes were detected during the fattening period, 30–90 kg (Walters and Sellwood, 1984). A larger study comprising 564 crossbred pigs (Swedish Landrace × Swedish Yorkshire) showed differences in performance both during early life (0–6 weeks) and during the fattening period, 24–100 kg (Edfors-Lilja *et al.*, 1986b). In this study, piglets with the receptor for the K88*ac* had a significantly poorer daily gain during the first three weeks of life, indicating a greater incidence of diarrhoea during early life in piglets with the receptor. However, for a time period with a lower total frequency of diarrhoea outbreaks, no difference in daily gain was found between the phenotypes. During the fattening period, pigs with the receptor had a higher daily gain of lean than pigs lacking the receptor, irrespective of performance during the piglet period. Thus it seems that the receptor phenotype has a better lean meat deposition.

The poorer daily gain of young pigs of the receptor phenotype, was recently confirmed in a Swiss study comprising 120 Swiss Landrace and 250 Large White pigs from different breeding centres (Gautschi and Schwörer, 1989). They found a poorer daily gain, up to 25 kg of pigs with receptor for K88*ab* and K88*ac*, which resulted in a delayed start in the performance test of two to three days.

At present, the function and significance of the receptor on a more basic level is not known, although it has been suggested that the receptor might be involved in the presentation of antigens in the gut (Newby and Stokes, 1984).

Sow performance

After oral challenge with *E. coli* K88, sows possessing the receptor secrete colostrum having antibacterial properties (Sellwood, 1982). The anti-adhesive antibodies to K88, including both IgA and IgM, were found in both colostrum and milk and protected the offspring in an outbreak of neonatal diarrhoea caused by *E. coli* K88 (Sellwood, 1984).

Table 24.1. Frequency of K88ac receptor phenotype pigs in different populations.

Country	Frequency of receptor pigs (%)	Breed	Number of animals	Reference
Australia	88[a]	Large White, Landrace	459	Snodgrass *et al.* (1981)
Netherlands	40	Not given	101	Bijlsma *et al.* (1982)
Sweden	41	Yorkshire × Landrace	564	Edfors-Lilja *et al.* (1986b)
Sweden	57	Landrace, Yorkshire, Hampshire	155	Edfors-Lilja and Lundeheim, personal communication
USA	50	Several breeds	345	Rapacz and Hasler-Rapacz (1986)
Great Britain	10–91	Not given		Walters and Sellwood (1982)
Finland	41[b]	Landrace	237	Atroshi *et al.* (1983)
Switzerland	22–47[b]	Hampshire, Landrace, Large White	468	Gautschi and Scwörer (1989)

[a]K88 type not given.
[b]K88*ac* and/or K88*ab*.

Table 24.2. Receptor phenotypes with regard to adherence of K88*ab*, *ac* and *ad*.

E. coli K88			Reference
ab	*ac*	*ad*	
+	+	+	1, 2 and 3
+	+	−	1, 2 and 3
+	−	−	3
+	−	+	1 and 3
−	+	+	3
−	+	−	2 and 3
−	−	+	1, 2 and 3
−	−	−	1, 2 and 3

Reference: 1. Bijlsma *et al.* (1982); 2. Rapacz and Hasler-Rapacz (1986);
3. Edfors-Lilja and Lundeheim, personal communication.

The highest incidence of diarrhoea will be seen if a boar of the receptor phenotype is mated to a sow lacking the receptor. Some of the offspring will then have the receptor and thus be susceptible to infection. However, they will not be protected by the colostrum, since the sow does not recognize the coliforms as pathogens and consequently does not produce antibodies against them (Gibbons *et al.*, 1977; Sellwood, 1979). If the sow is vaccinated parenterally before farrowing she will probably produce antibodies to K88 – though not of the prefered immunoglobulin class, secretory IgA. There are indications that pigs possessing the receptor have a more pronounced IgG response to K88 after a subcutaneous immunization, than have pigs lacking the receptor (Edfors-Lilja *et al.*, 1986a). It has also been suggested that sows having the receptor produce colostrum and milk containing cell membrane surface determinants with the receptor structure (Atroshi *et al.*, 1983). These membrane determinants may bind K88-positive *E. coli* and thus protect the piglets from infection. Receptor-like glycocompounds that can prevent adherence of *Vibrio cholerae* with receptors on target cells have been described in human milk (Holmgren *et al.*, 1983).

In a recent study the influence on piglet performance of different receptor phenotype matings were analysed (Welin-Berger, 1989). The number of piglets born alive was found to be lower, but the number of piglets born dead higher in litters after receptor-free parents.

Vaccination

After oral challenge with *E. coli* K88, sows lacking the receptor secrete colostrum with poorer antibacterial properties (Sellwood, 1982). Administration of large quantities of live K88-positive *E. coli* culture for a prolonged period stimulated a higher specific IgA response in sows susceptible to the vaccination strain, compared to sows lacking the receptor (Bijlsma *et al.*, 1987). However, the response of receptor-free sows to an oral vaccination under field conditions is less documented. A higher IgG response to K88 in pigs of the receptor phenotype after a subcutaneous immunization has been found (Edfors-Lilja *et al.*, 1986a).

In Sweden, it was observed that vaccination against neonatal diarrhoea had little effect in approximately 20% of vaccinated herds (Bergström, 1975). It can not be excluded that this was partly due to quantitative genetic differences between herds and/or in differences in receptor frequency. Quantitative genetic differences in antibody response to complex natural antigens are well documented in mice (for a review, see Biozzi *et al.*, 1979). An estimated heritability around 0.2 was found for antibody response to antigens such as sheep red blood cells, *Salmonella* antigens, bovine serum albumin and rabbit gamma globulin. Genetic differences in antibody response of the same magnitude have subsequently been demonstrated for chickens (Zijpp *et al.*, 1983), cattle (Lie, 1979) and pigs (Edfors-Lilja *et al.*, 1985; Huang, 1977; Rothschild *et al.*, 1984a,b).

Use of genetic resistance in breeding

The use of receptor-free boars may be beneficial in herds with diarrhoea problems, as suggested by Walters and Sellwood (1982). Any susceptible progeny produced would then be born to susceptible sows that secrete antibodies to K88 in their colostrum and milk (see above). Walters and Sellwood (1982) showed that the use of homozygous recessive boars would not only decrease the frequency of susceptible unprotected piglets to zero after one generation, but also increase the frequency of genetically resistant piglets rapidly. In a paper by Ollivier and Renjifo (1989) the utilization of genetic resistance in different breeding schemes was discussed. For instance, the use of a resistant terminal sire gives the lowest mortality in the progeny, irrespective of the frequency of the resistant gene in the dam line. However, if the terminal sire is homozygous for the receptor gene, mortality in the progeny increases with the frequency of the resistant gene in the dam line. In a three-way cross, the use of a homozygous susceptible maternal grand sire should then be recommended.

There are some results indicating that the receptor phenotype pigs have a better lean tissue growth (Edfors-Lilja *et al.*, 1986b). Selection for increased growth rate would then lead to an increased frequency of the receptor gene. Results presented by Chappuis *et al.* (1984) indicate that there may be a very low frequency of the receptor in some Chinese breeds.

Post-weaning diarrhoea

E. coli infections causing post-weaning diarrhoea and/or oedema disease are often due to other strains than those causing neonatal diarrhoea. The change in diet at weaning is often believed to be a predisposing factor to the infection, as discussed by Bertschinger *et al.* (1978/1979) and Svendsen (1979).

A genetic influence on the resistance to post-weaning diarrhoea and oedema disease was observed by Smith and Halls (1968). In a more recent study by Bertschinger *et al.* (1986), rather large sire differences in mortality due to *E. coli* enterotoxaemia were found. The mortality after inoculation with bacterial cultures, varied between half-sib groups from 0 to 64%. The heritability for enteric disorders recorded at a progeny test station was estimated to 0.59 by Lundeheim (1988). However, the common environmental effect was large, 0.66.

Conclusions

The use of K88 receptor-free pigs to increase resistance to neonatal diarrhoea might be beneficial. However, the function and significance of the receptor on a more basic level is not known. Also, there are some reports indicating that the inheritance of the receptor might be more complicated than believed earlier. The impact of the receptor on results of vaccination schemes is not completely outlined. Thus, with the present knowledge, neither of the phenotypes can be recommended with certainty for breeding.

Resistance to post-weaning diarrhoea might possibly be improved by breeding, although different environmental effects might be more important. If records of diarrhoea incidence are available, i.e. from progeny testing stations, they could be used in the selection of boars.

References

Atroshi, F., Schildt, R. and Sandholm, M. (1983) K88-mediated adhesion of *E. coli* inhibited by fractions in sow milk. *Zentralblatt für Veterinärmedizin Reihe B* 30, 425–33.
Bergström, C.G. (1975) Klinisk prövning av polyvalent colivaccin mot spädgrisenteriter. (Eng. summary). *Svensk Veterinärtidning* 27, 287–91.
Bertschinger, H.U., Eggenberer, E., Jucker, H. and Pfirter, H.P. (1978/1979)

Evaluation of low nutrient, high fibre diets for the prevention of porcine *Escherichia coli* enterotoxaemia. *Veterinary Microbiology* 3, 281–90.

Bertschinter, H.V., Munz-Müller, M., Pfirter, H.P. and Schneider, A. (1986) Vererbte Resistenz gegen Colienterotoxämie beim Schwein. *Journal of Animal Breeding and Genetics* 103, 255–64.

Bijlsma, I.G.W. and Bouw, J. (1987) Inheritance of K88 mediated adhesion of *Escherichia coli* to jejunal brush borders in pigs: a genetic analysis. *Veterinary Research Communication* 11, 509–18.

Bijlsma, I.G.W., Houten, M. van, Frik, J.F. and Ruitenberg, E.J. (1987) K88 variants K88*ab*, K88*ac*, K88*ad* in oral vaccination of different porcine adhesive phenotypes. Immunological aspects. *Veterinary Immunology and Immunopathology* 16, 235–50.

Bijlsma, I.G.W., Nijs, A. de, Meer, C. van der and Frik, J.F. (1982) Different pig phenotypes affect adherence of *Escherichia coli* to jejunal brush borders by K88*ab*, or K88*ad* antigen. *Infection and Immunity* 37, 891–4.

Biozzi, G., Mouton, D., Sant'Anna, O.A., Passos, H.C., Gennari, M., Reis, M.H., Ferreira, V.C.A., Heumann, A.M., Bouthillier, Y., Ibanez, O.M., Stiffel, C. and Siqueira, M. (1979) Genetics of immune responsiveness to natural antigens in the mouse. *Current Topics in Microbiology and Immunology* 85, 31–98.

Brinton, C.C., Fusco, P., Wood, S., Jayappa, H.G., Goodnow, R.A. and Strayer, J.G. (1983) A complete vaccine for neonatal swine colibacillosis and the prevalence of *Escerichia coli* pili on swine isolates. *Veterinary Medicine and Small Animal Clinics* 78, 962–66.

Chandler, D.S., Chandler, H.H., Luke, R.K., Tzipori, S.R. and Craven, J.A. (1986) Screening of pig intestines for K88 non-adhesive phenotype by enzyme immunoassay. *Veterinary Microbiology* 11, 153–61.

Chappuis, J.P., Duval-Iflah, Y., Ollivier, L. and Legault, C. (1984) *Escherichia coli* K88 adhesion: a comparison of Chinese and Large White piglets. *Génétique, Sélection et Evolution* 16, 385–90.

Duval-Iflah, Y., Guerin, G., Renard, C. and Ollivier, L. (1987) K88 colibacillosis in pig: Preliminary genetic results. *38th Meeting of European Association of Animal Production*, Lisbonne, Portugal. 6pp.

Edfors-Lilja, I., Gahne, B. and Petersson, H. (1985) Genetic influence on antibody response to two *Escherichia coli* antigens in pigs. II. Difference in response between paternal half-sibs. *Zeitschrift für Tierzüchtung und Züchtungsbiologie* 102, 308–17.

Edfors-Lilja, I., Lundström, K. and Rundgren, M. (1986a) Antibody response to *E. coli* K88 antigen in genetically different pigs. Abstract. *1st International Veterinary Immunology Symposium*, Guelph.

Edfors-Lilja, I., Petersson, H. and Gahne, B. (1986b) Performance of pigs with and without the intestinal receptor for *Escherichia coli* K88. *Animal Production* 42, 381–7.

Gautschi, C. and Schwörer, D. (1989) An analysis of the intestinal receptor for *E. coli* K88 in different Swiss pig breeds. *Animal Genetics* 20, Suppl. 1, 38–9.

Gibbons, R.A., Sellwood, R., Burrows, M., and Hunter, P.A. (1977) Inheritance of resistance to neonatal *E. coli* diarrhoea in the pig: Examination of the genetic system. *Theoretical and Applied Genetics* 51, 65–70.

Glock, R.D. and Whipp, S.C. (1986) In: Lehman, A.D., Straw, B., Glock, R.D.,

Mengeling, W.L., Penny, R.H. and Scholl, E. (eds), *Diseases of Swine*, 6th edn. The Iowa State University Press, Ames, Iowa, pp. 144–9.

Guinée, P.A.M. and Jansen, W.H. (1979) Behavior of *Escherichia coli* K antigens K88*ab*, K88*ac*, and K88*ad* in immunoelectrophoresis, double diffusion, and hemagglutination. *Infection and Immunity* 23, 700–5.

Holmgren, J., Svennerholm, A.-M. and Lindblad, M. (1983) Receptor-like glyco-compounds in human milk that inhibit classical and *El Tor Vibrio cholerae* cell adherence (hemagglutination). *Infection and Immunity* 39, 147–54.

Hu, Z. (1988) Studies of genetic and expression variations in susceptibility and resistance of swine enterocytes by enteropathogenic K88*ad Escherichia coli.* MSci thesis. University of Wisconsin, Madison, USA.

Huang, J. (1977) *Quantitative Inheritance of Immunological Response in Swine.* Dissertation, University of Hawaii. University Microfilms International, Ann Arbor, Michigan, USA.

Isaacson, R.E., Nagy, B. and Moon, H.W. (1977) Colonization of porcine small intestine by *Escherichia coli*: colonization and adhesion factors of pig enetero-pathogens that lack K88. *Journal of Infectious Diseases* 135, 531–9.

Jones, G.W. and Rutter, J.M. (1972) Role of the K88 antigen in the pathogenesis of neonatal diarrhoea caused by *Escherichia coli* in piglets. *Infection and Immunity* 6, 918–27.

Kearns, M.J. and Gibbons, R.A. (1979) The possible nature of pig intestinal receptor for the K88 antigen of *Escherichia coli*. *FEMS Microbiological Letters* 6, 165–8.

Lie, Ø. (1979) Genetic analysis of some immunological traits in young bulls. *Acta Veterinaria Scandinavica* 20, 372–86.

Lundeheim, N. (1988) Health disorders and growth performance at a Swedish pig progeny testing station. *Acta Agricultura Scandinavica* 38, 77–88.

Moon, H.W., Nagy, B., Isaacson, R.E. and Ørskov, I. (1977) Occurrence of K99 antigen on *Escherichia coli* isolated from pigs and colonization of pig ileum by K99 enterotoxigenic *E. coli* from calves and pigs. *Infection and Immunity* 15, 614–20.

Morris, J.A., Thorns, C., Scott, A.C., Sojka, W.J. and Wells, G.A. (1982) Adhesion *in vitro* and *in vivo* associated with an adhesive antigen (F41) produced by a K99 mutant of the reference strain *Escherichia coli* B41. *Infection and Immunity* 36, 1146–53.

Nagy, B., Moon, H.W. and Isaacson, R.E. (1977) Colonization of porcine intestine by enterotoxigenic *Escherichia coli*: selection of piliated forms *in vivo*, adhesion of piliated forms to epithelial cells *in vitro*, and incidence of a pilus antigen among porcine enterpoathogenic *E. coli. Infection and Immunity* 16, 244–52.

Newby, T.J. and Stokes, C.R. (1984) The intestinal immune system and oral vaccination. *Veterinary Immunology and Immunopathology* 6, 67–105.

Ollivier, L. and Renjifo, X. (1989) Utilization de la resistance genetique a la coli-bacillose K88 dans les schemas d'amelioration genetique du porc. *40th Meeting of European Association for Animal Production*, Dublin, Ireland. 16pp.

Rapacz, J. and Hasler-Rapacz, J. (1986) Polymorphism and inheritance of swine small intestinal receptors mediating adhesion of three serological variants of *Escherichia coli*-producing K88 pilus antigen. *Animal Genetics* 17, 305–21.

Rothschild, M.F., Chen, H.L., Christian, L.L., Lie, W.R., Venier, L., Cooper, M.,

Briggs, C. and Warner, C.M. (1984a) SLA complex and immune response. Breed and swine lymphocyte antigen haplotype differences in aggultination titres following vaccination with *B. bronciseptica. Journal of Animal Science* 59, 643–9.

Rothschild, M.F., Hill, H.T., Christian, L.L. and Warner, C.M. (1984b) Genetic differences in serum-neutralizing titers of pigs after vaccination with pseudorabies modified live-virus vaccine. *American Journal of Veterinary Research* 45, 1216–18.

Sellwood, R. (1979) *Escherichia coli* diarrhoea in pigs with and without the K88 receptor. *Veterinary Record* 105, 228–30.

Sellwood, R. (1980) The interaction of the K88 antigen with porcine intestinal epithelial cell brush borders. *Biochimica Biophysica Acta* 632, 326–35.

Sellwood, R. (1982) *Escherichia coli*-associated porcine diarrhoea: antibacterial activities of colostrum from genetically susceptible and resistant sows. *Infection and Immunity* 35, 396–401.

Sellwood, R. (1984) An intestinal receptor for the K88 antigen of porcine enterotoxigenic *Escherichia coli.* In: Boedecker, E.C. (ed.), *Attachment of Organisms to the Gut Mucosa*, vol. 2. CRC Press, Boca Raton, pp. 167–75.

Sellwood, R., Gibbons, R.A., Jones, G.W. and Rutter, J.M. (1975) Adhesion of enteropathogenic *Escherichia coli* to pig intestinal brush borders: The existence of two pig phenotypes. *Journal of Medical Microbiology* 8, 405–511.

Smith, H.W. and Halls, S. (1968) The production of oedema disease and diarrhoea in weaned pigs by oral administration of *Escherichia coli.* Factors that influence the course of the experimental disease. *Journal of Medical Microbiology* 1, 45–9.

Smith, H.W. and Linggood, M. (1971) Observations on the pathogenic properties of the K88, Hly and Ent plasmids of *Escherichia coli* with particular reference to porcine diarrhoea. *Journal of Medical Microbiology* 4, 467–85.

Snodgrass, D.R., Chandler, D.S. and Makin, T.J. (1981) Inheritance of *Escherichia coli* K88 adhesion in pigs: identification of non-adhesive phenotypes in a commercial herd. *Veterinary Record* 109, 461–3.

Söderlind, O. and Möllby, R. (1978) Studies on *Escherichia coli* in pigs. V. Determination of enterotoxigenity and frequency of O groups and K88 antigen in strains from 200 piglets with neonatal diarrhoea. *Zentralblatt für Veterinärmedizin. Rehie B* 25, 719–28.

Söderlind, O., Olsson, E., Smyth, C.J. and Möllby, R. (1982) Effect of parenteral vaccination of dams on intestinal *Escherichia coli* in piglets with diarrhoea. *Infection and Immunity* 36, 900–6.

Söderlind, O., Thafvelin, B. and Möllby, R. (1988) Virulence factors in *Escherichia coli* strains isolated from Swedish piglets with diarrhoea. *Journal of Clinical Microbiology* 26, 879–84.

Sojka, W.J. (1965) *Escherichia coli in domestic animals and poultry,* CAB International, Wallingford, Oxfordshire.

Stevens, A.J. (1963) Coliform infections in the young pig and a practical approach to the control of enteritis. *Veterinary Record* 75, 1241–6.

Stirm, S., Ørskov, F., Ørskov, I. and Birch-Andersen, A. (1967) Episome-carried surface antigen K88 of *Escherichia coli.* III. Morphology. *Journal of Bacteriology* 93, 740–8.

Svendsen, J. (1979) 'Enteric *Escherichia coli* in suckling pigs and in pigs at weaning. Aspects of pathogenesis, prevention and control.' Thesis, Swedish University of

Sweeney, E.J. (1968) *Escherichia coli* in enteric disease of swine: observations on herd resistance. *Irish Veterinary Journal* 22, 42–46.

Agricultural Sciences, Lund 6.

Valpotic, I., Dean, E.A. and Moon, H.W. (1989) Phenotyping of pigs for the presence of intestinal receptors mediating adhesion of enterotoxigenic *Escherichia coli*-bearing K88*ac* pilus antigen by ELISA. *Veterinarski Arhiv* 59, 161–75.

Walters, J.R. and Sellwood, R. (1982) Aspects of genetic resistance to K88 *E. coli* in pigs. *2nd World Congress on Genetics Applied to Livestock Production.* Madrid. Vol. VII, 362–7.

Walters, J.R. and Sellwood, R. (1984) Observations on the performance of pigs genetically resistant to K88 *E. coli. Proceedings of 8th International Pig Veterinary Society Congress,* Ghent. p. 74.

Welin-Berger, S. (1989) 'Tunntarmsreceptor för *E. coli* K88ac: Nedärvning och inverkan på smågrisars tidiga produktionsresultat.' (Eng. summary) Undergraduate thesis. Dept Animal Breeding and Genetics, Swedish University of Agricultural Sciences, Uppsala.

Zijpp, A. van der, Frankena, K., Bonenschanscher, J. and Nieuwland, M.G.B. (1983) Genetic analysis of primary and secondary immune responses in the chicken. *Poultry Science* 62, 565–72.

Section 7
Major Genes and Animal Disease

Several important groups of major genes that directly or indirectly impinge on animal disease are discussed with a view to minimizing their deleterious effects or to employing their advantages more fully. Some major genes are of interest because of their large effect on economic traits which is potentially useful but whose use is associated with deleterious side effects. These include the halothane gene in pigs, the double muscling gene in cattle and the fecundity gene in sheep, all of which are discussed in some detail. It is concluded that some have substantial economic advantages in particular circumstances, often when animals heterozygous at the locus in question are involved.

Other major genes are largely deleterious dredged from the shadows of the genome by over-intense concentration on traits presumed to be desirable. These include the spider gene that has recently been observed in North America in some sheep breeds.

Chapter 25

Spider and Other Major Genes in Farm Livestock

J.B. Owen

School of Agricultural and Forest Sciences, University of Wales, Bangor, Gwynedd LL57 2UW, UK

Summary

A special aspect of the genetic basis of disease is related to the deleterious effect of specific genes with major effects. Several major genes are considered ranging from the recent discovery of the spider gene in sheep in the USA to the exploitation of genes for muscle hypertrophy (double muscling) in cattle and sheep. Some of these genes have promising possibilities for exploitation because of their positive effects e.g. 'halothane' gene in pigs, double muscling and the fecundity gene in sheep. Means of maximizing their net positive value at the expense of their deleterious effects are considered. Other genes for dwarfism and giantism (spider genes) are deleterious by-products of extreme selection for limited objectives. The widespread nature of major genes and the way they arise is discussed.

It is concluded that many existing and hitherto uncovered major genes can be useful in the development of farm animals, possibly in heterozygous form. These genes are prime candidates for the application of molecular biology either for gene transfer or for the genotyping of potential stock.

Introduction

Recently there has been increasing interest in genes at a single locus that have a major effect on economic traits in farm animals. These genes, whilst not having the sole control of a trait (as in some of Mendel's pea traits) assert a major effect, often superimposed on a basal continuous (polygenic) variation. Currently topical examples are halothane gene (pigs), double muscling gene (mainly cattle), fecundity gene (sheep). These three examples

439

are notable in that they are genes at a single locus that have a large effect on a potentially important commercial trait. Other examples that can be described as 'major' genes have major and sometimes bizarre effects, e.g. dwarfism in cattle, known since early in the century and the spider gene in sheep in North America first reported around 1980. These genes are lethal or semilethal in their effects but mirror, in an extreme form the beneficial trait that was the aim of the breeders of the livestock in question.

Major genes, almost without exception, involve deleterious main or side-effects that are pertinent in a discussion of genetic effects on disease. These effects vary from the repercussion of the sheer magnitude of the phenotypic expression, e.g. of the fecundity or double muscling gene, on the physiological capability of the animal, to the gross deformity of limbs and other organs caused by the spider gene. In this chapter a survey of some of the most important and interesting major genes is attempted without attempting to cover all deleterious lethal genes in farm livestock. An hypothesis of how such conditions arise is suggested and the best use of sometimes potentially useful genetic variation, without incurring penalties in the health and welfare of animals, is discussed.

Genes associated with leanness

Several genes have been discovered that have large direct effects on the quality and proportion of lean (muscle) and of fatty tissue in farm animals. Two of these, the halothane gene of pigs and the double muscling gene of cattle, are described in greater detail by Hanset (1991) and Archibald (1991). These genes cause muscle hypertrophy and a reduction in the carcass fat content in the species concerned and there are some intriguing similarities in the genes for the two conditions. They are quite widespread between breeds within the species concerned. Double-muscling, long known in cattle (Oliver and Cartwright, 1969) has been observed in many breeds including Charolais, Belgian Blue and Welsh Black and is also known in sheep (first reported by Naerland, 1940). These genes not only cause substantial muscle hypertrophy but can also lead, particularly in the pig, to physical and chemical changes in muscle quality (pale soft exudative muscle). The latter effects are deleterious from the consumer quality viewpoint. Evidence for cattle show that the hypertrophied muscle is pale and tender but not necessarily exudative (Hanset, 1991). Other characteristics are also affected, notably rate of growth in the case of both genes and incidence of dystocia in the case of the double muscling condition. So severe is the dystocia problem in homozygous double muscled cows bred to carry homozygous calves that cesarean section is adopted as a routine herd procedure (Hanset, 1991).

An extreme form of leanness, associated with anorexia, has been observed in some strains of young female pigs (thin sow syndrome; Riley, 1989) and sheep (Lees, 1969). These young females may stem from strains subjected to intense selection for leanness, as in the case of the pigs that have shown susceptibility to this condition. In both pigs and sheep the females in question are subject to high output and restricted feed intake.

Genes associated with obesity

Some genes with major effects on body composition lead to obesity in laboratory rodents and in humans, but there are no reports of such conditions with widespread significance for farm animals. It is, however, possible that such genes may be involved in some pig breeds, possibly in the fixed state, but that their expression leads to no greater obesity than is expected of the pig, and therefore they are not categorized as remarkable single gene effects. The alleles of the 'leanness' genes could be regarded as 'obesity' genes with large effects, depending on which allele is associated with what is considered the 'norm' of body composition for the species.

The absence of high fat feeds and the necessity for energy expenditure in grazing, may preclude conditions conducive to the appearance of obesity genes in farm ruminants.

Genes affecting skeletal stature

Two types of gene have been observed in farm animals following intense selection for extremes of tallness or of compactness. The 'compact' beef breeds developed in the early twentieth century showed a significant frequency of genes for snorter dwarfism. This reached a frequency of 10% in some Hereford herds in the USA between 1945 and 1955 (Nicholas, 1987). An extreme manifestation of a similar gene is the dwarf bulldog gene present in the heterozygous state in the Dexter cattle breed (Hammond *et al.*, 1983). These genes illustrate effects which were deemed desirable in moderation, i.e. skeletal compactness or short legs in the heterozygotes, but disastrously lethal in the homozygous state.

At the other extreme selection for large-framed, tall animals seems to have been associated with the emergence of the spider syndrome in the Suffolk and Hampshire sheep breeds in North America. The condition described as hereditary chondrodysplasia has been reported in sheep breeder's journals in the USA since about 1980 although Vanek *et al.* (1986) were the first to report it in the scientific literature. The condition has not yet been reported from outside the North American continent, which accords with the general perception that American breeders have concentrated more on extreme body size than their counterparts elsewhere.

Affected animals (homozygous recessives) show several symptoms as shown in Figs 25.1, 25.2 and 25.3. These symptoms include angular deformities of the legs, chest wall deformities and roman nose (Rook, 1987; Rook *et al.*, 1986, 1988). Heterozygous carriers are characterized by very long legs and fine bones (Thomas *et al.*, 1988).

Genes affecting fecundity

The discovery of the Booroola gene in the Australian Merino, reported by Piper and Bindon (1982), was the first of several such genes reported for

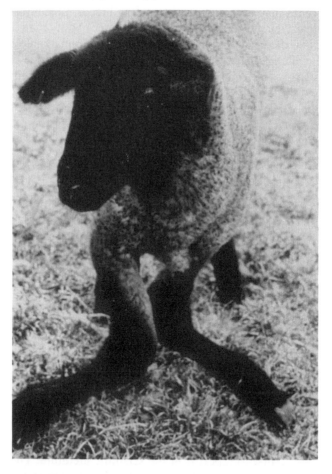

Fig. 25.1. View head-on of the long spindly legs of lamb with the spider syndrome.
Source: Dr Richard Cobb, University of Illinois.

Fig. 25.2. View of spider syndrome lamb from above.
Source: Dr Richard Cobb, University of Illinois.

sheep of several breeds (Owen *et al.*, 1990). This gene has a major effect on ovulation rate, varying in the range of 1–2 extra ova per copy of the gene, depending on the breed and age of ewe. Work is underway at Bangor to study the mechanism of the gene at molecular level with the purpose of enabling early genotype identification. It is possible that the effect arises from a defect of a gene coding for one of the ovulation regulatory mechanisms in the sheep.

The gene is of interest in the present context because its effect is so large that it puts the physiological mechanisms of the ewe, for sustaining the embryo and the lamb, under challenge (Fig. 15.4). As with many other genes with major effects, such as the leanness/muscling genes, the fecundity gene may best be exploited in practice in the heterozygous form (Owen and Whitaker, 1987). This seems particularly appropriate for ovulation rate since confining its expression to the heterozygous state can reduce the inflated variance in ovulation rate and litter size characteristic of flocks where the gene is segregating.

Associated with fecundity is lactation in the ewe and it is interesting to note that the occurrence of supernumerary teats in some breeds of sheep, particularly some of the prolific breeds, seems to be influenced significantly by a single gene (Davies, 1988). Such a gene, unlike some of those discussed so far, appears at first sight to be largely beneficial since it could

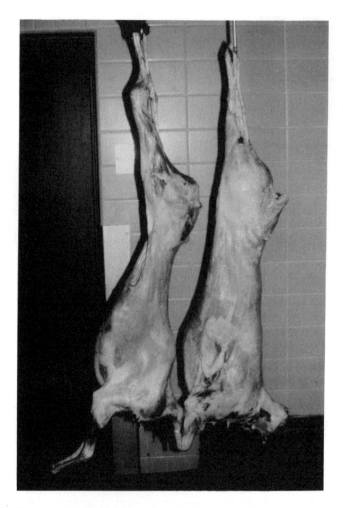

Fig. 25.3. Carcass of spider lamb compared with a control.
Source: Dr Richard Cobb, University of Illinois.

increase milking ability or, more important, reduce the stress and teat damage in the ewe suckling more than two lambs. However, it is not yet established that deploying such a gene is the most efficient way of increasing milk yield nor is it certain that even with extra teats that a ewe could more comfortably suckle more than two lambs at a time at the stage of highest demand.

a

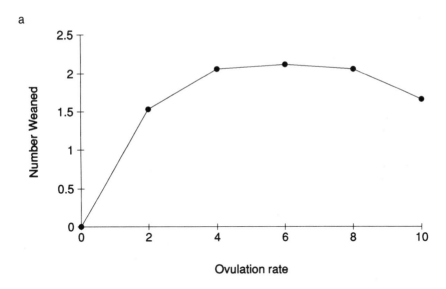

b

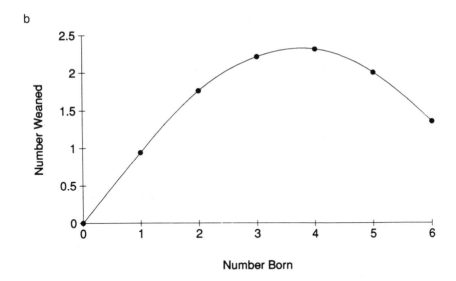

Fig. 25.4. (a) Number weaned and ovulation rate; (b) number weaned and number born.

Discussion

Occurrence of major genes

Many of the examples of major genes reviewed have been observed in animal populations subject to intense selection for a trait related to that coded for by the gene in question. Such selection is seen at its most intense in the screening of diverse genetic populations as in the formation of the Cambridge sheep breed (Owen *et al.*, 1986). In such situations the population in question is subjected to truncation with respect to the trait with the isolation of a subpopulation of small effective size. Following such truncation in the formation of the foundation group intense selection for the trait may then be continued for several generations. The consequence of such a selection procedure is that there is a high probability that the small founding subpopulation will carry any rare alleles with a major effect on the trait, that may be present in the large base population at low frequency. This would apply even to recessive genes with lethal or highly adverse effects in the homozygous form, provided the heterozygous individual showed some advantage for the selected trait. Once captured in the foundation group, the frequency of such a gene would quickly rise in the new population under the influence of selection.

Since mutations largely destabilize gene function it is possible that many of the genes in question arise from the malfunction of regulatory genes that evolved in the wild to prevent the expression of very extreme forms, e.g. high ovulation, extreme body composition or deviant skeletal anatomy.

Practical implications

Traditional breeders have generally employed balanced selection for all-round merit, with relatively lax selection for any individual trait. This strategy is probably the safest from the point of view of disease resistance and the value of native adapted stock subject to such selection should not be underestimated. However, as the result of disruptive selection as noted above, there are several major genes which have a potential usefulness, particularly as new methods of gene and genotype identification stem from advances in molecular genetics. Many major genes are useful in practice only when in the heterozygous form since two copies are normally too much.

Fortunately, genes at such loci often have a much reduced effect in the heterozygous state (i.e. they do not show full dominance). Heterozygous animals often show a better balance of advantage in the main effect over the disadvantageous side-effects. However, the efficient production of heterozygous animals depends on identification of the parent genotypes or the use of homozygous populations where the gene at the locus in question is fixed. This is where greater knowledge of the genome and the development of genotype 'fingerprinting' techniques will greatly increase the potential value of major genes so as to improve efficiency without incurring the health hazards involved in unbalanced genotypes.

References

Archibald, A.L. (1991) Inherited halothane induced malignant hyperthermia in pigs. In: Owen, J.B. and Axford, R.F.E. (eds), *Breeding for Disease Resistance in Farm Animals*. CAB International, Wallingford, pp. 449–66.

Davies, D.A.R. (1988) Breeding sheep with four teats in a flock in Devon, England. *Journal of Agricultural Science in Finland* 60, 620–1.

Hammond, J., Bowman, J.C. and Robinson, T.J. (1983) *Hammond's Farm Animals*, 5th edn. Edward Arnold, London.

Hanset, R. (1991) The major gene for muscular hypertrophy in the Belgian Blue cattle breed. In: Owen, J.B. and Axford, R.F.E. (eds), *Breeding for Disease Resistance in Farm Animals*. CAB International, Wallingford, pp. 467–78.

Lees, J.L. (1969) The reproductive pattern and performance of sheep. *Outlook on Agriculture* 6, 82–8.

Naerland, G. (1940) Forekommer dobeeltlaenderkarakteren hos andre husdyrarter enn storfe? Universell hyperplasi av stammens enn lemmenes muskulatur hos sau. (Does the 'Doppellender' condition occur in other species of domestic animals beside cattle?) *Skandinavisk VeterinarTidskrift* 30, 811–30.

Nicholas, F.W. (1987) *Veterinary Genetics*. Clarendon Press, Oxford.

Oliver, W.M. and Cartwright, T.C. (1968) *Double muscling in cattle. A review of expression, genetics and economic implication*, Technical Report No. 12. Texas A&M University Agricultural Experiment Station, p. 58.

Owen, J.B. and Ap Dewi, I. (1988) The Cambridge breed – its exploitation for increased efficiency of lamb production. *Journal of Agricultural Science in Finland* 60, 585–90.

Owen, J.B., Crees, S.R.E., Williams, J.C. and Davies, D.A.R. (1986) Prolificacy and 50 day lamb weight of ewes in the Cambridge sheep breed. *Animal Production* 42, 355–63.

Owen, J.B. and Whitaker, C.J. (1987) A comparison of cross-bred ewes raised from Welsh Mountain dams by three sire breeds: Cambridge, Border Leicester and Lleyn. *Journal of Agricultural Science* 109, 159–64.

Owen, J.B., Whitaker, C.J., Axford, R.F.E. and Ap Dewi, I. (1990) Expected consequences of the segregation of a major gene in sheep population in relation to observations on the ovulation rate of a flock of Cambridge sheep. *Animal Production* 51, 277–82.

Piper, L.R. and Bindon, B.M. (1982) Origins of the CSIRO Booroola. In: Piper, L.R., Bindon, B.M. and Nethery, R.D. (eds), *The Booroola Merino*, CSIRO Division of Animal Production, Melbourne, pp. 9–20.

Riley, J.E. (1989) Recent trends in pig production: the importance of intake. In: Forbes, J.M., Varley, M.A. and Lawrence, T.L.J. (eds), *The Voluntary Food Intake of Pigs*. British Society of Animal Production, Edinburgh, pp. 1–5.

Rook, J.S. (1987) Spider lamb syndrome. *Foreign Animal Disease Report* 15, 7–11.

Rook, J.S., Kopcha, M., Spaulding, K., Coe, P., Benson, M., Krehbiel, J. and Trapp, A.L. (1986) The spider syndrome: a report on one purebred flock. *Compendium on Continuing Education for the Practicing Veterinarian* 8, S402–S405.

Rook, J.S., Trapp, A.L., Krehbiel, J., Yamini, B. and Benson, M. (1988) Diagnosis of hereditary chondrodysplasia (spider lamb syndrome) in sheep. *Journal of the American Veterinary Medical Association* 193, 713–18.

Thomas, D.L., Cobb, A.R. and Waldron, D.F. (1988) Spider syndrome – a genetic defect found in American Suffolk sheep. *Proceedings, 3rd World Congress on Sheep and Beef Cattle Breeding* vol. 1, pp. 649–51.

Vanek, J.A., Alstad, A.D., Berg, I.E., Misek, A.R., Moore, B.L. and Limesand, W. (1986) Spider syndrome in lambs: a clinical and post-mortem analysis. *Veterinary Medicine* 81, 663–8.

Chapter 26

Inherited Halothane-Induced Malignant Hyperthermia in Pigs

Alan L. Archibald

AFRC Institute of Animal Physiology and Genetics,
Edinburgh Research Station, Roslin, Midlothian EH25 9PS,
UK

Summary

The porcine stress syndrome (PSS) manifests itself in sudden (stress) deaths and rapid post-mortem changes in skeletal muscle, which cause pale soft exudative meat (PSE). The deaths arise from uncontrolled malignant hyperthermic reactions, which can be triggered by handling, sexual intercourse, excessive ambient temperature and a number of chemical agents including the anaesthetic gas halothane. Sensitivity to halothane-induced malignant hyperthermia (MH) has been shown to be controlled by a recessive gene at a single autosomal locus (*HAL*).

Malignant hyperthermia in pigs also serves as a model for the inherited malignant hyperthermia syndrome in humans. Current research on the molecular genetics of malignant hyperthermia has two related objectives. First, the identification of informative marker loci in order to facilitate marker assisted selection for the desired *HAL* allele(s). Second, the identification of the *HAL* gene itself using the DNA markers as a starting point. Already two candidates for the *HAL* gene are under consideration – the ryanodine receptor (*RYR*) and the hormone-sensitive lipase (*LIPE*) genes. Both the *RYR* and *LIPE* genes map to the relevant regions of the porcine and/or human genomes. The candidature of the *RYR* gene is also supported by biochemical data concerning the ryanodine receptor protein isolated from stress-susceptible pigs.

If either of these genes proves to be the *HAL* gene, then the molecular nature of the MH mutations can be determined. Among the benefits to accrue would be improved tests for identifying genetically susceptible pigs or humans and a greater understanding of the aetiology of this genetic disease.

Introduction

Chemical-induction of malignant hyperthermia was first recognized in pigs about 25 years ago (Hall *et al.*, 1966; Harrison *et al.*, 1968). Subsequently, it became apparent that the breeds of pigs, that exhibited high frequencies of sudden (stress) deaths and of pale soft exudative (PSE) meat, were particularly prone to halothane-induced malignant hyperthermia (Harrison, 1972; Mitchell and Heffron, 1980). These three phenotypes of sudden (stress) deaths, pale soft exudative meat (PSE) and sensitivity to halothane-induced malignant hyperthermia can be considered to be manifestations of a general porcine stress syndrome (PSS). The deaths arise from uncontrolled malignant hyperthermic reactions, which can be triggered by handling, sexual intercourse, excessive ambient temperature and a number of chemical agents. The correlation between PSE and MH, however, is not complete (Mitchell and Heffron, 1980). Man is also afflicted by halothane-induced malignant hyperthermia. Although the human and porcine forms of this disorder are not identical, the similarities are sufficiently compelling to merit using porcine MH as a model for MH in humans.

MH has been extensively studied for several years (for reviews see Harrison, 1979; Louis *et al.*, 1990; Mitchell and Heffron, 1982; Webb *et al.*, 1982). In this chapter the genetic aspects of porcine MH, the development of a molecular genetic or 'reverse genetics' approach to studying MH and the contribution of linkage analysis to controlling the disease will be considered. Some of the biochemical and physiological research on MH will be described. Finally, the convergence of these two approaches in the identification of a candidate gene responsible for MH will be discussed.

Breeding and genetics

Differences in the frequencies of PSS phenotypes in different breeds indicated that PSS was a genetic disorder in pigs. Of the three phenotypes, only halothane sensitivity can be measured in the living animal. Eikelenboom and Minkema (1974) developed a predictive field test for susceptibility to PSS based on a short controlled exposure to the anaesthetic gas halothane. Although exposure to halothane can induce malignant hyperthermia and death, if the anaesthetic mask is removed immediately the first signs of muscle rigidity are observed, susceptible individuals generally recover. The development of this non-lethal test has been critical to subsequent studies of the genetic control of the disease.

Sensitivity to halothane-induced malignant hyperthermia has been shown to be controlled by a recessive gene at a single autosomal locus (*HAL*) (Mabry *et al.*, 1981; Minkema *et al.*, 1977; Ollivier *et al.*, 1975; Smith and Bampton, 1977). Carden *et al.* (1983) and Grashorn and Müller

(1985) have tested two-locus models with a susceptibility locus and a suppressor locus in attempts to find a genetic explanation for the incomplete penetrance observed by Ollivier *et al.* (1975) and Smith and Bampton (1977). Carden *et al.* (1983) concluded that a single and strictly recessive mode of inheritance may not be enough to explain their observations in the ABRO British Landrace lines selected for high and low incidence of halothane sensitivity. After further studies on these lines, Southwood *et al.* (1986, 1988) concluded that the degree of penetrance had been underestimated and that a single locus model provided the most appropriate explanation of halothane sensitivity.

As halothane sensitivity is a recessive trait, the standard halothane challenge test fails to distinguish between the homozygous resistant animals and heterozygous carriers. The alternative diagnostic tests are no better and are indeed generally less effective for genotyping (Webb *et al.*, 1987). Progeny testing, which offers the only reliable way of genotyping animals, is both expensive and time consuming. Therefore, there is a need for a cheap and accurate test which would facilitate the unambiguous identification of all three genotypes – *N/N, N/n* and *n/n*.

Although the halothane test does not allow the detection of carrier individuals, it is sufficient in combination with progeny testing to facilitate the selective removal of the HAL^n allele from pig populations. The halothane challenge test is relatively cheap, provides an immediate diagnosis of homozygous *HAL n/n* pigs and can be carried out by technicians after limited training. Given all these advantages it is surprising that the incidence of the HAL^n allele has not already been reduced to insignificant levels. Two factors have militated against the elimination of the HAL^n allele. First, the perceived advantages in lean meat production associated with the HAL^n allele has resulted in a variety of breeding programmes being designed to maximize the number of heterozygous (stress-resistant but lean) pigs in the slaughter generation (Simpson and Webb, 1989). It seems that these programmes are now being abandoned in favour of elimination of the HAL^n allele. Second, the incidence of the HAL^n allele in some breeds or populations, such as the Pietrain, mean that survival of the breed is currently incompatible with selection against HAL^n. I consider the HAL^n allele to be a disease gene, which should be eliminated in production stock but retained in experimental animals to serve as a model for human malignant hyperthermia.

Linkage relationships

By segregation analysis the *HAL* locus has been assigned to the best characterized linkage group in pigs (for a review see Archibald and Imlah, 1985). The structure of the *HAL* linkage group is summarized in

Figure 26.1(b). The *GPI* locus encodes the enzyme glucose-6-phosphate isomerase (EC 5.3.1.9), which is found in all tissues including erythrocytes. This locus was formerly known in the animal breeding community as *Phi* (phosphohexose isomerase). There are two common electrophoretic variants of the GPI protein (A and B) and a rare third variant, C, which has been found only in Eastern Bulgarian Mountain pigs (Van de Weghe *et al.*, 1988). The alleles at the *H* locus can be deduced from the serologically detected variants of the H erythrocyte antigens. It is not known whether the *H* locus gene product is the polypeptide backbone of the H antigens or a glycosyltransferase involved in the post-translational modification of the H antigens. The *H* locus is termed an open system by serologists, i.e. an animal, which is phenotypically Ha, may be genotypically H^a/H^- or H^a/H^a. The H^- allele may either be a genuine null allele or an allele, for which the serological reagents needed to detect its gene product have not been isolated. The *S(A-O)* locus, which was first postulated by Rasmusen (1964), influences the expression of the A-O erythrocyte antigens. The S^s allele, which suppresses A-O expression, is recessive and, therefore, genotyping animals for the *S* locus requires pedigree data as well as careful analysis of the A-O antigens. Although it is possible that the *S(A-O)* locus may be the porcine homologue of Secretor (*Se*), now known as *FUT2* in man, there is no evidence to suggest that the porcine *H* locus is homologous to the human H (*FUT1*) locus (Le Beau *et al.*, 1989). The *A1BG* locus encodes a protein (alpha1B-glycoprotein) of unknown function, which is found circulating in plasma. There are two common electrophoretic variants of this protein, which was formerly known as postalbumin-2, with the corresponding locus being *Po2*. The *PGD* locus encodes the ubiquitous enzyme 6-phosphogluconate dehydrogenase (EC 1.1.1.44). Electrophoretic variants of the PGD are controlled by two codominantly expressed alleles (*PGD*^{A} and *PGD*^{B}) at the *PGD* locus. A third variant PGD C and the

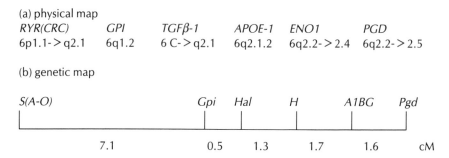

(a) physical map

RYR(CRC)	GPI	TGFβ-1	APOE-1	ENO1	PGD
6p1.1->q2.1	6q1.2	6 C->q2.1	6q2.1.2	6q2.2->2.4	6q2.2->2.5

(b) genetic map

S(A-O)		Gpi	Hal	H	A1BG	Pgd
	7.1		0.5	1.3	1.7	1.6

Fig. 26.1. (a) Physical map of the *HAL* region of porcine chromosome 6. (b) Genetic map of the *HAL* linkage group. Recombination distances calculated by maximum likelihood from pooled data in Archibald and Imlah (1985).

corresponding allele (*PGD^C*) have recently been described by Archibald and McTeir (1988).

These loci meet the requirements of genetic markers, as they are polymorphic and are linked to the locus of interest. Indeed, Gahne and Juneja (1985) have shown that a combination of the halothane test and haplotyping for *PHI, Po2* and *PGD* allows *HAL* genotypes to be predicted with an accuracy of 90–95% in Swedish Landrace and Yorkshire pigs. This approach has been used in Sweden to effect a breeding policy of significantly reducing the frequency of the *HAL^n* allele, and is one of the best examples of marker-assisted selection in livestock. However, the gene frequencies at these marker loci are not ideal for marker assisted selection in all pig populations. Furthermore, there are technical difficulties associated with the accurate determination of genotypes at the *S* and *H* loci.

In order to use linked genetic markers it is necessary to establish the phase of the linkage between alleles at the marker loci and alleles at the locus of interest. For the *HAL* locus this requires both halothane testing and marker genotyping parents and their offspring. To establish the linkage phase for the recessive *HAL^n* allele, a minimum of one halothane-sensitive parent or offspring is needed. Once the halotypes for any individual have been established, the haplotypes of its offspring can be deduced from their marker genotypes. A number of the deduced haplotypes will be in error, as a result of recombination disrupting the parental haplotypes. The genetic distance (recombination frequency) between the marker loci and the disease locus will determine this error rate. These genetic markers, therefore do not offer a direct test of the *HAL* genotypes of randomly chosen individuals. Such a direct test will only be possible once the defect in the *HAL^n* gene product or the underlying mutation in the gene has been identified.

Two complementary strategies are being pursued to identify the *HAL* mutation and so to reach a better understanding of this inherited disease. First, a biochemical and physiological approach is being used to compare susceptible and resistant pigs. Second, the linkage and genetic approach is being continued using molecular biological techniques. The present understanding of malignant hyperthermia is a synthesis of the data from these two approaches.

Biochemical approaches

One of the earliest signs of onset of an MH incident is muscle contractures and rigidity. The breeds of pigs most severely afflicted by the MH disorder yield leaner meat and appear better muscled. The pale soft exudative (PSE) meat, which is more prevalent in stress-susceptible pigs, arises from rapid post-mortem degenerative changes in skeletal muscle. It is not surprising, therefore that MH is regarded as a myopathy and that skeletal muscle has been the focus of most biochemical or physiological studies of malignant hyperthermia.

The comparison of a variety of muscle extracts/proteins from MH susceptible and MH resistant pigs and humans by both one- and two-dimensional polyacrylamide gel electrophoresis has failed to identify any consistent differences (Lorkin and Lehmann, 1983; Marjanen and Denborough, 1984; Sullivan and Denborough, 1982). The sampling procedures used may not have been sufficiently stringent to avoid confusing protein patterns arising from rapid post-mortem or post-biopsy degenerative changes caused by the MH lesion with putative causal differences in protein profiles. The animals sampled also may not always have been rigorously genotyped with respect to the *HAL* locus. Nevertheless, these data suggest that MH is not caused by the complete absence of, or a major change to a moderately abundant protein component of muscle.

The malignant hyperthermia reactions observed in humans and pigs appear to be caused by a failure of the regulation of intracellular Ca^{2+} levels in skeletal muscle, which lead to sudden and sustained increases in intracellular levels of Ca^{2+}. Calcium levels in muscle can be regulated by three subcellular membrane systems; the sarcolemma or plasma membrane, the mitochondria and the sarcoplasmic reticulum (Heffron, 1987). Although some studies implicated Ca^{2+} efflux from mitochondria as the trigger for MH (Cheah and Cheah, 1976, 1978, 1979) others failed to find differences between Ca^{2+} regulation by mitochondria from MH-susceptible individuals and controls (Heffron, 1984). Elevated phospholipase A2 activity has been implicated in the increased rates of Ca^{2+} efflux from mitochondria (Cheah and Cheah, 1981a; Cheah, 1984).

In recent years attention has focused on the regulation of calcium release from the sarcoplasmic reticulum (Kim *et al.*, 1984; Nelson, 1983; O'Brien, 1986; Ohnishi *et al.*, 1983). Studies by Louis and others suggest that the causal defect is in the Ca^{2+}-induced Ca^{2+} release channel (Louis *et al.*, 1990; Mickelson *et al.*, 1988, 1989; Fill *et al.*, 1990). The calcium-release channel is a 565 kDa protein, which is also known as the ryanodine receptor (RYR) from its affinity for the plant alkaloid ryanodine. Sarcoplasmic reticulum vesicles and single calcium release channels have been purified from pigs of differing malignant hyperthermia genotypes and studied in reconstituted lipid bilayer membranes. Calcium release channels from malignant hyperthermia susceptible (MHS) pigs are slower to close after a contractile event, thus raising cytoplasmic calcium (Fill *et al.*, 1990; Mickelson *et al.*, 1988). More recently, the peptides generated by trypsin digestion of the Ca^{2+} release channel proteins isolated from pigs of different *HAL* genotypes have been compared (Knudson *et al.*, 1990). Differences in the tryptic peptides were observed between the two *HAL* homozygous classes, with the heterozygotes having an intermediate pattern. The different patterns appear to correspond to different rates of trypsin cleavage, rather than polymorphisms at the cleavage sites and may represent subtle conformational differences between putative allelic forms of the calcium-release channel. These data support the earlier conclusion that the MH lesion does not correspond to an absence of function or a null mutation. Rather MH seems to be caused by a qualitative difference in the performance of a Ca^{2+} regulatory system, which under stress crosses a physiological Rubicon.

Molecular genetic approaches

Porcine malignant hyperthermia is a monogenic disease where the underlying lesion remains unknown, and as such MH resembles some of the inherited disorders of man, such as cystic fibrosis, Huntington's disease and human MH. A molecular genetic approach has been used successfully to isolate linked polymorphic markers (restriction fragment length polymorphisms, RFLPs) for the prediction of genotypes at disease loci and, for example in the case of cystic fibrosis, for the isolation and characterization of the disease gene itself (Kerem *et al.*, 1989; Riordan *et al.*, 1989; Rommens *et al.*, 1989). The case for applying this approach to the study of the porcine stress syndrome has been made previously (Archibald, 1987). Linkage between the *HAL* locus and a number of blood group and biochemical marker loci has been described above. However, the polymorphic information content (Botstein *et al.*, 1980) of these biochemical markers is limited and they do not provide a route to identifying the *HAL* gene itself.

Clones for the closest of the known markers, the glucose-6-phosphate isomerase gene (*GPI*) have been isolated and characterized (Archibald, 1989a,b; Chaput *et al.*, 1988; Davies *et al.*, 1987, 1988). These *GPI* clones have been used to search for DNA polymorphisms. Multiple allele *Pvu*II or *Sac*I polymorphisms (Davies *et al.*, 1988) and diallelic *Eco*RI, *Hin*dIII, *Not*I, *Taq*I and *Xba*I RFLPs (Brenig *et al.*, 1990a,b; Archibald and Bowden, in press) have been described and shown to be linked to the *HAL* locus. The *Eco*RI and *Xba*I polymorphisms are in complete linkage disequilibrium in the IAPGR halothane backcross population. Although these DNA polymorphisms have higher polymorphic information contents than the biochemical polymorphisms, they still have the same limitations as all linked markers as described earlier.

It would be useful for the molecular mapping of the *HAL* linkage group to have other cloned DNA markers. One of the other members of this linkage group, which could be readily cloned, is the gene for 6-phosphogluconate dehydrogenase (PGD). Partial cDNA clones have been isolated for rat PGD (Miksicek and Towle, 1983) and for human PGD (Kleyn, 1989). These human and rat clones have been used to isolate partial cDNA

clones encoding PGD (Archibald and Bowden, in press). The porcine
PGD cDNA has been used to probe pig DNA digested with a panel of 18
different restriction endonucleases. Although the DNA samples analysed
were isolated from pigs of four different PGD protein phenotypes, no
RFLPs were detected (Archibald and Bowden, in press).

The *HAL* linkage group has been mapped to porcine chromosome 6 as
a result of *in situ* hybridization studies using cloned porcine *GPI* and *PGD*
sequences (Chowdhary *et al.*, 1989; Davies *et al.*, 1988; Yerle *et al.*, 1990;
Yerle *et al.*, in press). *GPI* was initially mapped to 6p12- > q22 (Davies *et
al.*, 1988) and then more recently by combining *in situ* hybridization with
high resolution G-banded prometaphase chromosomes the *GPI* locus has
been localized more precisely to 6q1.2 (Yerle *et al.*, in press). The assignment of the *PGD* locus to 6q2.2- > 2.5 (Yerle *et al.*, 1990) allows the
orientation of the *HAL* linkage group to be deduced.

Comparative mapping

The comparative mapping of mammalian genomes offers opportunities for
identifying animal homologues of human genetic diseases. As noted earlier,
the aetiology of one such human genetic disorder (malignant hyperthermia
(MH)) suggests that it is the human equivalent of the porcine malignant
hyperthermia syndrome (Mitchell and Heffron, 1982). Indeed, as will be
seen later, comparative mapping data provide some of the critical evidence
in support of the strongest candidate gene for MH. It is surprising given the
extensive use of porcine MH as a model for MH in humans that the locus
responsible for malignant hyperthermia (*MH*) in man has only recently
been mapped to the q13.1 region of human chromosome 19 (McCarthy *et
al.*, 1990). The conservation of syntenic groups across mammalian species
had always suggested that human *MH* would map to chromosome 19 close
to the *GPI* locus.

An examination of the comparative map locations of the homologues of
the loci known to be linked to *HAL* could be beneficial in the search for
candidate genes for MH. The human and murine homologues of *GPI* and
PGD can be readily identified, as can the human equivalent of *A1BG*,
through the use of standardized nomenclature for these well-characterized
proteins. As noted before, Secretor (*Se* or *FUT2*) may be the human
homologue of the porcine *S(A-0)* locus. Whilst *GPI* and *A1BG* map to
human chromosome 19, as does *FUT2*, *PGD* maps to human chromosome
1 (Burns and Sherman, 1989; Le Beau *et al.*, 1989). In mice the *Gpi* and
Pgd loci map to chromosomes 7 and 4 respectively. Indeed, the linkage
groups around *GPI* and *PGD* represent two of the most highly conserved,
yet independent, syntenic groups in animals (Lalley *et al.*, 1989). The
TGFβ−1 locus which maps to chromosome 19q in man and chromosome 7

in mouse (Dickinson *et al.*, 1990; Fujii *et al.*, 1986) has also been localized to porcine chromosome 6cen- > q2.1 (Yerle *et al.*, 1990). Yerle *et al.* (in press) have also mapped the *APOE* and *ENO1* loci to chromosome 6 using cloned human DNA sequences (Fig. 26.1a).

The genes which have been physically mapped to porcine chromosome 6 and the chromosomal locations of their human and murine homologues are summarized in Fig. 26.1(a) and Table 26.1. From these data it can be concluded that the long arm of porcine chromosome 6 contains regions homologous to human 19q13.1- > q13.2 and human 1pter- > p36.13, which in turn are homologous to murine chromosomes 7 and 4 respectively.

As well as confirming that malignant hyperthermia in man and pigs are likely to be the result of mutations in homologous genes, comparative mapping data suggest that in mice the corresponding gene is likely to map to chromosome 7. With this knowledge it may be possible to develop a murine model of malignant hyperthermia by screening appropriate mutants or by targeting mutations in the murine gene through the use of embryo-derived stem cells (Capecchi, 1989). More generally, this comparison confirms the usefulness both of the human gene map as a guide for deriving a porcine gene map and of cloned human genes as probes for such mapping studies (Haley *et al.*, 1990).

Candidate genes

There are two lines of evidence, which support the hypothesis that the gene responsible for susceptibility to halothane-induced malignant hyperthermia encodes a Ca^{2+}-release channel protein known as the ryanodine receptor. First, there are biochemical data, which indicate that in stress-susceptible (MHS) pigs the sarcoplasmic Ca^{2+} release channel is defective and that in particular it is slow to close after a contracting event (Fill *et al.*, 1990; Louis *et al.*, 1990; Mickleson *et al.*, 1988). The second line of evidence comes from molecular genetic data. The isolation and characterization of cDNA clones encoding the rabbit and human ryanodine receptor will allow the contribution of this gene to MH to be evaluated (Takeshima *et al.*, 1989; Zorzato *et al.*, 1990). The human *RYR* cDNA clones have been used to reveal a number of RFLPs, which have been used in linkage analysis of a limited number of human MH pedigrees (MacLennan *et al.*, 1990). No recombinants were observed between the *RYR* RFLPs and the *MH* locus in 20 informative meioses. The assignment by physical mapping techniques of the human and porcine ryanodine receptor genes to human chromosome 19cen- > q13.2 and porcine chromosome 6p11- > q21 respectively enhances the likelihood that this gene is responsible for MH in man and pigs (MacLennan *et al.*, 1989, 1990; Harbitz *et al.*, 1990). The *RYR* locus has also been mapped to chromosome 7 in mouse (Cavanna *et al.*,

Table 26.1. Chromosomal location of human and murine homologues of genes mapped by physical methods to porcine chromosome 6.

	RYR (CRC)	GPI	TGFβ-1	APOE-1	ENO1	PGD
Homo sapiens	19q13.1	19q13.1	19q13.1	19q13.2	1pter->36.13	1p36.2->36.13
Mus musculus	7	7	7	7	4	4
Sus scrofa	6p1.1->q2.1	6q1.2	6 C->q2.1	6q2.1.2	6q2.2->2.4	6q2.2->2.5

1990) further underlining the homology of porcine chromosome 6, murine chromosome 7 and human 19q. The reported biochemical defect in the calcium release channel is not the first plausible explanation of MH, but it is the first such defective protein encoded by a gene in the relevant region of both the porcine and human genomes. Taken together these data suggest that not only is the ryanodine receptor gene a strong candidate for the MH locus but that the human and porcine malignant hyperthermia syndromes arise from mutations in homologous genes.

Although the combination of biochemical and molecular genetic evidence provide compelling support for the hypothesis that malignant hyperthermia is caused by mutations in the ryanodine receptor gene the hypothesis is not yet proven. The false dawns of the search for the gene responsible for cystic fibrosis in humans should be recalled at this juncture. The absence of recombinants between *MH* and DNA variants within the *RYR* gene is based on only 20 informative meioses. With a sample of such limited size one would not expect to observe recombination events between loci less than 5 cM (or approximately 5 megabases) apart. Malignant hyperthermia in humans appears to be more heterogeneous than in pigs and probably includes hyperthermia arising as a complication of other myopathies. This heterogeneity may also reflect greater diversity in the genetic backgrounds of humans than in artificially selected pig populations. Nevertheless, it is disquieting that families, which did not fit the standard criteria for autosomal dominant inheritance of MH, were excluded from the linkage analysis performed by MacLennan *et al.* (1990). Given the more outbred nature of human populations it is possible that several different mutations in what is a large gene (*RYR*) could give rise to similar but heterogeneous phenotypes. The *HAL* mutation in pigs by contrast could be a single event which has been widely disseminated by pig breeders pursuing gains in lean.

Levitt and others (1990) have nominated an alternative candidate for the *MH* locus in humans – the hormone-sensitive lipase gene. They support the candidature of this lipase again on both genetic and biochemical grounds. The hormone-sensitive lipase locus (*LIPE*) has been mapped to the *GPI-MH* region of human chromosome 19q (Holm *et al.*, 1988). However, the biochemical data linking hormone-sensitive lipase to susceptibility to malignant hyperthermia are not as robust as those for the Ca^{2+} release channel. Hormone-sensitive lipase, which is regulated by hormonal and neuronal factors, has a key role in the mobilization of the free fatty acids from stored triglycerides (Holm *et al.*, 1988). The elevated phospholipase activity in porcine and human MH skeletal muscle has not only been implicated in raising the Ca^{2+} efflux from mitiochondria but causes increased free fatty acid release from skeletal muscle triglyceride stores (Cheah and Cheah, 1981b; Fletcher *et al.*, 1989). These increased free fatty acid levels can disrupt the regulation of intracellular Ca^{2+} by the sarcoplasmic reticulum resulting in an abnormal net release of Ca^{2+} into the cytoplasm (Fletcher *et al.*, 1989, 1990). Levitt *et al.* (1990) speculate that catecholamine stimulation of, and a failure in the insulin activation of, the hormone-sensitive lipase in MH-susceptible individuals could lead to excess levels of free fatty acids thus disrupting muscle metabolism. However, no experimental data concerning hormone-sensitive lipase in humans or pigs of different *MH* genotypes are presented.

As well as susceptibility to malignant hyperthermia different HAL phenotypes in pigs are associated with differences in the lean and fat content of muscles (Simpson and Webb, 1989; Webb and Simpson, 1986). The mapping of the *APOE* gene, or the *TGFβ−1* gene, with its known effects on myogenesis and adipogenesis (Massagué, 1987), to the *HAL* linkage group may explain some of the body composition phenotypes associated with the *HAL* locus. However, if the hormone-sensitive lipase (*LIPE*) gene is the gene responsible for malignant hyperthermia as suggested by Levitt *et al.* (1990), then both HAL phenotypes might be explained by a single gene. Alternatively, *LIPE* may be responsible for the differences in fat content whilst *RYR* is responsible for MH. It may be possible to dissect these alternative hypotheses by site-directed manipulation of the relevant genes (Capecchi, 1989) once porcine embryo-stem cells become available (Notarianni *et al.*, 1990). If the different HAL phenotypes can be mimicked by independent mutations, then the pig breeders' dream of retaining the beneficial HAL phenotypes without the cost of the deleterious ones may be achievable through manipulation of the germline.

Conclusions

Increasingly, the halothane gene is regarded by pig breeders as a mutation, which should be eliminated from breeding populations. The controlled maintenance of the gene for its perceived benefits on lean content seems to be losing favour. Breeders working with breeds like the Pietrain, which have a very high incidence of the halothane gene, have less scope for eliminating the gene and perhaps need to focus their efforts on developing husbandry, slaughtering, and carcass processing regimes, which ameliorate the effects of the gene. Current research on molecular genetics of malignant hyperthermia may soon yield the identity of the MH mutations. Whilst such findings will provide a robust test for determining all three *HAL* genotypes in pigs and MH in humans, they will also confirm the validity and benefit of the pig as a model for MH in humans. Genetic engineering of the *HAL* gene to acquire its beneficial effects on lean, but without the deleterious effects of stress deaths and PSE meat may also become possible, not only in pigs but in other species reared for their meat.

Acknowledgements

The study of the molecular genetics of halothane-induced malignant hypthermia at IAPGR was supported with a grant (CSA946) from the UK Ministry of Agriculture Fisheries and Food.

References

Archibald, A.L. (1987) A molecular genetic approach to the porcine stress syndrome. In: Tarrant, P.V., Eikelenboom, G. and Monin, G. (eds) *Evaluation and Control of Meat Quality in Pigs.* Martinus Nijhoff. Dordrecht, pp. 343–57. pp. 343–57.

Archibald, A.L. (1989a) Linkage analysis of the halothane sensitivy locus in pigs. *Genetical Research* 53, 231.

Archibald, A.L. (1989b) Progress on the halothane gene in pigs. *Animal Genetics* 20, 332.

Archibald, A.L. and Bowden, L.M. (in press). Molecular genetic markers for halothane-induced malignant hyperthermia in pigs. Proceedings of the XXIIth International Conference on Animal Genetics 25–31 August 1990, *Animal Genetics* (in press).

Archibald, A.L. and Imlah, P. (1985) The halothane sensitivity locus and its linkage relationships. *Animal Blood Groups and Biochemical Genetics* 16, 253–63.

Archibald, A.L. and McTeir, B.L. (1988) A new allele at the Pgd locus in pigs. *Animal Genetics*, 19, 189–91.

Botstein, D., White, R.L., Skolnick, M. and Davis, R.W. (1980) Construction of a genetic linkage map in man using restriction fragment length polymorphisms. *American Journal of Human Genetics* 32, 314–31.

Brenig, B., Jürs, S. and Brem, G. (1990a) The porcine PHIcDNA linked to the halothane gene detects a HindIII and XbaI RFLP in normal and malignant hyperthermia susceptible pigs. *Nucleic Acids Research* 18, 388.

Brenig, B., Jürs, S. and Brem, G. (1990b) The porcine PHIcDNA linked to the halothane gene detects a NotI RFLP in normal and malignant hyperthermia susceptible pigs. *Nucleic Acids Research* 18, 388.

Burns, G.A.P. and Sherman, S.L. (1989) Report of the committee on the genetic constitution of chromosome 1. *Cytogenetics and Cell Genetics* 51, 67–90.

Capecchi, M.R. (1989) The new mouse genetics: altering the genome by gene targeting. *Trends in Genetics* 5, 70–6.

Carden, A.E., Hill, W.G. and Webb, A.J. (1983) The inheritance of halothane susceptibility in pigs. *Génétique Sélection et Evolution* 15, 65–82.

Cavanna, J.S., Greenfield, A.J., Johnson, K.J., Marks, A.R., Nadal-Ginard, B. and Brown, S.D.M. (1990) Establishment of the mouse chromosome 7 region with homology to the myotonic dystrophy region of human chromosome 19q. *Genomics* 7, 12–18.

Chaput, M., Claes, V., Portetelle, D., Cludts, L., Cravador, A., Burny, A., Gras, H. and Tartar, A. (1988). The neurotrophic factor neuroleukin is 90% homologous with phosphohexose isomerase. *Nature* 332, 454–7.

Cheah, K.S. (1984) Skeletal muscle mitochondria and phospholipase A_2 in malignant hyperthermia. *Biochemical Society Transactions* 12, 358–60.

Cheah, K.S. and Cheah, A.M. (1976) The trigger for PSE condition in stress-susceptible pigs. *Journal of the Science of Food and Agriculture* 27, 1137–44.

Cheah, K.S. and Cheah, A.M. (1978) Calcium movements in skeletal muscle mitochondria of malignant hyperthermic pigs. *FEBS Letters* 95, 307–10.

Cheah, K.S. and Cheah, A.M. (1979) Mitochondrial calcium efflux and porcine stress-susceptibility. *Experientia* 35, 1001–3.

Cheah, K.S. and Cheah, A.M. (1981a) Mitochondrial calcium transport and calcium activated phospholipase in porcine malignant hyperthermia. *Biochimica et Biophysica Acta* 634, 70–84.

Cheah, K.S. and Cheah, A.M. (1981b) Skeletal muscle mitochondrial phospholipase A and the interaction of mitochondria and sarcoplasmic reticulum in porcine malignant hyperthermia. *Biochimica et Biophysica Acta* 638, 40–9.

Chowdhary, B.P., Harbitz, I., Mäkinen, A., Davies, W. and Gustavsson, I. (1989) Localization of the glucose phosphate isomerase gene to the p12- > q21 segment of chromosome 6 in pig by in situ hybridization. *Hereditas* 111, 73–8.

Davies, W., Harbitz, I., Fries, R., Stranzinger, G. and Hauge, J.G. (1988) Porcine malignant hyperthermia carrier detection and chromosomal assignment using a linked probe. *Animal Genetics* 19, 203–12.

Davies, W., Harbitz, I. and Hauge, J.G. (1987) A partial cDNA clone for porcine glucosephosphate isomerase: isolation, characterization and use in detection of restriction fragment length polymorphisms. *Animal Genetics* 18, 233–40.

Dickinson, M.E., Kobrin, M.S., Silan, C.M., Kingsley, D.M., Justice, M.J., Miller, D.A., Ceci, J.D., Lock, L.F., Lee, A., Buchberg, A.M., Siracusa, L.D., Lyons, K.M., Derynck, R., Hogan, B.L.M., Copeland, N.G. and Jenkins, N.A. (1990) Chromosomal localization of seven members of the murine TGF-β superfamily suggests close linkage to several morphogenetic mutant loci. *Genomics* 6, 505–20.

Eikelenboom, G. and Minkema, D. (1974) Prediction of pale, soft, exudative muscle with a non-lethal test for halothane-induced porcine malignant hyperthermia syndrome. *Netherlands Journal of Veterinary Science* 99, 421–6.

Fill, M., Coronado, R., Mickelson, J.R., Vilven, J., Ma, J., Jacobson, B.A. and Louis, C.F. (1990) Abnormal ryanodine receptor channels in malignant hyperthermia. *Biophysics Journal* 50, 471–5.

Fletcher, J.E., Beech, J., Tripolitis, L., Hanson, S. and Rosenberg, H. (1990) Fatty acids modulate Ca^{2+}-induced Ca^{2+} release in skeletal muscle: implications for malignant hyperthermia. *FASEB Journal* A1209 (abstract 5477).

Fletcher, J.E., Rosenberg, H., Michaux, K., Tripolitis, L. and Lizzo, F.H. (1989) Triglycerides, not phospholipids, are the source of elevated free fatty acids in muscle from patients susceptible to malignant hyperthermia. *European Journal of Anaesthesiology* 6, 355–62.

Fujii, D., Brissenden, J.E., Derynck, R., and Francke, U. (1986) Transforming growth factor beta gene maps to human chromosome 19 long arm and to mouse chromosome 7. *Somatic Cellular and Molecular Genetics* 12, 281–8.

Gahne, B. and Juneja, R.K. (1985) Prediction of the halothane (Hal) genotypes of pigs by deducing Hal, Phi, Po2, Pgd haplotypes of parents and offspring: results from a large-scale practice in Swedish breeds. *Animal Blood Groups and Biochemical Genetics* 16, 265–83.

Grashorn, M. and Müller, E. (1985) Relationships between blood groups, isozymes and halothane reaction in pigs from a selection experiment. *Animal Blood Groups and Biochemical Genetics* 16, 329–35.

Haley, C.S., Archibald, A.L., Andersson, L., Bosma, A.A., Davies, W., Fredholm, M., Geldermann, H., Groenen, M., Gustavsson, I., Olliver, L., Tucker, E.M. and Van de Weghe, A. (1990) The Pig Gene Mapping Project – PiGMaP. *4th World Congress on Genetics Applied to Livestock Production.* XIII, 67–70.

Hall, L.W., Woolf, N., Bradley, J.W.P. and Jolly, D.W. (1966) Unusual reaction to suxamethonium chloride. *British Medical Journal* 2, 1305.

Harbitz, I., Chowdhary, B., Thomsen, P.D., Davies, W., Kaufman, U., Kran, S., Gustavsson, I., Christensen, K. and Hauge, J. (1990) Assignment of the porcine calcium release channel gene, a candidate for the malignant hyperthermia locus, to the 6p11-q21 segment of chromosome 6. *Genomics* 8, 243–8.

Harrison, G.G. (1972) Pale soft, exudative pork, porcine stress syndrome and malignant hyperpyrexia – an identity? *Journal of the South African Veterinary Association* 43, 57–63.

Harrison, G.G. (1979) Porcine malignant hyperthermia. *International Anesthesiology Clinics* 17, 25–61.

Harrison, G.G., Biebuyck, J.F., Terblanche, J., Dent, D.M., Hickman, R. and Saunders, J. (1968) Hyperpyrexia during anaesthesia. *British Medical Journal* 3, 594–5.

Heffron, J.J.A. (1984) Mitochondrial and plasma membrane changes in skeletal muscle in the malignant hyperthermia syndrome. *Biochemical Society Transactions* 12, 360–2.

Heffron, J.J.A. (1987) Calcium releasing systems in mitochondria and sarcoplasmic reticulum with respect to the aetiology of malignant hyperthermia: a review. In: Tarrant, P.V., Eikelenboom, G. and Monin, G. *Evaluation and Control of Meat Quality in Pigs* Martinus Nijoff, Dordrecht. pp. 17–26.

Holm, C., Kirchgessner, T.G., Svenson, K.L., Fredrikson, G., Nilsson, S., Miller, C.G., Shively, J.E., Heinzmann, C., Sparkes, R.S., Mohandas, T., Lusis, A.J., Belfrage, P. and Schotz, M.C. (1988) Hormone-sensitive lipase: sequence, expression, and chromosomal localization to 19 cent-q13.3. *Science* 241, 1503–6.

Kerem, B.-S., Rommens, J.M., Buchanan, J.A., Markiewicz, D., Cox, T.K., Chakravarti, A., Buchwald, M. and Tsui, L.-C. (1989) Identification of the cystic fibrosis gene: genetic analysis. *Science* 245, 1073–80.

Kim, D.H., Sreter, F.A., Ohnishi, S.T., Ryan, J.F., Roberts, J., Allen, P.D., Meszaros, L.G., Antoniu, B. and Ikemoto, N. (1984) Kinetic studies of Ca^{2+} release from sarcoplasmic reticulum of normal and malignant hyperthermia susceptible pig muscles. *Biochimica et Biophysica Acta* 775, 320–7.

Kleyn, P. (1989) Study of the PGD locus on chromosome 1p. *Genetical Research* 53, 233.

Knudson, C.M., Mickelson, J.R., Louis, C.F. and Campbell, K.P. (1990) Distinct immunopeptide maps of the sarcoplasmic reticulum Ca^{2+} release channel in malignant hyperthermia. *Journal of Biological Chemistry* 265, 2421–4.

Lalley, P.A., Davisson, M.T., Graves, J.A.M., O'Brien, S.J. Womack, J.E., Roderick, T.H., Creau-Goldberg, N., Hillyard, A.L., Doolittle, D.P. and Rogers, J.A. (1989) Report of the committee on comparative mapping. *Cytogenetics and Cell Genetics* 51, 503–32.

Le Beau, M.M., Ryan, D. and Rericak-Vance, M.A. (1989) Report of the committee on the genetic constitution of chromosomes 18 and 19. *Cytogenetics and Cell Genetics* 51, 338–57.

Levitt, R.C., McKusick, V.A., Fletcher, J.E. and Rosenberg, H. (1990) Candidate gene (scientific correspondence) *Nature* 345, 297–8.

Lorkin, P.A. and Lehmann, H. (1983) Investigation of malignant hyperthermia: analysis of skeletal muscle proteins from normal and halothane sensitive pigs by two dimensional gel electrophoresis. *Journal of Medical Genetics* 20, 18–24.

Louis, C.F., Gallant, E.M., Remple, E. and Mickelson, J.R. (1990) Malignant hyperthermia and porcine stress syndrome: a tale of two species. *Pig News and Information* 11, 341–4.

Mabry, J.W., Christian, L.L. and Kuhlers, D.L. (1981) Inheritance of porcine stress syndrome. *Journal of Heredity* 72, 429–30.

McCarthy, T.V., Healy, J.M.S., Heffron, J.J.A., Lehane, M., Deufel, T., Lehmann-Horn, F., Farrall, M. and Johnson, K. (1990) Localization of the malignant hyperthermia susceptibility locus to human chromosome 19q12–13.2. *Nature* 343, 562–4.

MacLennan, D.H., Duff, C., Zorzato, F., Fujii, J., Phillips, M., Korneluk, R.G., Frodis, W., Britt, B.A. and Worton, R.G. (1990) Ryanodine receptor gene is a candidate for predisposition to malignant hyperthermia. *Nature* 343, 559–61.

MacLennan, D.H., Zorzato, F., Fujii, J., Otsu, K., Phillips, M., Lai, F.A., Meissner, G., Green, N.M., Willard, H.F., Britt, B.A., Worton, R.G. and Korneluk, R.G. (1989) Cloning and localization of the human calcium release channel (ryanodine receptor) gene to the proximal long arm (Cen->q13.2) of human chromosome 19. *American Journal of Human Genetics* 45, Suppl. A205 (abstract 803).

Marjanen, L.A. and Denborough, M.A. (1984) Electrophoretic analysis of proteins in malignant hyperpyrexia susceptible skeletal muscle. *International Journal of Biochemistry* 16, 919–29.

Massagué, J. (1987) The TGF-β family of growth and differentiation factors. *Cell* 49, 437–8.

Mickelson, J.R., Gallant, E.M., Litterer, L.A., Johnson, K.M., Rempel, W.E. and Louis, C.F. (1988) Abnormal sarcoplasmic reticulum ryanodine receptor in malignant hyperthermia. *Journal of Biological Chemistry* 263, 9310–15.

Mickelson, J.R., Gallant, E.M., Rempel, W.E., Johnson, K.M., Litterer, L.A., Jacobson, B.A. and Louis, C.F. (1989) Effects of the halothane-sensitivity gene on sarcoplasmic reticulum function. *American Journal of Physiology* 257, C787–C794.

Miksicek, R.J. and Towle, H.C. (1983) Use of a cloned cDNA sequence to measure changes in 6-phosphogluconate dehydrogenase mRNA levels caused by thyroid hormone and dietary carbohydrate. *Journal of Biological Chemistry* 258, 9575–9.

Minkema, D., Eikelenboom, G. and van Eldik, P. (1977) Inheritance of M.H.S.-susceptibility in pigs. *Proceedings of the Third International Conference on Production Disease in Farm Animals.* Wageningen, The Netherlands, 1976, Pudoc, pp. 203–7.

Mitchell, G. and Heffron, J.J.A. (1980) The occurrence of pale, soft, exudative musculature in Landrace pigs susceptible and resistant to the malignant hyperthermia syndrome. *British Veterinary Journal* 136, 500–6.

Mitchell, G. and Heffron, J.J.A. (1982) Porcine stress syndromes. *Advances in Food Research* 28, 167–230.

Nelson, T.E. (1983) Abnormality in calcium release from skeletal sarcoplasmic reticulum of pigs susceptible to malignant hyperthermia. *Journal of Clinical Investigation* 72, 862–70.

Notarianni, E., Galli, C., Laurie, S., Moor, R.M. and Evans, M.J. (1990) Derivation of pluripotent, embryonic cell lines from porcine and ovine blastocysts. *4th World Congress on Genetics Applied to Livestock Production* XIII, 58–64.

O'Brien, P.J. (1986) Porcine malignant hyperthermia susceptibility: hypersensitive calcium-release mechanism of skeletal muscle sarcoplasmic reticulum. *Canadian Journal of Veterinary Research* 50, 318–28.

Ohnishi, S.T., Taylor, S. and Gronert, G.A. (1983) Calcium-induced Ca^{2+} release from sarcoplasmic reticulum of pigs susceptible to malignant hyperthermia. *FEBS Letters* 161, 103–7.

Ollivier, L., Sellier, P. and Monin G. (1975) Déterminisme génétique du syndrome d'hyperthermie maligne chez le porc de Piétrain. *Annales de Génétique et de Sélection Animale* 7, 159–66.

Rasmusen, B.A. (1964) Gene interaction and the A-O blood group system in pigs. *Genetics* 50, 191–8.

Riordan, J.R., Rommens, J.M., Kerem, B.-S., Alon, N., Rozmahel, R., Grzelczak, Z., Zielenski, J., Lok, S., Plasvsic, N., Chou, J.-L., Drumm, M.L., Iannuzzi, M.C., Collins, F.S. and Tsui, L.-C. (1989) Identification of the cystic fibrosis gene: cloning and characterization of complementary DNA. *Science* 245, 1066–73.

Rommens, J.M., Iannuzzi, M.C., Kerem B.-S., Drumm, M.L., Melmer, G., Dean, M., Rozmahel, R., Cole, J.L., Kennedy, D., Hidaka, N., Zsiga, M., Buchwald, M., Riordan, J.R., Tsui, L.-C. and Collins, F.S. (1989) Identification of the cystic fibrosis gene: chromosome walking and jumping. *Science* 245, 1059–65.

Simpson, S.P. and Webb, A.J. (1989) Growth and carcass performance of British Landrace pigs heterozygous at the halothane locus. *Animal Production* 49, 503–9.

Smith, C. and Bampton, P.R. (1977) Inheritance of reaction to halothane anaesthesia in pigs. *Genetical Research* 29, 287–92.

Southwood, O.I., Simpson, S.P. and Webb, A.J. (1986) Incomplete recessive inheritance and maternal effects on halothane sensitivity in British Landrace pigs. *Proceedings of the 3rd World Congress on Genetics Applied to Livestock Production* XI, 401–6.

Southwood, O.I., Simpson, S.P. and Webb, A.J. (1988) Reaction to halothane anaesthesia among heterozygotes at the halothane locus in British Landrace pigs. *Génétique Sélection Evolution* 20, 357–66.

Sullivan, J.S. and Denborough, M.A. (1982) The isolation and chemical characterisation of skeletal muscle microsomes from swine susceptible to malignant hyperpyrexia. *International Journal of Biochemistry* 14, 741–5.

Takeshima, H., Nishimura, S., Matsumoto, T., Ishida, H., Kangawa, K., Minamino, N., Matsuo, H., Ueda, M., Hanaoka, M., Hirose, T. and Numa, S. (1989) Primary structure and expression from complementary DNA of skeletal muscle ryanodine receptor. *Nature* 339, 439–45.

Van de Weghe, A., Yablanski, Ts., Van Zeveren, A. and Bouquet, Y. (1988) A third variant of glucose phosphate isomerase in pigs. *Animal Genetics* 19, 55–7.

Webb, A.J., Carden, A.E., Smith, C. and Imlah, P. (1982) Porcine stress in pig breeding. *Proceedings of the 2nd World Congress on Genetics Applied to Livestock Production* VI, 588–608.

Webb, A.J. and Simpson, S.P. (1986) Performance of British Landrace pigs selected for high and low incidence of halothane sensitivity. 2. Growth and carcass traits. *Animal Production* 43, 493–503.

Webb, A.J., Southwood, O.I. and Simpson, S.P. (1987) The halothane test in improving meat quality. In: Tarrant, P.V., Eikelenboom, G. and Monin, G. (eds), *Evaluation and Control of Meat Quality in Pigs*, Martinus Nijhoff, Dordrecht, pp. 297–315.

Yerle, M., Archibald, A.L., Dalens, M. and Gellin, J. (1990) Localization of PGD and TGFβ-1 to pig chromosome 6q. *Animal Genetics* 21, 411–17.

Yerle, M., Gellin, J., Dalens, M. and Galman, O. (1991) Localization on pig chromosome 6 of markers: GPI, APOE, ENO1 carried by human chromosomes 1 and 19 using in situ hybridization. *Cytogenetics and Cell Genetics* (in press).

Zorzato, F., Fujii, J., Otsu, K., Phillips, M., Green, N.M., Lai, F.A., Meissner, G. and MacLennan, D.H. (1990) Molecular cloning of cDNA encoding human and rabbit forms of the Ca^{2+} release channel (ryanodine receptor) of skeletal muscle sarcoplasmic reticulum. *Journal of Biological Chemistry* 265, 2244–56.

Chapter 27

The Major Gene of Muscular Hypertrophy in the Belgian Blue Cattle Breed

R. Hanset

Chaire de Génétique, Faculté de Médecine Vétérinaire, Rue des Vétérinaires, 45, B-1070 Brussels, Belgium

Summary

The results of an investigation on the genetic determination of the extreme muscled phenotype acquired by the Belgian Blue cattle breed are presented. The analyses of experimental and population data led to the conclusion that a major gene, partially recessive (symbol mh) is involved. The segregation of a single major gene was confirmed by linkage studies, a DNA fingerprint band co-segregating with the double-muscled condition in the progeny of a double heterozygous bull. Moreover, the muscled (or meaty) phenotype has been characterized from the anatomical, physiological and cellular points of view. There is a place in the field of animal production for the mh gene since its favourable effects may exceed the unfavourable ones if certain conditions are fulfilled regarding consumer requirements, market, farm structure, etc. The emergence and the expansion of the Belgian Blue breed are in contradiction with gloomy reports concerning the survival and the breeding of this extreme biological type.

Introduction

The extreme muscled phenotype acquired in a short period of time by the Belgian Blue cattle breed was a logical candidate for the possible disclosure of a gene with major effect. The genetic analysis that was undertaken was based on two kinds of data: experimental and population data. Linkage studies were recently added to the research programme. On the way, a biological profile of the condition could be drawn as well as a balance sheet of its favourable and unfavourable features.

Inheritance

Experimental data

F_1 cows from the cross Belgian Blue (BB) sires × Friesian cows were backcrossed to BB sires. The criterion of the muscle development was the total weight of the most important muscles in the half-carcase of calves slaughtered at the constant weight of 84 kg.

The distribution of this criterion in the backcross was clearly bimodal as might be expected in the case of the segregation of a gene with a large enough effect.

The two component distributions of this backcross are referred to as BC_1 and BC_2. Their means are given in Table 27.1, under the heading 'observed means' at the same time as the means of pure Friesian calves, pure BB calves, F_1 calves, all reared and slaughtered in the same conditions, thus giving five genetic types in total. The mean of the F_1 calves is similar to that of the BC_1 calves; the same is true for the mean of the pure BB calves and the mean of the BC_2 calves. The symbol mh was proposed for this locus; so, the three genotypes are written as follows: mh/mh; mh/+ and ++ (Hanset and Michaux, 1985a).

A monogenic model could be fitted to the data in order to estimate the gene effect. The metrical values, a, d and −a correspond to the three genotypes. The two breeds considered differed not only at the mh locus but probably also for many quantitative genes. Therefore, if the difference between the two homozygotes at the mh locus (mh/mh and +/+) is put equal to 2a and the difference due to the other genes is 2m, then the breed difference is:

$$BB - Friesian = 2m + 2a$$

Table 27.1. Expectations of the five genetic types (observed means, weighted least squares solutions, expected means).

Expectations	Observed means (kg)	Expected means (kg)
FR. $= \mu - m - a$	12.36 ($n = 5$)	12.36
B.B. $= \mu + m + a$	18.34 ($n = 30$)	18.38
F_1 $= \mu + d$	14.31 ($n = 7$)	14.50
$BC_1 = \mu + m/2 + d$	14.69 ($n = 28$)	14.64
$BC_2 = \mu + m/2 + a$	18.29 ($n = 32$)	18.25

Solution: $\mu = 15.369$; m = 0.266; a = 2.746; d = −0.865; residual standard deviation = 0.791.

FR = Friesian; B.B. = Belgian Blue; F_1 = first cross BB × Fr; BC = back cross.

Consequently, the expectations of the different means are:

Friesian = $\mu - m - a$; BB = $\mu + m + a$; $F_1 = \mu + d$; $BC_1 = \mu + m/2 + d$; $BC_2 = \mu + m/2 + a$ (if d = dominance deviation).

The different terms of the model (μ, m, a and d) were estimated by the method of weighted least squares.

The solutions are given in Table 27.1 as well as the means expected for the different genotypes. For the criterion considered, the gene effect is very important compared to the residual standard deviation. Furthermore, the mh gene behaves as a partial recessive, the dominance deviation -d- being negative; this means that the heterozygote is near the homozygous normal. Moreover, these two genotypes belong to the conventional conformation class in contrast to the homozygous mh/mh which exhibits the double-muscled conformation. This experiment extended over a few years and was completed in 1984. Since then, a research programme aimed at evaluating the performances of crossbred cows (F_1): BB × Friesian provided additional data on the segregation in the backcross of the two morphological types. The following frequencies were observed: 42 of the conventional type, 46 of the meaty type, in agreement with the frequencies expected if two alleles are segregating.

Population data

This segregation of a major gene of muscle development can also be revealed on progeny-test data collected in commercial herds where cows of the two morphological types coexist (dual-purpose or conventional and double-muscled or meaty). As practically all AI bulls are of the meaty type and therefore mh/mh, in these herds practically all cows of the conventional type must be heterozygous (mh/+). In fact, the evolution of the proportions of meaty calves born from dual-purpose and meaty cows (Fig. 27.1) shows that a big difference remains through the years as can be expected if the meaty cows are mh/mh and the dual-purpose ones are mh/+, the sires being mh/mh.

The hypothesis of a polygenic determinism with a threshold could be excluded. In fact, there is a break in the regression (line) of the proportion of meaty calves on the conformation score (value in francs per kilo) of the mothers, this break being situated at the level of the antimode of the bimodal distribution of this score (Hanset and Michaux, 1985b). This observation was repeated on 1,600 calvings recorded in the year 1988 (Fig. 27.2). The deviations from the regression line gave a $\chi^2_{17} = 44.82$, the hypothesis of linearity being rejected at the 1% level. Once more, the presence of a major gene was confirmed.

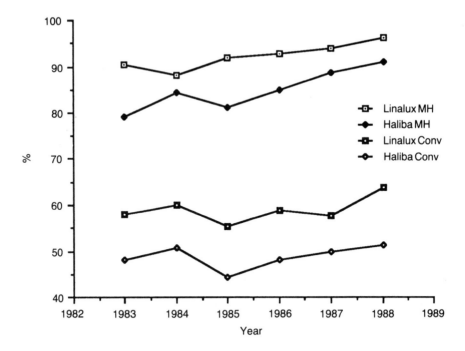

Fig. 27.1. Percentage of meaty calves according to the conformation of the dam (MH or Conv) and the region (Linalux and Haliba).

Linkage studies

The co-segregation of the gene of interest with a marker is a proof of linkage. This implies that at least one sire had to be found which was at the same time heterozygous at the mh locus and at as many marker loci as possible. Such bulls of the dual-purpose type are very few at present and are used only on dual-purpose cows. A battery of markers was used in this investigation. They were: one visible trait (coat colour), nine blood group loci, 12 biochemical markers, four RFLPs, five probes revealing multilocus DNA fingerprints (Georges *et al.*, 1990). Besides linkages that were revealed between classical markers, between these markers and fingerprint bands, and between fingerprint bands, co-segregation was discovered between a band and the mh allele in the progeny of a bull which was heterozygous for both. In this bull, the presence of the band (the alternative being its absence) and the mh allele were in the coupling phase. In 12 sire–dam pairs with absence of the band in the dam and a typical meaty progeny, the latter had received the band in each case.

% Meaty calves

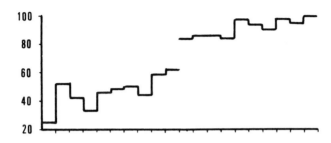

Nr of dams

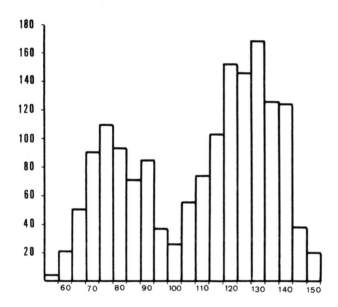

Fig. 27.2. Regression of the percentage of meaty calves on the conformation score of the dam.

The corresponding lodscore was 3.31 ($\alpha_{unil.}$ = 0.0002). The 95% confidence limits for θ, the coefficient of recombination, were 0 and 26.5%. The assumption of linkage is sufficiently strong to justify the cloning of the band and the development of a locus specific probe which would then become the first marker of the mh locus and would allow its chromosome assignment by *in situ* hybridization. Other markers will be hunted for, the aim being to come closer and closer to the mh locus.

Biological features of the homozygote mh/mh

Quantitative anatomy of the young

The data collected on the calves born from the backcross, slaughtered and dissected at the constant weight of 84 kg, allow comparison of the two biological types 'within sexes' (Tables 27.2 and 27.3). In these tables, R% represents the ratio (meaty/conventional − 1) × 100 and P, the level of significance of the difference between the two means. The gestation length is significantly increased: +2.4% for the males and +3.3% for the females whereas the increase of the birth weight is +16.5% and +25.4%. In the backcross now under way, these percentages are: +15.7% and 17.3%. This factor prevents natural calving.

On the other hand, many organs are hypotrophied and particularly the thymus, the spleen, the liver and the lungs. This is also true for the testis and the mammary gland. Table 27.3 shows that the muscle hypertrophy is about 25% on average but the hypertrophy of the individual muscles is heterogeneous ranging from 7.15% (*m. l. dorsi*) to 43.1% (*m. infraspinatus*). The amount of fat is dramatically reduced, mainly in the male. The hypotrophy of the bones is moderate compared to that of the other organs. A

Table 27.2. The main characteristics of the two biological types: MH (muscular hypertrophy or meaty) and Conv (conventional) at the adjusted weight of 84 kg.

	♂ MH	♂ Conv	R (%)	P(%)	♀ MH	♀ Conv	R (%)	P(%)
Gestation length (days)	285.9	279.2	+2.4	5.0	283.4	274.2	+3.3	1.0
Birth weight (kg)	49.4	42.2	+26.5	1.0	45.4	36.2	+25.4	0.1
Organs (kg)								
Skin	4.18	5.09	−17.9	0.1	4.15	4.93	−15.8	0.1
Liver	0.874	1.233	−29.1	0.1	1.022	1.204	−15.1	1.0
Heart	0.433	0.496	−12.7	1.0	0.412	0.484	−14.9	0.1
Lungs	0.693	0.884	−21.6	0.1	0.706	0.862	−18.1	0.1
Spleen	0.151	0.238	−36.5	0.1	0.172	0.249	−30.9	0.1
Thymus	0.123	0.269	−54.3	0.1	0.155	0.260	−40.4	0.1
Kidney (L.)	0.115	0.145	−20.7	1.0	0.119	0.152	−21.7	0.1
Kidney (R.)	0.112	0.141	−20.6	1.0	0.117	0.154	−24.0	0.1
Testes	0.013	0.018	−20.3	1.0				
Mammary gland					0.125	0.175	−28.8	10.0

Table 27.3. The main characteristics of the two biological types; MH (muscular hypertrophy or meaty) and Conv (conventional) at the adjusted weight of 84 kg.

	♂ MH	♂ Conv	R (%)	P(%)	♀ MH	♀ Conv	R (%)	P(%)
Muscles (kg) (half carcase)	18.51	14.91	+24.1	0.1	18.35	14.56	+26.0	0.1
Fat (kg)	0.44	0.74	−40.2	0.1	0.52	0.65	−20.1	10.0
Bones (kg)								
Scapulum	0.224	0.242	−7.4	1.0	0.208	0.225	−7.5	1.0
Humerus	0.450	0.489	−8.0	0.1	0.415	0.468	−11.3	0.1
Rad. Cub.	0.381	0.403	−5.5	5.0	0.342	0.369	−7.3	0.1
Femur	0.652	0.717	−9.1	0.1	0.609	0.683	−10.8	0.1
Tibia	0.455	0.478	−4.8	10.0	0.420	0.446	−5.8	5.0

centripetal gradient is apparent. The antagonism existing between the muscle development on the one hand and the development of other organs and the fatty tissues on the other hand is particularly striking.

A thick and protruding tongue can be observed at birth and this defect, generally temporary, can prevent proper suckling (Hanset and Michaux, 1978). At an older age, from 6–7 months, brachygnathism can appear (Hanset and Michaux, 1978). A strong selection is applied against these anomalies.

Physiology

At any age, the creatine content (measured in plasma or red cells) is significantly reduced whereas the creatinine content is clearly increased. These parameters are in relation with the muscle mass and they can be used to discriminate between the meaty type and other types as dual-purpose and dairy (Hanset and Michaux, 1986). Muscle samples taken from the seventh rib cut of bulls slaughtered at 12 months of age have shown: (1) similar DNA content; (2) nitrogen content slightly increased; (3) halving of the lipid content; (4) reduction in the hydroxyproline content ranging from 20 to 30% according to the muscle considered (Hanset *et al.*, 1982).

A delay in sexual maturity is observed in females (Table 27.4) and in males. At 12 months of age, the scrotal circumferences of Belgian Blue bulls and dual-purpose bulls of the same breed were 32 cm (± 2.2) and 35.3 (± 2.4) respectively and the testes weights were 436.3 (± 100) and 521.3 g (± 91) (Michaux and Hanset, 1981). If the volume of the ejaculate is reduced in the same proportion, the fertilizing ability does not seem altered.

Table 27.4.   Differences between the two biological types (meaty and conventional) expressed in real units, in phenotypic (σ_p) and genetic (σ_G) standard deviations.

	D	In σ_p	In σ_G
Production traits			
Feed conversion ratio	−0.56 kg	−0.80	−1.3
Killing-out percentage	+4.83 %	+3.20	+4.4
% lean in the 7th rib cut	+11.8 %	+3.9	+5.9
% fat in the 7th rib cut	−9.3 %	−3.5	−5.5
Selling price per kg liveweight	+40 BEF	+5.9	+10
Reproductive traits			
% culled for infertility			
I. Heifers	+3.51	0.12	0.69
II. First calvers	N.S.		
Non-return rate			
I. First gestation	−4.09	0.09	—
II. Second gestation	−4.91	0.10	1.34
No. of insemination/gestation			
I. First gestation	+0.09	0.11	—
II. Second gestation	0.11	0.10	1.34
Age at calving			
I. Heifers	+34.64	0.27	1.60
II. First calvers	+41.65	0.30	1.48
Calving interval	+9.17	0.18	1.08
Milk yield score			
I. First lactation	−0.33	0.45	1.67
II. Second lactation	−0.24	0.35	1.12

Regarding the hormonal status, Michaux *et al.* (1982) made the following observations on bulls: apart from a higher plasma content of Growth Hormone (GH) at 2.5 months of age, the plasma concentration of GH, Luteinizing Hormone (LH), testosterone and insulin are lower in the meaty type. These findings are to be interpreted with regard to the heavier birthweight, the smaller height at withers, a reduced feed intake, the smaller size of the testes and the seminal glands, and the quite different body composition. Moreover, the meaty animal is more susceptible to respiratory diseases mainly those of viral origin. An immunological deficiency does not seem to be involved but rather a reduced volume of the respiratory system (Gustin, 1989). A specific vaccine against Bovine Respiratory Syncytial Virus (BRSV) developed in Belgium has considerably reduced morbidity and mortality due to this cause.

Cellular characteristics

The muscular hypertrophy is due to a larger number (hyperplasia) of smaller muscle fibres (Hanset *et al.*, 1982). The connective tissue content is smaller as measured by the content of hydroxyproline, an amino acid specific to this kind of tissue. The tenderness of this meat is a well-known characteristic.

The muscle of the meaty animal has a propensity to develop an anaerobic type of metabolism, to be poorer in pigment and to show a quicker post-mortem glycolysis, these characteristics being associated with a paler meat. Nevertheless, this meat has a normal water-holding capacity (Van Hoof, personal communication).

This meat may not be compared to the PSE meat of the pig. Moreover, a susceptibility to halothane was not observed. Regarding the magnitude of their effect on muscle development, there is no possible comparison between these two genes.

Regarding the basic function of the gene, we are not yet beyond the stage of hypotheses (Hanset, 1986). An answer will be provided either by the identification of a particular protein or by the cloning and sequencing of the gene.

The mh gene in animal production

The demand, by an increasing number of consumers, for lean and tender meat, with the ensuing price differential, prompted Belgian breeders to select for animals with rounded outline and prominent muscles. This selection was not knowingly for a double-muscled condition as such but for a better muscle development on sound and fine animals. Of course the shape of these selected animals is reminiscent of that of a 'true' culard. But for Belgian breeders this latter term does not apply to their cattle, therefore they coined a new word 'viandeux' (meaty). The condition is far from being lethal. This selection has probably brought with it a major gene and poly-genes and in the process some of the undesirable effects of the major gene have been corrected (Hanset, 1982).

The price obtained for the meaty animals is particularly high, in any case higher than might be expected from the lean content of the carcase (Michaux *et al.*, 1983). In fact, this kind of carcase with about 80% lean (against 65% for the conventional type) provides a regrouping of some muscles on the one hand because of their size (hypertrophy) and on the other hand because of their tenderness. This regrouping is particularly apparent in the forequarter where many muscles of inferior quality are commercially upgraded into valuable cuts (Sonnet, 1982). 'Extensive cutting' is the name given to this special cutting system developed by the Belgian butchers for this type of carcase.

Such a high muscle development, coupled with large size is attainable because no constraint was put on calving performance. Since no chances are taken which could imperil the life of the cow and that of the calf, there is general recourse to caesarian section. This is performed by the veterinarian quickly after the beginning of the parturition and no pulling is tolerated in the herds where caesarian section is uniformly adopted. As a consequence, perinatal mortality is at a minimum level (Michaux and Hanset, 1986). This implies that, in the cow and her fetus, suffering is minimal and frequently less than in normally calved animals. In these conditions, a concern for animal welfare is not valid. On the other hand, from an economic point of view, it is worth mentioning that the selling-price of a meaty calf and the cost of the caesarian are in the ratio of ten to one.

The decrease in milk yield of about one-third is an undesirable trait; this is in a dual-purpose breed whose average milk yield was about 4,000 kg. Many cows are still milked but large herds have changed to a suckler herd system. In comparison to other continental breeds, the emergence of this extreme beef breed from a dual purpose breed has important consequences regarding milk yield, precocity and docility. The outcome would probably have been less favourable if the original gene pool had been that of a late maturing and poor milking breed.

The unfavourable effects on intrinsic female fertility as a consequence of the fixation of the extreme meaty type are considered in Table 27.4 (Hanset *et al.*, 1989a). In spite of the general effect of the gene on age at first calving, some large herds ($n > 80$) have an average age at first calving below 26 months (Michaux *et al.*, 1987).

In the same table, these effects are compared to the favourable effects on production traits. It is seen that the step forward regarding the carcase composition and its value on the Belgian market exceeds the step backward regarding the reproductive traits by several units. How otherwise can the boom of the breed with its 450,000 cows and 350,000 first inseminations be explained. This mere fact contradicts many reports stating that extreme muscling in cattle is incompatible with survival and breeding. The Belgian Blue breed has become one of the main assets of Belgian agriculture. It is used either in pure breeding or in crossbreeding with dairy cows (mainly Friesian). In crossing, the level of dystocia is similar to that of other large continental breeds (Cook, 1988; Liboriussen, 1982) and the crossbreds have a significantly improved carcase composition (Hanset *et al.*, 1989b; Michaux *et al.*, 1990).

The homozygosity for the mh allele does not represent the limit regarding muscular hypertrophy. In fact, despite the fixation at the mh locus a significant genetic variation still remains and at present, selection within the Belgian Blue breed continues and acts on a genetic variation of polygenic nature (Hanset *et al.*, 1987, 1989c, 1990).

Acknowledgement

This work is supported by the Institut pour l'encouragement de la Recherche Scientifique dans l'Industrie et l'Agriculture.

References

Cook, N.K. (1988) The Belgian Blue sire in crossing. Mimeo, Milk Marketing Board.

Georges, M., Lathrop, M., Hilbert, P., Marcotte, A., Schwers, A., Swillens, S., Vassart, G. and Hanset, R. (1990) The use of DNA fingerprints for linkage studies in cattle. *Genomics* 6, 461–74.

Gustin, P. (1989) Spécificités fonctionnelles du système respiratoire des bovins hyperviandeux. PhD Thesis, Faculté de Médecine Vétérinaie, Université de Liège.

Hanset, R. (1982) Muscular hypertrophy as a racial characteristic: the case of the Belgian Blue breed. In: King, J.W.B. and Ménissier, F. (eds), *Muscle Hypertrophy of Genetic Origin and its Use to Improve Beef Production*. Martinus Nijhoff, The Hague, pp. 437–49.

Hanset, R. (1986) Double-muscling in cattle. In: Smith, C., King, J.W.B., McKay, J.C. (eds), *Exploiting New Technologies in Animal Breeding*. Oxford Science Publications, pp. 71–9.

Hanset, R., Detal, G. and Michaux, C. (1989b) The Belgian Blue breed in pure- and crossbreeding. Growth and carcase characteristics. *Revue de l'Agriculture* 42, 255–64.

Hanset, R. and Michaux, C. (1978) Anomalies au niveau du maxillaire inférieur (brachygnathisme et déviation) chez le bovin culard. *Annales de Médecine Vétérinaire* 122, 37–44.

Hanset, R. and Michaux C. (1985a) On the genetic determinism of muscular hypertrophy in the Belgian White and Blue cattle breed. I – Experimental data. *Génétique Sélection Evolution* 17, 359–68.

Hanset, R. and Michaux C. (1985b) On the genetic determinism of muscular hypertrophy in the Belgian White and Blue cattle breed. II – Population data. *Génétique Sélection Evolution* 17, 369–86.

Hanset, R. and Michaux, C. (1986) Characterization of biological types of cattle by the blood levels of creatine and creatinine. *Journal of Animal Breeding and Genetics* 103, 227–40.

Hanset, R., Michaux, C. Dessy-Doize, C. and Burtonboy, G. (1982) Studies on the 7th rib cut in double-muscled and conventional cattle. Anatomical, histological and biochemical aspects. In: King, J.W.B. and Ménissier, F. (eds), *Muscle Hypertrophy of Genetic Origin and its Use to Improve Beef Production.* Martinus Nijhoff, The Hague, pp. 341–9.

Hanset, R., Michaux, C. and Detal G. (1989a) Genetic analysis of some maternal reproductive traits in the Belgian Blue cattle breed. *Livestock Production Science* 23, 79–96.

Hanset, R., Michaux, C., Detal, G., Boonen, F. and Leroy, P. (1990) Conformation et format dans la sélection du Blanc-Bleu Belge. Introduction d'un systéme de cotations linéaries. *Annales de Médecine Vétérinaire* 134, 197–204.

Hanset, R., Michaux, C., Leroy, P. and Detal, G. (1989c) Que peut-on encore attendre de la sélection en Blanc-Bleu Belge? *Annales de Médecine Vétérinaire* 133, 89–114.

Hanset, R., Michaux, C. and Stasse, A. (1987). Relationships between growth rate, carcase composition, feed intake, feed conversion ratio and income in four biological types of cattle. *Génétique Sélection Evolution* 19, 225–48.

Liboriussen, T. (1982) Comparison of sire breeds represented by double muscled and normal sires. In: King, J.W.B. and Ménissier, F. (eds), *Muscle Hypertrophy of Genetic Origin and its Use to Improve Beef Production.* Martinus Nijhoff, The Hague, pp. 637–43.

Michaux, C., Detal, G., Fernandez, J-J., Gielen, M., Lenoir, J-M., Thewis, A., Bienfait, J-M. and Hanset R. (1990) Growth and carcass characteristics of cross-bred bulls (Belgian Blue x Dairy) on different diets. *Revue de l'Agriculture* 43, 55–64.

Michaux, C., Detal, G. and Hanset, R. (1987) Age aux vêlages, intervalle de vêlage et taux de renouvellement à l'intérieur des troupeaux Blanc-Bleu Belge de type viandeux. *Annales de Médecine Vétérinaire* 131, 553–70.

Michaux, C. and Hanset, R. (1981) Sexual development of double-muscled and conventional bulls. I – Testicular growth. *Zeitschrift für Tierzüchtung und Züchtungsbiologie* 98, 29–37.

Michaux, C. and Hanset, R. (1986) Mode de vêlage et reproduction chez les génisses de race Blanc-Bleu Belge des types viandeux et mixte. *Annales de Médecine Vétérinaire* 130, 439–51.

Michaux, C., Van Sichem-Reynaert, R., Beckers, J-F., de Fonseca, M. and Hanset, R. (1982) Endocrinological studies on double-muscled cattle: LH, GH, testosterone and insulin plasma levels during the first year of life. In: King, J.W.B. and Ménissier, F. (eds), *Muscle Hypertrophy of Genetic Origin and its Use to Improve Beef Production.* Martinus Nijhoff, The Hague, pp. 350–67.

Michaux, C., Stasse, A., Sonnet, R., Leroy, P. and Hanset R. (1983) La composition de la carcasse de taureaux culards Blanc-Bleu Belge. *Annales de Médecine Vétérinaire* 127, 349–75.

Sonnet, R. (1982) Analytical study on retail cuts from the double-muscled animal. In: King, J.W.B. and Ménissier, F. (eds), *Muscle Hypertrophy of Genetic Origin and its Use to Improve Beef Production.* Martinus Nijhoff, The Hague, pp. 565–74.

Appendix

Abstracts of posters presented at the Symposium

Investigation of Mastitis Occurrence in Purebred and Crossbred Ewes.
C. Dario and G. Bufano, *Department of Animal Production, University of Studies-Bari, Italy.*

1544 lactations of the following genotypes were controlled: 280 Altamurana, 193 Leccese, 209 Sarda, Comisana and 674 F_1 (Sarda × Altamurana); all the lactations were recorded throughout 8 years (1979–1986) on an experimental farm located on the Barensis hills. Clinical mastitis occurred in a total of 119 ewes with an occurrence of 7.7% during the whole period.

Sardinian breed showed highest average milk yield (132l) as well as highest mastitis occurrence (17.7%), followed by Comisana both for milk yield (117l) and mastitis occurrence (9.5%). On the contrary Leccese breed, though resulting in lower milk yield (103l), showed a mastitis occurrence (9.3%) similar to the Comisana breed. Altamurana ewes had the lowest mastitis occurrence (1.43%) as well as milk yield (86l); whereas F_1 (Sarda × Altamurana), even though showing a noticeably increased milk yield vs pure breed, did not have a corresponding increase of mastitis occurrence 6.7%.

Paratuberculosis Incidence in Different Goat Breeds: Canary Island Breeds, a case of Genetic Resistance?
A. Molina, J.V. Delgado, R. Madueno, L. Morera and D. Llanes. *Departmento de Genetica, Instituto de Zootecnia, Facultad de Veterinaria, Universidad de Cordoba, 14005 Cordoba, Spain.*

Differential incidence of disease among breeds may be a first step in pointing out genetic implication to resistance to disease.

In Southern Spain we studied the incidence of paratuberculosis (Johne's disease) in Malagueña and Murciana-Granadina breeds by means of an ELISA test with two antigens (COA1, produced from infected goats and the commercial PPA-3[a] antigen). This test has a specificity of 93% and 95% and sensitivity of 83% and 86% for COA1, and PPA-3 respectively (Molina *et al.*, 1989).

The incidence varied from 10% (10% COA1, and 12% PPA-3) in Malagueña breed (*n*=242) to 40% (39.4% COA1, and 41.2% PPA-3) in Murciana-ü Granadina breed (*n*=251). No herds were found paraturberculosis free. The differential incidence may be explained by environmental and genetic factors.

The Canary Island breed may be a clearer case of genetic resistance, since these populations were paratuberculosis free (3.4% COA1, and 0.6% PPA-3; *n*=176).

An experimental herd (*n*=30) from the Canary Island breed was placed in Córdoba (Southern Spain) in the same sanitary, management and climatic conditions as the other breeds, and after two years, was analysed by the ELISA test and continued paratuberculosis free.

These results may show a genetic resistance of the Canary breed but new tests to discard environmental factors must be performed.

References

Molina A., Madueno R., Fernandez, J. and Llanes D. (1989) Incidence of serological positivity in goat paratuberculosis in Southern Spain by ELISA. *The Paratuberculosis Newsletter* 1, 4–5.

[a] Allied laboratories.

The Conformation Determination System (CDS): A Computerized Photogrammetric Technique for Measuring and Recording Bovine Skeletal Size and Angularity

Paul R. Greenough[1], P.A. Berg[1], Jos J. Vermunt[1] and J.A. Koehler *[1] The Western College of Veterinary Medicine* and *[2] Institute of Space and Atmospheric Studies, University of Saskatchewan, Saskatoon, Canada.*

The computer program described in this paper creates a set of bovine conformation measurements from an image stored on a 35 mm transparency.

The program, called CDS (Conformation Determination System), receives anatomical input from a digitizing pad onto which a photographic image of the subject has been projected. Anatomical landmarks had been identified and marked on the subject to facilitate their digitization. The landmarks are converted, using vector algebra, into a set of conformation values and stored into the program's database. The stored data are used to generate reports on size and angularity or changes in conformation of a single animal between examinations. Reports may also be produced on average conformation values for an entire breed/age/sex grouping.

Rigorous testing under laboratory conditions indicated that CDS-generated conformation values were accurate to within 2% of the expected values.

A Rationale for Selecting the Variables for Body Measurements of Cattle and the Precise Location of the Anatomical Landmarks

Paul R. Greenough and Jos J. Vermunt *Western College of Veterinary Medicine, University of Saskatchewan, Saskatoon, Canada.*

Internationally agreed criteria for defining body measurements in cattle do not appear in the literature. Also, the anatomical parts from which the measurements are taken have not been described. The descriptions used by the authors adhere closely to the terminology contained in the Nomina Anatomica Veterinaria while approximating to the locations most frequently used by the cattle industry. The anatomical points identified also are those that are compatible with photogrammetric and videogrammetric techniques of mensuration. Several definitions of body length appear in the literature; therefore the authors recommend that a standard should be established. Similarly the term 'hip height' is inappropriate and misleading; a clear recommendation is offered.

Selection of Sheep for Innate Resistance to Infection with *Haemonchus contortus*

T. Kassai[1], W.M.L. Hendrikx[3], É. Fok[1], P. Redl[1], Cs. Takáts[1], E. Takács[1], L. Fésüs[2], Ph. R. Nilsson[3], M.A.W. van Leeuwen[3], J. Jansen[3], W.E. Bernadina[3] and K. Frankena[3]. [1]*University of Veterinary Science, Budapest, Hungary,* [2]*Research Institute of Animal Husbandry, Herceghalom, Hungary and* [3]*Veterinary Faculty, State University of Utrecht, The Netherlands.*

It has been shown that resistance/susceptibility of sheep to gastrointestinal nematode infection is influenced by the genetics of the host. In an experiment involving 106 lambs the variation in host responsiveness to a single (day 0) or double (days 0 and 37) infection with 7,000 *H. contortus* L_3 was studied. The post-mortem worm count was considered to be the most direct measurement of the response. Lambs harbouring more than 500 or less than 50 adult worms were classified as low (LR) or high responders (HR), respectively.

Worm counts of a primary infection were found to be more subject to variation than those of a challenge. Adult worm counts at around 50 days after a secondary infection appear to allow LR and HR discrimination. However, to determine this parameter the animals have to be slaughtered. By regression analysis a strong relationship ($r=0.769$) was found between adult worm counts and faecal egg counts taken close to the time of slaughter. Haematocrit values ($r=0.028$), haemoglobin concentration and total serum protein values did not prove to be useful parameters in the selection of LR and HR individuals.

Although primary infection induced partial immunity to challenge, this was not associated with any apparent increase of serum IgG_1 and IgA antibody titres (measured by the ELISA test using crude somatic L_3 antigen). Mean antibody levels of LR lambs showed no difference from those of HRs. Thus serum IgG_1 and IGA levels are of no predictive value in identifying LR and HR lambs.

It is proposed that in living animals LR and HR selection can be based on individual faecal egg counts around 50 days after a secondary infection. In the offspring of HR parent sheep selected by faecal egg output analysis greater proportion of HR lambs did occur than in random bred population.

The Genetic Component of Footshape and Associated Foot Lameness in Dairy Cattle

J.B. Merritt and R.D. Murray *Department of Veterinary Clinical Science, University of Liverpool, Liverpool, UK.*

The foot shape of 500 dairy cattle, examined monthly as part of a study to determine the risk factors associated with foot lameness, was recorded on farm. Locomotion scoring was carried out at the same time.

The three components of the qualitative footscore describe: shape of the toe, height of the heel, and length of the toe. Standards were obtained using slaughterhouse material, and 'normals' were defined for each component.

Continuous variation in foot shape was observed both between cows on different farms and between animals under similar environmental and management conditions on the same farm. This is in accordance with previous findings, and probably reflects the genetic component of foot shape.

Cattle were grouped according to sire, and footshapes and the corresponding incidences of foot lameness were compared.

A Genetic Component of Lameness in Dairy Cattle

J.B. Merritt *Department of Veterinary Clinical Science, University of Liverpool, Liverpool, UK.*

The relationship between sire and the incidence of lameness was investigated in a group of animals in their 1st and 2nd lactations, on a 200-cow Cheshire dairy herd, as part of a three-year survey to study the risk factors associated with lameness in cattle.

The cows were examined monthly, both in the parlour, in order to assess foot shape, and whilst walking on concrete, in order to assess 'locomotion score' or lameness.

Data collected between January and June 1990 were subjected to simple univariate analysis. No significant association was found between bull and the incidence of lameness among daughters in this herd, or between bull and any of the aspects of foot shape observed. Abnormally shaped feet were, however, more often lame (chi-squared 5%) than those of normal shape.

This raises the question as to whether abnormally shaped feet are the original cause of lameness, or whether abnormality of horn growth is merely a secondary manifestation of the behaviour of an unhealthy quick.

It is interesting to note the generally high level of overgrowth of the heel of the outer claw (16%). This is considered by many to be a natural phenomenon, and one which will not necessarily lead to lameness, unless accompanied by environmental stress or disease of the quick. Also the variability in the incidence of corkscrew claw, though not significant, suggests in accordance with the findings of others, a higher degree of heritability for this trait.

Application of a Screening Programme for Genetic Resistance to Nematode Parasites in Commercial Sheep Stud Flocks in New Zealand

T.G. Watson and B.C. Hosking *MAF Technology, Ruakura Agricultural Centre, Private Bag, Hamilton, New Zealand.*

Sheep breeders and producers are being confronted with three significant issues that will change farming systems for the future. Resistance to members of two of the three action families of broad-spectrum anthelmintics appears to be developing rapidly in widely different geographic regions of New Zealand; resistance to ivermectin already appears to be established on goat farms; and there is increasing demand to minimize application of chemicals in animal production. One viable option to overcoming the effects of all three problems is to increase natural resistance of stock to infection through identification and use of 'elite' genotypes.

During 1987–88 a programme was initiated on a Romney sheep stud screening over 700 ram and ewe lamb progeny from over 15 single sire matings to investigate the practicality of developing single trait selection lines in commercial environments. Faecal nematode egg counts (FEC) were used to predict sire and progeny resistance to infection.

The protocol involves screening undrenched grazing lambs after weaning. Infections are acquired naturally from pasture until the mean FEC from selected monitor animals reaches 800–1000 epg. FECs are determined on two occasions when lambs are between 4 and 8 months of age. Anthelmintic therapy is given at commencement of the programme and after each sampling. Lamb progeny for inclusion in the programme are from single sire matings enabling collection of pertinent data such as sire, dam, breed, weaning weights, liveweight gains and fleece weights. These data can be used in a full assessment of breeding values over the time the programme is adopted and it is expected that they will assist development of single or multiple trait selection lines.

In 1980–90, through farmer awareness and interest, this study expanded to 10 farms and involved over 2,700 lambs and four sheep breeds (Romney, Perendale, Coopworth and Merino). Demand expressed by breeders over the last year suggests that these farm and sheep numbers may double or treble for the 1990–91 lambing season.

The results of the initial investigations will be given along with brief commentary with regards the difficulties encountered associated with application of the programme, costs to the farmer and estimated value to animal production will be discussed.

Initial Investigation on the Association Between BoLA and TB Infection

M. Longeri[1], M. Zanotti Casati[1] G. Ceriotti[1], M. Polli[1] and T.M. Gliozzi[1] [1]*Instituto di Zootecnica, Facoltà di Medicina Veterinaria, via Celoria 10, 20133 Milano, Italy* and [2]*I.D.V.G.A. – C.N.R., via Celoria 10, 20133 Milano, Italy.*

The correlation between TB-infected animals and BoLA antigens has been evaluated on 110 Holstein Friesian cattle. TB is becoming more and more serious in Italy with an increasing number of affected animals in some areas. Blood samples, collected from animals, were found to be positive and negative to tuberculin PPD skin test in the same farms and were tested for BoLA type. All the tested animals were also submitted to post mortem inspection to confirm the specificity of the allergic reactions.

The statistical analyses on the preliminary data suggest a possible negative effect of w15 and a positive one of w14 on cattle response to TB.

The Relationship Between Linear Assessment and Longevity in Dairy Cows

R. Broad, I. Ap Dewi and C.J.C. Phillips *School of Agricultural and Forest Sciences, University of Wales, Bangor, Gwynedd.*

Records for 45,015 British Friesian dairy cows from registered commercial herds were analysed to determine the relationship between longevity (age at culling) and linear assessments, on a scale from 1 to 9, of 16 traits. Thirty one, 34, 21, 11 and 3% of the cows had been culled at 1, 2, 3, 4 and over four years of age respectively. The percentage distributions of cows by trait values were approximately normal with the majority of animals (> 60%) having trait values between 4 and 6. The exceptions were the traits Rear Legs (rear view), Fore Udder Attachment and Udder Depth for which fewer than 60% of animals had trait values between 4 and 6. For these traits 36, 39 and 50% of animals respectively had trait values above 6.

The data were analysed using least squares analysis of variance in a model (Model 2 of Harvey, 1985) that included the effects on longevity of each trait and herd of origin. All traits were initially included in the model as linear and quadratic terms. The data were analysed further, using the same model, but excluding those traits that had a non-significant ($P>0.05$) effect on longevity in the initial analysis. Statistically insignificant ($P>0.05$) linear terms were retained in the model if the quadratic term for the trait had a significant ($P<0.05$) effect. In the initial analysis, Angularity, Rump Width, Rear Udder Width and Teat Placement (side view) had neither significant quadratic or linear terms and were not included in the second analysis.

The traits were ranked on the basis of their combined sums of squares for the linear and quadratic term, if included in the model. This analysis suggests the following ranking of traits, beginning with the one having the largest single effect on longevity, Teat Placement (rear view) > Body Depth > Stature (height at withers) > Udder Depth > Rear Legs (side view) > Chest Width > Fore Udder Attachment > Udder Support > Foot Angle > Rear Legs (rear view) > Rump Angle > Teat Length.

It is intended that further analysis be conducted to examine alternative statistical models and to calculate optimum values for each trait.

Reference

Harvey, W.R. (1985) *Users Guide to LFMLMW.* Ohio State University, Ohio, USA.

Index